Advances in Drug Delivery
Volume - I

Advances in Drug Delivery

Volume - I

Editors

Y. Madhusudan Rao

University College of Pharmaceutical Sciences,
Kakatiya University, Warangal, India

A.V. Jithan

Vaagdevi College of Pharmacy,
Warangal, India

PharmaMed Press
An Imprint of
Pharma Book Syndicate

PharmaMed Press
An imprint of Pharma Book Syndicate

An unit of BSP Books Pvt., Ltd.

4-4-316, Giriraj Lane,
Sultan Bazar, Hyderabad - 500 095.

Published by

PharmaMed Press
An imprint of Pharma Book Syndicate

An unit of BSP Books Pvt., Ltd.

4-4-316, Giriraj Lane, Sultan Bazar, Hyderabad - 500 095.
Phone: 040-23445605, 23445688; Fax: 91+40-23445611
E-mail: info@pharmamedpress.com
www.pharmamedpress.com/pharmamedpress.net

Printed at

Aditya Offset Process (I) Pvt. Ltd.
Hyderabad.

ISBN: 978-81-7800-254-5

Foreword

Advances in drug delivery is a very specialized area of pharmaceutics where in the principles and technologies applied in the formulation and development, are emerging and progressing. Currently no single book is available covering all the relevant and latest technologies developed.

Prof. Y. Madhusudhanrao and Dr. A.V. Jithan have been working and contributed considerably in the area and their work has been recognised well internationally. Their expertise is clearly reflected in the preparation of each and every chapter of the book. Various chapters presented in this book have great relevance to the Postgraduate students, scholars and teachers working in institutions and industry. Most of the topics reflect modern concepts and approaches in the design of novel drug delivery systems. The authors deserve praise and appreciation for their excellent piece of work and I hope this book will be very helpful for the users in the field.

Prof. R.Nagaraju

Institute of Pharmaceutical Technology

Sri Padmavati Mahila Visvavidyalayam (Women's University)

Tirupati, A.P.

Preface

Drug delivery is a broad term encompassing various means of achieving optimum drug reach to the target tissue, cell or the receptor. Several preformulation, formulation, biopharmaceutical, targeting and pharmacokinetic principles are applied in drug delivery. The book series entitled, "Advances in Drug Delivery" incorporates latest information regarding various subjects of drug delivery. In this volume 1f book series various aspects ranging from Sustained Release Dosage Forms to Chirality are covered.

Research and development in drug delivery is increasing at a rapid pace throughout the world. The need for increased efficiency of new therapies and reduction in future public health expenses will definitely bolster this area of research and development. In order to meet this demand, many well known and efficiently applied drugs will be reformulated in new drug delivery systems that can be value-added for optimized therapeutic activity. Further, several new molecules are being generated by medicinal chemists and their formulation is not any more empirical but it is now very systematic. These issues are not clearly dealt in various books available. One aim of this book is to enlighten pharmaceutical scientists all around the world with latest information on the topics which are involved in cutting edge growth of pharma research and industry. For instance, in Volume 1, a chapter on Chirality is included. This topic is being overlooked by various scientists around the world and we cited the importance of this burning topic in this first volume. Similarly we have included a chapter on P-gp and CYP3A limiting oral drug absorption. This chapter gives insight of drug metabolism and efflux occurring in GIT which play a key role in drug discovery and drug development process. Similarly, we have included many such chapters in this book. We anticipate that, more than ever before, the clinical success of future advances made by medical research will require an appropriate application of drug delivery technologies. This book caters to all those whose aim is to achieve higher objectives in drug delivery.

In next few volumes which are going to be released shortly similar topics like preformulation, pharmacogenomics, dendrimers, hydrogels etc., are going to be included along with transdermal drug delivery, solid lipid nanoparticles, polymeric nanoparticles, drug targeting, monoclonal antibodies and drug immunoconjugates, value added products like orodispersible tablets and films are going to be included.

We are very grateful to all the authors who have shared our enthusiasm and vision by contributing high quality book chapters, on time, keeping in tune with the original design and theme of this work. You will not be having this book in your hand without their dedication and sacrifice.

We encourage constructive criticism and will be glad to receive opinions from experts and readers so that coming volumes can be better shaped. The main objective of launching these volumes is to present vast and enormous knowledge to the researchers, postgraduates/ graduate teachers, besides young teachers and industry personnel.

Editors

Contributors

Dr. Kalyan Kumar Sen is currently the Principal, Gupta College of Technological Sciences, Asansol, West Bengal, India. Sri K.K Sen is an M.Pharm, Ph.D from Jadavpur University, Kolkata. He is having 22 years work experience in different fields of Pharmacy including industry, teaching etc.

Prof. Navin Sheth is currently a Professor and Head, Department of Pharmaceutical Sciences, Saurashtra University, Rajkot, Gujarat, India. He published several papers in national and international journals.

Shravan Kumar Yamsani is currently working as Associate Professor and PG guide at Department of Pharmaceutics, Vaagdevi College of Pharmacy, Warangal, Andhra Pradesh, India. He published several papers in national and international journals.

Bonepally Chandra Shekhar Reddy is currently working as Associate Professor and PG Guide at Department of Pharmaceutics, Vaagdevi College of Pharmacy, Warangal, Andhra Pradesh. He published several papers in national and international journals.

Mahesh R. Dabhi is currently working as an Assistant Professor and PG guide at Department of Pharmaceutical Sciences, Saurashtra University, Rajkot, Gujarat, India. He is currently working on one research project funded by University Grants Commission, New Delhi and another research project funded by Gujarat Council on Science and Technology, Government of Gujarat.

Milan Limbani is currently working at Alkem Laboratories Ltd., Daman, India. He completed his M. Pharm from Saurashtra University in 2010. He has so far handled several projects.

Palem Chinna Reddy is currently a UGC Major Research Project Fellow at University College of Pharmaceutical Sciences, Kakatiya University, Warangal, Andhra Pradesh, India. He is a recipient of several awards which include DST travel grant as well as AAPS travel grant to attend conferences in the USA.

Vamshi Vishnu Yamsani is currently working as an Associate Professor and PG guide at Department of Pharmaceutical Sciences, Vikas College of Pharmacy, Warangal, Andhra Pradesh, India. He published several papers in national and international journals.

Jagan Mohan Somgani is currently a Ph.D. scholar at University College of Pharmaceutical Sciences, Kakatiya University, Warangal, Andhra Pradesh, India.

Abheri Das Sarma is currently a Ph.D. scholar at Gupta College of Technological Sciences, Asansol, West Bengal, India.

Bandari Suresh is currently working as a scientist at Dr. Reddy's Research Laboratories, Hyderabad, Andhra Pradesh, India.

Contents

Chapter – 1

Controlled Release Products

Chapter – 2

Gastroretentive Drug Delivery Systems

Chapter – 3

Buccal Drug Delivery Systems

Chapter – 4
Multiple Emulsions

Chapter – 5
Nanoemulsions

Chapter – 6
Microspheres

Chapter –7

P-glycoprotein and CYP3A Limiting Oral Drug Absorption

Chapter –8

Liquid Crystals

Chapter –9

Chirality in Drug Development

1
Controlled Release Products

A.V. Jithan, B. Chandrasekhar Reddy and Y. Shravan Kumar
Vaagdevi College of Pharmacy, Warangal, India

1.1 Introduction

Controlled Release (CR) products are designed to maintain constant therapeutic plasma concentration of the drug within the therapeutic range of the drug over prolonged periods and offer minimum side effects. This can be achieved using a variety of delivery systems and also includes liposomes and drug-polymer conjugates. These products are designed to reduce the frequency of dosing by modifying the rate of drug absorption. Generally CR products administered by any route are designed such that rate of drug absorption should be equal to rate of drug elimination. Literally, the amount that is eliminated should be input into the compartment of interest at the same amount at any given time. These products can be administered by various routes including oral (peroral, buccal, sublingual), parenteral (IM, IV, SC, IP, IT, etc.), transdermal, respiratory, nasal, etc. Other miscellaneous routes such as intravaginal, rectal, etc. can also be used for CR products. Drug eluting stents is a novel concept in CR products. Examples of controlled release drug-eluting stents include Cypher® (reservoir) and TAXUS Express® (monolithic). The concept behind CR products was proposed long time ago. Early modified release products were often intramuscular/ subcutaneous injection of suspensions of insoluble drug complexes, eg. Procaine penicillin, protamine zinc insulin, insulin zinc suspension or injections of the drug in oil, eg. Fluphenazine decanoate. Advance in technology have resulted in novel modified release dosage forms.

Drug products that provide extended release first appeared as a major new class of dosage form in the late 1940's and early 1950s. Over the years, many terms (and abbreviations), such as sustained release(SR), sustained action (SA), prolonged action (PA), controlled release (CR), extended release (ER), timed release (TR), and long acting (LA), have been used by manufactures to describe product types and features. Although these terms often have been used interchangeably, individual products bearing these descriptions may differ in design and performance and must be examined individually to ascertain their respective features. In case of sustained release (SR) dosage forms the release of the active agent, although, is lower than in the conventional formulations, however, it is still substantially affected by the external environments into which it is going to be released. Controlled release (CR) systems provide drug release in an amount sufficient to maintain the therapeutic drug level over extended period of time, with the release profiles of predominantly controlled by the special technological construction and design of the system itself. The release of the active constituent is therefore, ideally independent of exterior factors. Extended release formulation is a controlled release formulation designed to produce even and consistent release of active ingredient. Extended release (ER) dosage forms are those which due to special technology of preparation provided, soon after a single dose administration, therapeutic drug levels maintained for 8-12 hours. Prolong or long action products are dosage forms containing chemically modified therapeutic substances in order to prolong biological half life. Although there are a variety of these dosage forms, the purpose of all such dosage forms is the same and the main aim is to achieve the controlled or sustained release of the drug. However, controlled release is the ideal release of the drug from delivery systems. In this case, the release is either a zero-order or first-order release. Timed-release, sustained release, prolonged release are less superior than controlled release. However, as of today, it is not practical to achieve controlled release with all the drugs and all the delivery systems. Most of the times the drug release is either sustained, prolonged or timed. However, all these types of releases can be called as CR release for its ideal nature and convenience. As a reason, in this book these formulations are called together as CR products. Plasma time profile of a drug following a CR form administration is shown in Figure 1.1.

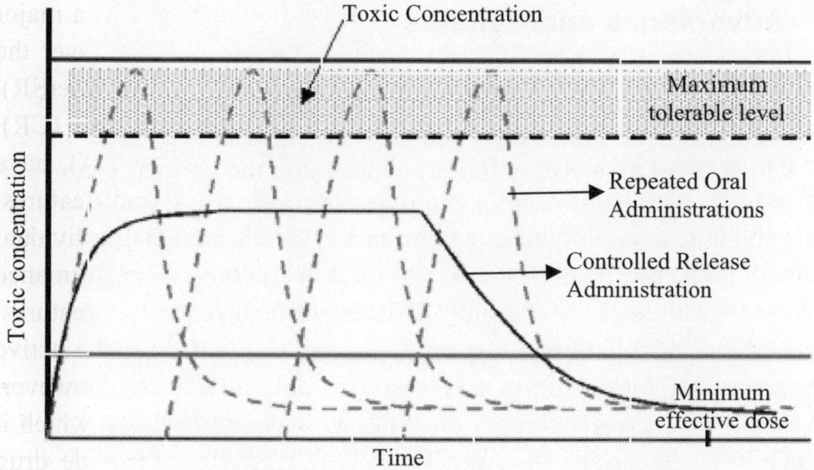

Figure 1.1 The plasma concentration time profile of a CR formulation.

The currently available CR products can be administered via various routes of administration. Based on this they could be classified into oral, parenteral, targeted, respiratory, vaginal and rectal. Parenteral CR products available in the market can release the drug upto five years. Some others release the drug for a period of two weeks to four weeks. On contrary, the currently available oral CR products or some products in clinical trials can release the drug, in case of short half-life for a period of 24 hours to seven days. Targeted CR products are generally administered at the site of the need and thereby release the drug for weeks. Several companies are now developing CR dosage forms to be administered by various routes other than the conventional routes mainly focusing on the respiratory tract. Recently, Acusphere, Inc. in the USA has conducted clinical trials on controlled release microsphere formulation of a leading asthma drug. This formulation releases the drug locally. Some of these formulations and the respiratory tract have been used in achieving the systemic sustained release of the drug. Similarly, vaginal and rectal route also have been investigated for CR products. The other CR formulations present in the market are transdermal dosage forms. Transdermal dosage forms are used either to achieve the drug levels in the skin tissue or the drug levels in the blood. These delivery systems sustain the release of the drug. They have been exhaustively investigated and ample information is available regarding these delivery systems. These CR delivery systems are dealt in a separate chapter in this book series.

1.2 Advantages and Benefits

All the conventional dosage forms excepting continuous IV infusion, release drugs largely according to first-order kinetics. Therefore immediately after administration and absorption, there is a higher plasma level which upon time will gradually reduce and the optimal therapeutic level is present for only a brief duration. On contrary, the CR systems release the drug at a constant rate (zero order) or at a predictably constant declining rate (first order) for a certain time period. This results in uniform concentration of drug in the plasma and tissue.

The high-peak blood concentration reached soon after administration of conventional formulations may result in adverse effects. For drugs with short biological half-life or with a clear relationship between concentration and response, it will be necessary to dose at regular, frequent intervals in order to maintain the concentration within the therapeutic range. Higher doses at less frequent intervals may result in higher peak concentrations with the possibility of toxicity. For drugs with wide margins of safety, this approach is satisifactory. Amoxycillin has a half-life of approximately one hour and a dosing frequency of eight hours. Therefore, there will be large fluctuations in the plasma levels with this drug in one dose. However, the drug has high therapeutic window and as a reason, such an administration is not a problem. On contrary, there are several examples available in the literature where such administration with low therapeutic window drugs is a problem. An example is hypotension in patients taking rapid-release nifedipine products. A CR nifedine product avoids the high initial blood concentration which causes the sudden hypotension and reflex tachycardia. Similarly, some times subtoxic levels can cause problems. Subtoxic levels due to conventional theophylline results in local irritation in the GIT. This happens whenever there is a consumption of conventional dosage form for this drug. Similarly, several examples can be seen in the market as well as literature. Gepirone, a psychoactive drug, is useful in the treatment of depressive and anxiety disorders. Pharmacologically, gepirone acts as a selective 5-HT_{1A} receptor partial agonist. Gepirone was originally developed by Bristol-Myers Squibb, but was out-licensed to Fabre-Kramer in 1993. In 2001, Organon filed a new drug application for gepirone extended release (ER). The need for ER formulation has been realized because of its short half-life which necessitated frequent administration. High peak concentrations seen at higher doses were associated with an increased incidence of adverse effects. Initial placebo-controlled clinical trials demonstrated that gepirone Immediate Release (IR) product improved symptoms of depression. The application for gepirone ER formulation is still pending with US Food and Drug Administration (FDA). On the other hand, the investigators concluded that gepirone ER appears safe and effective in short-term treatment of major depressive disorder and appears to be free of common side-effects.

CR products can significantly reduce the side effects associated with chronically administered drugs. Upon repeated administration of conventional products, the resulting pattern of drug concentration in plasma can vary widely and may cause inconsistent and undesired clinical effects. The saw tooth kinetic pattern that is noticed after repeated administration can result in the accumulation of the drug in tissues leading to undesirable side-effects especially after chronic administration. Several CR products for chronic drugs have been developed and the results clearly indicated a reduction in the side-effects. Also the subtherapeutic levels seen with every administration can result in times of ineffectiveness for this drug. This issue can be conveniently taken care using a CR dosage form. CR product of paroxetine delays the release of the active until the tablet has passed through the stomach; the drug is then released over four to five hours. Paroxetine CR was generally well-tolerated in clinical trials, as patients felt significantly less nausea than recipients of IR paroxetine. The effects of switching from IR carbamazepine formulations to an equal daily dose of carbamazepine ER capsules in epilepsy has shown significant benefits. Switching to ER formulation significantly improved patients adverse events and quality-of-life measures. It also improved seizure control. Risperidone is to date the only novel antipsychotic available as parenteral CR formulation (parenteral depot). Unlike the traditional esterification of conventional antipsychotics to achieve a long acting injectable formulation, long-acting risperidone is fabricated by a microsphere encapsulation process. This depot preparation significantly lowered several side-effects associated with the repeated oral administration of the drug. It also lowered the rate of reversible motor side effects when compared with oral therapy by constraining the peak levels below the moderate-to-severe threshold of reversible motor side-effects. In a study comparing regular alprazolam tablet, given four times a day and extended release alprazolam (XR) given once in the morning, drowsiness occurred more frequently with conventional alprazolam (86% of patients) than with the XR preparation (79%) or placebo (49%).

Several studies indicated that patient compliance also improves with CR products. Drugs with short half-life often need to be given at frequent intervals to maintain blood concentrations within the therapeutic range. An inverse correlation between the frequency of dosing and the patient compliance has been found. A reduction in the number of daily doses offered by CR products has the potential to improve patient compliance. In one study patient compliance in hypertensive outpatients between amlodipine (5 mg once daily) and slow release nifedipine (20 mg twice daily) were compared in an open, crossover. Four methods of assessment for patient compliance (pill count, taking compliance, days with correct dosing, timing compliance) were used. The results indicated that the compliance of the 320 hypertensive patients with once-daily amlodipine was markedly superior to twice-daily slow release nifedipine. Therapeutic coverage was also significantly better for amlodipine in the hypertensive patients. Amlodipine was better tolerated than nifedipine slow release. Patient compliance and therapeutic coverage with the calcium antagonist amlodipine given once daily was superior to slow release nifedipine bid in hypertensive outpatients recruited in general practice. Ideally, CR products are not significantly affected by the external environment, so that patient-to-patient variability is greatly reduced.

There is also a reduction in health care costs with CR products. The total cost of therapy of the controlled release product could be comparable or lower than the immediate-release product. With reduction in side effects, the overall expense in disease management also would be reduced.

1.3 Disadvantages and Problems

More complicated CR formulations may be more erratic in result. A sustained release product may contain a larger dose, i.e. the dose for two or three (or more) 'normal' dosing intervals. A failure of the controlled release mechanism may result in release of a large toxic dose. CR technology is more expensive. A growing number of new CR products have been submitted for regulatory approval. As mentioned previously, CR products have many advantages in safety and efficacy over immediate release drug products in that the frequency of dosing can be reduced, drug efficacy can be prolonged and the incidence and/or intensity of adverse effects can be decreased. However, some CR products developed have less clear rationale or are developed for active ingredients which are not appropriate for prolonged release dosage forms. In other cases, CR products are designed without full consideration of the basic properties of the drugs. Additionally, there is a big problem with in vitro-in vivo correlation. As a result, it is often difficult to evaluate whether a CR form is acceptable or not. Incomplete or undesirable prolonged release drugs may merely cause therapeutic confusion and, in addition may interfere with development and spread of good quality drugs.

1.4 Classification of CR Systems Based on Fabrication Techniques

Either modulation of dissolution of the active drug component or diffusion of the dissolved or solubilized species is the basic mechanisms for controlling drug release. Erosion of the polymer followed by drug reaching the dissolution medium also is one of the mechanisms. Some times all the four mechanisms may be involved in the release of the drug from the CR formulations. They may operate independently, together or consecutively. Several other mechanisms are also reported to play a role. These mechanisms apply to all the CR delivery systems administered by different routes of administration. Specific examples of drug release and mechanisms can be found in the literature regarding a particular marketed product, whether administered by oral route, parenteral route, local delivery or by any other route.

Purely diffusion, dissolution, erosion, osmotic pump type of mechanism, ion-exchange based drug release, smart materials etc. are known to be the main mechanisms utilized in the fabrication and drug release investigations. Based on these mechanisms CR products that exist include: 1. Diffusion controlled systems 2. Dissolution controlled systems 3. Dissolution and diffusion controlled release systems 4. Erosion products 5. Water penetration controlled systems 6. Drug covalently linked to polymer 7. Ion - exchange resin controlled release systems 8. Responsive Drug Delivery Systems.

1.4.1 Diffusion Controlled Products

There are two types of diffusion controlled systems: matrix controlled systems and reservoir controlled systems.

1.4.1.1 Matrix controlled systems

Matrix systems are also called as monoliths since the drug is homogeneously dispersed throughout a rate controlling medium. There are two types of matrix devices. 1. Insoluble matrix of rigid non-swellable hydrophobic materials. 2. Soluble swellable hydrophilic substances. In these systems the therapeutic agent is dispersed in either of the two above matrices. Materials used for case 1 (rigid matrix) are insoluble plastics such as polyvinyl chloride and fatty materials like stearic acid, bees wax, etc. Swellable matrix systems are generally composed of hydrophilic gums of natural (guar gum, tragacanth, karaya gum), semisynthetic (hydroxypropylmethylcellulose, carboxy methyl cellulose, xanthan gum) and synthetic (polyacrylamides) origin. The drug and matrix materials are granulated together and compressed into CR products. Examples include Plendil ER (Felodipine), Agon SR (Felodipine), Kapanol (Morphine sulphate) and Slow-K (Potassium chloride). TIMERx is a CR product based on agglomerated hydrophilic matrix consisting of xanthum gum and locust bean gum. Slofedipine XL (nifedipine) and Cystrin CR (oxybutynin) are other products developed using this technology. Drug release from insoluble nonswellable matrices involves penetration of fluid, followed by dissolution of the drug particles and diffusion through fluid filled pores. With case 2 matrices, the drug becomes available as the matrix swells or dissolves and the dissolved matrix then undergoes surface erosion with little or no bulk erosion. The surface area of the matrix decreases with time with a concomitant decrease in drug release. The diffusion depends on the solubility of the drug in the polymer. The drug may be either present below its solubility limit and dissolved in the polymer or present well above its solubility limit and dispersed in the polymer. The development of both matrices and reservoirs, which is discussed later, is a common place in pharmaceutical industries, as the technology is easy or enough experience has been gained. However, matrices have particular advantage over reservoirs in some cases and vice-versa is also true. For instance, Barochez et al., compared a monolithic matrix form with a reservoir device for a 80 mg highly soluble drug (1). At first, matrices were prepared containing 37.3 percent of a water soluble polymer: HPMC (Methocel or Metolose) and HEC (Natrosol). With such swelling agents, it was quite difficult to reach a zero order release. But industrial scale-up was easy, because the process uses only classical machines. With matrix formulations, variations between batches have been found very small and stability is good. Matrix formulations were also prepared using lipophilic matrix. Hard gelatin capsules were filled with a drug dispersion in Gelucire of different grades. The fabrication process is quite easy but at this time, information on stability is scanty. A third convenient way was a reservoir device, a tablet coated with an insoluble polymer film (Aquacoat ECD 30). A zero order release was obtained until 80 percent of drug released after 12 hours. But the coating was a very critical phase of the process. Slight disturbances in coating affected the drug dissolution rate. Additionally, there is a problem with the patients. The film may also be altered by the patient, who can break or crunch the tablet. In this case, all the drug is dissolved quasi instantaneously. It was suggested in this study that for all these reasons, the hydrophilic matrix was preferred, especially if a zero order is obtained. However, obtaining a zero-order drug release is not easy.

For instance, Mockel and Lippold, investigated the order of release of a drug from hydrocolloids (2). Matrices were manufactured by direct compression of a powder mixture of a polymer, e.g., methylhydroxypropyl cellulose (MHPC) or polyvinylalcohol (PVA), and a drug. The following factors that can influence the drug release mode were investigated at constant surface: (i) polymer solution viscosity, glass transition temperature, and swelling; (ii) drug concentration in the matrix and solubility; and (iii) conditions of release experiment (hydrodynamics). Only hydrocolloids with low viscosities yielded a zero-order release profile. In this case only the dissolution of the polymer appeared to control the drug release rate. Factors accelerating polymer dissolution resulted in higher release rates. Comparison of swollen and dry hydrocolloid matrices shows that the duration and kinetics of drug release were not controlled by the swelling front moving into the dry polymer, and water penetration and relaxation were not rate controlling. Therefore, the glass transition temperature had no effect on drug release from these hydrocolloids. The higher the hydrodynamic stress exerted on the eroding hydrocolloid, the faster the resulting drug release as a result of accelerated polymer dissolution. With hydrocolloids of very high viscosity the polymer dissolution is slow, and drug release from the swollen gel appears to be controlled by diffusion according to kinetics of the Higuchi type. Thus, different types of drug release can be obtained with different types of polymers. Additionally, the properties of the drug may also affect the release. Although a zero-order drug release is preferred, in the case of matrices very judicious selection of the polymers may be essential. On the other hand, zero-order drug release can be conveniently obtained from reservoir devices.

1.4.1.2 Reservoir controlled systems

A core of drug is coated with the water insoluble polymer. The polymer can be applied by coating or microencapsulation techniques. The drug release mechanism across the membrane involves diffusion of water through the membrane to the inside of the core, dissolution of the drug and then diffusion of the drug into the surrounding fluid. Materials used in such devices are hydroxypropyl cellulose, ethyl cellulose and polyvinyl acetate. The reservoir diffusion products are Plateau CAPS capsules (nicotinic acid), nio-bid (nitroglycerine), Nitrospan capsules (nitroglycerine), Brankadyl SR cap (theophylline). Some other examples of diffusion-controlled devices include drug-eluting stents such as Cypher® (reservoir) and TAXUS Express® (monolithic), intra-uterine contraceptives such as Progestasert® and Norplant ®, and various transdermal patches such as Nicoderm® and Transderm Nitro®

1.4.2 Dissolution-controlled Products

There are two types of products: A. Matrix dissolution controlled products and B. Reservoir dissolution-controlled systems.

1.4.2.1 Matrix dissolution products

In these systems, the drug is homogenously dispersed through out a rate controlling membrane. The drugs, which are highly water-soluble can also be formulated as CR products by controlling their dissolution rate. Slowly soluble polymers control the rate of dissolution of the drug. Waxes such as beeswax, carnauba wax and hydrogenated castor oil have been used. The wax embedded drug is generally prepared by dispersing the drug in the molten wax and congealing and granulating them.

1.4.2.2 Reservoir dissolution-controlled systems

In reservoir dissolution control system the drug particles are coated or encapsulated by one of the several microencapsulation techniques with slowly dissolving materials like cellulose derivatives, polyethylene glycols, polymethacrylates, waxes etc. The resulting reservoirs (coated beads, multi-particulate systems, pellets) may be filled as such in hard gelatin capsules (spansules) or compressed into tablets. The common multi-particulate systems are microparticles (microspheres or microcapsules), nanoparticles (nanospheres or nanocapsules), liposomes, etc. The dissolution rate of the drug depends upon the solubility and the thickness of the coating. By varying the thicknesses of the coat and its composition, the rate of drug release can be controlled. These products should not be chewed as the coating may be damaged. One of the advantages of encapsulated reservoir products is that the onset of absorption is less sensitive to stomach emptying. The entrance of the reservoir into the small intestine is usually more uniform than with non-disintegrating CR tablet formulations. An example of this type of product is fefol (Ferrous sulfate and folic acid).

1.4.3 Dissolution and diffusion controlled products (pore forming method)

In these system, the drug core is coated with a partially soluble membrane. Pores are thus formed due to dissolution of parts of the membrane, which permit entry of aqueous medium into the core and release of dissolved drug by diffusion. Using a mixture of ethylcellulose with polyvinylpyrrolidone or methyl cellulose, the latter material dissolves in water and forms pores in the insoluble ethylcellulose membrane. Based on the above preparation techniques, CR transdermal formulations, buccal drug delivery systems, nasal drug delivery systems (inhalers), oral gastroretentive systems and ocular inserts etc. can also be formulated.

1.4.4 Erosion Products

The release of a drug from these products is controlled by the rate of erosion of a carrier (polymer) matrix. Erosion of the polymer can be by dissolution or can be by a chemical means such as degradation of the polymer. The rate of release (amount of drug released from the dosage form per unit of time as defined by in vitro or in vivo testing) is determined by the rate of erosion. An example of such a formulation is Sinemet CR (carbidopa/levodopa). Several parenteral dosage forms also undergo this type of release mechanism. An implantable therapeutic system is fabricated by dispersing a loading dose of solid drug, in micronized form, homogenously throughout a polymer matrix made from bioerodible or biodegradable polymer, which is then molded into a pellet or bead-shaped implant. The controlled release of the embedded drug particles is made possible by the combination of polymer erosion through hydrolysis and diffusion through polymer matrix. The rate of drug release is determined by the rate of biodegradation, polymer composition and molecular weight, drug loading, and drug/polymer interaction. The rate of drug release from this type of drug delivery system is not constant and is highly dependent upon the rate process of polymer matrix erosion. It is exemplified by the development of biodegradable naltrexone pellets fabricated from poly(lactide/glycolide) copolymer for the antinarcotic treatment of opioid-dependent addicts. In addition to poly(lactide/glycolide) copolymer, several other biodegradable or bioerodible polymers, such as polysaccharide, polypeptide and homopolymer of polylactide or polyglycolide, can also be used to prepare biodegradable, implantable therapeutic systems. A biodegradable polymer, poly(ortho esters) possess an interesting example. This polymer contains linkages in the polymer backbone that are relatively stable at the physiological pH of 7.4, but become progressively unstable as the pH is lowered. The erosion rate of such polymers is controlled by the use of buffering agent, e.g., calcium lactate, which is physically incorporated into the polymer, and when in contact with water, produces a pH that activates the polymer to hydrolyze at a desired rate. If the polymer is maintained at a rather high hydrophobicity, then only the buffering agent in the surface layers is exposed to water and polymer hydrolysis will occur only in the surface layers. So, a constant (zero-order) rate of drug release is obtained.

1.4.5 Water penetration controlled systems

Some devices are designed using water as the main agent controlling the release of the drug. In these devices, the drug molecules cannot physically diffuse out of the device without water molecules diffusing in. There are generally two types of water penetration-controlled systems: 1. Swelling controlled systems. 2. Osmotic controlled systems.

1.4.5.1 Swelling controlled systems

These systems usually incorporate drugs in a hydrophilic polymer that is stiff or glassy when dry, but swells when placed in an aqueous environment. A typical oral capsule or pill is usually a swelling-controlled device. Although these devices are easy to manufacture, the release rates are often not steady. Swelling-controlled release systems are relatively new devices of the controlled release family of delivery devices for applications in pharmaceutical technology. To date, emphasis has been placed on swelling-controlled systems for release of a drug at a constant rate over a period of time (zero-order systems). Morita et al., investigated a swelling controlled system for a model drug (3). A novel controlled release system, the PVA swelling controlled release system, was evaluated in vitro and in vivo using emedastine difumarate. In the in vitro drug release study, the release profile of this system had almost zero-order kinetics. The effect of dissolution test conditions, which were paddle rotation speed, mechanical stress, and pH of the dissolution medium, on the release rate was very small. In an in vivo human bioavailability study of two formulations with a different release rate, the absorption rate was dependent on the release rate, and both formulations showed constant plasma levels of the drug for long periods. The variations of plasma concentration on the simulation of repetitive administration of the formulations at 24-h intervals were almost equal to the experimental value for the twice daily controlled release capsule currently on the market. It is concluded that the PVA swelling controlled release system is feasible for a long-acting preparation as a once-daily treatment.

1.4.5.2 Osmotic controlled systems

Reservoir systems have a drug core surrounded\coated by the rate controlling membrane. However factor like pH, presence of food and other physiological factor may affect drug release from conventional controlled release systems. Drug delivery systems utilizing the principles of osmosis have been very successful at addressing a wide variety of disease conditions. With principles first described in 1975, oral osmotic systems now address areas as diverse as cardiovascular disease to attention deficit disorder, with more than a dozen products marketed worldwide. Osmotic systems utilize the principle of osmotic pressure for the delivery of drugs. Drug release from these systems is independent of pH and other physiological parameter to a large extent and it is possible to modulate the release characteristic by optimizing the properties of drug and system. The oral osmotic pumps are now in advanced stage and the available products on this technology and number of patents granted in the last few years indicates its market success. Alza corporation of the USA was first to develop an oral osmotic pump and today also they are the leaders in this field with a technology by name OROS. The rate of release of drug in these products is determined by the constant inflow of water across a semi-permeable membrane into a reservoir, which contains an osmotic agent. The drug is either mixed with the agent or is located in a reservoir. The dosage form contains a small hole from which the dissolved drug moves out at a rate determined by the rate of entrance of water due to osmotic pressure. The rate of release is constant and can be controlled within tight limits yielding relatively constant blood concentrations. The rate of release can be modified by altering the osmotic agent and the size of the hole. An example of this type of product is Adalat Oros (Nifedipine).

For the first time in 1955 an Australian pharmacologist Rose and Nelson developed an implantable osmotic pump. Next quantum leap in osmotic dosage form came in1972 when Theuwes invented elementary osmotic pump. After that many have been invented which enable controlled delivery of almost all drugs. These devices can be now classified into implantable, oral osmotic pump, and other specific types. Different types of implantable osmotic pumps include: the Rose and Nelson pump, Higuchi Leeper pump, Higuchi Theuwes pump and implantable Miniosmotic pump. Different types of oral osmotic pumps include: single chamber osmotic pump (elementary osmotic pump), multichamber osmotic pump (push pull osmotic pump, osmotic pump with non expanding second chamber). Specific type osmotic pumps include: controlled porosity osmotic pump, bursting osmotic pump, Liquid OROS, delayed delivery Osmotic device, telescopic capsule, oros ct (colon targeting), sandwiched oral therapeutic system, osmotic pump for insoluble drugs, monolithic osmotic systems and OSMAT. The basic components of an osmotic pump include the drug, osmotic agent and semipermeable membrane. Drug generally incorporated has short biological half-life (2-6 hr), highly potent, and requires prolonged treatment. Examples of the drugs include nifedipine, glipizide and verapamil. Osmogents used for fabrication of osmotic dispensing device are inorganic or organic in nature a water soluble drug by it self can serve the purpose of an osmogent. Inorganic water-soluble osmogents include: magnesium sulphate, sodium chloride, sodium sulphate, potassium chloride, sodium bicarbonate. Organic polymer osmogents include: sodium carboxymethyl cellulose, hydroxypropylmethyl cellulose, hydroxyethylmethylcellulose, methylcellulose, polyethylene oxide, polyvinyl pyrollidine. The semi permeable membrane should be stable both to the outer/inner environment of the device. The membrane must be sufficiently rigid so as to retain its dimensional integrity during the operational lifetime of the device. The membrane should also be relatively impermeable to the contents of dispenser so that osmogent is not lost by diffusion across the membrane finally, the membrane must be biocompatible. Ideal properties of semi permeable membrane include: 1. The material must posses sufficient wet strength (-10^5) and wet modulus so as to retain its dimensional integrity during the operational lifetime of the device. 2. The membrane exhibit sufficient water permeability so as to retain water flux rate in the desired range. The water vapor transmission rates can be used to estimate water flux rates. 3. The reflection coefficient and leakiness of the osmotic agent should approach the limiting value of unity. Unfortunately, polymer membranes that are more permeable to water are also, in general more permeable to the osmotic agent. 4. The membrane should also be biocompatible. The substance forming a large part of the outer surface of the novel device of this invention is semipermeable, for example a material that is permeable to an external fluid such as water and the like while essentially impermeable to a selected product or other compounds in the device. This material can be non-erodible or bioerodible after a predetermined period of time and in each instance it is semi-permeable to solvent but not to solute and is suitable for construction of the outer layer of the device. Some materials which form semi-permeable membranes include cellulose acetate, cellulose nitrate, cellulose diacetate, cellulose triacetate, agar acetate, amylose triacetate, beta glucan acetate, beta glucan triacetate, cellulose acetate, cellulose acetate phthalate, polyurethanes, polyglycolic or polylactic acid and derivatives.

The following are the advantages of the osmotic pumps whether used for oral route or parenteral route. 1. They typically give a zero order release profile after an initial lag. 2. Deliveries may be delayed or pulsed if desired. 3. Drug release is independent of gastric pH and hydrodynamic condition. 4. They are well characterized and understood. 5. The release mechanisms are not dependent on drug. 6. A high degree of in-vitro and in vivo correlation is achieved. The following are the disadvantages: 1 Costly. 2. If the coating process is not well controlled there is a risk of film defects, which results in dose dumping 3. Size of the hole is critical.

Another example regarding the osmotic pumps, worth mentioning is Duros technology platform. This technology platform can be conveniently applied to any drug. This technological platform is mainly applied for controlled release parenteral delivery of the drugs. The ALZET® osmotic pump for animal research studies has been the basis for almost 7000 research studies. More recently, osmotic principles have been applied to human parenteral therapy, resulting in the development of the DUROS® technology. The potential of the DUROS technology as a platform for providing drug therapy was demonstrated by the Food and Drug Administration's approval in March 2000 of ALZA's Viadur product (leuprolide acetate implant), the first approved product to incorporate the DUROS implant technology. The Viadur implant delivers the GnRH analog leuprolide for 365 days for the palliative treatment of prostate cancer and has achieved good patient acceptance. The research studies performed with the ALZET osmotic pump demonstrated the breadth of applicability of parenteral delivery based on the principles of osmosis and that parenteral osmotic systems could be designed for effective site-directed delivery. DURECT Corporation was founded in 1998 to further develop the DUROS technology for advanced applications licensed from ALZA Corporation for selected fields in pain management and site-directed delivery. The DUROS technology is a miniature drug-dispensing system that operates like a miniature syringe and releases minute quantities of concentrated drug formulations in a continuous, consistent flow over months or years. The system is implanted under the skin and can be as small as 4 mm OD X 44 mm L or smaller. The system consists of an outer cylindrical titanium alloy reservoir. This reservoir has a high-impact strength and protects the drug molecules from enzymes, body moisture, and cellular components that might deactivate the drug prior to delivery. At one end of the reservoir is positioned the membrane, constructed from a specially designed polyurethane polymer. The membrane is permeable to water but substantially impermeable to ions. Positioned next to the membrane is the osmotic engine. The engine contains primarily NaCl, which is combined with other pharmaceutical excipients in tablet form. Next to the engine is the piston. The piston is made from elastomeric materials and serves to separate the osmotic engine from the drug formulation in the drug reservoir compartment. At the distal end of the titanium cylinder is the exit port. Exit ports can range from simple, straight channels to more complicated design configurations. The exit port design must be coupled to the rheological properties of the drug formulation. The drug formulation is contained in the drug reservoir compartment. The drug formulation may be either a solution or suspension. DUROS drug solutions can be both aqueous and non-aqueous in nature. DUROS drug formulations must exhibit stability at body temperature (37°C) for extended periods of time, usually ranging from 3 months to 1 year. Stable formulations have been developed for a number of drugs. It is possible to develop stable formulations of peptides and proteins, especially if suspension formulations are pursued. By formulating a non-aqueous suspension, the stability advantages of solids are exploited, and the absence of water greatly diminishes losses from hydrolytic degradation reactions. For many applications, the preferred site of implantation is subcutaneous placement in the inside of the upper arm. When implanted, a large, constant osmotic gradient is established between the tissue water and the osmotic engine. Osmosis is the movement of a solvent through a semi-permeable membrane from a region of low-solute concentration to a region of high-solute concentration; the osmotic engine provides a region of high NaCl concentration. The engine is specifically formulated with an excess of NaCl, such that solid NaCl is present throughout the delivery period. This results in a constant osmotic gradient throughout the delivery period. In response to the osmotic gradient, water is drawn across the membrane into the osmotic engine. The rate of water permeation is constant because of the constant osmotic gradient. Further, *in vivo* studies have shown that the rate is constant *in vivo*, confirming that the membrane is not fouled *in vivo*. The water imbibed into the osmotic engine expands its volume at a constant rate, thereby displacing the piston down the bore of the system at a controlled, steady rate. This displacement pumps drug formulation from the drug reservoir through the exit port and into the patient.

1.4.6 Drug Covalently Linked to the Polymer

Drug covalently linked to the polymer is now decisive in the design and preparation of control drug release formulations. There is an advantage with these systems compared when compared to other CR products. Ideally CR products should deliver a drug to specific site in a specific time and release pattern. Initially, constant or sustained drug release were the kinetics pursued by most of the CR products in order to avoid problems associated with conventional administration in chronic treatments. This concept has now evolved to the trend of developing CR systems that fits to the circadian rhythm by using the so-called stimuli responsive polymers or "intelligent" polymers. In this sense, the main advantages of polymers are their great versatility from the structural point of view, the possibilities to combine hydrophobic and hydrophilic components, as well as the interactions polymer-polymer, polymer-drug, polymer-solvent that offer many possibilities to design and prepare formulations with specific properties and functions. Other aspects that DDS covers are: the slow release of water soluble drugs, the improvement of the bioavailability of low soluble drugs, the delivery of two or more drugs from the same formulation, the possibility of having readily clearable polymer carriers, the control of the release of highly toxic drugs, and the improvement of the targeting to tissues or cells. Covalent polymer-drug conjugates are a special type of CR products where the drug or bioactive compound (peptides, proteins, growth factors, hormones, enzymes, etc.) is covalently linked to the macromolecular backbone through a physiologically labile bond. The possibility of linking any bioactive molecule to a macromolecular chain make polymeric conjugated systems very useful for applications not only related to medication, but also in fields as tissue engineering, biosensors, affinity separations, enzymatic processes, cell culture, etc.

1.4.7 Ion Exchange Resins

Drug release characteristics depend on the ionic state of the environment when a drug is contained in a resin. Thus, in the case of oral administration the drug release is determined by the ionic environment of the GIT. This principle can be conveniently applied to sustain the release of the drug. Thus, CR products can be developed using ion exchange resins. Because this approach of sustained release requires the presence of ions in solution, it would not be applicable to the skin, the external ear canal, or other areas with limited quantities of eluting ions. The subcutaneous and intramuscular routes, where the pool of available ions is more controlled, would appear better suited for this approach. With the GI tract appears to possess a rather constant ionic content, the variability in diet, water intake, and GI content composition make this constant ionic content unlikely. Nevertheless, oral product employing this principle are available in the market, especially for the prolonged release of the drug. Resins are water-insoluble materials containing anionic or cationic groups in repeating positions on the resin chain. The drug-charged resin is prepared by mixing the resin with drug solution either by repeated exposure of the resin to the drug in a chromatographic column or by keeping the resin in contact with drug solution for extended periods of time. The drug-resin is then washed to remove contaminant ions and dried to form particles or beads. When a high concentration of an appropriately charged ion is in contact with the ion-exchange group, the drug molecule is exchanged and diffuses out of the resin to the bulk solution. As is true with all CR products which involve diffusion processes, the area of diffusion and diffusional pathlength are important to the rate of diffusion. In addition, the amount of solvent in the matrix of the resin, as well as the structural rigidity of the resin also infuences the drug diffusion rate. For this reason, the porosity of the resin and the size of the bead or particle must be carefully controlled during the formulation process. The release rate can be further controlled by coating the drug-resin complex using one of the microencapsulation processes. Coated and uncoated drug-resin complexes may be mixed in certain ratios and filled into capsules with excipients or suspended in a palatable flavored vehicle containing suitable suspending agents. This has been shown to be a reliable technique to obtain desired release profiles. The release of drug from uncoated resin beads is expected to begin immediately while release from the coated form would be delayed depending on the type and thickness of the coat. Examples of ion-exchange resin type of product are Duromine containing the basic drug phentermine complexed onto an anionic resin and MS contin (Morphine sulfate) suspension which uses a polystyrene sulphonate resin.

1.4.8 Responsive Drug Delivery Systems

These devices are capable of releasing therapeutic agents by well-defined kinetics and have significant improvement over conventional CR systems. In these devices, the drug output is adjusted in reponse to physiological end. Responsive drug delivery systems can be classified as open- or closed-loop systems. Open-loop systems are also called pulse or externally regulated systems; the amount of drug released is not dependent on the environmental conditions the device is in. Among the most advanced externally-regulated devices are *mechanical pumps*, which dispense drugs from a reservoir outside the body via a catheter. Insulin-delivering pumps are commercially available with sophisticated control mechanisms and computers that can allow a programmed insulin delivery. Although these devices are not primarily made of polymers, the device-tissue interface can be expected to be polymeric. The rate of drug released can also be controlled and enhanced using external stimulants, like magnetism and ultrasound. In *magnetically-controlled drug delivery devices*, small magnetic spheres are embedded in a drug-containing polymer, which release a significant amount of drug when exposed to an oscillating field. Similarly, the release rate also increases when analogous drug-containing polymers are exposed to *ultrasound*. Ultrasound was found to enhance erosion and degradation of some biodegradable polymers, and it can also act as an on-off switch as in certain drug delivery systems.

In closed-loop systems, or self-regulated systems, the release is in direct response to the conditions detected, be it temperature, type of solvent, pH, or concentration, to name a few. Poly(N-isopropylacrylamide) is a well-known example of a *thermo-responsive polymer*: At its transition of 32°C, the polymer is soluble in water; but, as temperature is increased, the polymer precipitates and phase separates. Poly(ethylene glycol) and poly(propylene glycol) copolymers and poly(lactic acid) and poly(glycolic acid) copolymers also exhibit thermo-responsiveness. These polymers are useful in developing thermogelling systems (Atridox®); the drug is dissolved in the liquid form of the polymer at room temperature. When this mixture is injected in the body, the polymer turns into a gel, which eventually degrades and releases the drug molecules. Self-regulating insulin-delivery devices depend on the concentration of glucose in the blood to control the release of insulin. One system proposed immobilizing glucose oxidase (an enzyme) to a pH-responsive polymeric hydrogel, which encloses a saturated insulin solution. At high glucose levels, glucose is catalyzed by glucose oxidase and converts it to gluconic acid, thus lowering the pH. This decrease in pH causes the membrane to swell, forcing the insulin out of the device.

1.5 Design of Sustained Release Dosage Forms

The development of CR delivery systems stemmed from the need to reduce the dosing burden by releasing active substance at a passively controlled rate over an extended period of time. The advantages include reductions in fluctuations in drug concentration and adverse side effects, an increase in patient comfort and compliance, and, potentially, reduced healthcare costs. Since the early 1970s, CR delivery systems exploded in popularity. This led to the use of drug delivery technologies as a life cycle management tool and repatentability strategy employed by many pharmaceutical manufacturers for their leading brands. The rapid and continuous innovation behind CR technologies is impressive as evidenced by the creation of specialty drug delivery companies. CR dosage forms have been developed over three decades. They have increasingly gained popularity over other dosage forms in treating diseases. One of the first controlled release dosage form is the spansule introduced in 1950s. Spansule capsules were manufactured by coating a drug onto nonpareil particles and further coating with glyceryl stearate and wax. Subsequently, several formulations with different mechanisms of release have been introduced. Thus, the pharmaceutical companies have provided a variety of dosage forms and dosage levels of particular drugs, thus enabling the physician to control the onset and duration of drug therapy by selecting a suitable dosage form available.

Formulator should have knowledge of several important aspects before initiating the program of design of sustained release dosage form for a certain drug. To establish a procedure for designing sustained release dosage forms, it is useful to examine the properties of drug blood-level time profiles characteristics of multiple dosing therapy of immediate release forms. In general for drug therapy, selection of proper dose and dosage interval is a prerequisite to obtain a drug level pattern that will remain in therapeutic range. The minimum effective and maximum safe doses should be known. With sustained release dosage forms, the dosage interval is more compared to the conventional dosage forms. In general with oral CR dosage forms, the objective is to able to provide a sustained release plasma profile for upto 12 hours. With parenteral dosage forms, it depends on the need of the duration. Physicians and pharmacists were aware of plasma pattern of drug concentration versus time before even the CR formulations entered the market. Elimination of drug level oscillations can be achieved by administration of drug through constant-rate intravenous infusion. To design an efficacious sustained release dosage form, one must have a thorough knowledge of the pharmacokinetics of the drug chosen for this formulation. Either for conventional dosage form or for CR dosage form it is always assumed that drug blood levels are assumed to correlate with therapeutic effect and drug kinetics are assumed to be adequately approximated by a one-compartment model. In this case, the drug distribution is sufficiently rapid so that a steady state is immediately attained between the central and peripheral compartments, i.e., the blood-tissue transfer rate constants, k_{12} and k_{21} are large. Under the foregoing circumstances, the drug kinetics can be characterized by three parameters: the elimination rate constant (k_e) or biological half-life, the absorption rate constant (k_a) and volume of distribution (Vd). In case of the drug following a two compartment model Vc is the volume of the central compartment. In case of CR products, several parameters such as the loading dose, maintenance dose, rate of release of the maintenance dose should be calculated. To obtain a constant drug level, the rate of drug absorption must be made equal to its rate of elimination. Consequently, drug must be provided by the dosage form at a rate such that the drug concentration becomes constant at the absorption site. Detailed theoretic treatments of a number of sustained release dosage for designs have been reported. These systems include:

1. No loading dose with zero-order drug release.
2. No loading dose with first-order drug release.
3. Loading dose with zero-order drug release.
4. Loading dose with first-order drug release.

When loading dose is included, designs based on both immediate and delayed release of maintenance dose have been described. Any of the above designs is simple. For instance, Meka et al., (2008) developed a biphasic gastroretentive floating drug delivery system with multiple-unit mini-tablets based on gas formation technique to maintain constant plasma level of a drug concentration within the therapeutic window (4). The system consists of loading dose as uncoated core units, and prolonged-release core units are prepared by direct compression process; the latter were coated with three successive layers, one of which is seal coat, an effervescent (sodium bicarbonate) layer, and an outer polymeric layer of polymethacrylates. Another type of these formulations can be prepared using HPMC. HPMC matrix / mini-matrix systems have been formulated to achieve such "fast/slow" drug release patterns. These matrices are reported to contain the drug fraction for the extended release phase, while the drug fraction for immediate release was integrated into the matrix / minimatrix via an immediate releasing layer in a double-layer tablet system, incorporated into a release controlling coating over the matrices or incorporated into the voids between compressed mini-matrices. The equations useful in the calculations of the loading dose and maintenance dose were clearly described by Nicholas Lordi in the Theory and Practice of Industrial Pharmacy (Eds. Lachman, Liberman and Kanig, 3rd Edition) (5). Thus, when intended for acute or intermittent administration it is desirable to have an initial slug of drug rapidly absorbed followed by a slower maintenance component. For chronic administration, a zero order absorption rate is the theoretical goal, where the rate of systemic drug appearance, $R_{systemic}$, is given by (6):

$$R_{systemic} = F.R = CL.C$$

Where F is the drug bioavailability, R is the rate of drug administration, CL is total plasma clearance and C is the target steady-state plasma concentration. The amount of drug contained in each unit dosage form, D_{unit} is given by:

$$D_{unit} = CL.\ C.\ \tau/F$$

Where τ is a constant dosing interval. In practice, very few (if any) CR dosage forms result in zero-order systemic drug appearance. There is a fundamental dilemma encountered with any oral dosage form, but this is particularly true with CR forms. There are several factors that may affect the performance of the CR dosage form and this may contradict the zero-order systemic drug appearance. Zero-order systemic drug appearance is important on several occasions. One aspect of research about controlled-release delivery systems involves designing a system which produces steady-state plasma drug levels, which is also referred to as zero-order drug release kinetics. To meet this objective, numerous design variations have been attempted, and their major controlling mechanisms include diffusion/dissolution, chemical reactions, the use of osmotic pump devices, and multiple layer tablet designs, all of which incorporate numerous manufacturing steps and many associated drug release mechanisms. In general, drug release parameters (duration of drug release, release rate), expected steady state plasma drug concentrations and dosage form index are calculated and these are compared with theoretical controlled release parameters developed based on the pharmacokinetic characteristics of the drug (See reference 5). As mentioned before, the common CR approach is to combine a rapid-release dose fraction with a fraction having pseudo-first order release characteristics. When absorption is not rate limiting, the ideal approach to this situation is a zero-order drug delivery of the drug to the absorption site. Thus, when absorption is not the rate limiting step, the release of the drug from the dosage form becomes important. A zero-order drug release to the absorption site will result in steady-state plasma concentration. The fact that drug in a CR dosage form is not absorbed at a zero-order rate does not imply a faulty product. The goals of CR therapy can be achieved with first-order absorption. Gibaldi and Perrier have demonstrated that for many drugs, acceptable steady-state plasma level-time profiles may be obtained assuming absorption half-lives of about 3-4 hours (7). Their criterion of acceptability was a low $C_{av,max}/C_{av,\ min}$ quotient, where the plasma concentrations are the time averaged maximum and minimum steady state values, respectively. Theeuwes and Bayne have termed this ratio the "dosage form index" and used it successfully to compare acetazolamide CR products (8). It is widely held that zero-order in vivo release is the intent, but this is incorrect – the ultimate goal is zero-order systemic drug appearance. Zero-order in vivo release will produce zero-order systemic absorption only if: (1) the gut behaves as a one compartment model, i.e., its various segments are homogenous with respect to absorption; and/or (2) drug release rate is rate limiting for absorption. This has to be always kept in the mind. Just developing an in vitro zero-order drug release system may not be the only criteria. Scientists have successfully predicted plasma concentration-time profiles for drugs in oral CR dosage form based on in vitro dissolution and in vivo oral solution absorption profiles. This is under the implication that one of both of the above assumptions are mathematically valid under certain situations.

1.6 CR Formulations Used in Different Routes of Administration and the Salient Features

The goal of every drug delivery system is to deliver the precise amount of a drug at a pre-programmed rate to the desired location in order to achieve the drug level necessary for the treatment. A number of design options are available to control or modulate the drug release from a dosage form. Majority of the CR dosage form fall in the category of matrix, reservoir or osmotic system. This is true for all the routes of administration. In matrix system, the drug is embedded in polymer matrix and the release takes place by partitioning of drug into the polymer matrix and the release medium. In contrast, reservoir systems have a drug core surrounded\coated by the rate controlling membrane. However factors like pH, presence of food and other physiological factors may affect drug release from conventional controlled release systems. Osmotic systems utilize the principle of osmotic pressure for the delivery of drugs. Drug release from these systems is independent of pH and other physiological parameters to a large extent and it is possible to modulate the release characteristics by optimizing the properties of drug and system. Some of these issues are discussed in this section taking into consideration each route of administration.

1.6.1 Oral CR formulations

Over the last few years, consumers witnessed the wide spread and availability of a plethora of oral controlled release (CR) products in the marketplace. For example, by 1998, the U.S. Food and Drug Administration (FDA) approved 90 oral CR products for marketing.

From 1998 to 2003, in just five years, the FDA approved an additional 29 new drug applications that used CR technologies. Consequently, oral CR technologies are becoming more complex and encompassing multiple presentations. It is well recognized in the pharmaceutical industry that oral CR dosage forms can be defined based on release-profile characteristics or the underlying release- controlling mechanism.

The technologies behind oral drug delivery have emerged from the mainstream pharmaceutical industry and have become influential forces in their own right, as evidenced by the burgeoning "drug delivery companies" that are at the forefront of innovation and hold their own niche market. CR products evolved with simple matrix technology. Several research articles in the 1950s and 1960s reported simple matrix tablets or monolithic granules. For the first time in 1952, Smith Kline & French introduced the Spansule, a timed-release formulation that launched a widespread search for other applications in the design of dosage forms. The aim behind the development of these dosage forms was to achieve a constant release of the entrapped drug. As a reason products like Procardia XL were developed and became one of the blockbusters in the market for the past several years. On the basis of zero-order drug release concept, the zero-order osmotic delivery system used in Procardia XL was developed.

1.6.1.1 Currently marketed oral CR products

The development of technology in leaps and bounds and the availability of various polymers and the machinery available to prepare novel designs has currently resulted in the development of these oral CR products in a reproducible manner. The main oral drug-delivery approaches that are currently available include:

1. Coating technology using various polymers for coating tablets, nonpareil sugar beads, and granules.
2. Matrix systems made of swellable or nonswellable polymers.
3. Slowly eroding devices.
4. Osmotically controlled devices.

Two distinct drug release profiles, extended and delayed release, are achievable, and they can be used in various combinations to provide the desired release rate. Three delivery systems dominate today's market of oral CR products: matrix, reservoir, and osmotic systems. Release mechanisms from these dosage forms have been the subjects of extensive studies. Among them, diffusion plays a key role in both matrix and reservoir systems, whereas osmotic pressure is the predominant mechanism of drug release from osmotic systems and could also play a role in a reservoir system. Owing to technology accessibility, manufacturing, cost, and other considerations, diffusion-based CR products are used more widely than osmotic systems. For example, of the 29 CR products approved by the FDA between 1998 and 2003, 12 were based on matrix systems and 10 based on reservoir technologies compared with 2 osmotic tablets. In the design of single-unit, matrix-type controlled release dosage forms, conventional tablets are still popular. The advancement of granulation technology and the array of polymers available with various physicochemical properties (such as modified cellulose or starch derivatives) have made the development of novel oral controlled release systems possible. Matrix devices made with cellulose or acrylic acid derivatives, which release the homogeneously dispersed drug based on the penetration of water through the matrix, have gained steady popularity because of their simplicity in design. The drawback of matrix-type delivery systems is their first-order drug delivery mechanism caused by changing surface area and drug diffusional path length with time. This drawback has been addressed by osmotic delivery systems, which maintain a zero-order drug release irrespective of the pH and hydrodynamics of the GI tract. Multiparticulate systems are gaining favor over single-unit dosage forms because of their desirable distribution characteristics, reproducible transit time, and reduced chance of gastric irritation owing to the localization of drug delivery. Although several technologies for the production of microparticulate systems have been designed, thus far the mainstream technologies are still based on spray-drying, spheronization, and film-coating technology. Reservoir type of devices are very often mentioned. However, there is a problem of manufacture reproducibility and lack of safety.

1.6.1.2 Special features in the design of oral CR products

The design of oral CR products is based on several factors. The properties of the drug to be incorporated are important. Drugs with long elimination half lives are generally undesirable for CR products. Exception is for the use of this technology to prevent toxic effects due to a peaking effect or to reduce the dose. Also the pharmacological effect for some drugs is inherently sustained. For these drugs CR formulations may be redundant. Ex. 1. The drug binds to tissues (tissue bound ACE inhibitors). For these drugs, less frequent dosing is needed even though the drug may have a short half-life. 2. The drug that has irreversible effects (the inhibition of platelet cyclo-oxygenase by aspirin). 3. The relationship between response and plasma/blood concentrations is relatively flat or if the dose given results in concentrations which are in the plateau region of the dose-response relationship (thiazides in hypertension). 4. The drug is metabolized to pharmacologically active metabolite (s), which are more slowly cleared than the parent drug (quinapril, trandolapril and venlafaxine). Several pharmacokinetic parameters associated with the drug also have profound influence on the suitability of a drug for CR formulation. To avoid accumulation in the body, generally drugs with biological half-life between 2-6 hr is preferred. If a drug that undergoes extensive first-pass metabolism is incorporated in a CR product its bioavailability may be significantly impaired. If the absorption site is limited, absorption is likely to decrease and variable bioavailability will occur for CR dosage forms. Drugs which undergo non-linear elimination due to drug metabolism, saturation or other factors may not be good candidates for oral CR dosage forms. Drugs that undergo non-linear absorption are also not suitable candidates for oral CR dosage forms. Undesirable adverse reactions may develop by using CR dosage forms. Prior to its application to a suitable drug, the following factors should be clarified: 1. The clinical response should be correlated with blood-drug concentrations or tissue concentrations at the site of action. 2. There should be no induction or inhibition of the metabolizing enzymes by prolonged concentration of the drug in the blood; nor it should chance the casual change of pharmacological response or lead to possible tolerance or addiction for the drug. 3. There should be no interactions with other drugs due to protein binding. The major purpose for developing CR products of the drug is generally to maintain the blood concentration of the active ingredient at therapeutically effective levels. Therefore, it is desirable that average minimum effective concentration and optimal therapeutic concentrations be clarified for each drug by evaluating blood concentrations of the active ingredient or therapeutic moiety(s) including active metabolite(s) in relation to drug efficacy. The intra- and intersubject variations should be investigated for further confirmation of these levels. It is also desirable to investigate toxic blood drug concentrations. If the effective blood drug concentration is not known, estimates should be made from dose levels, blood concentrations, and clinical data based on the immediate release drug product. If effective blood drug concentration is unclear, the usefulness of the CR forms should be demonstrated by well-designed clinical studies.

Biopharmaceutical properties of the active ingredient incorporated into CR dosage form should be well known in rational formulation design. The following factors should be thoroughly investigated and understood: 1. location of major absorption sites or specificity in the site of absorption, 2. absorption rate, 3. the elimination half life of the drug,

4. whether absorption is non-linear due to the saturated drug absorption, first pass effects, or other reasons, 5. whether elimination is non-linear due to drug, 6. The effect of food, drugs likely to be used concurrently and physiological factors such as renal or hepatic function on the absorption, distribution, metabolism and excretion of the drug be studied and evaluated. In addition, the study effects of age, sex and smoking on the pharmacokinetics of the drug may be useful. Also, chemical and physicochemical properties of drugs, especially, pH- solubility characteristics should be known in advance. There are certain properties of the drug, which must be taken into consideration for the design of oral CR dosage form. The aqueous solubility and intestinal permeability of drug compounds are of paramount importance. A drug which has high solubility at intestinal pH and absorbed by passive diffusion has an ideal characteristics for fabrication of oral CR dosage form. A drug with high solubility and high permeability is also a best case for CR. Low soluble and low permeable drugs pose a worst case for oral CR formulations. A $p < 0.5 \times 10^{-6}$ mms^{-1} is not at all suitable for oral CR formulations. This is because once the drug is dissolved in the GIT, its permeability across the membrane becomes important. More than 90% absorption in vivo may be expected for compounds with permeability coefficient $p > 4 \times 10^{-6}$ mms^{-1}, whereas less than 20% absorption is expected when $p < 0.5 \times 10^{-6}$ mms^{-1}. A drug with no site-specific absorption characteristics is preferred. A drug with low aqueous (< 1 mg/ml) may already possess inherent sustained release potential. Generally, a dose preferable between 125-325 mg is suitable for oral CR dosage forms. However, this is slowly changing. Now technology is veering towards high dose oral CR formulations.

Physiological factors that affect the absorption should be thoroughly considered. The release of active ingredient from a CR formulation and its absorption are very much influenced by the physiological factors in the GIT. CR dosage forms are more susceptible to these factors than immediate release dosage forms. Therefore, the possible effects of the physiological factors should be fully considered for the dosage form design. The physiological characteristics of the gastrointestinal tract (the volume, composition, pH, surface tension and viscosity of the gastrointestinal content, and gastrointestinal motility) vary greatly from site to site. CR dosage forms remain in the gastrointestinal tract longer than conventional preparations. Therefore, physiological conditions of the gastrointestinal tract can affect the release of active ingredients of these forms much more than release from conventional forms. Noteworthily, gastric pH varies from acidic to neutral, and these variations can affect release of the active ingredient from the dosage form. These points should be considered when a formulation is being designed and assessed. If the drug is intended for use in a specific subpopulation, attentions should be paid to the specific physiology of the subpopulation.

The transit rate of a dosage form through the gastrointestinal tract is known to depend on the formulation properties such as size, form, specific gravity and adhesiveness of the preparation and physiological properties such as the length, size and motility of the gastrointestinal tract; and on the composition and volume of the gastrointestinal content. It is also affected by food, diseases, posture, and stress. The bioavailability of drugs often depends on the gastrointestinal transit rate of the dosage form. Therefore, the traveling characteristics of the dosage form through the gastrointestinal tract should be fully considered in designing advantageous dosage forms.

Desirable criteria of performance for CR dosage forms are: 1. duration of appropriate blood drug concentration for a sufficient time with minimal influence of food and physiological conditions of the gastrointestinal tract. 2. minimal contribution to intra- and intersubject variation. To select the best possible dosage form, all candidate forms should be fully tested for release characteristics. Moreover the pharmacokinetic profile should be evaluated in an appropriate species of animal or volunteer.

The release of the active ingredient from the preparation in the gastrointestinal tract is affected by many physiological factors including the mechanical force exerted by the digestive tract in relation to its movement, and the volume, composition, pH, surface tension, and viscosity of the gastrointestinal fluid. Therefore, the in vitro release behaviors should be investigated under as many conditions as possible to understand possible effects of gastrointestinal variables on in vivo release. To achieve stable blood concentrations, it is generally desirable to prepare prolonged release dosage forms whose release rates are minimally pH dependent. Therefore, release of the active ingredient should be evaluated at multiple levels of pH, such as 1.2, 4.0 and 6.8, representing typical gastrointestinal pH variation. Considering the variation in gastrointestinal motility; agitation rates should also vary more than 2 levels among 50, 100 and 200 rpm, when the paddle method is used, at an appropriate pH. If it is anticipated that the release rate is influenced by the wettability, ionic strength and composition of the test medium, their effects should also be investigated. It is also desirable to perform release tests using different kind of apparatus. On the other hand, taking into consideration the variation of mechanical stress in the gastrointestinal tract, the drug release from CR dosage forms containing an active ingredient with a narrow therapeutic window should be tested by the methods having a high mechanical stress, such as JP disintegration test method, the rotating flask method using beads and solubility simulator.

The specifications for drug releases should be established for quality control of CR dosage forms. Basically, it is desirable to employ the release tests which can predict the blood level profile of the drug as precisely as possible. It is also desirable to set the specification including sampling time and amount of drug to be released so as to show the release profile as accurately as possible. The tolerable range of the drug release change depending on the effect of the release rate on absorption or a related pharmacodynamic property (therapeutic window, toxicity or adverse reactions). Therefore, based on the relation between release rate and blood concentration or pharmacological effects, the tolerable range should be set within limits which do not allow great changes in blood concentrations or in clinical efficacy. The narrow tolerance limits should be set as much as possible to decrease the variation in drug release which will provide stable clinical effects. If the relation between the release rate and blood concentration is not clear, or if sufficient data are not available to prove the correlation, it is difficult to set rational specification. In such a case it is desirable to set specifications using the second method (paddle method) in the Japanese Pharmacopoeia at sampling time points of 20-40%, 40-60%, and more than 70% of the labeled amount of the active ingredient is released. If 100 rpm and 900 ml of test fluid was used for the paddle method, the tolerance ranges at 1st, 2nd and 3rd points should be set within 15%, 15% and 10% of the average release, respectively. At the 3rd sample point, only lower limit is acceptable instead of the tolerance range. The acceptance criteria of the drug release follow the criteria of dissolution or release tests of JP XI or USP XXI. Specimens for long term stability tests should be subject to dissolution testing and comply with the standards of the specifications.

As far as possible, the pharmacokinetics of the prolonged release dosage form should be compared with the immediate release product in healthy volunteers. Pharmacokinetic evaluation should be made, based on blood concentration data, except for the case that the concentrations of the active ingredient can be determined at the site of action whose effective concentrations are known. Data on drug concentration in the urine, saliva, or other body fluids will be accepted only when the concentrations of the active ingredient in the blood or at the site of action are correlated with that in these fluids. Unless the drug shows linear pharmacokinetics within the clinical dose range, the investigation should be made at two dose levels, high and low.

(i) ***Single dose study:*** The usefulness of the new CR dosage form given according to the dosage regimen should be evaluated by comparing the blood concentration with that of the immediate release dosage form or alternative forms such as solution or a powder; or with a prolonged release product which has already been approved, when better prolonged release characteristics are claimed. The parameters to be compared are AUC (zero to the final sampling time), AUC (0-∞), Cmax, the duration of the minimum effective concentration, or optimal effective concentrations of the active ingredient if these concentrations are known or can be estimated. It is desirable to determine the time to reach the minimum effective concentration or the optimal effective concentration, Tmax, absorption rate constant, elimination rate constant, clearance, extent of absorption and MRT and VRT by the moment analysis method.

(ii) ***Multiple dose study:*** Prior to a multiple dose study, a blood concentration profile at steady state for multiple dosing of both standard and test dosage forms should be simulated from the single dose pharmacokinetic trials. In the multiple dose studies, it should be ascertained that Cmax and Cmin at steady state are within the estimated ranges, and the usefulness of the prolonged release dosage form should be evaluated by comparing it with the reference product in 1) Cmax, 2) Cmin, 3) the difference between Cmax and Cmin or the ratio (dosage form index, Cmax/Cmin), and the duration of the minimum effective concentration or that of optimal effective concentration. For drugs with non-linear absorption or elimination, those with a narrow therapeutic window, or those which may cause severe adverse reactions, the blood concentration profile at steady state should be characterized by multiple dose studies. When multiple dose studies in healthy volunteers are not done, the usefulness of the prolonged release dosage form should be shown using the simulated parameters, where it is necessary to confirm that Cmax and Cmin are within the predicted range, by monitoring blood concentrations in clinical studies.

Factors which might affect the pharmacokinetics of a CR dosage form should be studied in which food is particularly an important factor because it is know to affect transit of dosage forms in gastrointestinal tracts, disintegration, and release of the drug. Therefore, the blood concentration profiles of the prolonged release dosage form should be compared between fasting and fed conditions. If a significant effects of food was observed, a special caution should be included in the dosage regimen (i.e. indication of drug administration only after meals), and it should be clarified whether the food effect was related to the drug itself or dosage forms by performing similar food studies using the drug solution or the immediate release product, although the studies are not needed when there is published evidence. In addition, as far as possible, it is desirable to clarify other factors of food (e.g., the volume and composition of meal, and intervals between food and drug administration) affecting the in vivo release and absorption. It is also desirable to investigate diurnal variations of pharmacokinetic parameters.

The clinical usefulness of the prolonged release dosage form should be shown comparing it with its already approved immediate release product or its already approved CR product (if a better CR dosage form is claimed). If the relation between the pharmacological effectiveness and blood concentration is unclear, the usefulness should be proved by the well-controlled clinical studies where the effective and toxic concentrations should be investigated by monitoring blood concentrations of the drug. The appropriate dosing regimen should be established during Phase I and II clinical studies in which it is recommended that the blood concentrations are monitored during Phase II clinical trials to establish a better dosing regimen. Factors that are to be considered in establishing dosing regimen include: 1. Overdose or dose dumping: Sustained release dosage forms might be more likely to produce significant adverse and toxic effects than immediate release dosage forms in case of overdose or dose dumping because of the higher doses of active ingredients which are absorbed over a prolonged time. Dose dumping, e.g. resulting from crushing by the teeth, may be another problem with prolonged release dosage forms. This is of particular concern for drugs with a narrow therapeutic window, and so studies are desired to establish preventive measures and actions to be taken in such cases. 2. Disease state: The physiological changes in gastrointestinal tract, liver, kidneys, or heart due to diseases often affect absorption, distribution and elimination of drugs and there is a possibility that CR dosage forms are particularly susceptible to the changes. In such cases, the dosing regimen should be studied and established as to reflect the pathological changes. 3. Combination therapy: If any other drug is used concurrently, it may affect the absorption, distribution, and elimination of the drug contained in the CR dosage form. As a result, blood concentrations of the drug may be changed, and this may affect the efficacy. The possible effect of drugs which might be used together in practice should be studied, and suitable indications and special warnings for the concurrent use of other drugs should be established. Dosing guidelines have to be set very well in advance. Recommendations for dosing conditions, frequency of dosing per day, and dose levels (initial dose, maintenance dose, dose adjustment for insufficient response, and the maximum tolerable dose) should be established, based on the available pharmacokinetic data during Phase II clinical studies. The action to be taken if toxic signs or adverse effects develop should also be specified in these guidelines. Detailed dosing guidelines including information about dose adjustment based on blood concentration monitoring or changes in renal clearance of each patient may be useful to maximize the therapeutic efficacy by making the utmost use of the advantages of the CR dosage form. It is desirable to set up corresponding detailed guidelines particularly for CR products containing A) drugs, blood concentrations of which may change strikingly by minimal changes in dose (drugs with non-linear absorption or elimination), B) drugs, the clearance and blood concentrations of which are susceptible to physiological conditions, age and so forth, C) drugs with a narrow therapeutic window, and D) drugs which might cause tolerance and/or severe adverse effects

1.6.1.3 Techniques of preparation and manufacture

Oral drug delivery is the largest and the oldest segment of the total drug delivery market. It is the fastest growing and most preferred route for drug administration. The fabrication of existing forms of orally administered pharmaceutical product is a relatively inexpensive and straightforward process. For example the process may involve insertion of the beneficial substance into a gelatin shell to form a capsule, or it may involve forming a drug and filler into a tablet. In general, lactose is the most common excipient in the tablet compression process. However, dose dumping may occur with hydrophilic drugs when incorporated as lactose tablets. One of the effective ways of prevention of dose dumping is to formulate into CR products. Materials currently used in known orally administered pharmaceuticals are relatively difficult to engineer in relation to control of rate and location of drug release. Variations, in gastrointestinal conditions, from human patient to human patient are not easily accommodated by existing orally administered products. The same methods with some modifications can be used to fabricate oral CR products. Along with the conventional techniques of preparation, there are several other improvements in the techniques and currently there are several other patented technologies.

As of today, tableting is the simplest, most common, and most economical method of processing an active agent into a drug-delivery product. Most of the times tablets are mainly intended for oral delivery of the active ingradient. However, some parenteral depot formulations such as tableted or compressed wafers are available. Examples of such tablet forms of controlled release implant depots intended for parenteral or local delivery is Gliadel Wafer. The tableting process involves feeding a metered amount (usually from 0.5 to 5 g) of a "granulation" (large particle blend) of an active agent to the dies of a tableting press where the granules are compressed into a tablet of the desired shape. Some manufacturers use intricate shapes for product identification. The premixed granulation contains the active and various excipients to modify the flow, compaction, die release, and dissolution characteristics of the tablet. Granulation is accomplished wet (using solvent) or dry (using no solvent) by high intensity mixers or fluidized-bed granulators. Although single-cavity presses can be used for specialized applications and testing, high-speed rotary presses capable of producing thousands of tablets per minute are typical. The pressures used in tablet compression are quite high, ranging from 50 to 500 mPa; therefore, the dies used in tableting are typically made with high-strength, surface hardened alloys. The same principles of tablet preparation and manufacture can be applied to controlled release oral tablets. These are much more sophisticated as the demand for controlled release has increased. Recently technology can produce bilayer tablets providing two different release rates for a drug. The technologies use different polymeric excipients in the tablet providing both immediate and sustained release. Various over-the-counter analgesics use this technology. Tablets composed of compacted microencapsulated beads have also been developed to provide both sustained release and protection from the irritating effects of certain drugs such as non-steroidal anti-inflammatory drugs. The drug is microencapsulated in a separate process, such as the fluidized-bed process, and then incorporated into the tableting formulation. Effervescent tablets are also produced to provide chewable and fast-dissolving tablets for patients who have difficulty swallowing. The drug is again typically incorporated into the effervescent tablet formulation as microencapsulated beads. Many tableted products are also subsequently coated to provide controlled release products. Coating usually takes place in a pan coater, which is a rotating drum, similar to a clothes drier, which tumbles the tablets in front of a spray nozzle for application of the coating. The coating is dried by a continuous flow of process air. The process air is often heated and solvent reclamation is used for organic-solvent based coatings. The batch sizes for pan coating can vary between 100 and 2,000 kg. Small laboratory models are also available. Various wax and enteric coatings are applied in this manner. For very small tablets, fluidized-bed coating may also be used.

The other types of oral SR dosage forms are available in the forms of capsules. By far the most versatile of all dosage forms is the gelatin capsule. Gelatin capsules can be filled with powders, small pellets or small tablets, liquids, or semisolids, as well as some combinations of these forms. The capsule also provides efficient taste-masking. The drug product is released by dissolution of the gelatin in the stomach. Gelatin capsules can be prepared from hard or soft gelatin; however, hard-gelatin capsules are more versatile for controlled drug delivery. Hard-gelatin capsules are available in several different sizes, and high-speed filling machinery, capable of filling 1,500 capsules per minute, is available. The machines typically consist of a drug hopper, a capsule hopper, a dose metering device, a dose chamber, a filling or tamping pin, a capsule tray, and a finished-capsule collection bin. The capsules are automatically opened, filled, and closed during the manufacturing process. Technologies have been developed whereby controlled release beads and mini-tablets are used to fill a gelatin capsule for convenient administration of an oral, controlled release dosage form. Examples of such products are the sustained release cold medications, where sustained release antihistamines, antitussives, and analgesics are first preformulated into extended release microcapsules or microspheres and then placed inside a gelatin capsule. Another example is enteric-coated lipase minitablets that are placed in a gelatin capsule for more effective protection and dosing of these enzymes.

Several factors are needed to be considered for the development of the oral SR dosage forms. Apart from those developments mentioned previously, several other techniques with different principles are also worthy to know. For instance, use of hydrophilic matrices for oral extended release of drugs is a common practice in the pharmaceutical industry. Hydrophilic matrices of high gelling capacity are of particular interest in the field of controlled release. Other matrix types of tablets with modifications and different other polymers also are in use. In the tablet matrix system, the tablet is in the form of compressed compact containing an active ingredient and tablet excipients such as filler, antiadherent and lubricant. The matrix may be compressed by direct compression of dried powder mixtures or granules. Unlike reservoir and osmotic systems, products based on matrix design can be manufactured using conventional processing and equipment. Second, development time and cost associated with a matrix system generally are viewed as favorable, and no additional capital investment is required. Lastly, a matrix system is capable of accommodating both low and high drug load and active ingredients with a wide range of physical and chemical properties. As with any technology, matrix systems come with certain limitations. First, matrix systems lack flexibility in adjusting to constantly changing dosage levels, as required by clinical study outcome. When a new dosage strength is deemed necessary, more often than not a new formulation and thus additional resources are expected. Furthermore, for some products that require unique release profiles (e.g., dual release or delayed plus extended release), more complex matrix-based technologies such as layered tablets (e.g., Allegra D) will be required. Nevertheless, we expect continued popularity of matrix systems because they have demonstrated success across a wide range of product profiles. With the growing need for optimization of therapy, matrix systems providing programmable rates of delivery become more important. Constant rate delivery always has been one of the primary targets of controlled release systems, especially for drugs with a narrow therapeutic index. Over the past 40 years, considerable effort has been and continues to be expended in the development of new delivery concepts in order to achieve zero-order or near-zero-order release. Examples of altering the kinetics of drug release from the inherent nonlinear behavior include the use of geometrical factors (cone shape, biconcave, donut shape, hemisphere with cavity, core in cup, etc.), erosion/dissolution control and swelling control mechanisms, nonuniform drug loading, and matrix-membrane combinations. Some of the systems are difficult or impractical to manufacture.

A typical reservoir system consists of a core (the reservoir) and a coating membrane (the diffusion barrier). The core contains the active ingredients and excipients, whereas the membrane is made primarily of rate-controlling polymer(s). A reservoir system normally contains many coated units (particulates) such as beads, pellets, and minitablets. Unlike single-unit tablets, the number of particulates in a reservoir system often is sufficient to minimize or eliminate the impact of any coating defect associated with a limited number of units. Another attractive feature of reservoir systems is that tailored drug release can be obtained readily by combining particulates of different release rates. An increasing number of products (e.g., Metadate CD and Ritaline LA) have been introduced using such a concept. Lastly, reservoir systems offer the flexibility of adjusting to varying dosage strengths without the need of new formulations. This is highly desirable during clinical development programs, where dose levels frequently are revised based on study outcome. Typically, special coating equipment such as the Wurster coater is required to apply the coating material uniformly. Scale-up of such processing can be challenging and may require changes in formulations between scales in order to maintain similar release characteristics. In addition, it has been recognized that dissolution of certain reservoir system–based products may change on storage. One way to minimize this problem is adding a curing step at the end of the coating process.

Dispensing systems for the delivery of compositions of matter are well known to the prior art. These systems generally delivery their composition by diffusion, for example, from an enclosed capsule or by diffusion from a multi-structured device having a wall formed of a known polymer permeable to the composition into a selected environment. However, there is a large category of compositions that cannot be delivered by the prior art delivery systems because of at least one feature inherent in these devices which adversely affects their rate of release from the system or substantially prevents the release of the composition from the system. For example, many compositions cannot be delivered from a diffusion controlled delivery system because their permeation rate through the rate controlling material comprising the system is too small to produce a useful effect, or in many instances the composition molecules are too big and will not diffuse through the rate controlling material forming the device. Also, there is an additional class of useful products that cannot be satisfactorily delivered by diffusion devices because of a particular chemical characteristic of the product. This additional class includes salts that because of their ionic character will not diffuse through most polymers and polymeric like materials and unstable polar compounds that cannot be formulated into a satisfactory composition suitable for storage and delivery from a prior art device. In all these situations, use of an osomotic pump for oral CR delivery is a solution. An osmotic drug (or other beneficial substance) delivery system comprises a multi-chamber compartment formed by an external shell and one or more chamber-dividing walls each with a small orifice, of a microporous material and overlayers of semipermeable membranes completely covering the outer shell of all but one chamber and substantially covering the outer shell of the remaining chamber. Osmotic agents, adjuvants, enzymes, drugs, pro-drugs, pesticides and the like are incorporated in the chambers covered by the semipermeable membrane, and external fluids that diffuse into those chambers form solutions and by osmotic pressure are forced through the orifice to the drug chamber to form a solution thereof and then through the exposed microporous shell to the exterior of the device at a rate controlled by the permeability of the semipermeable overlay and the osmotic pressure gradient across the shell.

Osmotic devices can be manufactured using a variety of methods. In the first method, manufacturing the device with an agent compartment and an osmogent compartment separated by a film, which film is movable from a rested to an expanded state is present. The device delivers agent by fluid being imbibed through the wall into the osmogent compartment producing a solution that causes the compartment to increase in volume and act as a driving force that is applied against the film. This force urges the film to expand against the agent compartment and correspondingly diminish the volume of this compartment, whereby agent is dispensed through the passageway from the device. While this device operates successfully for its intended use, and while it can deliver numerous difficult to deliver agents, its use is somewhat limited because of the manufacturing steps needed for fabricating and placing the movable film in the device. The osmotic device in an another method of manufacture comprises a semipermeable wall surrounding a compartment containing a beneficial agent that is insoluble to very soluble in an aqueous biological fluid and an expandable hydrogel. In operation, the hydrogel expands in the presence of external fluid that is imbibed into the device and in some operations mixes with the beneficial agents, thereby forming a dispensable formulation that is dispensed through the passageway from the device. This device operates successfully for its intended use, and it delivers many difficult to deliver beneficial agents for their intended purpose. Similarly, several techniques can be used in the manufacture of oral osmotic pumps. Based on the availability the following comprehensive information regarding osmotic pumps can be provided. There are two broad categories of osmotic pumps: elementary osmotic pumps and osmotic pumps having an expandable push layer or material. Elementary osmotic pumps are typically formed by compressing a tablet of an osmotically active drug (or an osmotically inactive drug in combination with an osmotically active agent or osmagent) and then coating the tablet with a semipermable membrane which is permeable to an exterior aqueous-based fluid but impermeable to the passage of drug and/or osmagent. One or more delivery orifices may be drilled through the semipermeable membrane wall. Alternatively, orifice(s) through the wall may be formed in situ by incorporating leachable pore forming materials in the wall. In operation, the exterior aqueous based fluid is imbibed through the semipermeable membrane wall and contacts the drug and/or salt to form a solution or suspension of the drug. The drug solution or suspension is then pumped out through the orifice as fresh fluid is imbibed through the semipermeable membrane.

1.6.2 Parenteral CR Formulations

The parenteral administration route is still the most effective and common form of delivery for macromolecules (such as peptides and proteins), for active drug substances with metabolic liabilities (i.e. drugs for which the bioavailability is limited by high first pass metabolism effect or other physico-chemical limitations) and for drugs with a narrow therapeutic index (i.e. several anticancer drugs - where slow infusion is the best way to control the exact pharmacokinetic into the blood). Moreover, at the same time, this administration route is the least preferred by patients because of the discomfort and inconvenience that it causes. For this reason, whatever drug delivery technology that can reduce the total number of injections throughout the drug therapy period is truly advantageous not only in terms of compliance, but also for the potential to improve the quality of the therapy. Such reduction in frequency of drug dosing is achieved, in practice, by the use of specific formulation technologies that guarantee that the release of the active drug substance happens in a slow and predictable manner. For several drugs, depending on the dose, it may be possible to reduce the injection frequency from daily to once or twice monthly or even less frequently. In addition to improving patient comfort, less frequent injections of drugs in the form of depot formulations smoothes out the plasma concentration-time profiles by eliminating the hills and valleys. Such smoothing out of the plasma profiles has the potential to not only boost the therapeutic benefit, but also to reduce unwanted events and side effects. Continuous release profiles are suitable to generate on 'infusion like' plasma level time profile in the systemic circulation without the necessity of hospitalization. This can be achieved by IV infusions which need hospitalization. However, recently several novel delivery systems for parenteral delivery of drugs have been developed.

Drug delivery systems that can precisely control drug release rates or target drugs to a specific body site, although a relatively recent technology have had an enormous medical and economic impact. New drug delivery systems impact nearly every branch of medicine and annual sales of these systems are far in excess of 10 billion dollars. However, to intelligently create new delivery systems or to understand how to evaluate existing ones, much knowledge is needed. New approaches in treating diseases such as alcoholism, cancer, heart disease, and infectious diseases using parenteral sustained release systems are needed to be examined. Delivery of vaccines, contraceptive agents, anticalcification agents, orthopedic agents, and veterinary agents is the need of the hour. Novel polymeric materials including polyanhydrides, chitosan polyesters, polyphosphates, polyphosphazenes, hydrogels, bioadhesive materials, and poly(ortho esters) are being evaluated for their utilization in parenteral sustained release dosage forms. Extensive characterization approaches including differential scanning calorimetry, gel permeation chromatography, spectroscopy, X-ray photoelectron spectroscopy, X-ray powder diffraction, and surface characterization are being explored. New areas related to drug delivery such as gene therapy, blood substitutes, food ingredients, and tissue engineering come into the perview of parenteral sustained release dosage forms. An analysis of various routes of administration for parenteral delivery should be investigated. Different controlled release designs such as osmotic pumps, pendent-chain systems, membrane systems, nanoparticles, and liposomes needs examination. Further, patents, regulatory issues, manufacturing approaches, economics, in vitro-in vivo correlations, pharmacokinetics, release kinetics, assays, diagnostics, and related issues are considered for investigations.

Controlled and sustained release parenteral drug delivery systems include liposomes, microspheres, suspensions, gels, emulsions, and implants. They are generally used to improve the therapeutic response by providing appropriate dosing strategies (this may be constant or pulsatile release). Such systems can be considered safer than conventional parenteral dosage forms since less drug is required and since the drug may be targeted to the in vivo site, avoiding high systemic levels. Due to the lower dosing frequency and simpler dosage regimes, patient compliance can be improved with these dosage forms. For example, microspheres and larger implantable devices can be used to modify release over periods of months to years. Liposomes may achieve targeted delivery both by passive and active means following intravenous administration and are utilized to target toxic drugs, such as anti-cancer agents, to avoid systemic side effects. Perhaps the most complex of the controlled drug delivery systems are the human parenteral systems.

Biodegradable microsphere and implantable-rod systems which deliver peptides for treatment of prostate cancer have been developed and approved in several countries. Implantable osmotic pumps are used in laboratory animals to conveniently evaluate the controlled delivery of active agents under a variety of conditions. Implantable silicone rods have also been developed and marketed for delivery of steroidal hormones. Prior to the development of these dosage forms, parenteral sustained release dosage forms in the forms of drug suspensions were known to the pharmaceutical industry. These are the first controlled release parenteral dosage forms. While for a series of sparingly soluble active drug substances, such as steroids, sterile aqueous, oleaginous suspensions or oily solutions are formulation approaches that allow an extended duration of action (in these cases the release of the active from the injected formulations is governed quite exclusively by the dissolution kinetic of the active drug substance, which can last up to several months according to the administered dose and the physico-chemical properties of the drug). However, for the majority of the drug substances, such as peptides and macromolecules, it is mandatory to utilize specific drug delivery technologies that can tailor and govern the release profile of the active drug substance from the formulation itself. As the reason various parenteral sustained release dosage forms other than those utilizing drug suspensions was developed. Now, we see this area is attractive research area for a pharmaceutical scientist as well as several such CR products are now available in the market.

A controlled release parenteral dosage form is usually selected when there are problems associated with oral delivery (eg, gastric irritation, first pass effects or poor absorption) and a need for extended release and/or targeted delivery (eg, rapid clearance, toxic side effects). The CR dosage form selected may be dependent on the desired effect (eg, long term localized release) as well as compatibility of the drug with the manufacturing process. Examples of applications for CR parenteral delivery include: fertility treatment, hormone therapy, protein therapy, infection treatments (antibiotics and antifungals), cancer therapy, orthopedic surgery and post-operative pain treatment, chronic pain treatment, vaccination/immunization, treatment of CNS disorders, and immunosupression. Approved CR parenteral products are listed in Table 1.1.

Table 1.1 Approved Parenteral CR Products.

Trade Name	Active Ingredient	Approval Date*
Suspension Products:		
Depo-Medrol	Methylprednisolone	pre-1982
Depo-Provels	Medoxyprogesterone	pre-1982
Celestone Soluspan	Betamethasone	pre-1982
Insulin	Lente Unltralente NPH	pre-1962
Plenaxis	Abarelix	2003
Microsphere Products:		
Lupron Depot	Leuprolide	1989
Sandostatin LAR	Octreotide	1998
Nutropin Depot	Somatropin	1999
Trelstar Depot	Triptorelin	2000
Plenaxis	Abarelix	2003
Suspension Products:		
Depo-Medrol	Methylprednisolone	pre-1982
Depo-Provels	Medoxyprogesterone	pre-1982
Celestone Soluspan	Betamethasone	pre-1982
Insulin	Lente Unltralente NPH	pre-1962
Plenaxis	Abarelix	2003
Liposome Products:		
Doxil	Doxorubicin	1995
Daunoxome	Daunorubicin	1996
Ambisome	Amphotericin B	1997
Depocyt	Cytarabine	1999
Lipid Complex Products:		
Ambelcet	Amphotericin B	1995
Amphotec	Amphotericin B	1997
Implant Products:		
Norplant	Levonorgestrel	1990
Gliadel	Carmustine	1996
Zoladex	Goserelin	1989
Viadur	Leuprolide	2000

*Approval dates refer to the date of approval by US FDA.

Although apart from polymeric devices, several dosage forms that can sustain the drug levels in the systemic circulation which include osmotic pumps, silicone devices, so far, only drug delivery devices based on polymers and copolymers deriving from lactic acid (LA) enantiomers, glycolic acid, and e-caprolactone (abbreviated as PLA, PGA, and PCL, respectively) have been commercialized. The prospective applications include devices to treat cancer, drug addiction, and infection, as well as drugs for contraception, vaccination, and tissue regeneration. A number of products are commercially available such as Decapeptyl'", Lupro Depots, Zoladexv, Adriamycinv, and Capronorv. Different types of implants are now available or are under active research investigations. Large-size implants require surgery, whereas needle-like implants can be injected subcutaneously (s.c.) or intramuscularly (i.m.) using a trochar. Microparticles can also be injected s.c. and i.m. and behave as tiny implants. Intravenous (i.v.) injection is possible with microparticles, the size of which must be below 7 *um* to avoid lung capillary embolization. However, microparticles can be taken up very rapidly by the macrophages of the reticuloendothelial system to finally end up in Kupffer cells in the liver. Although nanoparticles have been proposed to overcome the size limitation imposed by capillary beds; they can also be taken up by macrophages. Stealth nanoparticles with a surface covered by a brush of poly(ethylene oxide) (PEO) have been proposed to avoid macrophage uptake (9). In any event, nanoparticles, as well as microparticles, can hardly leave the vascular compartment. Recently, colloidal particles have been considered in the form of macromolecular micelles of amphiphilic biblock copolymers (10) or of aggregates of hydrophilic polymers bearing hydrophobic side chains (11). These systems can serve as a drug carrier via physical entrapment of a lipophilic drug within the hydrophobic microdomains formed by the core of micelles or aggregates. Macromolecular prodrugs, where a drug molecule is temporarily attached to a polymeric carrier, have also been proposed (12). Last but not least, the next century might see the development of polymeric drugs, because any synthetic polymer can advantageously interact with elements of living systems, such as molecules, cell membranes, viruses, and tissues (13). Liposomal formulations also are helpful as sustained release parenteral dosage forms. Liposomes, lipid based drug carrier vesicles have recently emerged as a new technology in pharmaceutical sciences. Liposomes are composed of nontoxic, biodegradable lipids, in particular of phospholipids. Attempts have been made to prepare liposomes from nonphospholipid components which have the potential to form lipid bilayers that are more durable than conventional liposomes. Currently, both conventional and non-phospholipid liposomes are rapidly becoming accepted as pharmaceutical agents which improve the therapeutic value of a wide variety of compounds. Liposome drug delivery systems are reviewed in detail in another chapter of this book series and as a reason it is not discussed in detail here. In general, liposomes are advantageous in that they can provide controlled release of an entrapped drug, reduce side effects by limiting the concentration of free drug in the bloodstream, alter the tissue distribution and uptake of drugs in a therapeutically favorable way, and make therapy safer and more convenient by reducing the dose or frequency of drug administration. Liposomes generally have been known to improve formulation feasibility for drugs, to provide prolonged sustained release, to reduce toxicity and to improve the therapeutic ratio, to prolong the therapeutic effect after each administration, to reduce the need for frequent administration, and to reduce the amount of drug needed and/or absorbed by the mucosal or other tissue. Advantages such as decreased toxicity and degradation, use of smaller doses, the possibility of targeting the liposome towards a specific site, and reducing side effects of a liposome-bound drugs over the use of a free or polymer-bound drugs is well documented. These dosage forms are not only popular for human use but several of these dosage forms are currently used or being developed for veterinary purposes.

The majority of implants used in veterinary medicine are compressed tablets or dispersed matrix systems in which the drug is uniformly dispersed within a nondegradable polymer. Drug release from dispersed matrix systems involves dissolution of the drug into the polymer, followed by diffusion of the drug through the polymer, and partitioning from the surface of the polymer into the surrounding aqueous environment. Implants are available to increase weight gain and feed conversion efficiency in food-producing animals. These implants are typically prepared in a manner similar to tablets. One controlled-release implant consists of a cylindrical core of silicone, surrounded by an outer layer of estradiol-loaded silicone. A range of implants is available to enhance reproductive performance in breeding animals. These include ear implants containing norgestomet dispersed in polyethylene methacrylate or silicone, a biocompatible tablet implant containing deslorelin (a GnRH agonist) for use in mares that does not require removal, and a sustained-release pellet of melatonin, which is implanted in the ear of ewes to enhance breeding performance. Testosterone pellets are available for implanting in the ears of wethers at doses of 70-100 mg every 3 mo for the prevention of ulcerative posthitis. It is worth mentioning briefly about development of the various biodegradable and other polymers that are used in the development of these parenteral sustained release dosage forms.

During the past 50 years, synthetic polymers have changed the everyday life of humans due to the possibility of covering a wide range of properties by modification of macromolecular structures and introduction of additives (fillers, plasticizers, etc.). In the meantime, surgeons and pharmacists tried to use these materials as biomaterials. About 30 years ago, distinction was made between permanent and temporary therapeutic uses. The former requires biostable polymeric materials, and the main problem is resistance to degradation in the body. In contrast, the latter needs a material only for a limited healing time. In this regard, degradable polymers became of great interest in surgery as well as in pharmacology. The first degradable synthetic polymer was poly(glycolic acid) (PGA), which appeared in 1954. This polymer was first discarded because of its poor thermal and hydrolytic stabilities, which precluded any permanent application. Later on, people realized that one could take advantage of the hydrolytic sensitivity of PGA to make polymeric devices that can degrade in a humid environment and, thus, in a human body. This led to the first bioabsorbable suture material made of a synthetic polymer. It is worth noting that terminology is one of the sources of confusion in the field. Nowadays, people tend to use the word *degradable* as a general term and reserve *biodegradable* for polymers that are biogically degraded by enzymes introduced in vitro or generated by surrounding living cells. The possibility for a polymer to degrade and to have its degradation by-products assimilated or excreted by a living system is thus designated as *bioresorbable*. Most of the degradable and biodegradable polymers identified during the past 20 years have hydrolyzable linkages, namely ester, orthoester, anhydride, carbonate, amide, urea, and urethane in their backbone. The ester bond-containing aliphatic polyesters are the most attractive because of their outstanding biocompatibility and versatility regarding physical, chemical, and biological properties. The main members of the aliphatic polyester family are numerous. Only a few have reached the stage of clinical experimentations as bioresorbable devices in drug delivery. This is primarily due to the fact that being degradable or biodegradable is not sufficient. Many other prerequisites must be fulfilled for clinical use and commercialization.

Many classes of cross-linked polymer gels display phase transition characteristics i.e. abrupt change in swollen volume in response to small environmental changes like pH, light, temperature, intensity, electric field, ionic strength, and even specific stimuli like glucose concentration. Drugs containing charged hydrogel networks have been recognized as useful matrices for delivering drugs because their volume, consequently, deliver drug solution in response to external pH variation. Such hydrogels have been applied in glucose sensitive insulin releasing devices, an osmotic insulin pump and site specific delivery in the gastrointestinal tract .The polymeric devices are generally classified into the following categories: 1. Diffusion controlled devices: Monolithic devices and Reservoir devices 2. Solvent controlled devices: Osmotically controled devices and Swelling controlled devices 3. Chemically controlled devices: Bioerodible system and Drug polymer conjugates. Biodegradable polymer may be classified based on the mechanism of release of the drug entrapped in it:

Natural: albumin, starch, dextran, gelatin, fibrinogen, hemoglobin.

Synthetic: Poly ethyl-poly (alkyl cynoacrylates), poly amides, Nylon, 6-10 nylon-cynoacrylates, poly butyl - 6-6, poly acryl amides, poly amino acid, poly urethane. Aliphatic poly esters are poly (lactic acid) poly (lactide – co glycolide), poly (glycolic acid), poly caprolactone, polydihydroxy butyrate, poly hydroxy butyrate co-valently cross linked protein, hydrogel. Biodegradable polymers investigated for controlled drug delivery are 1. Poly lactide / poly glycolide polymers. 2. Poly anhydrides. 3. Poly caprolactone 4. Poly orthoesters 5. Psuedo polyamino acid 6. Poly phosphazenes 7. Natural polymers.

Interest in poly lactide material has been generated due to its considerable chemical, biological and mechanical characteristic. Most of polylactide materials developed so far are designed to deliver drugs to the systemic compartment. Also local drug delivery is a possibility in this case one attempts to achieve high drug concentration at the site of implantation without exposing non affected tissue to the drug. Implants are used as depot formulations either to limit high drug concentrations to the immediate area surrounding the pathology or to provide sustained drug release for systemic therapy. Clinically, implant systems have been used in situations where chronic therapy is indicated, such as hormone replacement therapy and chemical castration in the treatment of prostate cancer. Biodegradable materials, such as polylactic acid co-glycolic acid, are of course preferred as this removes the need for surgical removal of the implant after treatment has ended. However, non-biodegradable materials do provide therapeutic levels of drug for up to one year in vivo. One of the reasons for the popularity of the lactide/ glycolide material in drug delivery system is their relative ease of fabrication into various types of delivery systems: Micro particles (Microspheres and microcapsules), implants and fibers. Aliphatic poly esters undergo biodegradation by bulk erosion the lactide/glycolide polymer chains are cleaved by hydrolysis to the monomeric acids and are eliminated from body through Krebs cycle, Primarily as carbon dioxide and in urine. Very little difference in observed in the rate of degradation at different body sites as the hydrolysis rate is dependent only on significant changes in temperature and pH or presence of catalysts. The role of enzyme in the biodegradation of the polymers has been still unclear. Lactide glycolide polymers show wide range of hydrophilicity which makes them versatile in designing controlled release system. The use of different PLGA polymers for the development of parenteral SR dosage forms for desired period is given in Table 1.2.

Table 1.2 Biodegradation of lactide/glycolide polymers.

Polymer	Months
Polylactide	18-24 months
Poly dl-lactide	12-16 months
Poly glycolide	2-4 months
PLGA 50:50	2 months
PLGA 85:15	5 months
PLGA 90:10	2 months

Some pictures of microspheres under Scanning Electron Microscope are provided in Figure 1.2(a) and 1.2(b).

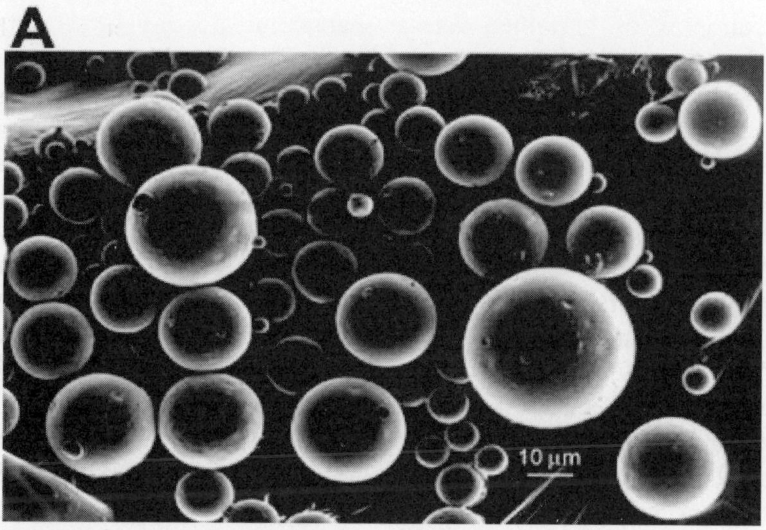

Figure 1.2(a) Characteristics of PLGA microspheresA: Picture of rhGDNF-loaded microspheres analyzed by scanning electron microscopy. Scale bar in A represents 10 µm.

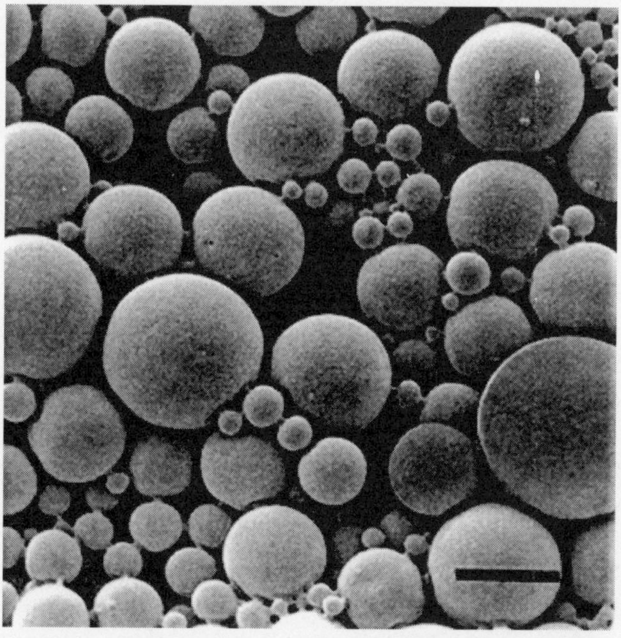

Figure 1.2(b) Scanning electronic photomicrograph of PLGA microspheres obtained by the multiple emulsion W/O/W method. Bar = 5 µm.

Examples of sustained release parenteral dosage forms and the drawbacks and need for different dosage forms is illustrated with the following drug. Buprenorphine is a semi-synthetic opioid analgesic with mixed agonist-antagonist properties (14). Besides being 20-40 times more potent than morphine, one of its main advantages is that the dose does not need to be increased during chronic administration. Buprenorphine can be in various forms such as sublingual tablets (0.2 mg) for the treatment of moderate, severe acute, and chronic pain, or as a pre-operative medication. Sublingual tablets containing 0.4, 2 and 8 mg of the drug are used for the treatment of opioid addiction. Alternatively, it is available as an injection (0.3 mg/mL) for IV, IM, intrathecal, and epidural administration as an analgesia in cases of severe acute pain and as a pre-medication. Recommended doses are 200-600 μg by IV or IM injection every six to eight hours, 30-45 μg intrathecally or 100-300 μg epidurally every six to twelve hours or 400 μg sublingually every six to eight hours. Different injectable SR dosage forms for this drug are available. For example, U.S. Pat. No. 6,495,155 discloses an injectable slow-release partial opioid agonist and/or opioid antagonist formulation in a poly (D, L-lactide) excipient with a small amount of residual ethyl acetate (15). The microparticles are under 125 μm in diameter and can be readily injected intramuscularly to provide at least about 0.5 ng/ml of drug over an extended period of time (28-60 days). The formulations are provided for use in the treatment of alcoholics and heroin addicts. Additionally, a subcutaneous depot product (Norvex®) exists wherein buprenorphine microcapsules consisting of buprenorphine base and biodegradable PLA-PGA polymer are disclosed. A study has found that buprenorphine propionate when prepared as a depot had a long-lasting analgesic effect, which was 7.5-fold longer than the traditional dosage form of buprenorphine in saline preparation, following IM injection in rats. The long lasting effect of IM depot of buprenorphine propionate is reported to be due to a slow release of buprenorphine propionate from its oil vehicle. They have subsequently synthesized and formulated other depots of buprenorphine esters, buprenorphine enanthate and decanoate. The buprenorphine decanoate in oil produced a 14-fold longer duration of action than buprenorphine HCl in saline. U.S. Pat. No. 6,335,035 discloses the preparation of a sustained release delivery system using a polymer matrix containing a drug for use in treating acute or chronic conditions (16). The drug is dispersed within a polymer matrix solubilized or suspended in a polymer matrix. The polymer matrix is composed of a highly negative charged polymer material such as polysulfated glucosoglycans, glycosaminoglycans, mucopolysaccharides and mixtures thereof, and a nonionic polymer such as carboxymethylcellulose sodium, hydroxyethyl cellulose, hydroxypropyl cellulose, and mixtures thereof.

Although the use of emulsions and suspensions for drug delivery is not uncommon and has been used in other analgesics, there are also problems associated with them. However, sustained release injectable buprenorphine formulations that exist in the prior art utilize more complicated systems such as microparticles or prodrugs in an oil vehicle. More particularly, the manufacturing of microparticles involves utilizing complex and costly processes with the use of organic solvents. Additionally, it can be difficult to achieve sterility of microparticles and other oil solutions because terminal sterilization is not always possible. In addition to these disadvantages, it is difficult to appropriately control the release of a drug such as buprenorphine in an injectable dosage form in order to achieve the desired onset and duration of analgesic effects in the target species. Accordingly, there continues to be a need for reasonably simpler and more practical formulations for sustained release of buprenorphine. To weed out the problems associated with the existing parenteral SR dosage forms, one embodiment, an oil-in-water buprenorphine formulation including buprenorphine and a surfactant that emulsifies the buprenorphine in oil, wherein the buprenorphine release is controlled by varying the oil concentration and/or pH of the emulsion has been disclosed.

1.6.2.1 Currently marketed parenteral CR products

The majority of sustained release technologies are designed for the use of biodegradable materials and a great number of them are specifically based on the use of polyesters, such as poly-lactides (PLAs) and poly-lactide-co-glycolide (PLGAs) copolymers. These polymers, once injected in the body, undergo random, non-enzymatic, hydrolytic cleavage of the ester linkages to form lactic acid and glycolic acid, which are normal metabolic compounds, eliminated via the tri-carboxylic acid cycle as carbon dioxide and water. The kinetic degradation of the polymers, and thus the release profile of the active drug substance incorporated into them, strongly depends on the selected polymer and on its physico-chemical properties. As an example, PLA/PLGAs bio-degradation kinetic is governed by a series of variables that interplay, among them the lactide-to glycolide ratio (the higher the glycolide content, the quicker the degradation rate), molecular weight of the polymer, end-caps of the polymer (i.e. acidic or neutral), drug to polymer ratio, solubility and hydrophobicity of the incorporated active drug substance and the formulation manufacturing method. Active drug substances are released from these formulations via a combination of diffusions through the polymeric matrix and from holes that are created by the erosion of the matrix itself. As drug particles diffuse out of the matrix, the exposed polymer is hydrolyzed, solubilized and released as monomers. New drug/matrix surface is thus exposed, and the process of diffusion and erosion continues. Several manufacturing procedures can be applied to obtain such formulation systems. The manufacturing method that will be chosen and tailored to the specific drug substance needs to take into consideration the stringent requirements needed for parenteral-acceptable products, such as: citing only some key features; using approved excipients (in this case no further insights on the toxicological characterization of the excipients will be needed); needing to produce particle size distribution or a rod/implant that can be delivered via convenient needles (19 gauge and above); needing to have high encapsulation/loading efficiency and manufacturing yields (this is particularly valid for costly biotechnological products); limiting as much as possible the use of organic solvents into the manufacturing process and in the final product; being relatively easy to scale-up and reliable in terms of batch to batch consistency and reproducibility; and being sterilizable or manufacturable through an aseptic process.

These currently available delivery systems have the following salient features: 1. No surgical removal of depleted system is required as it is metabolized in non toxicological by product. 2. The drug release from this system can be controlled by following a. Diffusion of drug through the polymer, b. Erosion of the polymer surface with concomitant release of physically entrapped drug, c. Cleavage of covalent bond between the polymer bulks or at the surface followed by diffusional drug loss, d. Diffusion controlled release at the physically entrapped drug with bio adsorption of the polymer until drug depletion. Long acting parenteral drug formulation are designed, ideally to provide slow constant, sustained, prolonged action.

The other sustained release parenteral formulations include infusion pumps. Implantable infusion pumps were introduced for continuous drug delivery in 1969. Several implantable pumps have been approved by the FDA, and many more are being tested and used clinically around the world. The Infusaid Models 100, 100, and 550 (Pfizer Infusaid, Inc., Norwood, MA) were FDA approved starting in 1982 for the delivery of various cancer therapeutics, including floxuridine (FUDR), fluorouracil (5-FU), methotrexate sodium (MTX), and cisplatin (CDDP). They have also been approved for the delivery of morphine sulfate for the treatment of pain resulting from incurable cancer. The Infusaid pumps consist of a flexible bellows containing the drug solution that is surrounded by a rigid chamber filled with a charging fluid (usually a volatile chlorofluorocarbon). Vapor pressure from the charging fluid expels the drug solution by compressing the internal drug reservoir. The release of the drug solution is regulated by a capillary flow restrictor or a valve/accumulator combination. The reservoir can be refilled by percutaneous puncture of a needle, which reexpands the reservoir and increases the pressure in the adjacent charging fluid chamber, causing the fluid to recondense. The pump is then ready for its next infusion cycle. Medtronic, Inc. (Minneapolis, MN) received FDA approval for its SychroMed Infusion System in 1988 for the delivery of CDDP, FUDR, doxorubicin hydrochloride (DOX), and MTX and received FDA approval in 1996 for the intrathecal delivery of morphine sulfate for the treatment of pain. The SynchroMed pump delivers a drug solution from a percutaneously refillable reservoir via a peristaltic pump mechanism that is driven by a lithium battery with a life of 1 to 3 years. The pump can be programmed with an external computer and a magnetic field telemetry link, allowing for more complex delivery regimens, including those based on circadian rhythms. The doctor has the option of tailoring the drug dosage, flow rate, and dosage schedule to best fit the needs of the patient, while the patient is able to receive treatments untethered.

1.6.2.2 Special Features in the Design of Parenteral CR Products

The development of parenteral sustained release dosage forms is achieved after taking several factors into consideration. The route of administration, the release rate both in vitro and in vivo, duration of action needed, all should be considered when a parenteral sustained release dosage form for a particular drug is designed. Route of administration is a very important aspect. Not all the types of dosage forms can be administered by all the routes of administration. The needed release of the drug can be achieved after using a particular polymer, particular dosage form and particular geometry. These are prepared using a variety of approaches. Implantable large devices such cylinders, pellets, slabs, discs, and films thicker than 0.1 mm are usually prepared by compression molding an intimate polymer-drug mixture. However, there is risk of thermal degradation. Tubings and needle-like implants can be obtained by extrusion. The temperature of compression molding or extrusion depends on the morphological characteristics of the polymer. For crystalline polymers, the processing temperature has to be above the melting temperature *(Tm)*' .

In the case of amorphous polymers, temperatures above the glass transition *(Tg)* are usually sufficient. Prior to processing, the polymer should be thoroughly dried to prevent thermal and/or hydrolytic degradation. Thin films can be prepared by casting a polymer-drug solution. Hollow fibers of highly crystalline PLA100 can be prepared by using a "dry-wet" coagulation spinning process. The use of different spinning systems (i.e.,different solvent-nonsolvent pairs and with or without additive) leads to hollow fibers with varying asymmetric membrane structures. PCL fibers containing tetracycline hydrochloride, with an outer diameter of 0.5 mm, have been prepared by melt spinning at 161°C (17). When organic solvents are used, the elimination of residual solvents is of major importance because they can generate toxicity regardless of the polymer matrix. Implantable mesh sheets were also reported. Implantable systems provide various advantages such as prolonged release of drugs, reproducibility of drug release profiles, and ease of fabrication. However, implantation of such systems requires surgery with risk of infection. Some commercial examples of these types of preparation are illustrated here. Currently, two controlled-release systems have been approved by the FDA to treat cancer. Both act as depots for the sustained release of anticancer peptides. The first, Zoladex® (Zeneca Pharmaceuticals), contains goserelin acetate, a synthetic decapeptide analogue of luteinizing hormone-releasing hormone (LH-RH) dispersed in a poly(lactide-coglycolide) rod. Zoladex® is indicated for the palliative treatment of advanced prostate cancer and breast cancer. One formulation contains 3.6 mg of goserelin acetate and is designed to last 4 weeks; the second formulation contains 10.8 mg of the peptide and is designed as a 3-month implant. The 4-week rods are 1 mm in diameter, and the 3-month rods are slightly larger, with a diameter of 1.5 mm. The 3-month implant is indicated for prostate cancer patients only. The rods are supplied in a special syringe that is used to implant the delivery system subcutaneously in the upper abdominal wall. In a randomized, prospective clinical trial comparing radiation therapy alone to radiation therapy combined with Zoladex® implants, survival of patients with locally advanced prostate cancer was improved with the addition of Zoladex® to the treatment. Both groups of patients received 50 Gy of radiation to the pelvis over 5 weeks and an additional 20 Gy over the following two weeks. Patients in the combined therapy group received the 4-week implants (Zoladex® 3.6 mg) starting on the first day of radiotherapy and again every 28 days for 3 years. In the combined therapy group, 85% of the patients were disease free at 5 years compared with only 48% of the radiotherapy-only group. Lupron Depot® (TAP Pharmaceuticals, Inc.) has also been FDA approved for the treatment of prostate cancer. The formulations are supplied as lyophilized microspheres that are resuspended in a diluent for intramuscular injections every 1, 3, or 4 months.

The amount of the drug incorporated into the delivery systems is also important. This is decided on the entire dose of the drug. The drug is slowly released from the injection site and is absorbed into the blood stream. Thus, in this case drug levels are absorption limited. This definitely depends on the polymer and the characteristics of the drug also. The design should be such that there should be a strict manufacturing control, less pharmacokinetic pitfalls, predictions of injection site residues should be easy, and safety of injection residues should be high. Ideal parenteral sustained release formulation should be administered once, give long term effects as desirable, be easy to manufacture and possess enough shelf-stable properties.

The physicochemical and biopharmaceutical properties of the drug can have a tremendous impact on its bioavailability and, hence, on its efficacy and toxicity profile. Thus, understanding these parameters is often tantamount to the selection and development of the optimum dosage form. Several physicochemical factors controlling the delivery of a bioactive agent to the host:

1. Size and Geometry of the delivery system.
2. Local pH of the host site.
3. Hydrophilic or hydrophobic nature of the active agent.
4. Solubility of the active in the local environment.
5. Solubility of the active in the delivery matrix.
6. Permeability of the delivery matrix to water.
7. Permeability of the delivery matrix to the active agent.
8. Biostability of the delivery matrix.
9. Concentration gradient across the delivery system.
10. Drug loading.
11. Polymer characteristics such as glass transition temperature and molecular weight etc.

12. Morphological characteristics such as porosity, tortuosity, surface area and shape of the system.
13. Chemical interaction between drug and the polymer.
14. External stimuli such as pH, ionic strength, thermal and enzymatic action.

The release of the drug can be modified using a variety of approaches (18). Microspheres of etoposide prepared by oil/oil suspension and solvent evaporation technique using polylactide (PLA) of molecular weight 50,000 Da were divided into size ranges of less than 75 μm, 75 to 180 μm, and 180 to 425 μm by passing through series of standard sieves, and their drug release was evaluated. Particles that are less than 75 μ showed faster release rates compared with larger size fractions. The difference in the rate of release is attributed to the difference in the surface area. Alterations in drug release rates therefore could be attained by simple mixing of different size fractions of microspheres. Drug loading is another important factor that effects the release of the drug from the delivery system. The release of the drug can be controlled by achieving suitable drug loading. The rate of diffusion will be higher for drugs with higher aqueous and polymer solubility, as well as for those not chemically interacting with the polymer. Higher drug loading will mean higher amounts of drug present on the surface or proximal to the surface that will lead to higher initial release. In addition, the rate of pore formation can be higher on drug depletion because the drug-polymer ratio is higher. An example illustrates the effect of leuprolide acetate loading on the physicochemical properties and in vitro drug release of PLGA microspheres. Formulations *A* and *B* with 11.9 and 16.3 percent of drug loads were prepared by a solvent-extraction-evaporation method. Higher drug incorporation resulted in a substantial increase in specific surface area and a decrease in bulk density. When observed under the scanning electron microscope, higher-drug-loaded microspheres showed a higher surface porosity. This resulted in higher initial release from microspheres with higher drug incorporation. Another important maneuver that can be performed to obtain the desired release is the selection of the polymer with desired molecular weight. To date, the largest body of literature exists on polyesters such as poly(DL-lactide) or poly(DL-lactidecoglycolide). These polyesters are available from a variety of vendors on a commercial scale with varied molecular weights and monomer ratios of lactide and glycolide. They are also available with acid end groups to impart higher hydrophilicity. Addition of low-molecular-weight poly(DLlactide) (MW 2000 Da) increases drug release from a biodegradable poly(DL-lactide) (MW 120,000 Da) drug delivery system. Bodomeier et al. found that the duration of action could be varied over a range of several hours to months by varying the amount of low-molecular-weight poly(DL-lactic acid) (18). Degradation of these polymers occurs by hydrolysis of ester linkages causing random scission and mainly depends on the polymer concentration, ratio of comonomers, and hydrophilicity. Drug solubility is another important feature that can have a significant influence on the drug release. In case of macromolecular drugs, a major portion of drug is released by polymer degradation and erosion, and a small portion is released by the diffusion mechanism. Polypeptides usually have limited solubility in the polymer, which greatly prevents their diffusion. In addition, the aqueous channels present in the delivery system could be too narrow or tortuous for these macromolecules. Reports from the literature indicate that the drug release is multi-or triphasic, which is characterized by higher initial release (can be termed *burst* in some cases), a lag phase where minimal amount of drug is released, and finally, release of drug at a higher rate until depletion. Similarly, the influence of the other factors is profound and can be found from various literature sources.

In the design of parenteral CR products, the drug release studies and the in vivo absorption determination are one of the main criteria. Current uses of in vitro release testing include: 1) formulation development, to include assessment of dose-dumping and in vivo stability (e.g., Stealth-type liposomes, which should remain stable without significant drug release until uptake at the target site in vivo); 2) quality control to support batch release, 3) evaluation of the impact of manufacturing process changes on product performance, 4) substantiation of label claims; and 5) compendial testing. Although in vitro release testing of CR parenterals is primarily utilized for quality control purposes, in vitro release tests should be developed with regard to clinical outcomes (bio-relevance). The rationale for this understanding is that the ultimate purpose of quality control testing is to ensure the clinical performance, i.e., efficacy and safety of the product. In order to achieve in vivo relevance, physiological variables at the site need to be considered including: body temperature and metabolism (both can significantly affect blood flow), muscle pH, buffer capacity, vascularity, level of exercise, as well as volume and osmolarity of the products. Any tissue response, such as inflammation and/or fibrous encapsulation of the product may need to be considered. In vitro release methods should be designed based on in vivo release mechanisms. With this understanding, the following general approaches for in vitro test method design are important: 1) identification of release media and conditions that result in reproducible release rates; 2) preparation of formulation variants that are expected to have different biological profiles; 3) testing of formulation variants in vitro as well as in vivo; and 4) modification of in vitro release methods to allow discrimination between formulation variants that have different in vivo release profiles. The relevance of sink conditions is also very relevant in in vitro test design for CR parenterals, considering that sink conditions may not exist at a particular in vivo site. General agreement was that sink conditions should be used for in vitro testing for quality control purposes provided that the study design allowed for discrimination between formulation variants with different in vivo release profiles. However, it can be argued that non-sink conditions may be necessary if the purpose of the in vitro test is to establish in vitro-in vivo correlation (IVIVC). Although IVIVC is not utilized at present for CR parenterals, with sufficient bio-relevance built into the in vitro tests to support an IVIVC it may allow subsequent waiver of in vivo studies

Although IVIVC may not be possible for all CR parenteral products, it is an important area for research. The principles used in IVIVC of oral extended-release products may be applied to parenterals with appropriate modification, justified on a scientific basis. IVIVC modeling and measurements may be different for different types of products (e.g. targeted release versus extended release products). Similarly, in vitro release methods and media are likely to vary depending on the product and should be developed based on in vivo relevance. For example, in vitro cellular tests may be acceptable as long as they are reproducible and can be validated. Similarly, in vivo measurements may vary and may include plasma concentrations, efficacy/safety data, surrogate endpoint data, as well as tissue concentrations. In vitro and in vivo measurements must be justified scientifically. In the case of some products, such as liposomes, it may be necessary to measure in vivo concentrations of both free and encapsulated drug. Models that represent multiple processes (e.g., physical and biological) should be considered, as appropriate. The use of animals was considered to be acceptable to prove that an in vitro release system is discriminating. However, the use of animal models was considered inappropriate to prove an IVIVC for regulatory purposes.

Instead, bio-relevance should be developed using clinical data. Nevertheless, IVIVC modeling using animal data would be suitable for "proof of principle" for initial research purposes. Research in this area should be encouraged, possibly coordinated through Product Quality Research Initiative (PQRI).

The issue of data variability with respect to IVIVC should consider the following salient features:

- Increase the number of dosage units or individuals.
- Variability may be acceptable as long as its source can be estimated and a valid IVIVC is obtained.
- If the source and importance of the variability can be determined, it may be possible to minimize it.

Fibrous encapsulation, may affect release in vivo and this needs to be considered in establishing an IVIVC. However, these types of tissue response may be difficult to simulate in vitro.

In the development of in vitro release methods, animal data may be used to obtain tissue distribution and pharmacokinetic information. Plasma levels may not be the best measure of in vivo behavior for CR parenteral products intended for local delivery or targeted release, and therefore, the use of animal models to investigate in vivo product performance should be more exhaustive. More extensive bio-data can be obtained using animal models, including tissue levels at the local site. Animal models were considered to be invaluable and serial tissue samples might be used to compare product performance before and after manufacturing changes for CR parenterals with tissue-specific delivery. Although data will be useful in initial development, ultimately human data must be used to establish an IVIVC. As this could be important issue, the use of just in vitro testing may eventually become the reality for product approval. A brief overview of in vitro release testing methodologies suitable for parenteral CR products is very relevant at this current juncture.

Current USP apparatus for in vitro release testing are designed for oral and transdermal products and may not be optimal for controlled release parenteral products. USP apparatus 1 (basket) and 2 (paddle) were designed for immediate- and modified-release oral formulations. USP apparatus 1 and 2 suffer from problems with sample containment and although this can be overcome by use of a sinker for monolithic depots and dialysis tubing to contain dispersed systems (such as, microspheres), these solutions in themselves create additional problems. For example, microsphere aggregation due to confinement in the dialysis tubing, and un even dissolution from the sides of monolithic depots associated with the sinker device. Violation of sink conditions may also result from confinement within dialysis tubing. Another concern is the large volume required with apparatus 1 and 2, which may not be relevant for small volume parenterals injected subcutaneously and intramuscularly. USP apparatus 5 (paddle over disc), 6 (cylinder) and 7 (reciprocating holder) were designed for the transdermal route and do not offer any advantages for parenteral delivery systems. USP apparatus 3 (reciprocating cylinder) and 4 (flow through cell) were designed for extended release oral formulations. These latter two methods may be the most relevant to CR parenterals and may be suitable following appropriate modifications. Some researchers have noted evaporation problems with apparatus 3. Alternative apparatus, such as small sample vials and vessels, with and without agitation, are currently used for CR parenterals. However, problems are associated with these alternative apparatus, including lack of sink conditions and sample aggregation. USP apparatus 4 can be currently described as the most suitable USP apparatus for controlled and sustained release parenterals. This apparatus allows flexibility in volume, sample cell, flow rate and can be modified for specific product applications (such as avoidance of aggregation problems and of potential violation of sink conditions). There are some disadvantages with USP 4 apparatus. It is considered not robust under extreme conditions applied for accelerated testing. Examples of robustness problems were O-ring failure and filter blockage leading to variable flow rates, as well as polymer migration resulting in valve problems. These problems were product specific and they could be overcome by suitable method alteration (e.g. solvent change) and apparatus modification with parts that could with stand the desired operating conditions (such as high temperature).

1.6.2.3 Techniques of Preparation and Manufacture

Scale-up of a manufacturing process to prepare a polymer for controlled delivery applications is somewhat different than for most commodity applications. For most commodity applications, the specifications of the final polymer are far less stringent than for polymers used in controlled delivery with far fewer regulatory requirements. And often there are numerous final products that can be made from a particular grade of such a polymer. Thus, the market size for a particular type and grade of a polymer used to make commodity products is typically very large. Here, cost is normally a key factor. On the other hand, final controlled delivery products are usually high value-add and their performance in the intended application usually requires very exacting specifications. Therefore, the specifications of all raw materials, including any polymers used to manufacture these products, are also very exacting. Also, the market for a controlled release product is typically a relatively small niche market requiring much smaller volumes of raw materials. Thus, for many controlled release applications the polymers used are custom made in relatively small size lots. A production lot of such polymers is therefore often very small by comparison and may range in size from < 1 kg to 25 kg, especially for biodegradable controlled drug delivery products for human use. Production quantities of such polymers are therefore usually prepared in small-scale stirred batch reactors ranging in size from 1 to 50 gallons in a clean-room environment with close adherence to applicable regulatory guidelines. There are, however, some polymers used for human drug delivery that are prepared in relatively large quantities. Examples are the cellulose ethers and esters that are prepared in large-scale batch reactors. Typical batch sizes are 1,000-2000 kg. These polymers are used in enteric-coated oral formulations and currently represent one of the largest volume polymeric raw materials used in human controlled drug delivery. Several techniques are available for the preparation of various parenteral SR dosage forms. These are discussed in detail here.

Techniques based on Extrusion and Compression

Extrusion: Extrusion is a process used for melting, blending, and forming a polymeric material into a desired finished product. Post forming operations such as orientation, pressing, or final molding may also be coupled with extrusion. Rod, tubing, film, channels, and filaments are examples of shapes that can be continuously extruded. Coatings and coextruded shapes (two different polymers extruded through a die and combined into the final product shape) of all of the above can also be produced. Extruders are also used for compounding and pelletizing materials to be later molded by various processes. An extruder consists of a heated barrel having one or more rotating extrusion screws through it. The single screw variety is the most common. The screw turns and the material moves forward through the extruder in a fashion similar to the action of a progressive-cavity pump. The barrel is often vented to remove volatiles (residual monomers, solvents, moisture, and entrapped air), thus preventing defects in the finished product. A screen pack and breaker plate (support for the screen) for filtering the material are located at the barrel exit. The pressure is measured and controlled at the exit by a feedback control loop to the screw-rotation-speed controller. Heated dies are attached to the end of the extruder to form the polymeric material into the desired shape. A melt-metering pump usually precedes the forming die to provide precise flow control. Extruders are sized by barrel diameter and can be as small as one-half inch to as large as about 8 inches; although, for most controlled delivery applications the smaller size machines will most likely be used. The length of the flighted portion of the screw to the inside diameter of the barrel determines the available surface area of the barrel and the average residence time of the material. The barrel and screw are designed of materials suitable for the temperature, pressure, and chemical aggressiveness of the material being extruded. Typical process pressures are <35 mPa; however, pressures up to 70 mPa are not uncommon. The materials used are usually surface-hardened, high-strength alloys. They are sometimes chromeplated for added corrosion resistance. Barrel liners of highly corrosion-resistant materials are also available. Extruders are usually heated by external electrical heater bands controlled in various zones over the length of the barrel. Heating is controlled through a feedback controller loop which actuates an electrical contacter to activatethe heating elements. Because of the high shear forces involved in extrusion, heat can begin to build once the materials begin to be extruded, and it often becomes necessary to cool the extruder. Extruders are typically cooled by passing a heat transfer fluid through internal cores or jackets in the barrel, or cooling coils surrounding the barrel. Cooling is controlled through a feedback controller loop which actuates a valve to the circulating heat-transfer fluid system.

The geometry of the screw can vary considerably depending on the material and the final product desired. The screw can have a constant pitch or "lead" or variations in screw pitch or lead beginning larger at the feed throat (point of introduction of material) and getting progressively smaller as the material progresses through and exits the barrel. The later is usually used when an intense mixing action is desirable. However, each application typically has its own requirements. The major or outer diameter of the screw is as close as possible to the barrel diameter to prohibit material from passing over the screw flights. The minor or "root" diameter of the screw will typically vary in the first screw type already described, having a smaller minor or root diameter at the feed throat and getting progressively larger and becoming constant as the later portion of the barrel (metering section) is approached. This provides a deep "channel" on material entry to accommodate incoming unmelted and uncompressed material and floods the entry zone with material to prevent starving of the extruder. Most screws are also bored, at least in the feed section, to provide entry of heat-transfer fluids for cooling. The screen pack is located at the exit of the barrel and consists of several screens of different mesh size to filter rough contaminants from the melt. The finest mesh screen is located in the middle of progressively coarser mesh screens. The ones preceding it are for progressively finer filtering of the melt, and the ones after it are for support. The breaker plate is a thick metal plate with numerous large (-1/8-inch diameter) holes located past the screen pack and before the melt-metering pump and the forming die; it serves to support the screen pack and equilibrate the melt pressure to the pump. The metering pump is a positive displacement pump, which is controlled separately from the extruder screw to provide precise flow to the forming die.

Injection molding: Injection molding is probably the most widespread molding technique for quickly and easily forming a polymer melt into a finished product. In this process, the mold is split to allow part removal. It is kept closed during injection by an appropriate clamping force. The mold is filled by forcing a precompounded (containing all additives), molten polymer formulation into the mold. The injection molding machine may be a simple piston (ram) injector design, or a more complex reciprocating-screw design. Either type consists of a feed hopper attached to a heated cylindrical barrel with an injection nozzle attached to the end of the barrel. The reciprocating screw or ram are typically capable of applying 70-140 mPa of pressure to the melt during the injection cycle. This is the operation sequence of the reciprocating-screw machine:

1. With the reciprocating screw forward, unmelted material is fed from the hopper.
2. The material is then plasticated and forced to the front of the barrel by the rotating screw, which simultaneously moves backward against a hydraulic cylinder (reciprocates) as the front of the barrel fills.
3. The mold clamp is released, the mold opens, and the part, formed on the previous cycle, is ejected.
4. The mold closes and the clamp pressure is reapplied.
5. The screw moves forward, as a ram, injecting the melt into the mold, and remains forward to begin the next cycle.

The mold temperature is kept warm but held at a suitable solidification temperature for the material being injected. Too cold a mold can lead to material freezing before the mold is filled. Materials used to construct injection molding equipment are similar to those used to build extruders. The molds must also be constructed of rugged materials to avoid both the erosive and corrosive forces of material flow. The major advantages to injection molding are speed and the ability to simultaneously form multiple complex geometric parts. These are the disadvantages: 1. The high temperature and shear require the polymers and actives to be very stable. 2. The process wastes materials in runners and sprues. 3. The mold and equipment costs are high. 4. Mold erosion occurs from material flow. 5. Runners and sprues are difficult to clean from the final products. 6. The directional flow patterns, inherent in the process, can leave residual internal stresses in the part. Some examples of controlled delivery products produced by extrusion and injection molding are antibiotic periodontal fibers and insecticidal collars and ear tags. Extrusion and injection molding processes are very efficient and available equipment is capable of producing several pounds to several hundred pounds of final product per hour.

Compression Molding: Compression molding is usually used to process thermosetting polymers and is a simple and economical process. However, it can also be used to process thermoplastic materials where it is advantageous in producing the product and provided that appropriate molding conditions are used. In compression molding, the material to be molded is placed in the preheated mold in the form of a loose powder or prill. An excess of several percent is usually added to the mold to ensure complete filling. The mold is closed and sufficient pressure (several mPa) is applied to force the material into the mold cavity. The pressure is dependent on the flow characteristics of the molding material and the complexity of the part being molded. Excess material is forced out of the mold as flash or through a vent. For thermosetting materials the pressure is maintained long enough for the part to cure. The molds are generally heated and cooled by passing a heat transfer fluid through internal cores; however, internal or external electrical heating elements can also be used if precautions are taken to avoid hot spots. The advantages of compression molding are several: 1. Waste is low because no runners or sprues are used. 2. The final parts are easier to clean because no runners or sprues are used. 3. Mold erosion from material flow is minimized. 4. Residual internal stresses are low because of the short, multidirectional flow patterns of the material. 5. The mold and equipment costs are low. 6.

Low process temperatures can be used. There are some disadvantages to compression molding: 1. It is not suitable for intricate parts because the flow is minimal. 2. Because of polymer viscoelasticity, thermoplastic parts are difficult to mold without distortion. 3. It is best suited for fairly thin products.

Solvent Processing of Polymeric Controlled Delivery Products

Solvent Casting of Films: Solvent casting has been an established process for the preparation of polymeric films for decades. The polymer and soluble or dispersable additives are first dissolved and dispersed in a suitable solvent. The solution is then cast onto a continuous, release-coated belt or web-supported film and passed through an oven to drive off the solvent. The solvent is usually reclaimed. The dried film is continuously removed from the belt and wound as it passes from the oven. Care must be taken in design of the process line to ensure that it is particularly suited to the product being manufactured. Consideration should be given to the flow characteristics of the casting solution, the evaporation rate of the solvent, and the changing flow characteristics of the polymer solution throughout the drying process to ensure a uniform film. The thermal stability of the polymer and any additives or active agents must also be considered. Solvent casting of films containing active agents for controlled delivery can be advantageous to melt processing if the active is thermodynamically unstable. However, because of the high initial process investment and the inherent difficulty in process control, solvent casting should only be considered for manufacturing controlled drug delivery devices when absolutely necessary. Solvent casting of films is often used as a part of the overall manufacturing process in manufacturing transdermal patch products.

In Situ-Forming Implant Depots: The first in situ forming implant depot formulation to be approved for delivery of a drug in humans is ATRIDOX. The product is designed for controlled delivery of an antibiotic for treatment of periodontal disease. It is supplied as an injectable liquid. When injected into the periodontal cavity, the formulation sets forming a drug-delivery depot that delivers the antibiotic doxycycline to the cavity. ATRIDOX is based on the ATRIGEL® drug-delivery technology developed by Atrix Laboratories, Inc. The technology consists of a biodegradable polymer, a bioactive agent, excipients, and additives dissolved in a water-soluble, bioabsorbable, and biocompatible solvent. The formulations may be liquids, pastes, or putties; however, the liquid injectable is the preferred form for most applications. Products based on the ATRIGEL technology are typically made using a high-intensity mixer to dissolve the polymer and dissolve or disperse the active agent and any excipients. Another company, Matrix Pharmaceutical, Inc., has developed a proprietary biodegradable gel matrix consisting of purified bovine collagen for the targeted, intralesional delivery of chemotherapeutic agents. The system is prepared by dispersing chemotherapeutic agents in an aqueous solution of a protein such as purified bovine collagen. The chief advantage of this system is that the bovine collagen gel allows one to optimize the retention and release of the drug at a targeted injection site, thus minimizing systemic toxicity. The therapeutic effect of the delivery is further enhanced by the inclusion of a vasoconstrictor such as epinephrine. The tumor or other diseased tissues are exposed to high concentrations of the drugs for prolonged periods of time because of the site specific and sustained release of the drug. In one study, cisplatin either in a single solution (CDDP suspension) or within the Matrix gel (cisplatin/epinephrine gel) are injected into a mouse tumor (100 mm"). The free cisplatin was cleared from the site within 1 h, whereas the gel retained the drug between 24 to 72 h. One product based on this concept, AccuSite (fluorouracil/epinephrine) Injectible Gel for the treatment of recurrent genital warts has been approved in seven European countries and has completed Phase III studies in the United States. Here the gel matrix is a viscous, aqueous gel consisting of fluorouracil (30 mg/mL) and epinephrine (90.1 mg/mL) and other inactive buffering and osmotic excipients, within a purified bovine collagen matrix gel. This technology is covered by several U.S. and European patents. Another Matrix product, IntraDose Injectable Gel, has advanced into a second level of Phase II clinical trials in the U.S. for treatment of inoperable liver cancer. The product consists of a dispersion of cisplatin and epinephrine in a purified bovine collagen matrix. The protein gel matrix is a simple yet effective delivery system that permits direct delivery of chemotherapeutic agents to a tumor and provides a high local concentration of drug while minimizing systemic toxicity. The major disadvantage of the system is that it cannot provide long-term release of the drug (usually less than 1-3 days) because the drug can easily diffuse out of the gel matrix.

Encapsulation: Encapsulation, especially microencapsulation (particles ranging in size from a few to several hundred micrometers), is a process whereby particles of an active agent are surface coated to provide changes in the physicochemical properties of the active agent. There are many different processing techniques used depending on the desired properties of the final product, the properties of the agent being coated, and the properties of the coating material. The term microsphere is often used synonymously with microcapsule; however, a distinction should be made between the two terms as they are used in controlled delivery, because the final products produced and their release characteristics are quite different. Microcapsules are essentially discontinuous microspheres where the active core material is completely covered with a nonactive surface coating. The coating thickness may be varied depending on the characteristics desired in the final product. The surface coating of a microcapsule sequesters the active and serves as a protectant and/or a sustained release, rate-controlling membrane. The release mechanism of the active agent is usually mediated by diffusion of the active agent through the coating. On the other hand, microspheres are micrometer-sized homogeneous, monolithic spheres containing the active agent dispersed in a nonactive matrix material. The matrix material is often biodegradable. And in this case, the release mechanism of the active agent is usually mediated by degradation of the matrix. Although the final products are quite different, some of the processes used to prepare microspheres and microcapsules are very similar. Some of the more common processes used to form microspheres and microcapsules are as follows:

1. Spray-drying.
2. Fluid-bed coating.
3. Phase separation.
4. Solvent evaporation.
5. Solvent extraction.
6. Cryogenic solvent extraction.

Several proprietary processes also exist.

Spray-Drying: Spray-drying is a process that transforms the feed material from a fluid state into a dried particulate form by spraying the feed into a hot drying medium. It is a one-step, continuous, particle-drying process. The feed material can be in the form of a solution, suspension, emulsion, or paste. The resulting product can be powdered, granular, or agglomerated particles, depending upon the physical and chemical properties of the feed material, the drier design, and its operation. Spray-drying is used in all major industries where particle drying is required, ranging from food and pharmaceutical manufacturing to chemical industries such as mineral ores and clays. Spray-drying involves atomization of the feed into a drying medium, resulting in the evaporation of the solvent and the formation of dried particles. Atomization is a process that breaks up the bulk liquid into millions of individual spray droplets. The energy necessary for this process is supplied by centrifugal force (rotatory atomizer), pressure (pressure nozzle), kinetic (two-fluid nozzle), or sonic vibration (ultrasonic nozzle). The selection of the atomizer type depends on the nature of the feed and the desired characteristics of the final product. For all atomizer types, increasing the amount of energy available for atomization results in smaller droplet sizes. If the atomization energy is held constant and the feed rate is increased, larger particles result. Atomization also depends upon the fluid properties of the feed material, where higher viscosity and surface tension result in larger droplet sizes at the same atomization energy. In most cases, air is used as the spray-drying medium; however, dry nitrogen can be used for moisture-sensitive compounds. Contact with the spray-drying medium causes evaporation of the solvent (water or organic solvent) from the droplet surfaces. The evaporation is rapid due to the vast surface area of the droplets in the spray. The manner in which the spray contacts the drying medium is an important design factor. It influences droplet behavior; and therefore, has a great effect on the properties of the dried product. Contact with the spray-drying medium is determined by the position of the atomizer in relation to the drying air inlet. Co-current flow (the product and air pass through the dryer in the same direction), counter-current flow (the spray and air enter the dryer at the opposite direction), and mixed flow driers are available. The selection of the appropriate design is based on the required particle size, the required dried particle form and the temperature to which the dried particle can be subjected. For example, if a fine-particle product (mean size 20-120 μm) is required, but a low product temperature must be maintained at all times during the drying operation, a co-current, rotatory-atomizer spray-drier is selected. Product separation from the drying air follows completion of the drying stage. Primary separation of the dried product takes place at the base of the drying chamber. Small fractions can be recovered in separation equipment such as a cyclone. Spray-drying is a useful method for the processing of pharmaceuticals since it offers a means for obtaining powders with predetermined properties, such as particle size and shape. In addition a number of formulation processes can be accomplished in one step in a spray-drier; these include encapsulation, complex formation, and even polymerization. Spray-drying and spray-congealing processes can be used for preparing microparticles for controlled release applications. In the spray-congealing process, no solvent is used. The feed, which consists of the coating and core materials, is fed to the atomizer in the molten state. Microparticles form when the droplets meet cool air in the drying chamber and congeal. Oil-soluble vitamins, such as A and D, have been microencapsulated by spray-drying an emulsion of the oil in a gum arabic or gelatin solution. Spray-drying has also been used in the preparation of polymer-coated microcapsules for the purposes of taste masking. Biodegradable microparticles have also been prepared by spray-drying. PLA and PLGA microspheres have been prepared from solutions or suspensions of a number of drugs dissolved or dispersed in methylene chloride. Microcapsules of progesterone and PLA were formed with diameters of less than 5 μm. Crystallization of the drug occurred in the aqueous phase when microspheres were prepared by a solvent evaporation method, but spray drying avoided this problem. The main difficulty encountered in preparing spray-dried microcapsules is the formation of polymer fibers as a result of inadequate forces to disperse the feed liquid into droplets; the successful atomization into droplets is dependent on both the type of polymer used and the viscosity of the spray solution.

Fluid-Bed Coating: Fluid-bed coating is a process whereby particulates are suspended in a column of heated air or inert gas while a solution or emulsion of a polymer or other film-forming coating material is applied to the particles through spray nozzles. High-quality microcapsule products are economically produced by this process. Typical products are taste-masked drugs, enteric-coated drugs, and sustained-release drugs. Fluid-bed coating is a complex process consisting of three major operations: fluidization, atomization, and drying. The coating chamber has a high volume of flow to suspend, agitate, and dry the coated particles. The spraying nozzles can be located at various positions in the coating chamber providing top, bottom, and side or tangential spraying of the particles. Bottom spraying, or Wurster coating as it is often called, is the most common technique used for encapsulation particles as small as 30-40 μm.

Phase Separation: The phase separation process, or coacervation process as it is sometimes called, involves: 1. Preparing an organic solution of a water-insoluble, matrix material (usually polymeric), 2. Addition of an aqueous solution of the active agent or dispersion of particulates of an active agent to the organic solution with vigorous agitation, 3. Introduction of a coacervating agent or event for the matrix material to the matrix solution/active agent emulsion or dispersion 4. Depending on the means of coacervation, the coacervated matrix solution/active agent emulsion or dispersion is then added to an appropriate hardening agent to extract the excess matrix solvent, 5. collecting, washing, and drying of the final product Coacervation may be brought about by various means. It can be induced by:

- A change in the temperature of the system.
- A change in the pH of the system.
- A change in electrolyte balance.
- Addition of nonsolvents.
- Addition of other materials which are incompatible with the polymer solution.

Solvent Evaporation: It is the most widely used manufacturing technique for biodegradable microspheres. The microsphere formation process consists of three stages:

1. Droplet formation.
2. Droplet stabilization.
3. Microsphere hardening.

First, a dispersed phase containing the polymer is emulsified in an immiscible continuous phase containing a stabilizing agent. The second phase involves the diffusion of the solvent from the emulsion droplet into the continuous phase and its subsequent evaporation. Simultaneous inward diffusion of the nonsolvent into the droplet causes polymer precipitation, microsphere formation, and hardening. Depending on the nature of the two phases, the process may be termed oil-in-water (o/w) or water-in-oil (w/o) method. The solvent evaporation process requires the use of a surfactant to stabilize the dispersed-phase droplets formed during emulsification and inhibit coalescence. Surfactants are amphipathic in nature and therefore align themselves at the droplet surface, thereby promoting stability by lowering the free energy at the interface between the two phases. Furthermore, the creation of a charge or steric barrier at the droplet surface confers resistance to coalescencing and microsphere flocculation. Surfactants employed in the o/w process tend to be hydrophilic in nature and by far poly(vinyl alcohol) is the most widely used. The emulsification systems used for microparticle production have included both low- and high-speed mechanical stirring, sonication, and microfluidization. The particle size and size distribution can be controlled by the emulsification speed and mixing vessel design. Following emulsification, the removal of remanent solvent and complete microsphere hardening is usually accomplished by gentle agitation of the suspension. After evaporation of the solvent, the final stage of the emulsification-solvent evaporation process is the isolation of microspheres from the dispersed phase containing surfactant. This has generally been achieved by centrifugation and filtration, and it is usually followed by a further cleaning process in which the particles are washed several times with distilled water. The microspheres are finally dried using lyophilization or fluid-bed drying. There are several other techniques of preparing microspheres. Information on these systems can be obtained from research literature.

The limiting factor with regard to melt process of implant preparation for drug delivery is of course the heat stability of the active agent. Most of the lactide/ glycolide are injection molded at temperatures between 140°C and 175° C, hence they are not suitable for thermo labile drugs. Monomers levels greater than 2-3% by weight often cause substantial degradation of lactide/ glycolide copolymer in injection molding operation. Drug loaded fibers of both monolithic and reservoir types using lactide/ glycolide polymers have been reported. Monolithic formulation can readily be produced with melt extrusion using the blend of the active agent and polymer extruded under pressure at the lowest possible temperature. Reservoir or coaxial fiber can be produced from the glycolide/ lactide polymers by two important methods. ·Melt spinning technique in which the drug was introduced during the spinning process as a suspension or solution in a suitable lumen fluid. Dry wet phase process for poly lactide fibers, in which the drug must be added to the hollow fiber after the fibers are produced.

1.7 Testing of CR Systems

For oral controlled release dosage forms, apart from various test parameters, in vitro dissolution testing and in vivo bioequivalence studies are mandatory for approval. Dissolution testing should be conducted on 12 individual dosage units of the test and products. The potential for pH dependence of drug release from an extended release product is well recognized. Dissolution profiles should therefore be generated in aqueous media of the following pH ranges: 1 - 1.5, 4 - 4.5, 6 - 6.5, and 7 - 7.5.

Early sampling times of 1, 2, and 4 hours should be included in the sampling schedule to provide assurance against premature release of the drug (dose dumping) from the formulation. The general dissolution conditions to be followed are shown below:

1. Apparatus USP XXII Apparatus 1 (rotating basket) for capsules USP XXII Apparatus 2 (paddle) for tablets.
2. Rotation Speed 100 rpm (basket) 50 and 75 rpm (paddle).
3. Temperature 37 + 0.5ºC.
4. Units To Be Tested 12.
5. Dissolution Medium 900 ml of aqueous media of various pH.
6. Sampling Schedule 1, 2, 4 hours, and every two hours thereafter, 12 until 80% of the drug is released.
7. Tolerances as established.
8. Content Uniformity testing of the test product lot should be performed as described in the USP XXII.

In vivo bioequivalence studies recommended for approval for extended release formulations are designed to document that: The drug product meets the extended release claim made for it. The drug product does not release the active drug substance at too rapid a rate (dose dump). Performance is equivalent to that claimed following single doses and dosing to steady state. The impact of food on the *in vivo* performance on the bioavailability has to be assessed. The above objectives are generally met by the following three *in vivo* studies: 1. A single dose, randomized, two-period, two-treatment, two-sequence crossover study under fasting conditions. 2. A single dose, randomized, three-treatment, three period, six sequence, crossover, limited food effects administered under fasting conditions with those of the test products administered immediately after a standard breakfast. And 3. A multiple dose, steady state, randomized, two treatment, two-period, two-sequence crossover study under fasting conditions for the test formulation. For safety reasons, this study may be performed in the non-fasting state.

The different tests employed with other CR products are different. The compendial and the modified flow-through cell have been used successfully for implants and microparticulate formulations. The compendial flow-through apparatus is modified with regard to the inner diameter to suit the special properties for testing parenterals—that is, a low volume of fluid is used in the acceptor compartment. The flow rate of the medium has to be set very slow. Use of High Pressure Liquid Chromatography (HPLC) pumps may be considered to provide the necessary accuracy and precision at very low flow rates. In this case, the flow-through system may need to be redesigned with small internal diameter tubing. Intermittent flow might also be an option. Static or rotating bottles have also been used for in vitro release testing. As tests are often run over a long time period (eg, several weeks to months), measures have to be taken to compensate against evaporation. Suitable preservatives may be added to prevent microbial contamination. Standard preservatives, including cetylammonium bromide, benzalkonium chloride, parabens, phenol derivatives, and mercury salts, along with appropriate concentrations to be used, are listed in many pharmaceutical textbooks. The selection has to be based on criteria such as compatibility with the active pharmaceutical as well as other formulation ingredients and the pH of the test medium. Issues with these compounds include their ionization properties, physicochemical interactions, and analytical interferences. 0.1% sodium azide has also been used as preservative, but because of safety concerns, it cannot be generally recommended. The composition of the medium should take into consideration the osmolarity, pH, and buffer capacity of the fluids at the site of administration, which are usually assumed to resemble those of plasma (or muscle) but with lower buffer capacity. However, the main challenges with this type of dosage form are to determine the appropriate duration of the test and the times at which samples are to be drawn in order to characterize the release profile adequately. An in vitro release test for assessing the quality and for process control of liposome drug products is important, but the challenge remains to develop and identify a reliable method that can characterize drug release from the product.

These CR formulations can be characterized for various properties. To characterize the release from the dosage form adequately, one must generate a drug release profile in which release (dissolution) values are determined as a function of time. This multipoint characterization has been in place for modified release oral dosage forms for some time and is also recommended for slower dissolving immediate release products. Because many of the dosage forms discussed here are complex in composition and release mechanism, a multipoint drug release test will be required to characterize release from the drug product in general and to test for possible alterations in the release profile during storage.

Multipoint tests may also be needed for batch release testing in order to confirm acceptable batch-to-batch consistency. Typical cases where multipoint tests are likely to be needed include transdermal patches, semisolid preparations, chewing gums, implants, microparticulate formulations, solid solutions, solid dispersions, and liposomes. The experimental test conditions should be discriminating enough ("mild" conditions) to detect manufacturing variables that may affect biopharmaceutical product performance. Test conditions that may not be able to discriminate adequately among products/batches with different in vivo release profiles include those with very high agitation/flow rates, the use of strongly alkaline solutions to dissolve poorly soluble acids, and the use of very high surfactant concentrations to create sink conditions, to name but a few. As for solid oral dosage forms, development of in vitro dissolution/release tests and specifications for novel/special dosage forms should take into account relevant bioavailability or clinical data. However, expectations with respect to the quality and/or level of in vitro/in vivo correlation should not be set as high as for solid oral dosage forms, because of the higher level of complexity and data variability for novel/special dosage forms.

Ideally, physiological conditions at the site of administration should be taken into account when selecting the in vitro dissolution/release test conditions. The complexity of the release mechanism of some novel/special dosage forms and the lack of knowledge about the conditions under which release occurs in vivo make it difficult to design physiologically based tests in all cases, but it should be possible to conceive a test that can detect the influence of critical manufacturing variables, differentiate between degrees of product performance, and to some extent characterize the biopharmaceutical quality of the dosage form. As the release mechanism and site of application vary dramatically among the novel/special dosage forms, the experimental test conditions have to be tailored according to the conditions at the site of administration (eg, temperature of the test) and the release mechanism (eg, chewing gums will require different agitation rates than suspensions). Within a given category, it may be necessary to have product type-specific dissolution tests (eg, separate tests for lipophilic and hydrophilic suppositories), and in some cases for products containing the same drug and administered in the same type of novel/special dosage form but with a different release mechanism (analogous to the range of tests available in the USP for theophylline extended release dosage forms).

The quality control tests employed has a variety of applications. A specific value of in vitro dissolution/drug release testing is recognized in its application as a batch-to-batch quality control test and its value in evaluation and approval of SUPAC. SUPAC-SS defines the levels of changes with respect to component and composition, site of manufacturing, scale of manufacturing, and process and equipment changes. In vitro drug release is used to ensure product sameness for semisolid dosage forms under SUPAC-related changes. The same principles can easily be extended to other dosage forms where the product sameness can be ensured by profile comparison between prechange and postchange products using an appropriate in vitro test and profile comparison (eg, for transdermal patches). In addition to this, the dissolution/drug release test can also be used for providing biowaivers for lower strengths of a product from a given manufacturer, once the higher strength is approved based on the appropriate bioavailability/ bioequivalence test procedure. Even though less experience is available with novel/special dosage forms than is available with conventional dosage forms, in vitro/in vivo correlations have been established. In such cases it is legitimate to use in vitro dissolution as a surrogate for the in vivo performance of a drug product, as long as the rate-limiting step is the release of the drug from the formulation; regulations should also support this. Because of the typically higher variability of in vivo and in vitro data in the case of many novel/special dosage forms, expectations about the quality and level of in vitro/in vivo correlations might have to be adjusted in comparison to those for conventional dosage forms. It is worth noting that in general, an in vitro dissolution/release test is expected for each novel/special dosage form regardless of whether the intended effect is systemic or nonsystemic (eg, topical semisolid dosage forms), for formulation development, for investigations to support postapproval changes, and for batch-to-batch quality control. It has to be noted, however, that because of the specific formulation design, because of potential (physicochemical) interactions between the dosage form and the physiological environment at the site of administration, and because of the necessary design of in vitro dissolution equipment for novel/special dosage forms, dissolution/release data in vitro might be more strongly influenced by test or equipment parameters or less predictable for in vivo release than is usually experienced for conventional dosage forms. Therefore, a scientifically sound assessment of the relevance and validity of an in vitro dissolution test should affect the final decision about the application of the test and the specifications set for batch-to-batch quality control.

1.8 Conclusions

Controlled release systems are in high demand as of today. This is simply because of the commercial aim of a pharmaceutical company. Pharma companies are now aiming to improve productivity and lower the risks associated with new drug candidates. Often simply developing a new generation of medicines, often proprietary, formulation and delivery technologies saves companies millions of dollars. These new technologies can modulate release profiles to achieve dosing that enhances patient outcome, compliance and safety. Growth in controlled release formulations is driven in part by the dramatic increase in generic drugs and generic drug companies. These off-patent APIs often suffer from sub-optimal pharmacokinetics and unpleasant side effects, issues which can be typically be addressed by controlled release platforms. As generic drug suppliers search for ways to increase the marketability and positioning of their products, the number of controlled release technology formulations entering the marketplace will accelerate. Understanding controlled release technology through a quantitative and mechanistic approach by examining the common principles shared by several dosage forms regardless of the shape of the dosage form or the route of administration is always helpful. Thus, designing controlled release products in a systematic and encyclopedic manner rather than using trial and error approach is the most beneficial approach. Some basic concepts of controlled release dosage forms have been comprehensively covered in this chapter.

References

1. Barochez B, Lapeyre F and Cuine A. Oral sustained release dosage forms comparison between matrices and reservoir devices. Drug Dev and Ind Pharm, 15(6-7):1001-1020 (1989).

2. Mockel JE and Lippold BC. Zero-order drug release from colloid matrices. Pharm Res, 10(7):1066-1070 (1993).

3. Morita R, Honda R and Takahashi Y. Development of oral controlled release preparations, a PVA swelling controlled release system (SCRS). II. In vitro and in vivo evaluation. J Control Rel, 68(1):115-120 (2000).

4. Meka L, Kesavan B, Kalamata VN, Eaga CM, Bandari S, Vobalaboina V, Yamsani MR. Design and evaluation of polymeric coated minitablets as multiple unit gastroretentive drug delivery systems for furosemide. J Pharm Sci, 98(6):2122-2132 (2008).

5. Lordi NG. Sustained release dosage forms in The Theory and Practice of Industrial Pharmacy, 3rd edition (Eds. Leon Lachman, Herbert A. Lieberman, Joseph L. Kanig), pp. 430-456 (1986).

6. Boxenbaum H. Pharmacokinetic determinants in the design and evaluation of sustained release dosage forms. Pharm Res, 1(2): 82-88 (1984).

7. Milo Gibaldi and Donald Perrier. Pharmacokinetics, 2nd edition (Marcel Dekker, New York), pp. 188-194 (1982).

8. Theeuwes F and Bayne W. Dosage for index: an objective criterion for evaluation of controlled drug delivery systems. J Pharm Sci, 66(10):1388-1392 (1977).

9. Vittaz M, Bazile D, Spenlehaue G, Verrecchia T, Veillard M, Puisieux F and Labarre D. Effect of PEO surface density on long-circulating PLA-PEO nanoparticles which are very low complement activators. Biomaterials, 17(16):1575-1581 (1996).

10. Benahmed A, Ranger M, Leroux J. Novel polymeric micelles based on the amphiphilic diblock copolymer poly (N-vinyl-2-pyrrolidone)-block-poly (DL-lactide). Pharm Res, 18(3):323-328 (2004).

11. Lysenko EA, Trusov AN, Chelushkin PS, Bronich TK, Kabanov AV, and Zezin AB. Mixed micelles based on cationic and anionic amphiphilic diblock copolymers containing identical hydrophobic blocks. Polymer Sciences Series A, 51(6):606-615 (2009).

12. Ouchi T and Ohya Y. Macromolecular prodrugs. Progress in polymer science, 20(2):211-257 (1995).

13. Thatte S, Datar K and Ottenbrite RM. Perspectives on: polymeric drugs as delivery systems. J Bioactive and Compatible Polymers, 20(6):585-601 (2005).

14. Buprenorphine. Http: //en/Wikipedia.org/wiki/Buprenorphine (dated:28/02/2010).

15. Tice TR, Staas JK, Ferrell TM and Markland P. Injectable opioid partial agonist or opioid antagonist microparticle compositions and their use in reducing consumption of abused substances. US patent 6495155 (2002).

16. Drizen A, Rothbart P, Nath GM. Sustained release delivery system. US patent 6335035 (2002).

17. Goodson JM, Holborow D, Dunn RL, Hogan P and Dunham S. Monolithic tetracycline-containing fibers for controlled delivery to periodontal pockets. J Periodontol, 54(10):575-579 (1983).

18. Bodomeier R, Oh K and Chen H. The effect of the addition of low molecular weight poly (DL-lactide) on drug release from biodegradable poly (DL-lactide) drug delivery systems. Int J Phar, 51:1-8 (1989).

2
Gastroretentive Drug Delivery Systems

Y. Madhusudan Rao[1] and B. Suresh[2]

[1]*University College of Pharmaceutical Sciences, Kakatiya University Warangal, India*

[2] *Dr. Reddys Laboratory, Hyderabad, India*

2.1 Introduction

Drug delivery systems are becoming increasingly sophisticated as pharmaceutical scientists acquire a better understanding of the physicochemical and biological parameters pertinent to their performance. The word 'new' in relation to drug delivery system (DDS) may be a search for something out of necessity. The objective is to minimize the disadvantages associated with existing dosage form (DF) and optimizing therapy. Despite tremendous advancements in drug delivery, the oral route remains the preferred route for the administration of therapeutic agents because of the low cost of therapy and ease of administration lead to high levels of patient compliance. More than 50% of drug delivery systems available in the market are oral drug delivery systems (Source: International Market Study, America- Drug Delivery based products, 1998). Conventional oral dosage forms provide a specific drug concentration in systemic circulation without offering any control over drug delivery. Thus the oral administration of a medication by means of controlled release drug delivery systems (CRDDS) should ideally enable the drug release at a predetermined, predictable and controlled rate (1, 2) to obtain the required plasma levels and to keep them steady for a prolonged period of time. Figure 2.1 represents the theoretical basis and significance of designing controlled release drug delivery systems. Unfortunately, this ideal therapeutic target cannot be achieved systemically inspite of the progresses accomplished today in formulation and control of drug release kinetics from such type of dosage forms. The main limitations come from the inter subject and intra subject variability of gastrointestinal transit time and from the non-uniformity of drug absorption throughout the alimentary canal.

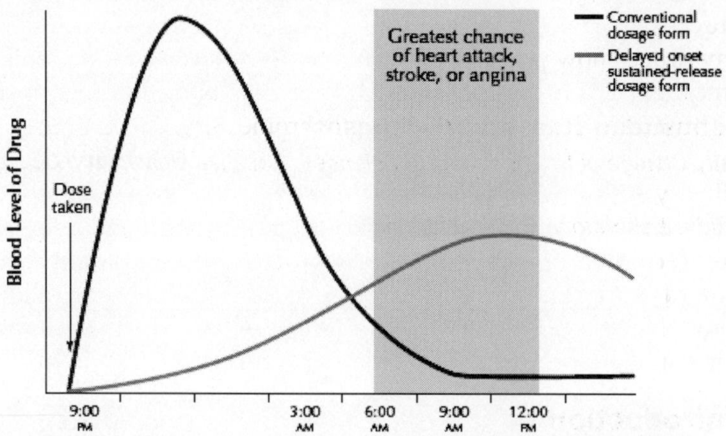

Figure 2.1 IR Vs SR / CRDDS in Hypertension Therapy.

(Adapted from Prisant L.M., Chronobiol. Hypertens. Med. Crossfire. 1999; 1:1-16.)

An important requisite for the successful performance of oral CRDDS is that the drug should have good absorption throughout the gastrointestinal tract (GIT), preferably by passive diffusion, to ensure continuous absorption of the released drug (3). A major constraint in oral controlled drug delivery is that not all drug candidates are absorbed uniformly throughout the GIT. Some drugs are absorbed in a particular portion of the GIT only or are absorbed to a different extent in various segments of the GIT (4, 5). Such drugs are said to have an absorption window, which identifies the drug's primary region of absorption in the GIT. An absorption window exists because of physiological, physicochemical, or biochemical factors. The pH-dependent solubility and stability level of a drug plays an important role in its absorption. A drug must be in a solubilized and stable form to successfully cross the biological membrane, and it will experience a pH range from 1 to 8 as it travels through the GIT. Because most drugs are absorbed by passive diffusion of the un-ionized form, the extent of ionization at various pH levels can lead to nonuniform absorption or an absorption window. The presence of certain enzymes in a particular region of the GIT also can lead to regional variability in the absorption of drugs that are substrates of those enzymes (6).

Perorally administered drugs are absorbed by passive diffusion processes and by non passive means. Drugs absorbed by active and facilitated transport mechanisms show higher regional specificity because of the prevalence of these mechanisms in only certain regions of the GIT (3). Many drugs show poor Bioavailability (BA) because of the presence of enzymes and efflux pumps. Intestinal metabolic enzymes—primarily, Phase I metabolizers such as cytochrome P450 (CYP3A)—are abundantly present in the intestinal epithelium. Their activity decreases longitudinally along the small intestine, with levels rising slightly from the duodenum to the jejunum and declining in the ileum and colon. This nonuniform distribution of CYP3A causes regional variability in the absorption of drugs that are substrates of that enzyme. In addition, carriers involved in the secretion of organic molecules from the blood into the intestinal lumen may affect drug absorption (7).

An example of such a secretory transporter is P-glycoprotein (P-gp), which is present in the villus tip of enterocytes and has the capacity to interact with a vast variety of drugs. P-gp sends the absorbed drug from the cytoplasm of the enterocyte back to the intestinal lumen, thus reducing the drug's BA. In the human body, these intestinal enzymes and efflux transporters work in coordination to protect against the entry of toxic agents into blood circulation.

Drugs having site-specific absorption are difficult to design as oral CRDDS because only the drug released in the region preceding and in close vicinity to the absorption window is available for absorption (8). After crossing the absorption window, the released drug goes to waste with negligible or no absorption (Figure 1.2(a)). This phenomenon drastically decreases the time available for drug absorption after its release and jeopardizes the success of the delivery system. All these above limitations could be overcome, for various judiciously selected drugs, by prolonging the gastric residence time of the pharmaceutical dosage form i.e., the development of gastroretentive drug delivery systems.

Dosage forms that can be retained in the stomach for prolonged and predictable period of time are called gastroretentive drug delivery systems (GRDDS) (9). Therefore the real issue in the development of oral GRDDS is not just to prolong the delivery of drugs for 12 hours or more, but to prolong the presence of DDS in the stomach or upper GI tract until the entire drug is released. Thus GRDDS can improve the controlled delivery of drugs that have an absorption window by continuously releasing the drug for a prolonged period of time before it reaches its absorption site (10) (Figure 2.2(b)) thus ensuring its optimal BA (11, 12). Within the context of this book chapter GRDDS formulation technologies and biopharmaceutical factors will be considered to design an effective system for optimizing the bioavailability and specifically modifying the release of drugs that are absorbed in the proximal intestine.

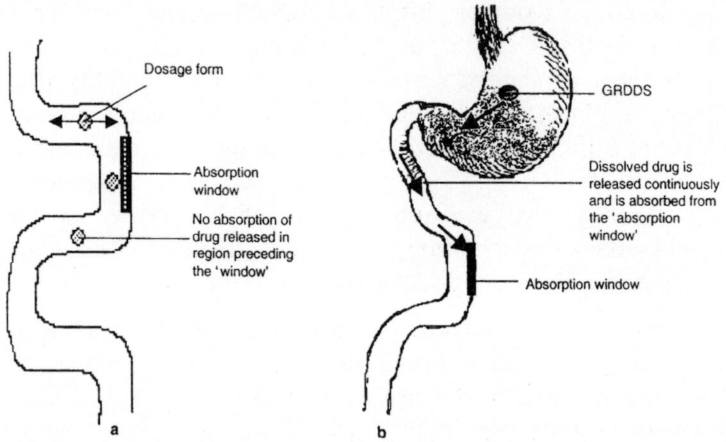

(a) Conventional dosage form (b) Gastroretentive drug delivery system

Figure 2.2 Absorption scenario for a controlled release dosage form versus a gastroretentive delivery system assuming that the drug is mainly absorbed in the proximal intestine.

2.2 Basic Anatomy and Physiology of Stomach

Stomach Anatomy

The design of efficient gastroretentive drug delivery systems can be achieved by thorough understanding of anatomy and physiological behavior of stomach. Stomach is recognized as a depot for (sustained release) gastroretentive drug delivery systems both in human and veterinary applications. The stomach is roughly shaped like the letter J; the medial concave side is known as the lesser curvature and the lateral convex side is known as the greater curvature (13) (Figure 2.3). The stomach can be divided into five parts. The cardia is an ill-defined region beginning at the gastroesophageal junction and extending into the first to 2 - 3 cm of the stomach. The fundus is the second section of stomach that lies superior to the gastroesophageal junction. The corpus or body of the stomach and fundus act as reservoir of undigested material. The distal third of the stomach, the antrum, finally, the pylorus is a narrow 1 to 2 cm long channel connecting the stomach to the duodenum. Pyloric antrum is a major site of mixing actions and act as pump for gastric emptying (14).

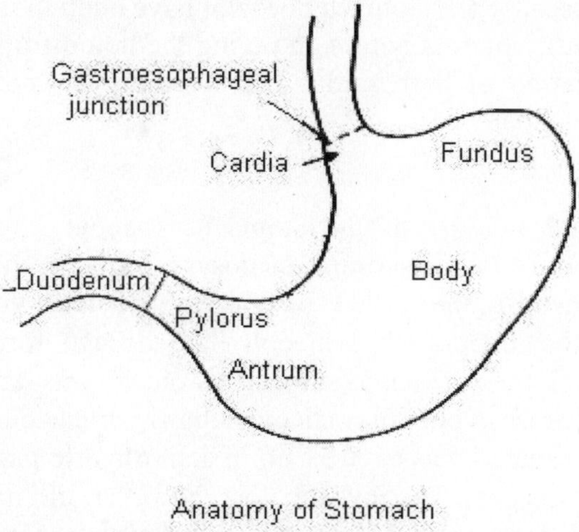

Figure 2.3 Diagrammatic representation of the stomach.

Physiology

Factors such as pH, nature and volume of the gastric secretions and gastric mucosa play a major role in drug release and absorption. Table 2.1 lists the anatomical and physiological features of the GIT.

Table 2.1 Morphology, Physiology and transit time in each segment of GIT.

Segment	Function	Size (Diameter X length) cm	Surface Area (m^2)	pH	Transit time depends on type of food	
					Liquid	Solid
Stomach	Digestion of food	15 × 20	3.5	1-3.5	10-30 min	<1 to 3 hrs
Duodenum	Neutralization of acids	3-5 × 20-30	2	4-6.5	Short	Minutes to hours
Jejunum	Absorption of nutrients	3-5 × 240	180	5-7	3 hr ±1.5 hrs	4 hr ± 1.5 hrs
Ileum	Absorption of nutrients	3-5 × 360	280	6-8		

pH

Environmental pH affects the performance of orally administered drugs. The pH of the stomach in fasted conditions is about 1.5 to 2, and in fed condition, usually it is about 2 to 6. A large volume of water administered with an oral dosage form increases the pH of the stomach initially. This change occurs because the stomach does not have enough time to produce a sufficient quantity of acid before emptying the liquid from the stomach. Thus the dissolution of both acidic and basic drugs would be affected with pH changes (15).

Volume

The resting fluid volume of the stomach is about 25-50ml. Gastric volume is important for dissolution of dosage forms in vivo. In a study done to demonstrate the effect of gastric pH and volume on the absorption of controlled-release theophylline dosage form in humans, gastric fluid volume of each subject as such was measured. The experiment was done in both normal and achlorhydric subjects. The mean volume ± SD estimated was 61 ± 51 ml in achlorhydric patients and 98 ± 38ml in normal subjects. The investigators, however, did not observe that such a volume difference could affect in vivo dissolution of drugs as each dose was administered with 180ml water (15).

Gastric Mucosa

Simple columnar epithelial cells line the entire mucosal surface of the stomach. Mucous, parietal, and peptic cells are present in the body of the stomach. These cells are associated with different functions. The parietal cells secrete acid whereas the peptic cells secrete precursor for pepsin. The surface mucosal cells secrete mucus and bicarbonate. They protect the stomach from digestion by pepsin and from adverse effects of hydrochloric acid (HCl). As mucus has a lubricating effect, it allows chyme to move freely through the digestive system (15).

Gastric Secretion

Acid, pepsins, gastrin, and mucus are the main secretions of the stomach. Normal adults produce a basal secretion up to 60ml with approximately 4 mmol of hydrogen ions every hour. The volume of this secretion can go beyond 200ml and 15 to 50 mmol hydrogen ions when stimulated. Pure parietal secretion is a mixture of hydrochloric acid and potassium chloride (15). Histamine stimulates acid secretion through the H_2-receptors located on gastric mucosa. Another potent stimulator of gastric acid is the hormone gastrin. Peptides, amino acids, and distention of stomach stimulate HCl release. The mean thickness of mucus in human stomach is about 140 μm. It is continuously digested from the surface. Generally, it takes 4 to 5 hours for mucus turnover. Mucus protects the gastric mucosa from pepsin and acid in the stomach (15). In the stomach and duodenum, the mucus layer provides a stable unstirred layer that maintains a pH gradient from the acid lumen to near neutral at the mucosal surface. In addition, it provides a physical barrier that prevents access by luminal digestive enzymes to the epithelium. The mucus layer also acts to protect the delicate epithelium from mechanical forces that result from contractile activity. Mucus consists of gel-forming glycoproteins (1-10% by weight) and water. The glycoprotein portion is over 70% carbohydrate by weight and is negatively charged (15).

Mucus is composed of two layers; the inner layer firmly adheres to the gastric mucosa while the outer layer (luminal side) loosely adheres to the inner layers. In the stomach, the loosely adherent layer is between 40 and 60% of the total mucus thickness. The outer mucus layer is in a state of continuous flux as a result of erosion by luminal protcases and mechanical shear and replacement by secretion of new mucus. Under normal physiological conditions, mucus secretion is equal to its erosion and so maintains an adequate protective layer. Polymers with excellent in vitro mucoadhesion properties have been developed based on the hypothesis that if an oral dosage form can be made to stick to the mucosal surface of the stomach or duodenum, gastric or duodenal residence time can be prolonged (16).

Effect of Food on Gastric Secretion

On average, the daily intake of a normal adult is 2±1 kg of food and drink. In response to this stimulus, the gut secretes an additional 5 liters of fluids. The volume produced within the first hour of eating can be twice that of the meals. A distinct pH gradient exists in the stomach after a meal. The contents of the body of the stomach are neutralized but the antrum remains relatively acidic in nature. This ingestion of food is the major stimulus to acid and enzyme secretions in the stomach. This effect is more pronounced if the meal has high protein content, as the protein content has a maximal buffering capacity. A meal can increase the pH of the stomach to 3 or 5, and foods such as milk can raise it over pH.

Gastric Motility

The stomach produces coordinated movements of the gastric contents due to three layers of smooth muscles. These layers are outer longitudinal muscle layer, inner circular layer, and an oblique layer (15).

The process of gastric emptying occurs both during fasting and fed states; however the physiological behavior of stomach differs markedly between fasting as well as in fed states. During fasting state an inter-digestive series of electrical events take place, which cycle through stomach and intestine every 2 to 3 hours (17). This activity is called the interdigestive myoelectric motor complex (IMMC) or migrating myoelectric cycle (MMC) (18). The MMC is organized into alternating cycles of activity and quiescence and can be subdivided into basal (phase I), preburst (phase II), and burst (phase III) intervals. Phase IV is a transient phase between phases I and III (19) Figure 2.4.

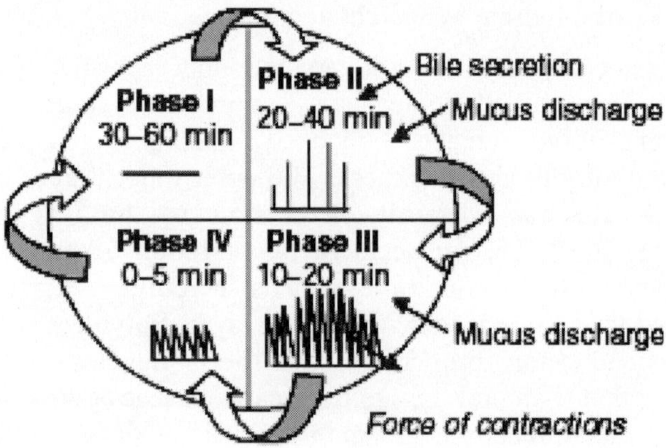

Figure 2.4 Motility patterns of GIT in the fasted state.

Phase I - This lasts for 30 to 60 min with rare contractions.

Phase II - This lasts for about 20 to 40 min with intermittent contractions that intensify over time.

Phase III - This phase lasts for 10 to 20 minutes with intense contractions resulting in complete emptying of stomach. This is also known as the housekeeper wave.

Phase IV - This lasts for 0 to 5 minutes, which occurs between burst phase and basal phase of two consecutive cycles.

In fed conditions, only one phase is present. This phase is present as long as there is food in the stomach. It consists of regular and frequent contractions. These contractions are not severe as those in the third phase of fasted motility pattern. During fed state onset of IMMC is delayed resulting in slow down of gastric emptying (20). This type of motility is known as postprandial motility. The particular phase during which a dosage form is administered influences the extent and rate of drug absorption as well as the performance of CRDDS and GRDDS as far as their residence within the GIT is concerned (21). When CRDDS are administered in the fasted state, the MMC may be in any of its phases, which can significantly influence the total gastric residence time and transit time in the GIT. This assumes even more significance for drugs that have an absorption window because it will affect the amount of time the dosage form or the liberated drug spends in the region preceding and around the absorption window. The less time spent in that region, the lower the degree of absorption. Therefore, the design of GRDDS should take into consideration the resistance of the dosage form to gastric emptying during phase III of the MMC in the fasted state and also to continuous gastric emptying through the pyloric sphincter in the fed state. This means that GRDDS must function quickly after administration and able to resist the onslaught of physiological events for the required period of time (19).

2.3 Factors Influencing Gastric Retention Time (GRT) of Dosage Forms

There are several factors that can affect gastric emptying (and hence GRT) of an oral dosage form. These factors include: density, size, shape of dosage form, concomitant intake of food and drugs such as anticholinergic agents (e.g. atropine, propantheline), opiates (e.g. codeine), and prokinetic agents (e.g. metoclopramide, cisapride) and biological factors such as gender, posture, age, body mass index, and disease states (e.g. gastrointestinal diseases, diabetes, Crohn's disease) (11, 22).

Density of Dosage Form

Density of dosage form determines the location of system in GIT. Dosage forms having density lower than that of the gastric fluids experience floating behavior and greater gastric retention. A density <1.0gm/ml is required to exhibit the floating property. The floating tendency of the dosage form usually decreases as a function of time, as the dosage form gets immersed into the fluid content of the stomach, as a result of the development of hydrodynamic equilibrium. Generally, the denser systems settle to the bottom of the stomach at a faster rate than the less dense systems, hence experiencing faster gastric emptying (23).

Size of Dosage Form

The size of the dosage form can also determine how long a dosage form is retained in the stomach. The mean gastric residence times of the non-floating dosage forms are highly variable and greatly dependent on their size, which could be small, medium, or large. In fed conditions, the smaller units get emptied from the stomach during the digestive phase and the larger units during the housekeeping waves. In most cases, the larger the size of the dosage form, the greater the gastric retention time, because the larger size would not allow the dosage form to quickly pass through the pyloric antrum into the intestine (11, 22). In an attempt to determine the GRT for non disintegrating tablets of different sizes, the longest gastric emptying time was observed for 13-mm tablets (171 ± 13min), followed by 11-mm (128±17min) and 7-mm (116±19min) tablets. These results depend on the fact that the aperture of the resting pylorus is 12.8 ± 7mm, which is considered as an important factor for GI transit of dosage forms (19). In another study done to compare the effect of size on the gastric retention time of floating (F) and non-floating (NF) units using γ-scintigraphy, it was found that F units with a diameter equal to or less than 7.5mm had longer GRTs compared to NF units. However, the GRTs were similar for F and NF units having a larger diameter of 9.9mm. Thus F units, which remain buoyant on gastric contents, are protected against gastric emptying during digestive phases. On the other hand, NF units lie in the antrum region and are repelled during the digestive process by peristalsis (11). On the other hand some investigators consider that there is no statistically significant difference in the extent of gastric residence in the case of floating systems of different sizes (19).Thus intragastric buoyancy or the position of the floating dosage forms away from the pyloric sphincter is considered more important compared to the size in determining the gastric residence time of such systems. The problems such as all or nothing emptying of dosage form (24) obstruction in the gastrointestinal tract made single unit floating systems unreliable and irreproducible in prolonging gastric residence time. However the multiple unit floating systems developed are reliable and reduce inter subject variability in absorption and lower the probability of dose dumping (25). One important fact to be kept in mind is that the gastroretentive drug delivery systems should dissolve or erode after drug release to the size that allows ejection of such delivery systems from the stomach.

Food Intake and Nature of Food

The nature of food such as caloric content, volume, viscosity of food along with frequency of feeding has profound effect on gastric retention of the dosage form. Usually presence of food in the stomach increases GRT of the dosage form and increases drug absorption by allowing it to stay at the absorption site for a longer time (26). The GRT of both non-floating and floating single units is shorter in fasted individuals and is prolonged after a meal. Khosla et al, reported a delayed emptying of 3, 4, and 5 mm diameter tablets when taken with breakfast in humans (21). The rate of gastric emptying primarily depends on the caloric contents of the ingested meal. It does not differ for proteins, fats, and carbohydrates as long as their caloric content is the same; the nutritive density of a meal helps to determine the rate of gastric emptying. Generally an increase in acidity, osmolarity, and caloric value slows down gastric emptying. Muller–Lissner et al, 1981 reported that in an *in-vivo* evaluation of floating drug delivery systems (FDDS), a GRT of 4-10 hours could be achieved after a fat and protein meals (27). In a study done on a bilayer-floating capsule of misoprostol, the mean GRT was 199±69 minutes after a single light meal (breakfast). However, after a succession of meals, the data showed remarkable prolongation of the mean GRT, to 618±208 minutes (28). In another study on the effect of fed state after a single meal all the floating units had a floating time (FT) of about 5 hours and GRT was prolonged by about 2 hours over the control. However, after a succession of meals, most of the floating units showed a FT of about 6 hours and a GRT prolonged by about 9 hours over the control though a certain variability of the data owing to mixing with heavy solid food ingested after the dosing was observed (29). Obviously, if the gastroretentive properties of a floating dosage form are independent of meal size, it can be suggested that the dosage form will be suitable for patients with a wide range of eating habits.

Interestingly, most of the studies related to the effect of food on GRT of floating drug delivery systems share a common viewpoint that food intake is the main determinant of gastric emptying, while specific gravity has only a minor effect on the emptying process. Stated otherwise, the presence of food, rather than buoyancy, is the most important factor affecting GRT and floating does not necessarily increase GRT. Thus it may be concluded that floating systems rely on the presence of a meal along with the system's ability to expand and enlarge in size as a mean to retard their emptying (11).

Effect of Age, Gender, Posture and diseased state

The biological factors such as age, body mass index (BMI), gender, posture and disease states (diabetes, Crohn's disease) influence the gastric emptying. Elderly persons and females showed slower gastric emptying, stress increases gastric emptying while depression slows it down. In a study done on both genders, it was found that gastric emptying in women is slower than in men regardless of weight, height, body surface area and even when hormonal changes due to menstrual cycle are normalized (11).

The investigators also studied the effect of posture on GRT and found no significant difference in the mean GRT for individuals in upright, ambulatory and supine states. On the other hand in a comparative study in humans, the floating and non-floating systems behaved differently (26). In the upright position, the floating systems floated to the top of the gastric contents showing prolonged GRT. But the non-floating systems showed an opposite trend and sank rapidly into the gastric contents. In the upright position, the non-floating units settled to the lower part of the stomach and underwent faster emptying as a result of the peristaltic contractions, and the floating units remained away from the pylorus. However, in a supine position, the floating units were emptied faster than non-floating units of similar size (30). Apart from the upright position, the posture of the individual in the ambulatory state may have different effects on the gastric emptying of the dosage form. When the individual rests on the left side, the floating of the dosage form will be toward the pyloric antrum; when the individual rests on the right side, the floating of the dosage form will be in the opposite direction. Thus gastric emptying is slower in individuals resting on their right side (22).

The GRT of the dosage forms may vary also with age due to the changes in physiology with increasing age and the hormonal responses responsible for gastric emptying. It has been demonstrated that gastric retention is prolonged in the elderly, especially in individuals 70 years or older (31).

Shape of the Dosage Form

The shape of the dosage form is one of the factors that affect its GRT. It is considered to be one of the formulation variables or factors affecting GRDDS performance. Six shapes (ring, tetrahedron, cloverleaf, string, pellet, and disk) were screened *in vivo* for their gastric retention. The tetrahedrons (each leg 2cm long) and rings (3.6cm in diameter) exhibited nearly 100% retention at 24 hours where as on the other hand cloverleaves (2.2-3.2 cm in diameter) exhibited 40-67% retention; discs (2.2-3.2 cm diameter). 67%; string (12cm × 2mm × 2mm / 24cm × 2mm × 2mm), 0%; and pellets (4mm) 0% retention at 24 hours (32). The specific shapes examined in the mentioned *in vivo* study are shown in Figure 2.5.

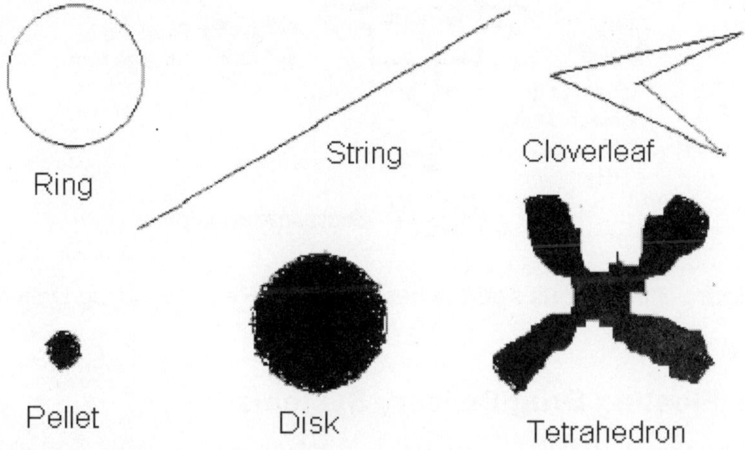

Figure 2.5 Delivery system shapes examined for their retention ability in the stomach.

2.4 Strategies for Gastroretentive Drug Delivery Systems

Over last two to three decades the need for GRDDS has led to extensive efforts in both academia and industry towards the development of such drug delivery systems. The numerous gastro retentive drug delivery systems that have been designed to prolong the gastric residence time (33-36). These drug delivery systems are beneficial for absorption window drugs, drugs degraded in colon, drugs with poor solubility and stability in alkaline conditions and drugs administered for local action in stomach. The different gastroretentive drug delivery systems resulted due to these efforts include floating dosage systems (35, 20), expandable and swellable systems (37, 38), modified systems (39, 40), mucoadhesive systems (41-43), superporous hydrogel systems (44), magnetic systems (45, 46) raft forming systems (47,48) and high-density systems (49). The general approaches for gastric retention are given in Figure 2.6.

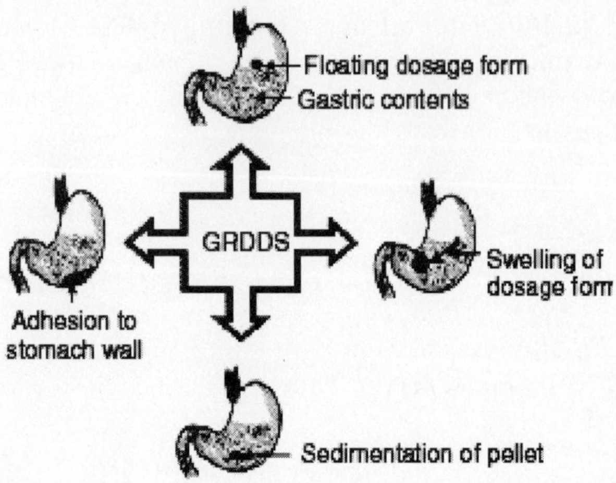

Figure 2.6 Various approaches of Gastro Retentive Drug Delivery Systems.

2.4.1 Floating Drug Delivery Systems

Among the GRDDS floating drug delivery systems (FDDS) or hydrodynamically controlled systems are widely investigated to prolong the gastric residence time. These floating dosage systems remain buoyant on the stomach fluid thereby increasing the gastric retention time. The various buoyant systems developed are single unit floating systems (Tablets, Capsules, laminated systems, raft forms, etc) and multiple unit floating systems (Hollow microspheres, multiple unit beads, granules, pellets etc). Most of the floating systems reported in literature are single-unit systems, such as hydrodynamically balanced system (HBS) and floating tablets. Based on the mechanism of buoyancy, two distinctly different technologies, i.e., noneffervescent and effervescent systems have been utilized in the development of FDDS.

2.4.1.1 Non effervescent/ Swellable FDDS

The most commonly used excipients in non-effervescent/swellable FDDS are gel-forming or highly swellable cellulose type hydrocolloids, polysaccharides, and matrix forming polymers such as polycarbonate, polyacrylate, polymethacrylate and polystyrene. One of the approaches to the formulation of such floating dosage forms involves intimate mixing of drug with a gel-forming hydrocolloid, which swells in contact with gastric fluid after oral administration and maintains a relative integrity of shape and a bulk density of less than unity within the outer gelatinous barrier (50). The air trapped by the swollen polymer confers buoyancy to these dosage forms. In addition, the gel structure acts as a reservoir for sustained drug release since the drug is slowly released by a controlled diffusion through the gelatinous barrier.

Sheth and Tossounian, 1978 developed a HBS capsule (Figure 2.7) containing a mixture of a drug and hydrocolloids. Upon contact with gastric fluid, the capsule shell dissolves; the mixture swells and forms a gelatinous barrier thereby remaining buoyant in the gastric juice for an extended period of time (51).

Javed Ali et al, 2007 developed a hydrodynamically balanced system of metformin as a single unit floating capsule. Various grades of low-density polymers were used for the formulation of this system. They were prepared by physical blending of metformin and the polymers in varying ratios. Capsules prepared with HPMC K4M and ethyl cellulose gave the best *in vitro* percentage release and was taken as the optimized formulation. By fitting the data into zero order, first order and Higuchi model it was concluded that the release followed zero order release with fickian diffusion mechanism. The in vivo studies on rabbits showed buoyancy for 5 h. The comparative pharmacokinetic study was performed by administration of the optimized HBS capsules and immediate release capsules, showed increase in AUC in optimized HBS capsules of metformin when compared with immediate release formulation (52).

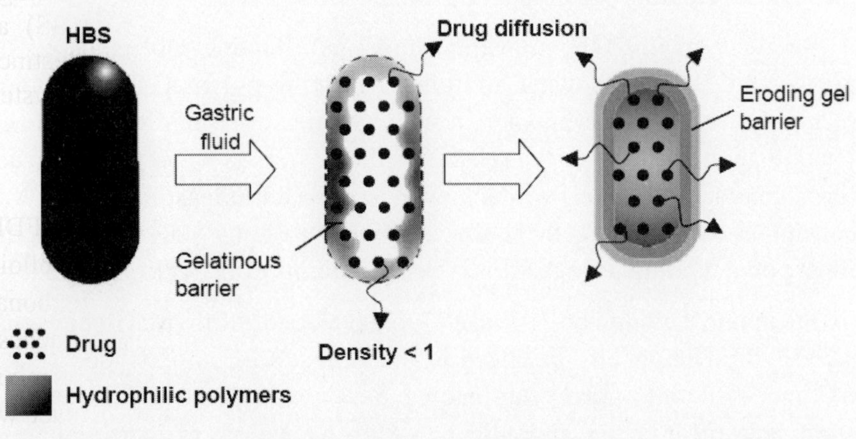

Figure 2.7 Working of Hydrodynamically balanced system.

Nur and Zhang, 2000, developed floating tablets of captopril using HPMC K4M, K15M and carbopol 934P. In vitro buoyancy studies revealed that tablets of 2 kg/cm^2 hardness after immersion into the floating media floated immediately and tablets with hardness 4 kg/cm^2 sank for 3 to 4 minutes and then came to the surface. Tablets in both cases remained floating for 24 hours. The tablet with 8 kg/cm hardness showed no floating capability. It was concluded that the buoyancy of the tablet is governed by both the swelling of the hydrocolloid particles on the tablet surface when it contacts the gastric fluids and the presence of internal voids in the center of the tablet (porosity). A prolonged release from these floating tablets was observed as compared to the conventional tablets and a 24-hour controlled release from the dosage form of captopril was achieved (53).

Ushomaru et al, 1987 developed sustained release composition for a capsule containing mixture of cellulose derivative or a starch derivative that formed a gel in water and higher fatty acid glyceride and/or higher alcohol, which was solid at room temperature. The capsules were filled with the above mixture and heated to a temperature above the melting point of the fat components and then cooled and solidified (54).

Bolton and Desai, 1989 developed a non-compressed sustained release tablet that remained afloat on gastric fluids. The tablet formulation comprised 75% of drug and 2% to 6.5% of gelling agent and water. The non-compressed tablet had a density of less than 1 and sufficient mechanical stability for production and handling (55).

Streubel et al, 2003 prepared single-unit floating tablets based on polypropylene foam powder and matrix-forming polymer. Incorporation of highly porous foam powder in matrix tablets provided density much lower than the density of the release medium. A 17% wt/wt foam powder (based on mass of tablet) was achieved in vitro for at least 8 hours. It was concluded that varying the ratios of matrix-forming polymers and the foam powder could alter the drug release patterns effectively (56).

Sheth and Tossounian 1979a, 1979b developed SR floating tablets that were hydrodynamically balanced in the stomach for an extended period of time until the entire drug-loading dose was released. Tablets were comprised of an active ingredient, 0–80% by weight of inert materials, and 20–75% by weight of one or more hydrocolloids such as methylcellulose, HPC, HPMC, hydroxyethylcellulose, and sodium carboxy methylcellulose, which upon contact with gastric fluid provided a water impermeable colloid gel barrier on the surface of tablets. These tablets could be layer-approach or composite systems (Figure 2.8) (57, 58).

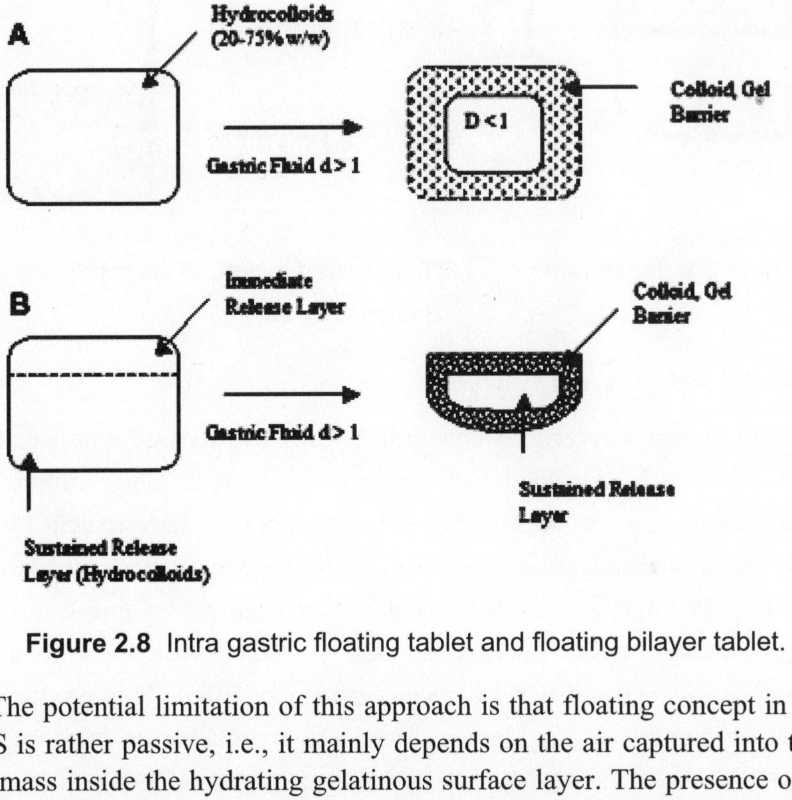

Figure 2.8 Intra gastric floating tablet and floating bilayer tablet.

The potential limitation of this approach is that floating concept in an HBS is rather passive, i.e., it mainly depends on the air captured into the dry mass inside the hydrating gelatinous surface layer. The presence of a small amount of fatty material, added to impede wetting, also aids buoyancy. Because of this passivity, the buoyancy of an HBS largely depends on the characteristics and amount of hydrophilic polymers used. Many investigators have tried other combinations of hydrophilic polymers (e.g. agar, alginic acid) and hydrophobic material (e.g. oil and porous calcium silicate) to improve HBS property. Since it was difficult to achieve both good buoyancy and a desirable release property, a modified version of an HBS was developed; the double-layered floating systems were proposed as a means to optimize floating capabilities and drug release profile separately (Figure 2.9).

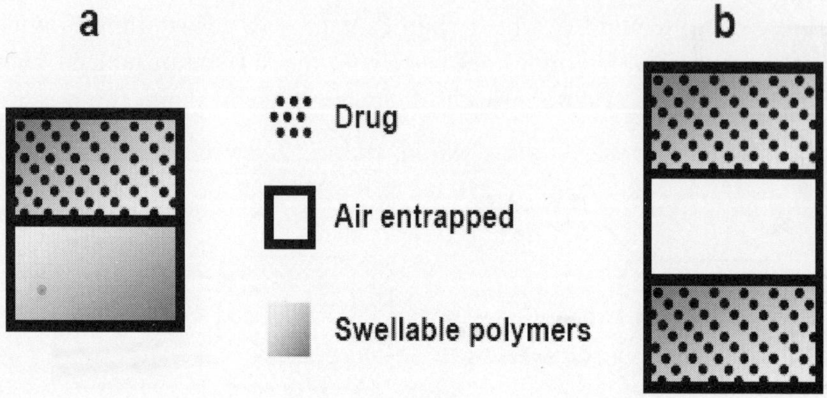

Figure 2.9 Improvement in HBS proposed by Oth et al and Krogel and Bodmeier

2.4.1.2 Effervescent systems

These buoyant delivery systems utilize matrices prepared with swellable polymers such as Methocel or polysaccharides and effervescent components, e.g., sodium bicarbonate and citric or tartaric acid (59) or matrices containing chambers of liquid that gasify at body temperature (60-62). The matrices are fabricated so that upon arrival in the stomach, carbon dioxide is liberated by the acidity of the gastric contents and is entrapped in the gellified hydrocolloid (Figure 2.10). This produces an upward motion of the dosage form and maintains its buoyancy. A decrease in specific gravity causes the dosage form to float on the chime (59). The carbon dioxide generating components may be intimately mixed within the tablet matrix, in which case a single-layered tablet is produced (63), or a bilayered tablet may be compressed which contains the gas generating mechanism in one hydrocolloid containing layer and the drug in the other layer formulated for a SR effect (64). Further refinements involve coating the matrix with a polymer which is permeable to water, but not to CO_2. This concept has also been exploited for floating capsule systems (65), prepared floating capsules by filling with a mixture of sodium alginate and sodium bicarbonate.

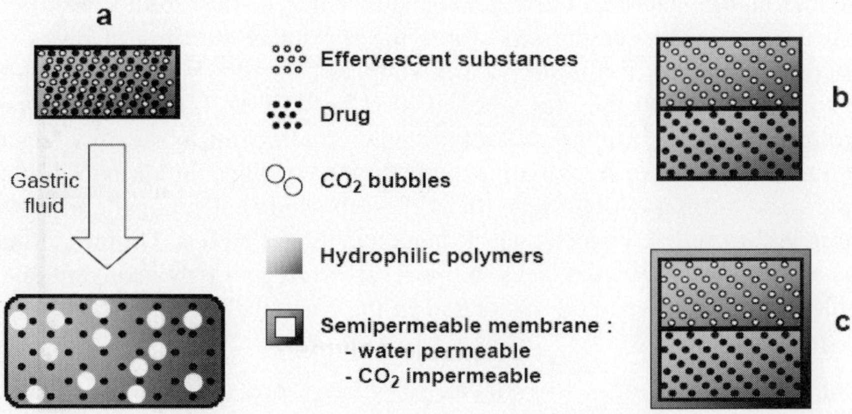

Figure 2.10 Gas generating systems (a) single layer system (b) Bilayer systems b) with out (c) with semipermeable membrane.

Yang et al, 1999 developed a swellable asymmetric triple-layer tablet with floating ability to prolong the gastric residence time of triple drug regimen (tetracycline, metronidazole, and clarithromycin) in Helicobacter pylori–associated peptic ulcers using hydroxy propyl methyl cellulose (HPMC) and poly ethylene oxide (PEO) as the rate-controlling polymeric membrane excipients (66). The design of the delivery system was based on the swellable asymmetric triple-layer tablet approach. Hydroxypropylmethylcellulose and poly ethylene oxide were the major rate-controlling polymeric excipients. Tetracycline and metronidazole were incorporated into the core layer of the triple-layer matrix for controlled delivery, while bismuth salt was included in one of the outer layers for instant release. The floatation was accomplished by incorporating a gas-generating layer consisting of sodium bicarbonate: calcium carbonate (1:2 ratios) along with the polymers. The in vitro results revealed that the sustained delivery of tetracycline and metronidazole over 6 to 8 hours could be achieved while the tablet remained afloat. The floating feature aided in prolonging the gastric residence time of this system to maintain high localized concentration of tetracycline and metronidazole.

Ozdemir et al, 2000 developed floating bilayer tablets with controlled release for furosemide. The low solubility of the drug could be enhanced by using the kneading method, preparing a solid dispersion with β cyclodextrin mixed in a 1:1 ratio. One layer contained the polymers HPMC 4000, HPMC 100, and CMC (for the control of the drug delivery) and the drug. The second layer contained the effervescent mixture of sodium bicarbonate and citric acid. The in vitro floating studies revealed that the lesser the compression force the shorter is the time of onset of floating, i.e. when the tablets were compressed at 15 MPa, these could begin to float at 20 minutes whereas at a force of 32 MPa the time was prolonged to 45 minutes. Radiographic studies on 6 healthy male volunteers revealed that floating tablets were retained in stomach for 6 hours and further blood analysis studies showed that bioavailability of these tablets was 1.8 times that of the conventional tablets. On measuring the volume of urine the peak diuretic effect seen in the conventional tablets was decreased and prolonged in the case of floating dosage form (67).

Fassihi and Yang, 1998 developed a zero-order controlled release multilayer tablet composed of at least two barrier layers and one drug layer. All the layers were made of swellable, erodible polymers and the tablet was found to swell on contact with aqueous medium. As the tablet dissolved, the barrier layers eroded away to expose more of the drug. Gas evolving agent was added in either of the barrier layers, which caused the tablet to float and increased the retention of tablet in a patient's stomach (68).

Talwar et al, 2001 developed a once-daily formulation for oral administration of ciprofloxacin. The formulation was composed of 69.9% ciprofloxacin base, 0.34% sodium alginate, 1.03% xanthum gum, 13.7% sodium bicarbonate, and 12.1% cross-linked poly vinyl pyrrolidine. The viscolysing agent initially and the gel-forming polymer later formed a hydrated gel matrix that entrapped the gas, causing the tablet to float and be retained in the stomach or upper part of the small intestine (spatial control). The hydrated gel matrix created a tortuous diffusion path for the drug, resulting in sustained release of the drug (temporal delivery) (69).

Baumgartner et al, 2000 developed a matrix-floating tablet incorporating a high dose of freely soluble drug. The formulation containing 54.7% of drug, HPMC K4 M, Avicel PH 101, and a gas-generating agent gave the best results. It took 30 seconds to become buoyant (70). In vivo experiments with fasted state beagle dogs revealed prolonged gastric residence time. On radiographic images made after 30 minutes of administration, the tablet was observed in animal's stomach and the next image taken at 1 hour showed that the tablet had altered its position and turned around. This was the evidence that the tablet did not adhere to the gastric mucosa. The MMC (phase during which large non disintegrating particles or dosage forms are emptied from stomach to small intestine) of the gastric emptying cycle occurs approximately every

2 hours in humans and every 1 hour in dogs but the results showed that the mean gastric residence time of the tablets was 240 ± 60 minutes (n = 4) in dogs. The comparison of gastric motility and stomach emptying between humans and dogs showed no big difference and therefore it was speculated that the experimentally proven increased gastric residence time in beagle dogs could be compared with known literature for humans, where this time is less than 2 hours.

Moursy et al, 2003 developed sustained release floating capsules of nicardipine HCl. For floating, hydrocolloids of high viscosity grades were used and to aid in buoyancy sodium bicarbonate was added to allow evolution of CO_2. In vitro analysis of a commercially available 20-mg capsule of nicardipine HCl (MICARD) was performed for comparison. Results showed an increase in floating with increase in proportion of hydrocolloid. Inclusion of sodium bicarbonate increased buoyancy. The optimized sustained release floating capsule formulation was evaluated in vivo and compared with MICARD capsules using rabbits at a dose equivalent to a human dose of 40 mg. Drug duration after the administration of sustained release capsules significantly exceeded that of the MICARD capsules. In the latter case the drug was traced for 8 hours compared with 16 hours in former case (71).

However these systems were shown to float during in vitro tests as a result of the generation of CO_2 that was trapped in the hydrating gel network on exposure to an acidic environment. However, gas-generating systems suffer from the disadvantage of not floating immediately after swallowing of the dosage form because the gas generation takes some time to occur. Thus, these dosage forms might be cleared from the stomach in an unpredictable manner especially if taken during phase III of MMC. The problems such as all or nothing emptying of dosage form, obstruction in the gastrointestinal tract made single unit floating systems unreliable and irreproducible in prolonging gastric residence time.

2.4.1.3 Multiple Unit Floating Systems

For conventional oral sustained- or prolonged-release dosage forms, multiple units are more advantageous than single units because they disperse widely and uniformly along the GIT and could lessen intra- and inter-subject variability in terms of bioavailability. For gastroretentive systems, multiple units may have the advantage of avoiding all-or-nothing emptying and increase the probability that some of the dosage will remain in the stomach. However the multiple unit floating systems (MUFS) developed are reliable and reduce inter subject variability in absorption and lower the probability of dose dumping.

Various multiple unit floating dosage systems developed utilizing numerous polymers and drugs are presented in Table 2.2. Different multiple unit floating dosage systems reported by different preparation methods are categorized into floating microspheres, floating alginate beads, floating resin beads, floating granules and pellets etc. Considering the advantages of MUFS the design of floating multiple matrix tablets with immediate release component and sustained release component were studied.

Significance of multiple unit floating systems

- Improved patient compliance and ease of administration.
- Improved drug absorption and bioavailability for absorption window drugs.
- Controlled and prolonged delivery of drugs.
- Local delivery of drug in stomach (Helicobacter pylori infection)
- Treatment of gastro-esophageal reflux.
- Avoids dose dumping.
- Reduced inter subject variability of drug absorption.
- Minimizes gastric irritation by slow and controlled release of drug.
- Reduced drug loss.
- Ease of manufacture with simple equipment.

2.4.1.4 Gas-generating Systems

Gas generation has also been used for multiple-unit systems; a conventional matrix sustained release beads were coated with a dual effervescent layer followed by swellable membrane coating as seen in Figure 2.11.

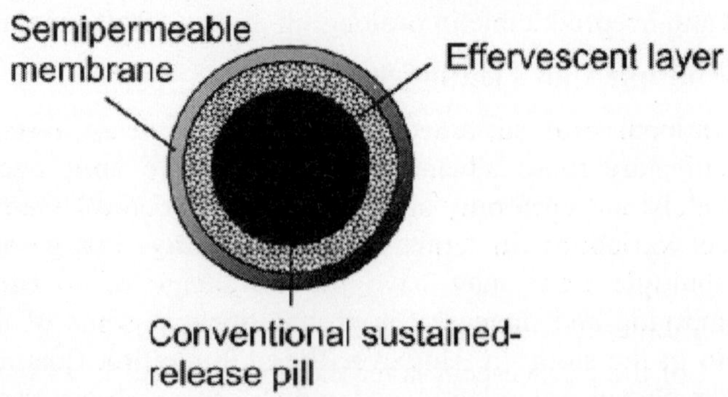

Figure 2.11 A multiple-unit oral floating dosage system.

The multiple-unit floating pellets in Figure 2.9 consisted of sustained release pills as seeds surrounded by double layers. The inner layer is a gas-generating layer containing both sodium bicarbonate and tartaric acid. The outer layer is a swellable membrane layer containing mainly polyvinyl acetate and purified shellac. Moreover, the gas-generating layer is divided into two sublayers to avoid contact between sodium bicarbonate and tartaric acid. Sodium bicarbonate is contained in the inner sublayer and tartaric acid in the outer layer. When the system is immersed in a buffer solution at 37°C, it sinks once in the solution and forms swollen pills, like balloons, with a density much lower than 1g/ml. The reaction is due to carbon dioxide generated by neutralization in the inner gas-generating layers with the diffusion of water through the outer swellable membrane layers (Figure 2.12).

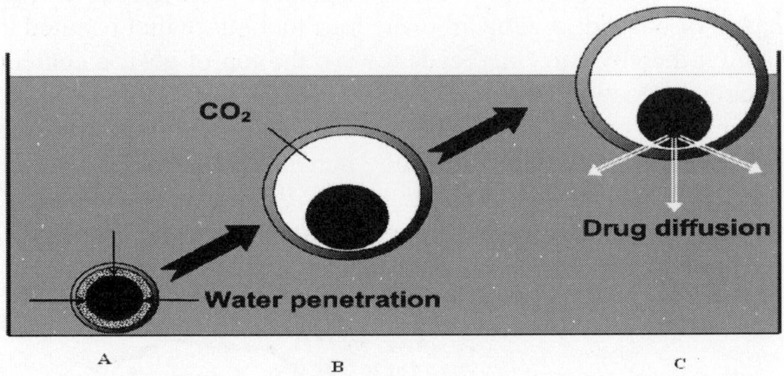

Figure 2.12 Stages of floating mechanism of a multiple unit oral dosage form: (a) penetration of water; (b) generation of CO_2 and floating; (c) dissolution and diffusion of drug.

The multiple unit oral floating system, mentioned in Figures 2.11 and 2.12, was found to float completely within 10 minutes and approximately 80% remained floating over a period of 5 hours irrespective of pH and viscosity of the test medium. The preparation processes of the multiple-unit system was threefold: "granulation" of the sustained-release pill; the first granulation is conducted to coat the sustained-release pills with barrier membrane layers then this was followed by granulation of "coating 1" of the effervescent layer then by "coating 2" of the swellable membrane layer (72).

Ion exchange resin beads

The property of ion exchange resins of exchanging ions when exposed to gastric fluid could be exploited in preparation of controlled release floating resin beads. These systems consist of resin beads, which are loaded with bicarbonate and a negatively charged drug that is bound to the resin. The resultant beads are encapsulated in a semipermeable membrane (Eudragit® RS) to overcome rapid loss of carbon dioxide. After exposure to the acidic medium in the stomach, chloride ions can pass through the membrane and exchange with bicarbonate. This exchange releases carbon dioxide, which is trapped inside the membrane, causing the resin particles to float. (Atyabi et al 1996 developed a floating system using ion exchange resin that was loaded with bicarbonate by mixing the beads with 1 M sodium bicarbonate solution (73). The loaded beads were then surrounded by a semi-permeable membrane to avoid sudden loss of CO_2. Upon coming in contact with gastric contents an exchange of chloride and bicarbonate ions took place that resulted in CO_2 generation thereby carrying beads toward the top of gastric contents and producing a floating layer of resin beads (Figure 2.13). The in vivo behavior of the coated and uncoated beads was monitored using a single channel analyzing study in 12 healthy human volunteers by gamma radio scintigraphy. Studies showed that the gastric residence time was prolonged considerably (24 hours) compared with uncoated beads (1 to 3 hours).

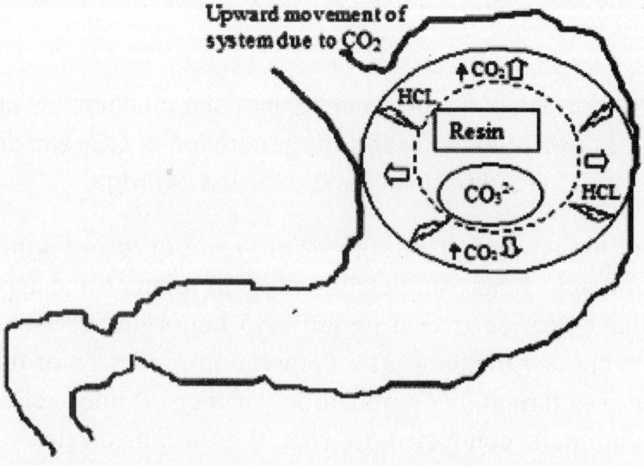

Figure 2.13 Pictorial presentation of floating effervescent ionic resin bead.

Umamaheswari et al, 2003 prepared cellulose acetate butyrate (CAB) coated cholestyramine microcapsules as an intragastric floating drug delivery system by emulsion solvent evaporation method (74). This process involves loading resin beads with bicarbonate by mixing resin beads with sodium bicarbonate solution for 15 min and slurry was stirred for 6 h to allow adsorption of bicarbonate on to resin. The beads were filtered, washed with deionized water and dried overnight at 40°C. Then drug was loaded on to bicarbonate charged beads by suspending in drug solution and stirred continuously for 12 h to form drug-resin complexes. Thus the complexes formed were filtered and washed with water to remove free uncomplexed drug followed by drying at 50°C. The drug resin complexes were encapsulated with CAB. This includes dispersion of complexes in 10% polymer solution that was emulsified in liquid paraffin containing magnesium stearate and sorbitan sesquioleate at 1500 rpm and stirring continued until all polymer solvent evaporated. Thus microcapsules formed were washed with hexane and dried under vacuum.

2.4.1.5 Hollow Microspheres

Hollow microspheres are considered one of the most promising buoyant systems because they possess the unique advantages of being multiple-unit systems and better floating properties as a result of the central hollow space inside the microspheres. The general techniques involved in their preparation include simple solvent evaporation and solvent diffusion and evaporation. The drug release and better floating properties mainly depend on the type of polymer, plasticizer, and solvents employed for the preparation. Polymers such as polycarbonate, Eudragit S, and cellulose acetate are used in the preparation of hollow microspheres, and the drug release can be modulated by optimizing the polymer quantity and the polymer-plasticizer ratio (22).

The type of excipients used in the preparation of microspheres especially the water-soluble substances, will have a profound effect on drug release because of their pore-forming (channeling) properties and rapid hydration of microspheres, facilitating rapid drug diffusion. The incorporation of hydrophilic substances such as polyethylene glycol and sucrose result in increased drug release from the microspheres, and sometimes the lag phase associated with the release from the microspheres can also be eliminated. According to the free volume theory, the diffusion occurs by localized activated jumps from the preexisting cavities to the next cavity. The nature and type of plasticizer used in the microspheres preparation also influences the drug release properties. The incorporation of plasticizer in the preparation results in decreased attractive forces between the polymer chains and increased mobility of the chains. An increased plasticizer concentration also reduces the total amount of polymer needed in the microspheres, resulting in less hindrance for the drug release (22).

*Yasua proposed the free volume theory to describe the relationship between the solute diffusivity in gel matrix and the matrix hydration, as diffusivity (D) increases whrn hydration (H) increases. The theory takes the form of $\ln D = \ln D_0 - K_r(1/H - 1)$ where D_0 is the diffusivity of the solute in pure solvent medium and K_r is a constant characteristic of the solute and solvent molecules. According to this theory, the plot of ln D versus $1/H - 1$ will be linear with the slope $-k_r$ and intercept $\ln D_0$[(D)].

Kawashima et al, 1991; 1992 and Sato et al, 2003 proposed hollow microspheres (so-called 'microballoons') based on Eudragit S (an enteric polymer), containing the drug in the polymeric shell (24,75,76). The preparation procedure and mechanism of microballoon formation is schematically illustrated in Figure 2.14. A solution of polymer and drug in ethanol/methylene chloride is poured into an agitated aqueous solution of polyvinyl alcohol. The ethanol rapidly partitions into the external aqueous phase and the polymer precipitates around methylene chloride droplets. The subsequent evaporation of the entrapped methylene chloride leads to the formation of internal cavities within the microparticles. However, according to Lee et al, 1999, many drugs fail to be released in significant amounts from this type of microparticle at the low pH of the stomach because of the shell consisting of an enteric polymer (77). To overcome this restriction, non-volatile oil (78) or highly porous calcium silicate powder pre-loaded with drug (79) can be added to the dispersed phase, or Eudragit S/RL mixtures used (80). The research group of Kawashima also prepared hollow microspheres based on blends of Eudragit S and other hydrophilic or hydrophobic polymers, such as Eudragit L, hydroxypropyl methylcellulose phthalate (HPMCP), HPMC or ethylcellulose (81). Incorporation of HPMC within the outer shell showed promising results concerning the control of drug release from the system at the pH of gastric fluids. With increasing HPMC content, riboflavin release was accelerated; however, the floating ability of the microspheres decreased. Importantly, the performance of these riboflavin-containing microballoons was also studied in vivo (82, 83). Upon oral administration to healthy volunteers, the intra-gastric behavior was investigated by γ-scintigraphy and the urinary excretion of riboflavin was monitored. In the fed state, the microballoons were retained in the stomach for up to 5 h. Interestingly, microspheres with good floating properties but low in vitro drug-release rates showed lower urinary excretion of riboflavin in the time period of 4–8 h after administration than microspheres with worst floating properties and high in vitro drug-release rates. Thus, it is important to choose an appropriate balance between the floating properties and drug-release rates with this type of gastroretentive drug delivery system.

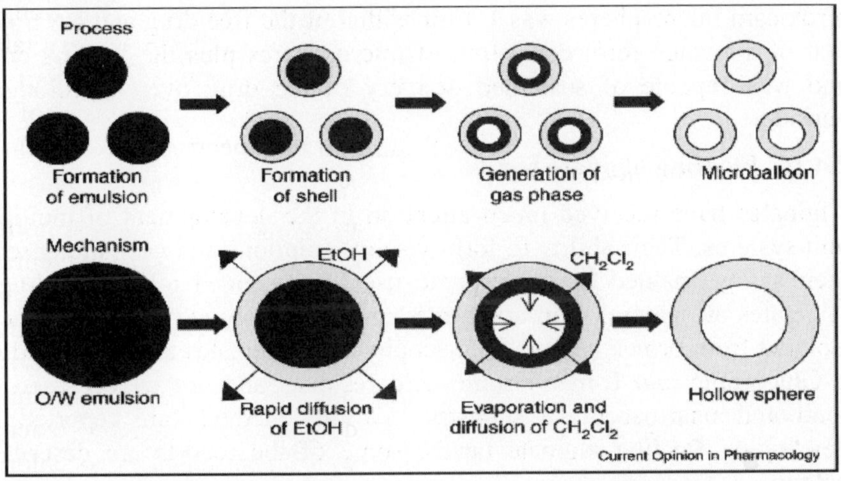

Figure 2.14 Preparation and Mechanism of Microballoon formation.

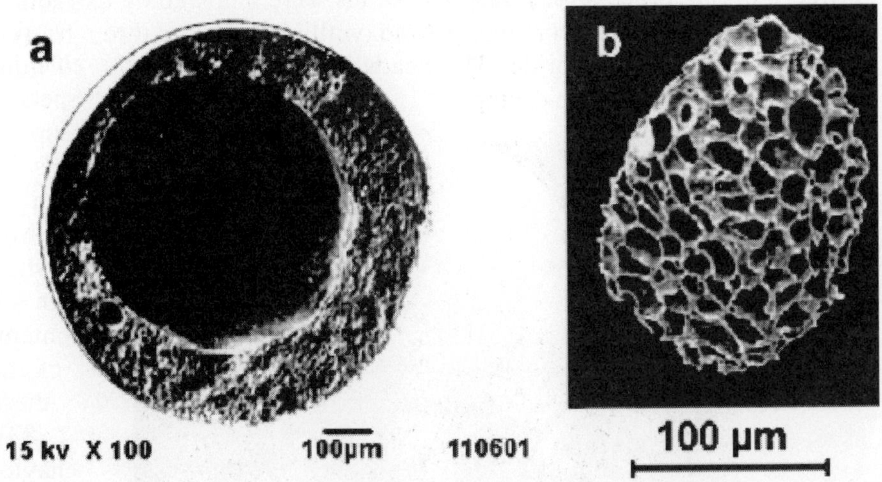

Figure 2.15 (a) Microballoons from sato et al (b) Foam particles from Streubel et al.

Joseph et al, 2002 developed a floating dosage form of piroxicam based on hollow polycarbonate microspheres (84). The microspheres were prepared by the solvent evaporation technique. Encapsulation efficiency of 95% was achieved. In vivo studies were performed in healthy male albino rabbits. Pharmacokinetic analysis from plasma concentration vs time plot revealed that the bioavailability from the piroxicam microspheres was 1.4 times that of the free drug and 4.8 times that of a dosage form consisting of microspheres plus the loading dose and was capable of sustained delivery of the drug over a prolonged period.

2.4.16 Floating alginate beads

Alginates have received much attention in the development of multiple unit systems. Their ability to form viscous solutions and gels, along with their safety enabled the widespread use for sustained release of drugs. Alginates are natural, non-toxic, biodegradable polysaccharide polymers isolated from brown seaweed (Phacophycae). Alginates are composed of L-Glucoronic and L-mannuronic acid residues and are widely used in food and pharmaceutical industry. Various methods are reported in developing floating alginate beads, some of the reports are described below:

Whitehead et al, 2000 developed floating alginate beads (Figure 2.16) of Amoxycillin using sodium alginate, and calcium chloride with the aid of freeze drying process (85). Alginate beads were prepared by extruding sodium alginate solution containing Amoxycillin tri hydrate through 21G needle in to calcium chloride. The beads formed were left for 30 min before filtration and freeze-drying.

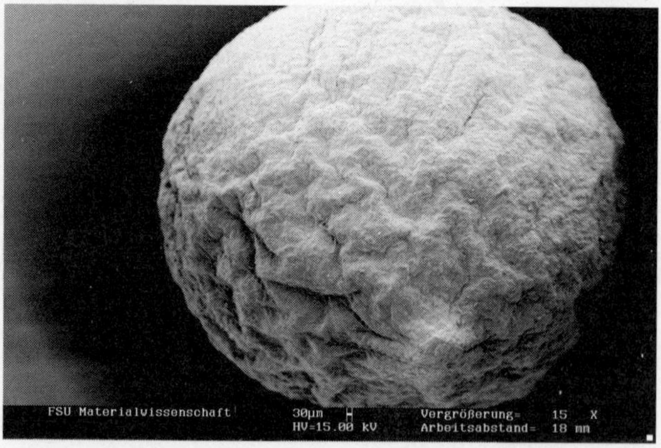

Figure 2.16 Freeze dried floating alginate bead.

Fassihi et al developed floating hollow beads by ionic gelation method followed by freeze-drying (68). They had prepared drug-encapsulated calcium pectinate and calcium pectinate alginate beads by preparing dispersion of pectin, drug and soluble / insoluble excipients or dispersion of pectin and alginate solution with drug and added dropwise through 16 guage needle into calcium chloride solution by peristaltic pump at a rate of 2ml/min. The beads formed were cured at 5°C for 24 h, washed with water and dried in freeze dryer for 72 h.

Sriamornsak et al, 2004 prepared floating oil-entrapped calcium pectinate gel beads by emulsion gelation method (86). These beads were prepared by dissolving pectin in water into which different amounts of oils were added that is homogenized at 3000 rpm for 5 min. Varying amounts of metronidazole was dispersed in an emulsion of oil and pectin mixture and extruded using syringe into calcium chloride solution with gentle agitation at room temperature. The beads formed were separated, washed with water and dried at 37°C for 12 h; these drug-loaded beads were coated with acrylic polymer by air suspension method.

Thus, floating alginate beads can be used for delivering high concentrations of drugs to gastric mucosa, taking advantage of floating property. These floating beads are particularly effective for site-specific controlled release of antibacterial agents effective against harmful stomach bacteria such as H. pylori.

2.4.1.7 Expandable/Swellable drug delivery systems

The diameter of pyloric sphincter was explored in developing gastroretentive drug delivery systems to retain in the stomach by increasing its size either by expansion or swelling. The expansion can be achieved by swelling or by unfolding of the dosage form in the stomach. If the dosage form can attain a size larger than that of the pyloric canal diameter, it can be retained in the stomach for a prolonged period of time. The expandable GRDDS are usually based on three configurations: a small ('collapsed') configuration which enables convenient oral intake; expanded form that is achieved in the stomach and thus prevents passage through the pyloric sphincter; and finally another small form that is achieved in the stomach when retention is no longer required i.e. after the GRDDS has released its active ingredient, thereby enabling evacuation (87-89). Swelling usually occurs because of osmosis. Unfolding takes place due to mechanical shape memory i.e. the GRDDS is fabricated in a large size and is folded into a pharmaceutical carrier e.g. a gelatin capsule, for convenient intake. In the stomach, the carrier is dissolved and the GRDDS unfolds or opens out, to achieve extended configuration.

These systems are sometimes referred to as plug-type systems because they tend to remain lodged within the pyloric canal.

Mamajek and Moyer, 1980, designed a GRDDS that comprised of an envelope from an elastic or non-elastic non-hydratable polymeric membrane, which is drug and body fluid permeable (Figure 2.17) (38). The envelope contains a drug reservoir and an expanding agent i.e. a swellable resin or hydrocolloid which causes expansion by osmotic pressure. Such devices of sizes larger than 1.5 × 31 cm were retained in the dog stomach for prolonged periods of time, typically more than 12 h.

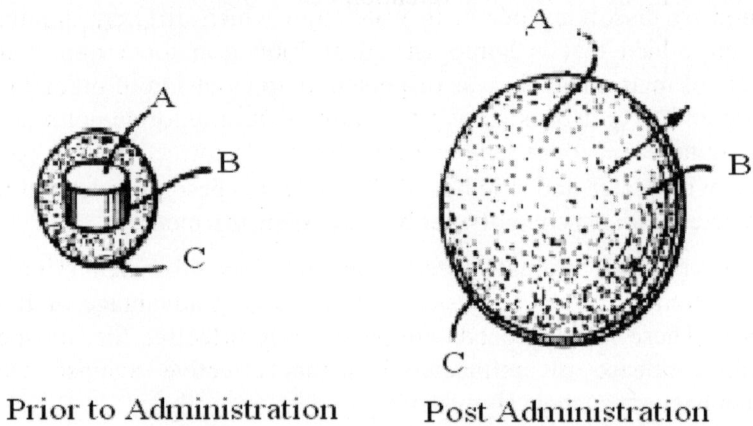

Figure 2.17 Prior to administration (a) Drug reservoir (b) Swellable expanding agent (c) and the whole enclosed by elastic outer polymeric envelope. Post administration Pressure of the expanding agent (b) swells the elastic polymer (c). Drug is released from the dosage form through the elastic polymeric envelope (c) as indicated by the arrow

Kagan et al, 2006 had developed an accordion pill, a novel controlled release gastroretentive unfolding dosage form, to increase the bioavailability of riboflavin in humans (90). It consists of planar polymeric sheets folded like an "accordion" into a standard gelatin capsule and therefore is termed "accordion pill" (AP). The two principal components of the accordion pill are the firm frame and dimensions of the dosage form contributes gastroretentivity; and the inner layer that consist of a polymeric matrix, loaded with the active compound, and controls the rate of drug release. Previously, they had demonstrated both in preclinical studies (91) and in humans (89) that the combination of elasticity and geometric properties of the frame is the key for ensuring proper gastric retention of the AP. The compositions of the inner matrix and the frame polymers can be modified separately to provide independently desired gastric retentivity and control of drug release rate.

Klausner et al, 2002, proposed unfolding multilayer, polymeric films based on a drug-containing shellac matrix as the inner layer, covered on both sides with (outer) shielding layers composed of hydrolyzed gelatin, Eudragit S, glycerin and glutaraldehyde (91). The system is optionally framed with rigid polymeric strips composed of L-poly-lactic acid and ethylcellulose, or ethylcellulose-triethylcitrate (Figure 2.18). Two factors influenced the in vivo gastric retention behavior: the dimensions and the mechanical properties of the films. Such dosage forms placed into gelatin capsules were administered to beagle dogs.

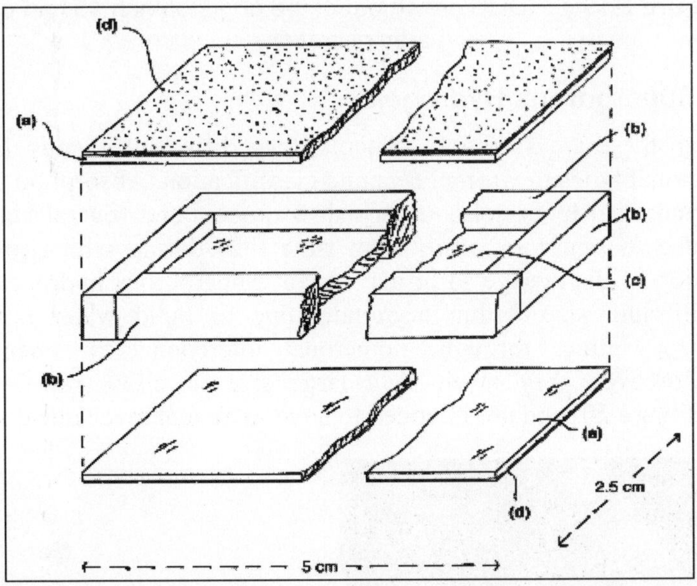

Figure 2.18 Schematic presentation of GRDDS proposed by Klausner et al.

(a) Shielding layer (b) Rigid frame (c) Polymer drug matrix (d) Anti adhering layers

A novel unfolding CR-GRDDS of levodopa administered to beagle dogs for in vivo evaluation, the unfolding GRDDS drawn out of the dog stomach 15 min post administration was given in Figure 2.19 (Klausner et al, 2003).

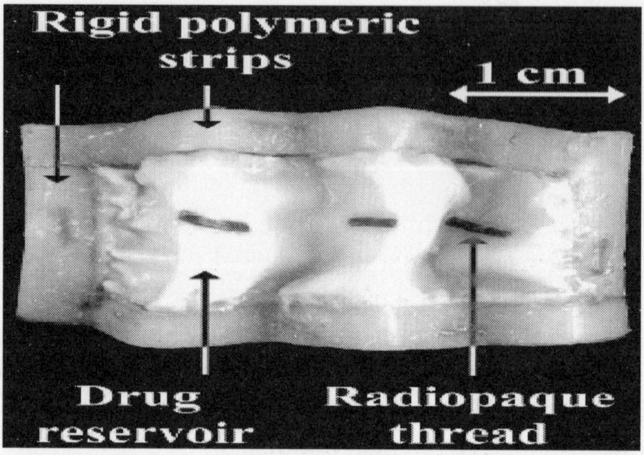

Figure 2.19 GRDDS drawn out of the dog stomach 15 min post administration.

2.4.2 Superporous Hydrogels

Although these are swellable systems, they differ sufficiently from the conventional types to warrant separate classification. Absorption of water by conventional hydrogel is a very slow process and several hours may be needed to reach an equilibrium state (92) during which premature evacuation of the dosage form may occur. Superporous hydrogels, swell to equilibrium size within a minute, due to rapid water uptake by capillary wetting through numerous interconnected open pores (93). Moreover, they swell to a large size (swelling ratio ~100 or more) (Figure 20) and are intended to have sufficient mechanical strength

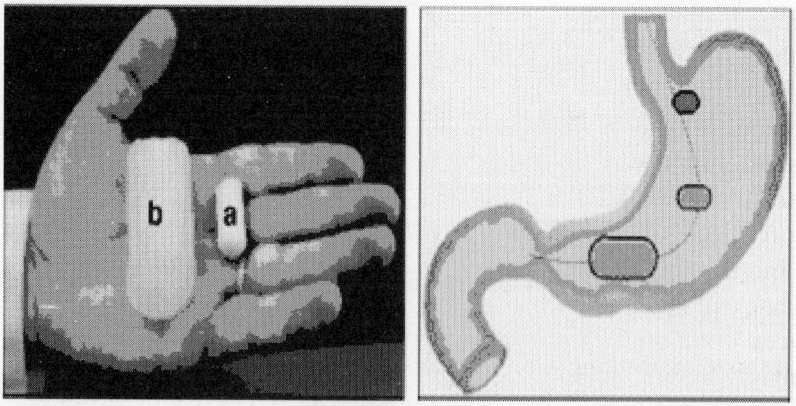

Figure 2.20 On the left, supreporous hydrogel in its dry (a), and water swelled state (b).

to withstand pressure by gastric contraction. This is achieved by co-formulation of a hydrophilic particulate material, Ac-Di-Sol (croscarmellose sodium). In vivo studies with dogs showed that under fasting condition, the superporous hydrogel composite (i.e. containing Ac-Di-Sol) remained in the stomach for 2–3 h. This time increased to >24 h after feeding, even though the fed condition was maintained only for a few hours. After several hours (~30 h), fragmentation occurred and the composite was rapidly cleared.

On the left, schematic illustration of the transit of superporous hydrogel.

2.4.3 Mucoadhesive or bioadhesive systems

Bio/mucoadhesive systems bind to the gastric epithelial cell surface, or mucin, and extend the GRT by increasing the intimacy and duration of contact between the dosage form and the biological membrane. The concept is based on the self-protecting mechanism of the GIT. Mucus secreted continuously by the specialized goblet cells located throughout the GIT plays a cytoprotective role. Mucus is a viscoelastic, gel-like, stringy slime comprised mainly of glycoproteins. The epithelial adhesive properties of mucin are well known and have been applied to the development of GRDDS through the use of bio/mucoadhesive polymers. The adherence of the delivery system to the gastric wall increases residence time at a particular site, thereby improving BA (94).

A bio/mucoadhesive substance is a natural or synthetic polymer capable of adhering to a biological membrane (bioadhesive polymer) or the mucus lining of the GIT (mucoadhesive polymer). The characteristics of these polymers are molecular flexibility, hydrophilic functional groups, and specific molecular weight, chain length, and conformation. Furthermore, they must be nontoxic and non-absorbable, form non-covalent bonds with the mucin–epithelial surfaces, have quick adherence to moist surfaces, easily incorporate the drug, and offer no hindrance to drug release, have a specific site of attachment, and be economical. The best bio/mucoadhesive is a slightly cross-linked poly acrylic acid, which is commercially available as polycarbophil and Carbopol®. Polycarbophil and Carbopol® are poly acrylic acid loosely linked with divinylglycol and sucrose, respectively. The binding of polymers to the mucin–epithelial surface can be subdivided into three broad categories: hydration-mediated adhesion, bonding-mediated adhesion, and receptor-mediated adhesion (95).

An unresolved issue related to bio/mucoadhesive systems is the attachment site of the system in the gut wall. The systems can attach both to the mucus layer and the epithelial surface of the stomach, the mucus can be found not only on the surface of the lumen but also within the lumen (called the soluble mucus) (96). Hence, it is difficult to understand how mucoadhesive systems identify the designated attachment site. However these bioadhesive systems does not seem to be a very feasible solution as the bond formation is uncontrolled and prevented by the acidic environment and thick mucus present in the stomach. High turnover of mucus adds to the difficulties in retaining a bioadhesive system bonded to the gastric mucosa.

2.4.4 Magnetic Systems

This system is based on a simple idea: the dosage form contains a small internal magnet, and a magnet placed on the abdomen over the position of the stomach. Ito et al. 1990 used this technique in rabbits with bioadhesives granules containing ultrafine ferrite (γ-Fe_2O_3). They guided them to the oesophagus with an external magnet (~1700 G) for the initial 2 min and almost all the granules were retained in the region after 2 h.

Fujimori et al. (1994) formulated a magnetic tablet containing 50% w/w ultra ferrite with hydroxypropylcellulose and cinnarizine (97). In beagle dogs, the tablet remained in the stomach for 8 h by the application of a magnetic field (1000 to 2600 G). Absorption of cinnarizine was sustained and the area under the plasma concentration–time-curve values (AUC_{0-24} h) increased . Groning et al. (1996) developed a method for determining the gastrointestinal transit of magnetic dosage forms under the influence of an extracorporeal magnet, using a pH-telemetering capsule (Heidelberg capsule). Small magnets were attached to the capsule and administered to humans. Using an extracorporeal magnet, gastric residence time of the dosage form was >6 h compared with 2.5 h control. Two years (Groning et al., 1998) later, the same group proposed oral acyclovir depot tablets with internal magnets (98). In vivo human studies showed that, in the presence of an extracorporeal magnet, the plasma concentrations of acyclovir were significantly higher after 7, 8, 10 and 12 h. Furthermore, the mean AUC_{0-24} h was ~2800 ng h/mL with the external magnet and ~1600 ng h/mL without. Although these systems seem to work, the external magnet must be positioned with a degree of precision that might compromise patient compliance.

2.4.5 Use of Passage Delaying Agents

Lipid vehicles, mainly fatty acids, can reduce the motility of the stomach and the intrinsic rate of gastric emptying. Thus the use of passage-delaying excipients, e.g. triethanolamine myristate, to delay gastric emptying has been proposed. Preliminary in-vivo results, nevertheless, depict a major problem related to the highly variable inter-subject reaction to such controlling excipients. Another approach is using passage-delaying drugs, e.g. propantheline, which is generally considered undesirable because of potential side effects. In both cases, trying to deliberately slow down the normal motor activity of the stomach is a risk. Gastric motility laziness is, in fact, an unwanted side effect often associated with various situations. Thus one should be very careful when considering the use of any passage-delaying substance that indiscriminately affects the gastric emptying mechanism.

2.5 Drugs Explored for GRDDS

Gastroretentive dosage forms may offer many advantages for a wide range of drugs. Table 2.3 lists some examples of drugs that can be best delivered using such devices.

Table 2.3 Drugs that can be delivered using gastric retention.

- Drugs acting locally in stomach (e.g. antacids, antibiotics for bacteria – based ulcers)
- Drugs that are absorbed primarily in the stomach (e.g. albuterol)
- Drugs, such as weak bases, that are poorly soluble in an alkaline pH
- Drugs that have a narrow window for absorption from GI tract (Drugs absorbed mainly from proximal small intestine. i.e. riboflavin, levodopa, furosemide, ranitidine hydrochloride)
- Drugs that are absorbed rapidly from the GIT (e.g. amoxicillin)
- Drugs that show potential degradation in the intestinal and colon pH (e.g. captopril)

Gastroretentive dosage forms would provide the best results for drugs that may act locally in the stomach or that may be absorbed primarily in the stomach. For many drugs that are absorbed mainly from the upper small intestine (i.e. drugs with absorption windows), controlled release formulations in the stomach would result in improved bioavailability. Improved bioavailability is also expected for drugs that are absorbed readily upon controlled release in GIT. Such drugs can be best delivered by slow release from the stomach. With the total GI transit duration being increased, a greater exposure of drug to its absorption site(s) may be accomplished and the relative bioavailability will consequently be increased.

The total transit time also becomes more predictable, delivery systems can thus be more reliably programmed to provide complete release of the drug before the dosage form reaches non-optimal absorption sites. Prolonged gastric retention not only enables us to achieve control over time, but also over space by maintaining the delivery systems positioned at a steady site and thus properly addressing the issue of drug delivery.

Numerous drugs have been explored for development of GRDDS. The list of drugs utilized in the development of single and multiple unit floating systems is given in Table 2.3.

Table 2.3 Some of the drugs formulated as single and multiple unit forms of floating drug delivery systems.

Type	Drug	Reference
Tablets	Chlorpheniramine maleate	Deshpande et.al., 1997
	Theophylline	Yang and Fassihi .1996
	Furosemide	Ozdemir et.al., 2000.
	Ciprofolxacin	Talwar et.al., 2001
	Pentoxyfillin	Baumgartneret.al., 2000
	Captopril	Nur et.al., 2000
	Acetylsalicylic acid	ShethandTossounian 1979
	Nimodipine	Wu et.al., 1997
	Amoxycillin trihydrate	Hilton and Deasy 1992
	Verapamil HCl	Chen ,Hao. 1998
	Isosorbide di nitrate	Ichikawa etal 1991
	Sotalol	Cheuh et.al., 1995
	Atenolol	Rouge et.al., 1998
	Isosorbide mono nitrate	Chitnis et.al.,1991
	Acetaminophen	Phuapradit 1989 PhuapraditandBolton1991
	Ampicillin	Gupta 1987
	Cinnarazine	Machida et.al., 1989
	Diltiazem	Machida et.al., 1989
	Florouracil	Watanbe et.al., 1993
	Piretanide	Rouge et.al., 1998
	Prednisolone	Inouye et.al., 1988
	Riboflavin- 5′ Phosphate	Ingani et.al., 1987

Contd...

Type	Drug	Reference
Capsules	Nicardipine	Moursy et.al., 2003
	L- Dopa and benserazide	Erni et.al., 1987
	hlordiazepoxide HCl	Shethand Tossounian1984
	Furosemide	Menon et.al., 1994
	Misoprostal	Oth et.al., 1992
	Diazepam	Gustafson et.al., 1981
	Propranlol	Khattar et.al., 1990
	Urodeoxycholic acid	Simoni et.al., 1995
Micros-pheres	Verapamil	Soppimath et.al., 2001
	Aspirin, griseofulvin, and p-nitroaniline	Thanoo et.al., 1993
	Ketoprofen	El-Kamel et.al., 2001
	Tranilast	Kawashima et.al., 1991
	Ibuprofen	Kawashima et.al., 1992
	Terfenadine	Jayanthi et.al., 1995
Granules	Indomethacin	Hilton and Deasy. 1992
	Diclofenac sodium	Malcolm et.al., 1987
	Prednisolone	Inouye et.al., 1989
Films	Drug delivery device	Harrigan et.al., 1977
	Cinnarizine	Gu et.al., 1992
Powders	Several basic drugs	Dennis et.al., 1992

2.6 Evaluation of Gastroretentive Drug Delivery Systems

Any drug product must be evaluated to ensure its performance characteristics and to control batch-to-batch quality. In addition to routine tests for general appearance, hardness, friability, drug content, weight variation, uniformity of content, and drug release, the gastroretentive performance of GRDDS must be evaluated (99). Gastroretentive performance includes floating/buoyancy property and lag time, and *in vivo* visualization for GI retention parameters using X-ray and gama scintigraphy. Apart from that, the multiple unit systems are characterized for surface morphology and physical state of the drug in the dosage form. The in vitro floating and dissolution performance of some single and multiple unit GRDDS are given in the Table 2.4.

Table 2.4 *In vitro* floating and dissolution performance of reported GRDDS.

Drug (polymers used)	Floating media/Dissolution medium and method	Reference
Pentoxyfylline (HPMC K4M)	500ml simulated gastric fluid pH 1.2 (with out enyzme), USP XXIII dissolution apparatus	Baumgartner et al.,2000
Amoxicillin (Calcium alginate)	900ml of deaerated 0.1 M HCl pH 1.2, USP XXII apparatus	Whitehead et al., 2000
Ketoprofen (Eudragit RL & S100)	20 ml simulated gastric fluid pH 1.2 (with out enyzme), 50 mg microparticles in 50 ml beaker % of floating is calculated. 900ml of 0.1N HCl or pH 6.8 Phosphate buffer, USP apparatus 1	El-Kamel et al., 2001
Verapamil (Propylene foam, Eudragit RS, ethyl cellulose)	30ml of 0.1N HCl pH 1.2, floatation was studied by placing 60 particles in 30 ml flask	Streubel et al., 2002
Theophylline (HPMC K4M, PEO)	0.1 N HCl pH 1.2, USP XXIII dissolution apparatus II	Yang and Fassihi 1996
Furosemide (HPMC 4000 & 100, CMC, PEG)	Gastric fluid pH 1.2, flow through cell flow rate 9ml/min	Choi et al., 2002
Piroxicam (Polycarbonate)	900ml dissolution medium in USP apparatus II	Mitra ., 1984
Ampicillin (Sodium alginate)	500ml distilled water, JP XII disintegration test medium pH 1.2 and pH 6.8 in JP XII dissolution apparatus with paddle	Katayama et al., 1999
Aspirin, Griseofulvin, p-nitro aniline (Polycarbonate, PVA)	500ml simulated gastric and intestinal fluid in 1000ml Erlenmeyer flask shaken in bath incubator for dissolution	Thanoo et al., 1993

Contd...

Drug (polymers used)	Floating media/Dissolution medium and method	Reference
Diclofenac (HPC-L)	0.1 g granules in 40 ml purified water,% granules were calculated, Triple flow cell followed by 900ml distilled water in JP XII with paddle	Yuasa et al., 1996
Sulphiride (Carbopol 934P)	500ml JP XII disintegration test medium pH 1.2 and pH 6.8 in JP XII dissolution apparatus	Kohri et al., 1995
Amoxicillin trihydrate (HPC)	500-1000ml citrate/phosphate buffer or HCl solution pH 1.2	Hilton and Deasy. 1992
Tranilast, Ibuprofen (Eudragit S)	900ml JP XI disintegration test medium pH 1.2 and pH 6.8 in USP XXI dissolution apparatus II	Kawashima et al., 1992
Isardipine (HPMC)	300 ml of artificial gastric fluid in a beaker, suspended in water bath with magnetic stirrer and 500/1000ml of 0.1 M HCl and surfactant lauryl sulfate dimethyl ammonium oxide	Mazer et al., 1988
Verapamil (HPC-H, HPC-M, HPMC K15)	Water in USP XXIII dissolution apparatus II 50rpm	Cheuh et al., 1995
Tetracycline, metronidazole, bismuth salt (HPMC K4M, PEO)	900ml of 0.1 M HCl (pH 1.8) in USP dissolution apparatus during floating was observed	Yang et al., 1999
Calcium carbonate (HPMC K4M, E4M and carbopol)	900ml simulated gastric fluid (pH 1.2), a continuous floating system was conceived, upward floating force measured by balance and data transmitted to online computer	Choi et al., 2002

The various parameters that need to be evaluated for their effects on GRT of buoyant formulations can mainly be categorized into following different classes:

Galenic parameters: Diametric size, flexibility and density of matrices.

Control parameters: Floating time, dissolution, specific gravity, content uniformity, hardness and friability (if tablets).

In case of multi particulate drug delivery systems, differential scanning calorimetry (DSC), particle size analysis, flow properties, surface morphology, and mechanical properties

Geometrical parameters: Shape.

Physiological parameters: Age, sex, posture and food.

The test for buoyancy and *in vitro* drug release studies are usually carried out in simulated gastric and intestinal fluids maintained at 37°C. In practice, floating time is determined by using the USP disintegration apparatus containing 900ml of 0.1N HCl as a testing medium maintained at 37°C. The time required to float the HBS dosage form is noted as floating or floatation time (100).

Dissolution tests are performed using the USP dissolution apparatus. Samples are withdrawn periodically from the dissolution medium, replenished with the same volume of fresh medium each time, and then analyzed for their drug contents after an appropriate dilution.

Burns *et al* (101) developed and validated an *in vitro* dissolution method for a floating dosage form, which had both rapid release and SR properties. The method, although based on the standard BP (1993)/ USP (1990) apparatus 2 methods, was modified such that paddle blades were positioned at the surface of dissolution medium. The results obtained with this modified paddle method showed reproducible biphasic release dissolution profiles when paddle speeds were increased from 70 to 100 rpm and the dissolution medium pH was varied (6.0-8.0). The dissolution profile was also unaltered when the bile acid concentration in the dissolution medium was increased from 7 to 14 m *M*. The specific gravity of FDDS can be determined by the displacement method using analytical grade benzene as a displacing medium (102).

The system to check continuous floating behavior contains a stainless steel basket connected to a metal string and suspended from a sartorius electronic balance. The floating object is immersed at affixed depth into a water bath, which is covered to prevent water evaporation. The upward floating force could be measured by the balance and the data transmitted to an online PC through RS232 interphase using a sarto wedge program. A lotus- spread sheet could automatically pick up the reading on the balances. Test medium used in floating kinetics measurements was 900 ml simulated gastric fluid (pH 1.2) maintained at 37°C, data was collected at 30 sec interval; baseline was recorded and subtracted from each measurement. Dissolution basket had a holder at the bottom to measure the downward force.

The specific gravity of FDDS can be determined by the displacement method using analytical grade benzene as a displacing medium. The initial (dry state) bulk density of the dosage form and the changes in floating strength with time should be characterized prior to *in vivo* comparison between floating (F) and nonfloating (NF) units. Further, the optimization of floating formulations should be realized in terms of stability and durability of the floating forces produced, thereby avoiding variations in floating capability that might occur during *in vivo* studies.

Resultant weight test: An *in vitro* measuring apparatus has been conceived to determine the real floating capabilities of buoyant dosage forms as a function of time. It operates by measuring the force equivalent to the force F required to keep the object totally submerged in the fluid (103).

This force determines the resultant weight of the object when immersed and may be used to quantify its floating or nonfloating capabilities. The magnitude and direction of the force and the resultant weight corresponds to the vectorial sum of buoyancy ($\mathbf{F}_{bouy}$) and gravity ($\mathbf{F}_{grav}$) forces acting on the object as shown in the equation

$$\mathbf{F} = \mathbf{F}_{buoy} - \mathbf{F}_{grav}$$

$$F = d_f gV - d_s gV = (d_f - d_s)\, gV$$

$$F = (df - M / V)\, gV$$

in which **F** is the total vertical force (resultant weight of the object), g is acceleration due to gravity, d_f is the fluid density, d_s is the object density, M is the object mass, and V is the volume of the object, by convention, a positive resultant weight signifies that the force **F** is exerted upward and that the object is able to float, whereas a negative resultant weight means that the force **F** acts downward and that the object sinks (100).

The crossing of the zero base line by the resultant weight curve from positive toward negative values indicates a transition of the dosage form from floating to nonfloating conditions. The intersection of lines on a time axis corresponds to the floating time of the dosage form.

The *in vivo* **gastric retentivity** of a floating dosage form is usually determined by gamma scintigraphy or roentgenography. Studies are done both under fasted and fed conditions using F and NF (control) dosage forms. It is also important that both dosage forms are non disintegrating units, and human subjects are young and healthy.

The tests for floating ability and drug release are generally performed in simulated gastric fluids at 37°C.

In vivo gastric residence time of a floating dosage form is determined by X-ray diffraction studies, gamma scintigraphy, (104) or roentgenography (105).

2.7 Marketed Products of GRDDS

The intensive research work carried in last three decades resulted in following gastroretentive commercial products (Table 2.5). And other suitable products are under development by different companies like Depomed (metformin).

Table 2.5 List of marketed dosage forms of GRDDS.

Product	Type	Active ingredient	Manufacturer	Ref. No
Madopar	HBS	Levodopa and benserzide	Roche	Erni., 1987
Valrelease	HBS	Diazepam	Hoffman LaRoche	Sheth and Tossounian 1984
Topalkan	Raft forming	Alginic acid, + Al and Mg salts	Enska	Degtiareva et al., 1994
Liquid gavison	Raft forming	Alginic acid and sodium bicarbonate	Glaxo smithkline	Washington et al., 1986
Almagate floatcoat	Raft forming	Antacid	-	Fabregas et al., 1994
Cifran OD	HBS	Ciprofloxacin	Ranbaxy	-

In-Vivo Evaluation

γ-Scintigraphy

γ-Emitting radioisotopes compounded into CR-DFs has become the state-of-art for evaluation of gastroretentive formulation in healthy volunteers. A small amount of a stable isotope e.g. Sm, is compounded into DF during its preparation. The main drawbacks of γ-scintigraphy are the associated ionizing radiation for the patient, the limited topographic information, low resolution inherent to the technique and the complicated and expensive preparation of radiopharmaceuticals (26).

Radiology

This method is the state of art in preclinical evaluation of gastroretentivity. Its major advantages as compared to γ-scintigraphy are simplicity and cost. However, use of X-ray has declined due to strict limitations, regarding the amount of exposure and it's often requirement in high quantity. A commonly used contrast agent is barium sulphate (106)

Gastroscopy

It comprises of peroral endoscopy, used with a fibereoptic and video systems. It is suggested that gastroscopy may be used to inspect visually the effect of prolonged stay in stomach milieu on the FDDS. Alternatively, FDDS may be drawn out of the stomach for more detailed evaluation (107).

Ultrasonography

Ultrasonic waves reflected substantially different acoustic impedances across interface enable the imaging of some abdominal organs (108). Most DFs do not have sharp acoustic mismatches across their interface with the physiological milieu. Therefore, Ultrasonography is not routinely used for the evaluation of FDDS. The characterization included assessment of intragastric location of the hydrogels, solvent penetration into the gel and interactions between gastric wall and FDDS during peristalsis.

Magnetic Resonance Imaging (MRI)

In the last couple of years, MRI was shown to be valuable tool in gastrointestinal research for the analysis of gastric emptying, motility and intra gastric distribution of macronutrients and drug models. The advantages of MRI include high soft tissue contrast, high temporal and spatial resolution, as well as the lack of ionizing irradiation. Also, harmless paramagnetic and supra magnetic MR imaging contrast agents can be applied to specifically enhance or suppress signal of fluids and tissues of interest and thus permit better delineation and study of organs (109).

2.8 Case study (Acta Pharm, Suresh Bandari et al., 2010)

Formulation and evaluation of multiple tablets as biphasic gastroretentive floating drug delivery system for fenoverine

Fenoverine is an anti-spasmodic agent that restores smooth muscle motility and relieves the distressing symptoms associated with irritable bowel syndrome and primary dysmenorrhoea.

In the present study biphasic GRDDS of fenoverine as model drug comprising of immediate release loading dose tablet (LDT) and sustained release floating multiple matrix tablets (FMMT) were developed and characterized for *in vitro* performance.

Experimental

Materials

Fenoverine and SPASMOPRIV™ 200 mg capsules were generous gifts from Eurodrugs, India. HPMC K4M and HPMC 100 LV were obtained from ISP, India. Sodium bicarbonate, citric acid anhydrous, magnesium stearate, was purchased from S.D. Fine-Chem Ltd, India. Purified talc was purchased from E. Merck, India. All other chemicals used were of analytical grade.

Methods

Preparation of biphasic gastroretentive drug delivery system. - The total dose of fenoverine for twice daily sustained release formulation was calculated based on pharmacokinetic data. The loading dose and maintainance dose was considered as 74mg and 126mg of fenoverine respectively. The total dose of GRDDS was 200mg.

The preliminary formulations were studied to optimize the drug to polymer ratio and effervescent composition (data not shown). Then the FMMT were prepared with optimized concentration of effervescent composition. All formulation ingredients of FMMT and LDT were sifted through 420 μm aperture size homogeneously blended in mortar separately and directly compressed using 6 mm flat punches on rotary compression machine (Riddhi, Ahmedabad, India). The GRDDS comprising of LDT and FMMT equivalent to 200 mg of fenoverine were placed in 0 size hard gelatin capsules and evaluated. The qualitative and quantitative composition of FMMT and LDT was shown in Table.1.

Physical characterization of the blend and compressed FMMT and LDT: Physical properties such as bulk density (ρb), tapped density (ρt), compressibility index and the angle of repose of final blend of FMMT were determined.

Compressed multiple tablets were characterized for mass and thickness variation ($n = 20$). Crushing strength ($n = 6$) was measured with Monsanto tester, friability ($n = 6$), with Roche type friabilator. The drug content in each formulation was determined by triturating 20 tablets in a mortar and powder equivalent to average weight was added in100 ml of 0.1N hydrochloric acid, followed by shaking for 30 minutes. The sample was analyzed by reported HPLC method.

Density of tablet: The density of the tablets was calculated from tablet height, diameter and weight using equation D (g/cm^3) = $W/(m/2)^2 \times \pi \times h$.

W is the weight of a tablet, m is the diameter of tablet, π is the circular constant, and h is the height of a tablet.

Floating behavior: Buoyancy lag time and duration of buoyancy were determined in the USP dissolution apparatus II with 0.1 mol L^{-1} HCl and Simulated Gastric Fluid (enzyme free SGF). The time between introduction of the tablet in to the media and its buoyancy on the media was taken as buoyancy lag time and the time of duration of the tablet remain buoyant was observed visually.

In vitro *drug release studies:* The release of fenoverine from FMMT and biphasic GRDDS was studied using USP dissolution apparatus II (Labindia, India). The dissolution media were 900 ml 0.1 mol L^{-1} HCl and SGF (enzyme free) maintained at 37 ± 0.5^0C with rotation speed of 50 rpm. Aliquots of 5 ml was collected at predetermined time intervals, filtered through a 0.45-μm membrane filter and replenished with equivalent volume of fresh medium. Drug content in the dissolution samples were determined by HPLC after suitable dilutions.

Drug release modeling: The suitability of several equations, which are reported in the literature to identify the mechanism(s) for the release of fenoverine, was tested with respect to the release data. The data for analysis was taken to Q12 (drug released up to 12 h) excluding the lag time for all models.

Further the release profiles were compared by statistically derived mathematical index similarity factor (*f*2) using theoretical profile as reference. The dissolution efficiency *(DE)*, at 2, 6 and 8h and time taken to release x% drug (t_x%) i.e., $t_{50\%}$, $t_{75\%}$. were also determined from the dissolution profiles for each formulation. The two dissolution profiles are considered to be similar, if *f*2 value is more than 50 (between 50 and 100).

Stability studies: To assess the drug and formulation stability were carried. Optimized formulation kept in the humidity chamber (LabTop, India) maintained at 40 °C and 75% relative humidity for 3 months. At the end of studies, the formulation was subjected to a drug assay, floating behavior and *in vitro* dissolution studies. For the comparison of release profiles of initial and stability samples, "similarity factor" *f*2, was calculated.

Results and Discussion

Physical properties of final blend and compressed FMMT and LDT: The floating multiple matrix tablets of fenoverine were prepared by effervescent technique using methocel (K4M, K100LV), the combination of two different grades of HPMC was utilized to obtain the desired buoyant characteristics and to maintain the integrity of the FMMT for 12 h. The blend of the final batches showed bulk density of 0.55 ± 0.08 gcm^{-3} and tapped density of 0.64 ± 0.10 gcm^{-3} indicating desirable flow properties. Further the angle of repose determined showed value of <30°, percent compressibility below 15 indicating good flow properties of the formulations.

The physical characteristics of loading dose tablet and floating multiple matrix tablets are shown in Table II. The mass and thickness of the all FMMT were uniform and hardness of tablets was found to be between 4.1 and 5.30 $kgcm^{-2}$ indicating sufficient crushing strength. The friability was below 1 % for all formulations, indicating good mechanical resistance of the tablet. The drug content varied between 41.9 to 43.1 mg in all FMMT indicating content uniformity of the prepared batches.

Buoyancy and density: The density of FMMT formulations was 1.202, 1.195, 1.225 and 1.234 gcm^{-3} respectively. Though the density of formulated tablets was bigger than 1.0 gcm^{-3} they remained buoyant for

12 h with lag time of 4.16 to 5.22 minutes. The optimized concentration of effervescent mixture (sodium bicarbonate and citric acid) aided in the buoyancy of all tablets.

In vitro release studies: Ideally, a sustained release formulation should release the required quantity of drug with predetermined kinetics in order to maintain an effective drug plasma concentration. To achieve this, the delivery system should be formulated so that it releases the drug in a predetermined and reproducible manner. The release of fenoverine from biphasic GRDDS was analyzed by plotting cumulative percent drug released against time. Figures. 1 a and b show the *in vitro* drug release profile of fenoverine in 0.1 mol L^{-1} HCl and SGF (enzyme free) from biphasic GRDDS. The biphasic GRDDS remains buoyant in dissolution media with liberation of FMMT and LDT after 10-15 minutes. There was an initial burst release of fenoverine from the gastroretentive system which can be attributed to the LDT, and release of fenoverine from FMMT was sustained over 12 h compared with conventional commercial capsules which released complete drug within 30 minutes. It was found that the initial burst release of 47.2 ± 1.8 in GRDDS studied in 0.1 mol L^{-1} HCl and 41.5 ± 1.6 in SGF (enzyme free) during the first hour. The drug release from biphasic GRDDS was found to be 102.4 ± 1.6 % in 0.1 mol L^{-1} HCl and 98.5 % ± 2.2 in SGF (enzyme free) respectively at the end of 12 h. The drug release from FMMT was sustained for prolonged period of time due to viscous nature of hydroxyl propyl methyl cellulose matrix through which drug to be diffused. Replacement of HPMC 100LV grade with HPMC K4M aided to increase the drug release within 12 h and maintained the integrity and buoyancy of FMMT formed.

Drug release pattern from GRDDS: The results of kinetic models for fenoverine release from biphasic GRDDS are shown in Table III. The coefficient of regression *(R^2)* was used as indicator of the best fitting, for each of the models considered. The results reveal that all formulations of biphasic GRDDS, best fits in the zero-order model, in both 0.1 mol L^{-1} HCl and SGF (enzyme free). The mechanism of drug release from these tablets was found to be diffusion as seen from high R^2 values of Higuchi model.

The optimized GRDDS formulation was selected based on the similarity factor (f2) values of all GRDDS and other dissolution parameters such as dissolution efficiency (DE) at 2, 6 and 8 h and time taken to release x% drug (t_x%), $t_{50\%}$, $t_{75\%}$ that are shown in Table IV. The similarity factor (f2) of GRDDS3 when compared with the theoretical release profile in 0.1 mol L^{-1} HCl was observed to be 91.5 which was greater than the other formulations and in SGF 61.4 which was well above 50 indicating that the drug release pattern was similar with theoretical release profile. The other independent model parameters such as dissolution efficiency (DE) also revealed that the drug release profile from GRDDS3 was similar with theoretical profile.

Influence of media on drug release: The fenoverine release in 12 h from optimized biphasic GRDDS3 in 0.1 mol L^{-1} HCl and SGF (enzyme free) was found to be 97.7% and 94.4 % respectively. Statistical analysis was carried out using t-test and there was no statistically significant difference ($p < 0.05$) in drug release between the two media. However the *in vitro* release behavior of GRDDS was also compared with the theoretical (predictive) profile and found to be quite similar.

Stability studies: In view of the potential utility of the formulation, stability studies were carried out at 40 °C and 75% RH for three months (for accelerated testing) to assess their long-term stability. After storage, the formulation was subjected to a drug assay, floating behavior and *in vitro* dissolution studies. The analysis of the dissolution data (Figure 2.2), after storage for three months showed no significant change in release pattern indicating the two dissolution profiles are considered to be similar (*f*2 value was more than 50). The other parameters evaluated were comparable to initial values.

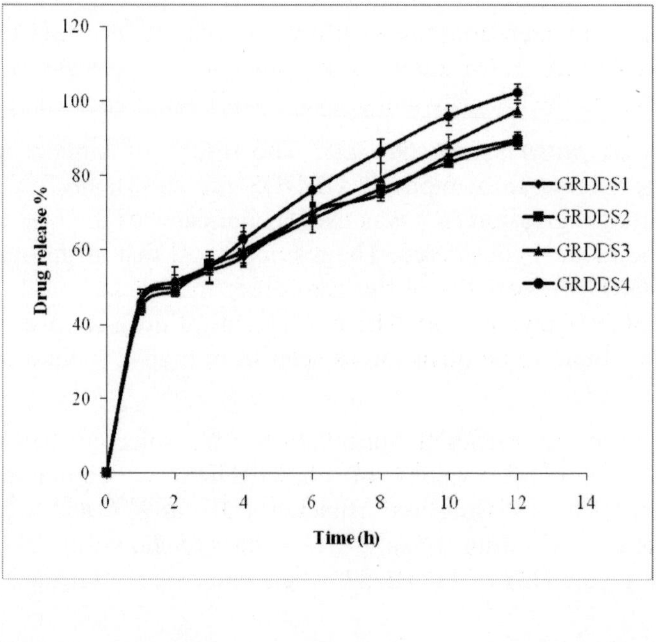

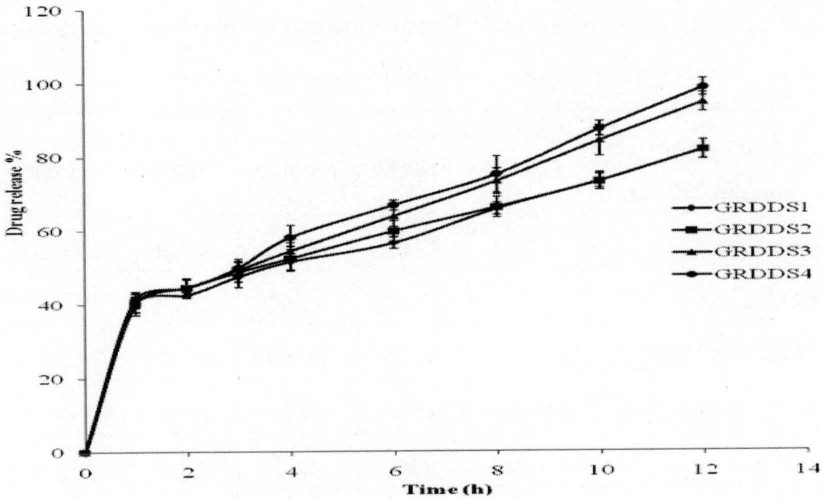

(b) SGF (enzyme free).

Figure 2.1 Dissolution profiles of fenoverine from biphasic GRDDS in (a) 0.1 mol L^{-1} HCl (b) SGF (enzyme free).

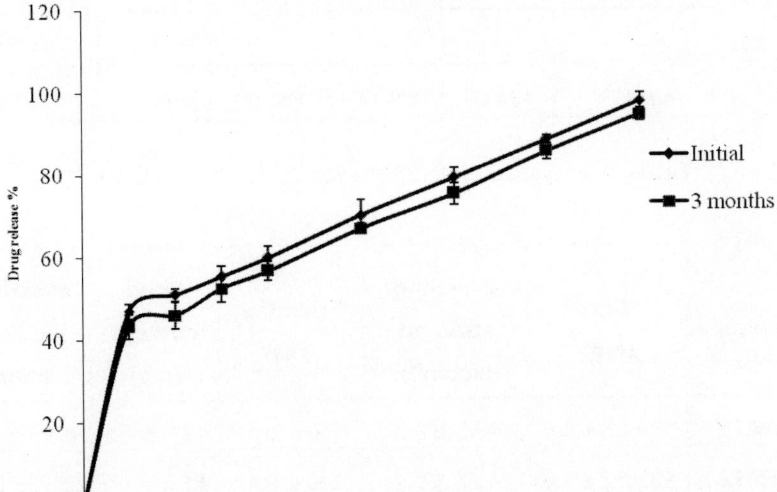

Figure 2.2 Dissolution profile of GRDDS3 after storage at 40 ^0C and 75% RH.

Table I Composition of biphasic Fenoverine gastroretentive drug delivery system.

Ingredients (mg per tablet)	FMMT1	FMMT2	FMMT3	FMMT4	LDT
Fenoverine	42.00	42.00	42.00	42.00	74.00
HPMC K 4M	63.00	47.25	31.50	15.75	-
HPMC K 100 LV	-	15.75	31.50	47.25	-
Sodium bicarbonate	20.00	20.00	20.00	20.00	-
Citric acid	5.00	5.00	5.00	5.00	-
Talc	2.00	2.00	2.00	2.00	2.00
Magnesium stearate	2.00	2.00	2.00	2.00	2.00
Cross caramellose sodium	-	-	-	-	13.00
Lactose	-	-	-	-	44.00
Total tablet mass	134.00	134.00	134.00	134.00	135.00

Table II Physical characteristics of multiple tablets.

Formulation	Mass (mg)	Crushing strength ($kgcm^{-2}$)	Friability (%)	Drug content (mg)	Floating lag time (min)
FMMT1	134.26 ± 1.0	4.05 ± 0.3	0.8 ± 0.1	42.3 ± 1.1	4.16 ± 0.3
FMMT2	133.52 ± 1.6	4.25 ± 0.4	0.6 ± 0.1	41.9 ± 1.5	5.22± 0.7
FMMT3	135.44 ± 1.7	5.10 ± 0.1	0.5 ± 0.1	42.6 ± 0.3	4.93± 0.3
FMMT4	136.01 ± 2.2	4.50 ± 0.2	0.6 ± 0.1	43.1 ± 0.9	4.74± 0.6
LDT	134.24 ± 2.8	5.30 ± 0.3	0.5 ± 0.1	74.3 ± 1.6	-

Table III In vitro release kinetics of fenoverine biphasic gastroretentive drug delivery Systems

Formulation	Q_{12} release[a]	R^2		
		First Order	Zero Order	Higuchi
0.1 mol L^{-1} HCl				
GRDDS1	89.70 ± 0.104	0.975	0.990	0.980
GRDDS2	89.10 ± 0.100	0.965	0.990	0.994
GRDDS3	97.69 ± 0.104	0.991	0.999	0.974
GRDDS4	102.44 ± 0.103	0.970	0.991	0.983
SGF (Enzyme free)				
GRDDS1	81.72 ± 1.125	0.992	0.996	0.964
GRDDS2	81.62 ± 0.437	0.986	0.998	0.980
GRDDS3	94.41 ± 0.173	0.991	0.999	0.970
GRDDS4	98.45 ± 0.254	0.983	0.996	0.969

[a] Drug released in 12 h. Values represent mean ± SD

Table IV Dissolution parameters and f_2 factor of GRDDS.

Formulation	$t_{50\%}$ (h)	$t_{75\%}$ (h)	DE (%)			*f2* factor
			2 h	6 h	8 h	
0.1 mol L^{-1} HCl						
Theoretical release	2.0	7.0	50	70	80	-
GRDDS1	2.0	8.0	55.93	78.24	83.57	80.4
GRDDS2	2.1	7.6	54.97	76.64	85.81	85.0
GRDDS3	**1.7**	**7.0**	**52.38**	**71.95**	**81.12**	**91.5**
GRDDS4	1.6	5.8	50.24	74.14	84.45	69.9
SGF (Enzyme free)						
GRDDS1	3.6	10.4	52.09	69.16	80.43	50.1
GRDDS2	3.3	10.5	54.55	73.11	80.97	53.0
GRDDS3	**2.95**	**8.3**	**47.16**	**67.49**	**77.47**	**61.4**
GRDDS4	3.0	8.0	45.13	67.79	76.20	68.0

DE – Dissolution efficiency

References

1. Chien YW. Controlled and Modulated-Release Drug Delivery Systems, in Encyclopedia of Pharmaceutical Technology, J. Swarbrick and J.C. Boylan, Eds. (Marcel Dekker, Inc.,New York, 1990), pp. 280–313 (1990).
2. Chien YW. Oral Drug Delivery Systems, in Novel Drug Delivery Systems, Chien, Y.W., Eds. Marcel Dekker, New York, pp. 139–196 (1992) .
3. Ritschel WA, and Kearns GL. Absorption/Transport Mechanisms, in Handbook of Basic Pharmacokinetics....including Clinical Applications, Ritschel W.A and Kearns, G.L. Eds. (American Pharmaceutical Association, Washington, DC,), p. 63 (1999).
4. Harder S, Furh U and Bergmann D. Ciprofloxacin Absorption in Different Regions of the Human GIT, Investigation With the Hf Capsule, Br. J. Clin. Pharmacol. 30 (1), 35–39 (1990).
5. Rouge N, Buri P and Doelker E. Drug Absorption Site in the Gastrointestinal Tract and Dosage Forms for Site-Specific Delivery, Int. J. Pharm. 136 (1), 17–139 (1996).
6. Chungi VS, Dittert LW Smith RB. Gastrointestinal Sites for Furosemide Absorption in Rats, Int. J. Pharm. 4, 27–38 (1979).
7. Benet LZ and Cummins CL. The Drug Efflux-Metabolism Alliance: Biochemical Aspects. Adv. Drug. Del. Rev. 50 (Supplement 1), S3–S11 (2001).
8. Drewe J, Beglinger C and Kissel T. The Absorption Site of Cyclosporin in Human GIT, Br. J. Clin. Pharmacol. 33 (1), 39–43 (1992).
9. Cremer K. Drug Delivery: Gastro-Remaining Dosage Forms, Pharm. J. 259 (108) (1997).
10. Kohri N. Improving the Oral Bioavailability of Sulpiride by Gastric-Retained Form in Rabbits, J. Pharm. Pharmacol. 48, 371–374 (1996).
11. Singh BN, and Kim KH. Floating Drug Delivery Systems: An Approach to Oral Controlled Drug Delivery via Gastric Retention, J. Controlled Rel. 63 (1–2), 235–259 (2000).

12. Gardner CR. Gastrointestinal Barrier to Oral Drug Delivery, in Directed Drug Delivery, Borchardt. R.T. Repta, A.J. and Stella, V.J. Eds. (Human Press, New Jersey), 61–82 (1985).

13. Daniel R, Clayburgh, Jerrold R, and Turner. Encyclopedia of gastroenterology, Elsevier science (2004).

14. Desai S. A Novel Floating Controlled Release Drug Delivery System Based on a Dried Gel Matrix Network [master's thesis]. Jamaica, NY, St John's University (1984).

15. Deshpande AA, Rhodes CT, Shah NH and Malick AW. Controlled-release drug delivery systems for prolonged gastric residence: an overview. Drug Dev. Ind. Pharm. 22 (6), 531–539 (1996).

16. Hou S, Cowles VE and Brener B. Gastric Retentive Dosage Forms: A Review Crit. Rev. Ther. Drug Carrier Syst., 20(6), 461-497 (2003).

17. Vantrappen GR, Peeters TL and Janssens J. The secretory component of interdigestive migratory motor complex in man. Sc. and J Gastroenterol. 14, 663 - 667 (1979).

18. Wilson CG and Washington N. The stomach: its role in oral drug delivery. In: Rubinstein MH, ed. Physiological Pharmaceutical: Biological Barriers to Drug Absorption. Chichester, UK: Ellis Horwood; 47 – 70 (1989).

19. Chawls G, Gupta P, Koradia V and Bansal A. Gastroretention – A means to address regional variability in intestinal drug absorption. Pharm. Technol., July, 50-68 (2003).

20. Desai, S and Bolton S. A floating controlled release drug delivery system: in vitro- in vivo evaluation. Pharm Res. 10, 1321 – 1325 (1993).

21. Khosla R, Feely LC and Davis SS. Gastrointestinal Transit of Non-Disintegrating Tablets in Fed Subjects, Int. J. Pharm. 53 (1), 107–117 (1989).

22. Hari Vardhan Reddy L and Murthy RSR. Floating dosage systems in drug delivery. Crit. Rev. Ther. Drug Carrier Syst., 19(6), 553-585 (2002).

23. Bardonnet PL, Faivre V, Pugh WJ, Piffaretti JC and Falson F. Gastroretentive dosage forms: Overview and special case of Helicobacter pylori. J. Control. Rel. 111, 1-18 (2006).

24. Kawashima Y, Niwa T, Takeuchi H, Hino T and Ito Y. Preparation of multiple unit hollow microspheres (microballoons) with acrylic resin containing tranilast and their drug release characteristics (in vitro) and floating behavior (in vivo), J. Control. Rel., 16 279–290 (1991).

25. Rouge N, Leroux JC, Cole ET, Doelker E and Buri P. Prevention of the sticking tendency of floating minitablets filled into hard gelatin capsules, Eur. J. Pharm. Biopharm. 43, 165–171 (1997).

26. Van Gansbeke B, Timmermans J, Schoutens A and Moes AJ. Intragastric positioning of two concurrently ingested pharmaceutical matrix dosage forms. Nucl. Med Biol 18, 711-718 (1991).

27. Muller-Lissner SA, Will N, Muller-Duysing W, Heinzel F, and Blum Al. A floating capsule with slow release of drugs: a new method of oral drug medication. Dtsch Med Wochenschr 106, 1143-1147 (1981).

28. Oth M, Franz, M, Timmermans J and Moes AJ. The bilayer floating capsule: a stomach- directed drug delivery system for misoprostol. Pharm Res. 9, 298-302 (1992).

29. Iannuccelli V, Coppi G, Sansone R and Ferolla G. Air compartment multiple-unit system for prolonged gastric residence. Part II. In vivo evaluation, Int. J. Pharm. 174, 55–62 (1998).

30. Timmermans J and Moes AJ. Factors controlling the buoyancy and gastric retention capabilities of floating matrix capsules: new data for reconsidering the controversy, J. Pharm. Sci. 83, 18–24 (1994).

31. Mojaverian P, Vlasses PH, Kellner PE and Rocci ML Jr. Effects of gender, posture, and age on gastric residence time of an indigestible solid: pharmaceutical considerations, Pharm. Res. 10, 639–644 (1988).

32. Cargill R, Caldwell LJ, Engle K, Fix JA, Porter PA and Gardner CR. Controlled gastric emptying. I. Effects of physical properties on gastric residence times of non disintegrating geometric shapes in beagle dogs. Pharm. Res., 5 (8), 533–536 (1988).

33. Groning R and Heun G. Oral dosage forms with controlled gastrointestinal transit, Drug Dev. Ind. Pharm. 10, 527–539 (1984).

34. Moes AJ. Gastroretentive dosage forms, Crit, Rev. Ther. Drug Carrier Syst. 10, 143–195 (1993).

35. Deshpande AA, Shah NH, Rhodes CT and Malick W. Development of a novel controlled-release system for gastric retention. Pharm. Res. 14, 815– 819 (1997).

36. Hwang S, Park H and Park K. Gastric Retentive Drug Delivery Systems Crit. Rev. Ther. Drug Carrier Syst., 15 (3), 243-284 (1998).

37. Urquhart J and Theeuwes F. Drug delivery system comprising a reservoir containing a plurality of tiny pills, US Patent 4, 434, 153, February 28, (1984).

38. Mamajek RC and Moyer ES. Drug-dispensing device and method, US Patent 4207890, June 17, (1980).

39. Fix JA, Cargill R and Engle K. Controlled gastric emptying. III. Gastric residence time of a non-disintegrating geometric shape in human volunteers. Pharm Res. 10, 1087-1089 (1993).

40. Kedzierewicz F, Thouvenot P, Lemut J, Etienne M, Hoffman A and Maincent P. Evaluation of peroral silicone dosage forms in humans by gamma-scintigraphy. J Control Rel. 58, 195-205 (1999).

41. Lenaerts VM and Gurny R. Gastrointestinal Tract- Physiological variables affecting the performance of oral sustained release dosage forms. In: V.M. Lenaerts, R. Gurny eds. Bioadhesive Drug Delivery Systems, CRC Press, Boca Raton, FL (1990).

42. Lehr CM. Bioadhesion technologies for the delivery of peptide and protein drugs to the gastrointestinal tract. Crit. Rev. Ther. Drug Carrier Syst. 11, 119–160 (1994).

43. Ponchel G and Irache JM. Specific and non-specific bioadhesive particulate systems for oral delivery to the gastrointestinal tract, Adv. Drug Del. Rev. 34, 191–219 (1998).

44. Chen J, Blevins WE, Park H and Park K. Gastric retention properties of superporous hydrogel composites. J. Control. Rel. 64 (1-3), 39-51 (2000).

45. Fujimori J, Machida Y and Nagai T. Preparation of a magnetically responsive tablet and confirmation of its gastric residence in beagle dogs, STP Pharma Sci. 4, 425– 430 (1994).

46. Rudiger G, Berntgen M and Georgarakis M. Acyclovir serum concentrations following peroral administration of magnetic depot tablets and the influence of extracorporal magnets to control gastrointestinal transit. European J. of Pharm. and Biopharm, 46, 285–291 (1998).

47. Washington N, Washington C, Wilson CG and Davis SS. What is liquid Graviscon? A comparison of four international formulations. Int J Pharm. 34, 105-109 (1986).

48. Fabregas JL, Claramunt J, Cucala J, Pous R and Siles A. In vitro testing of an antacid formulation with prolonged gastric residence time (Almagate flot coat). Drug Dev Ind Pharm. 20, 1199-1212 (1994).

49. Devereux JE, Newton JM and Short MB. The Influence of Density on the Gastrointestinal Transit of Pellets, J. Pharm. Pharmacol. 42 (7), 500–501 (1990).

50. Hilton AK and Deasy PB. In Vitro and In Vivo Evaluation of an Oral Sustained-Release Floating Dosage Form of Amoxycillin Trihydrate, Int. J. Pharm. 86 (1), 79–88 (1992).

51. Sheth PR and Tossounian JL. Sustained release pharmaceutical capsules, US Patent 4, 126, 672, November 21, (1978).

52. Javed A, Shweta A, Alka A, Anil KB, Rakesh KS, Roop KK and Sanjula B. Formulation and development of hydrodynamically balanced system for metformin: In vitro and in vivo evaluation Eur. J. Pharm. Biopharm. 67, 196-201 (2007).

53. Nur AO and Zhang JS. Captopril floating and/or bioadhesive tablets: design and release kinetics. Drug Dev Ind Pharm. 26, 965-969 (2000).

54. Ushomaru K, Nakachimi K and Saito H. Pharmaceutical preparations and a method of manufacturing them. US patent 4702918. October 27, (1987).

55. Bolton S and Desai S. Floating sustained release therapeutic compositions. US patent 4 814 179. March 21, (1989).

56. Streubel A, Siepmann J, Bodmeier R. Floating matrix tablets based on low density foam powder: effect of formulation and processing parameters on drug release. Eur J Pharm Sci. 18, 37-45 (2003).

57. Sheth PR and Tossounian JL. Sustained- Release Tablet Formulations, US Patent No.4, 140, 755, (1979a).

58. Sheth PR and Tossounian JL. Novel sustained release tablet formulations, US Patent 4, 167, 558, September 11, (1979b).

59. Rubinstein A and Friend DR. Specific delivery to the gastrointestinal tract, in: A.J. Domb (Ed.), Polymeric Site-Specific Pharmacotherapy, Wiley, Chichester, pp. 282–283 (1994).

60. Ritschel WA. Targeting in the gastrointestinal tract: new approaches, Methods Find. Exp. Clin. Pharmacol. 13, 313–336 (1991).

61. Michaels AS. Drug delivery device with self actuated mechanism for retaining device in selected area, US Patent 3, 786, 813, January 22, (1974).

62. Michaels AS, Bashwa JD and Zaffaroni A. Integrated device for administering beneficial drug at programmed rate, US Patent 3, 901, 232, August 26, (1975).

63. Hashim H and Li Wan Po A. Improving the release characteristics of water-soluble drugs from hydrophilic sustained release matrices by in situ gas-generation, Int. J. Pharm. 35, 201–209 (1987).

64. Ingani HM, Timmermans J and Moes AJ. Conception and in vivo investigation of peroral sustained release floating dosage forms with enhanced gastrointestinal transit, Int. J.Pharm. 35 (1987) 157–164 (1987).

65. Stockwell AF, Davis SS and Walker SE. In vitro evaluation of alginate gel systems as sustained release drug delivery systems, J. Control. Release 3 167–175 (1986).

66. Yang L, Esharghi J and Fassihi R. A new intra gastric delivery system for the treatment of helicobacter pylori associated gastric ulcers: in vitro evaluation. J Control Rel. 57, 215-222 (1999).

67. Ozdemir N, Ordu S and Ozkan Y. Studies of floating dosage forms of furosemide: in vitro and in vivo evaluation of bilayer tablet formulation. Drug Dev Ind Pharm. 26, 857-866 (2000).

68. Fassihi R and Yang L. Controlled release drug delivery systems. US patent 5 783 212. July 21, (1998).

69. Talwar N, Sen H and Staniforth JN. Orally administered controlled drug delivery system providing temporal and spatial control. US patent 6 261 601. July 17, (2001).

70. Baumgartner S, Kristel J, Vreer F, Vodopivec P and Zorko B. Optimisation of floating matrix tablets and evaluation of their gastric residence time. Int J Pharm. 195, 125-135 (2000).

71. Moursy NM, Afifi NN, Ghorab DM and El-Saharty Y. Formulation and evaluation of sustained release floating capsules of Nicardipine hydrochloride. Pharmazie. 58, 38-43 (2003).

72. Ichikawa M, Watanabe S and Miyake Y. A new multiple unit oral floating dosage system. I: Preparation and in vitro evaluation of floating and sustained-release characteristics. J. Pharm. Sci., 80(11), 1062-1066 (1991).

73. Atyabi F, Sharma HL, Mohammed HAH and Fell JT. In vivo evaluation of a novel gastro retentive formulation based on ion exchange resins. J. Control Rel. 42, 105-113 (1996).

74. Umamaheshwari RB, Subheet J and Jain NK. A New Approach in Gastroretentive Drug Delivery SystemUsing Cholestyramine, Drug Delivery, 10, 151–160 (2003).

75. Kawashima Y, Niwa T, Takeuchi H, Hino T and Itoh Y. Hollow microspheres for use as a floating controlled drug delivery system in the stomach. J. Pharm. Sci. 81, 135-140 (1992).

76. Sato Y, Kawashima Y, Takeuchi H and Yamamoto H. Physicochemical properties to determine the buoyancy of hollow microspheres (microballoons) prepared by the emulsion solvent diffusion method. Eur J Pharm Biopharm. 55, 297-304 (2003).

77. Lee JH, Park TG and Choi HK, Development of oral drug delivery system using floating microspheres. J Microencapsul. 16, 715-729 (1999).

78. Lee JH, Park TG, Lee YB, Shin SC and Choi HK. Effect of adding non-volatile oil as a core material for the floating microspheres prepared by emulsion solvent diffusion method. J Microencapsul. 18, 65-75 (2001).

79. Jain SK, Awasthi AM, Jain NK and Agrawal GP. Calcium silicate based microspheres of repaglinide for gastroretentive floating drug delivery: preparation and in vitro characterization. J Control Rel. 107, 300-309 (2005).

80. El-Kamel AH, Sokar MS, Al Gamal SS and Naggar VF. Preparation and evaluation of ketoprofen floating oral delivery system. Int J Pharm. 220, 13-21 (2001).

81. Sato Y, Kawashima Y, Takeuchi H and Yamamoto H. In vitro evaluation of floating and drug releasing behaviors of hollow microspheres (microballoons) prepared by the emulsion solvent diffusion method. Eur J Pharm Biopharm. 57, 235-243 (2004).

82. Sato Y, Kawashima Y, Takeuchi H, Yamamoto H and Fujibayashi Y. Pharmacoscintigraphic evaluation of riboflavin-containing microballoons for a floating controlled drug delivery system in healthy humans. J Control Rel. 98, 75-85 (2004a).

83. Sato Y, Kawashima Y, Takeuchi H and Yamamoto H. In vitro and in vivo evaluation of riboflavin-containing microballoons for a floating controlled drug delivery system in healthy humans. Int J Pharm. 275, 97-107 (2004b).

84. Joseph NJ, Laxmi S, Jayakrishnan A. A floating type oral dosage from for piroxicam based on hollow microspheres: in vitro and in vivo evaluation in rabbits. J. Control Rel. 79, 71-79 (2002).

85. Whitehead L, Collete JH and Fell JT. Amoxycillin Release from a floating dosage form based on alginates. Int. J. Pharm. 21, 45-49 (2000).

86. Sriamornsak P, Thirawong N and Putkhachorn S. Morphology and buoyancy of oil entrapped calcium pectinate gel beads, AAPS, 6(3), article 24 (2004).

87. Klausner EA, Lavy E, Barta M, Cserepes E, Friedman M and Hoffman A. Novel gastroretentive dosage forms: evaluation of gastroretentivity and its effect on levodopa absorption in humans, Pharm. Res. 20 (9), 1466–1473 (2003).

88. Klausner EA, Eyal S, Lavy E, Friedman M and Hoffman A. Novel levodopa gastroretentive dosage form: in-vivo evaluation in dogs J. Control Rel. 88, 117–126 (2003).

89. Klausner EA, Lavy E, Friedman M and Hoffman A. Expandable gastroretentive dosage forms, J. Control. Release 90 (2), 143–162 (2003).

90. Kagan L, Lapidot N, Afargan M, Kirmayer D, Moor E, Mardor Y, Friedman M and Hoffman A. Gastroretentive Accordion Pill: Enhancement of riboflavin bioavailability in humans J. Controlled Rel. 113, 208–215 (2006).

91. Klausner EA, Lavy E, Stepensky D, Friedman M and Hoffman A. Novel gastroretentive dosage forms: evaluation of gastroretentivity and its effect on riboflavin absorption in dogs, Pharm. Res. 19 (10), 1516–1523 (2002).

92. Groning R and Heun G. Oral dosage forms with controlled gastrointestinal transit, Drug Dev. Ind. Pharm. 10, 527–539 (1984).

93. Chen J, Blevins WE, Park H and Park K. Gastric retention properties of superporous hydrogel composites. J. Control. Rel. 64 (1-3), 39-51 (2000).

94. Wilding IR, Davis SS and Hagan DT. Targeting of Drugs and Vaccines to the Gut, in Pharmac. Ther., C.J. Hawkey, Eds. 98–124 (1994).

95. Park K and Robinson JR. Bioadhesive Polymers as Platforms for Oral-Controlled Drug Delivery:Method to Study Bioadhesion, Int. J. Pharm. 19 (1), 107–127 (1984).

96. Lehr CM and Hass J. Developments in the Area of Bioadhesive Drug Delivery Systems, Expert Opinion on Biological Therapy 2, 287–298 (2002).

97. Fujimori J, Machida Y and Nagai T. Preparation of a magnetically responsive tablet and confirmation of its gastric residence in beagle dogs, STP Pharma Sci. 4, 425– 430 (1994).

98. Rudiger G, Berntgen M and Georgarakis M. Acyclovir serum concentrations following peroral administration of magnetic depot tablets and the influence of extracorporal magnets to control gastrointestinal transit. Eur. J. of Pharm. and Biopharm, 46, 285–291 (1998).

99. Fabregas JL, Claramunt J, Cucala J, Pous R and Siles A. In vitro testing of an antacid formulation with prolonged gastric residence time (Almagate flot coat). Drug Dev Ind Pharm. 20, 1199-1212 (1994).

100. Chawla G, Gupta P, Koradia V and Bansal AK. Gastro retention: A means to address regional variability in intestinal drug absorption. Pharm.Tech, 50 – 68 (2003).

101. Burns SJ, Corness D, Hay G, Higginbottom S, Whelan I, Attwood D and Barnwell SG. Development and validation of an in vitro dissolution method for a floating dosage form with biphasic release characteristics, Int J Pharm, 121, 37-44 (1995).

102. Singh RN and Kim KH. Floating drug delivery systems: An approach to oral controlled drug delivery via gastric retention, J Control Release, 63 235-259 (2000).

103. Timmermanns J and Moes A. How well do floating dosage forms float? Int. J. Pharm., 62(3), 207 – 216 (1990).

104. Timmermans J, Gansbeke VB and Moes AJ. Assessing by gamma scintigraphy the in vivo buoyancy of dosage forms having known size and floating force profiles as a function of time. Vol I. Proceedings of the 5th International Conference on Pharmacy Technology. Paris, France: APGI: 42-51 (1989).

105. Babu VBM and Khar RK. In vitro and In vivo studies of sustained release floating dosage forms containing salbutamol sulphate. Pharmazie; 45:268-270 (1990).

106. Horton RE, Ross FGM and Darling GH. Determination of the emptying-time of the stomach by use of enteric-coated barium granules, Br Med J, **1,** 1537-1539 (1965).

107. Jao F, Edgren DE and Wong PS. Gastric retention dosage form havin multiple layers, Int Application WO0038650 (July 6, 2000).

108. Hendee WR. Textbook of Diagnostic Imaging II, vol 1, edited by C E Putman & C E Ravin (W B Saunders, Philadelphia), 1-6 (1994).

109. Steingotter A, Weishaupt D, KunzMader P, Legsfeld K, Thumshirn HM Boesinger P, Fried M and Schwizer W. Magnetic resonance imaging for the in vivo evaluation of gastric retentive tablets, Pharm Res, 20 (2003).

3
Buccal Drug Delivery Systems

P. Chinna Reddy, Y. Vamshi Vishnu and Y. Madhusudan Rao
University College of Pharmaceutical Sciences,
Kakatiya University, Warangal,Andhra Pradesh, India.

3.0 Introduction

Over the decades, peroral delivery has been the popular route of administration for the majority of therapeutic agents targeting systemic delivery. Oral administration generally leads to 'transmucosal' absorption in the gastrointestinal tract; however, this enteral route of delivery subjects compounds to extensive presystemic elimination, which may include gastrointestinal degradation, metabolism, or first-pass clearance via the liver. This first pass hepatic metabolism has often resulted in low systemic bioavailability, short duration of therapeutic activity, formation of inactive therapeutic agents and, at times, toxic metabolites (1). Parenteral routes, such as intravenous or intramuscular, unlike oral delivery, permit therapeutic agents to gain direct entry into the systemic circulation, and therefore reach the intended site of action more rapidly, with complete bioavailability often achievable with intravenous administration. Unfortunately, parenteral drug administration entails certain hazards, requires specialized equipment, often requires close medical supervision of the medication, high costs, inconvenience and poor patient compliance make it less conducive for long term therapy. Transdermal and transmucosal drug delivery offers attractive alternative routes for administration of drugs and may avoid the significant drawbacks of peroral and parenteral administration. These routes of administration bypass first pass metabolism by delivering drug directly into the systemic circulation.

Systemic transmucosal delivery of therapeutic agents via the mucosal epithelium lining of accessible body cavities, such as the oral cavity (buccal), nose (nasal), rectum (rectal), and vagina (vaginal), have received renewed interest within the last two decades. These routes have numerous advantages over the peroral drug delivery, such as bypassing hepatic first- pass clearance, and therefore potentially improving systemic bioavailability. In addition, these transmucosal routes may eliminate the disadvantages of parenteral administration. As a result, nasal, rectal, and vaginal routes of delivery have gained the attention of many researchers.

Although transdermal delivery has been demonstrated to be very useful for controlled administration of highly potent lipophilic drugs to maintain a sustained plasma drug concentration within a therapeutically effective range, it often requires an undesirable lag time for the attainment of therapeutic concentrations. Also, the skin permeation rate of hydrophilic and large drug molecules is often too low to be therapeutically useful. Nasal administration has received increased attention for the delivery of compounds that are not absorbed orally and to decrease the time of the onset of action of a particular drug. In particular, with the availability of proteins and peptides from advanced biotechnology, the development of intranasal drug delivery systems has become even more vital. However, long-term exposure of the nasal mucosa to exogenous compounds may lead to irreversible damage to the action of its cilia (2). The rectal, vaginal, and ocular mucosa possess various advantages for drug delivery, poor patient acceptability associated with these routes makes them useful only for local drug delivery. The oral cavity, however, is a highly accepted route for both local and systemic drug delivery (3). Indeed, buccal delivery is increasingly being considered to be the preferred route for many drug classes (4). Table 3.1 compares issues of various routes of drug delivery.

Table 3.1 Comparison of different routes for systemic drug delivery

Issues	Nasal	Buccal	Vaginal/Rectal	GI Tract	Dermal
Accessibility	++	++	+	+	+++
First pass clearance	+++	+++	++	+	+++
Acceptability	++	+++	+	+++	+++
Surface area	+	++	++	+++	+++
Onset of action	+++	++	+++	+++	+
Robustness	+	++	++	++	+++
Duration	++	+++	+++	+	++
Permeability	++	++	+++	+++	+
Vascular drainage	+++	+++	+++	+++	+
Surface environment	++	+++	+	+	++
+ = Not favorable; ++ = Intermediate; +++ = Very favorable					

Buccal drug delivery offers several advantages over the oral route and other alternative routes of drug administration. The membranes that line the oral cavity are readily accessible, robust, and exhibit fast cellular recovery following local stress or damage and is very used to being exposed to various exogenous compounds. Properly constructed oral transmucosal drug delivery systems are easy and painless to administer and well accepted by the patient. Precise localization of the dosage form is possible and there is the ability to terminate delivery when required (e.g. because of toxicity). Apart from these applications the following advantages are present with the buccal drug delivery system.

Advantages

(i) It is richly vascularized and more accessible for the administration and removal of a dosage form.

(ii) Buccal drug delivery has a high patient acceptability compared to other non-oral routes of drug administration.

(iii) Harsh environmental factors that exist in oral delivery of a drug are circumvented by buccal delivery.

(iv) Avoids acid hydrolysis in the gastrointestinal (GI) tract and by passing the first-pass effect.

(v) Moreover, rapid cellular recovery and achievement of a localized site on the smooth surface of the buccal mucosa

(vi) Buccal delivery allows use of drugs with short half-lives

(vii) Controls plasma concentrations of potent drugs and can interrupt drug input quickly in case of toxicity.

(viii) Reduces multiple administration and improvement in patient compliance

(ix) Useful for pediatric and geriatric patients

Apart from the numerous advantages of the buccal drug delivery, it is having the following limitations:

Limitations

(i) Low permeability of the buccal membrane: specifically when compared to the sublingual membrane.

(ii) Smaller surface area. The total surface area of the membranes of the oral cavity available for drug absorption is 170 cm^2 of which ~50 cm^2 represents non-keratinized tissues, including the buccal membrane.

(iii) The continuous secretion of saliva (0.5–2 L/day) leads to subsequent dilution of the drug.

(iv) Swallowing of saliva can also potentially lead to the loss of dissolved or suspended drug and, ultimately, the involuntary removal of the dosage form.

(v) Not suitable for high dose/blood concentration drugs

(vi) Limited absorption of high-molecular-weight drugs

3.1 Oral Mucosa

3.1.1 Anatomy of the Oral Mucosa

Light microscopy reveals several distinct patterns of maturation in the epithelium of the human oral mucosa based on various regions of the oral cavity. Three distinctive layers of the oral mucosa are the epithelium, basement membrane, and connective tissues. The oral cavity is lined with the epithelium, below which lies the supporting basement membrane. The basement membrane is, in turn, supported by connective tissues (Fig. 3.1). The epithelium, as a protective layer for the tissues beneath, is divided into (a) non-keratinized surface in the mucosal lining of the soft palate, the ventral surface of the tongue, the floor of the mouth, alveolar mucosa, vestibule, lips, and cheeks, and (b) keratinized epithelium which is found in the hard palate and non-flexible regions of the oral cavity (5). The epithelial cells, originating from the basal cells, mature, change their shape, and increase in size while moving towards the surface. The thickness of buccal epithelium in humans, dogs and rabbits has been determined to be approximately 500–800μm (6).

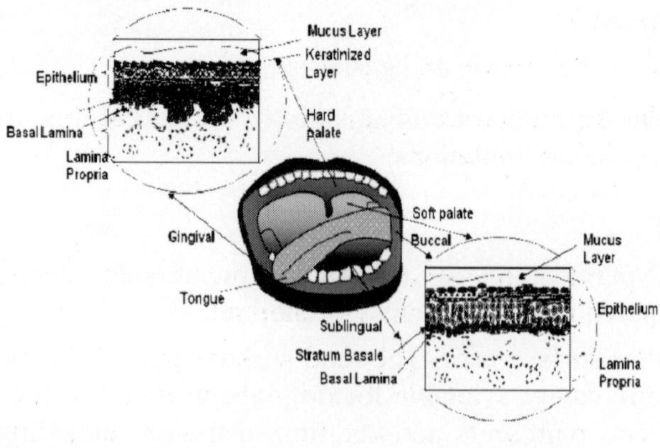

Figure 3.1 Schematic representation of the oral cavity.

The basement membrane forms a distinctive layer between the connective tissues and the epithelium. It provides the required adherence between the epithelium and the underlying connective tissues and functions as a mechanical support for the epithelium. The underlying connective tissues provide many of the mechanical properties of oral mucosa. The buccal epithelium is classified as a non keratinized tissue (7). It is penetrated by tall and conical-shaped connective tissues. These tissues, which are also referred to as the lamina propria, consist of collagen fibers, a supporting layer of connective tissues, blood vessels, and smooth muscles (1). A schematic representation of the tissue components of the oral mucosa is shown in Figure 3.2. The rich arterial blood supply to the oral mucosa is derived from the external carotid artery. The buccal artery, some terminal branches of the facial artery, the posterior alveolar artery, and the infraorbital artery are the major sources of blood supply to the lining of the cheek in the buccal cavity (8).

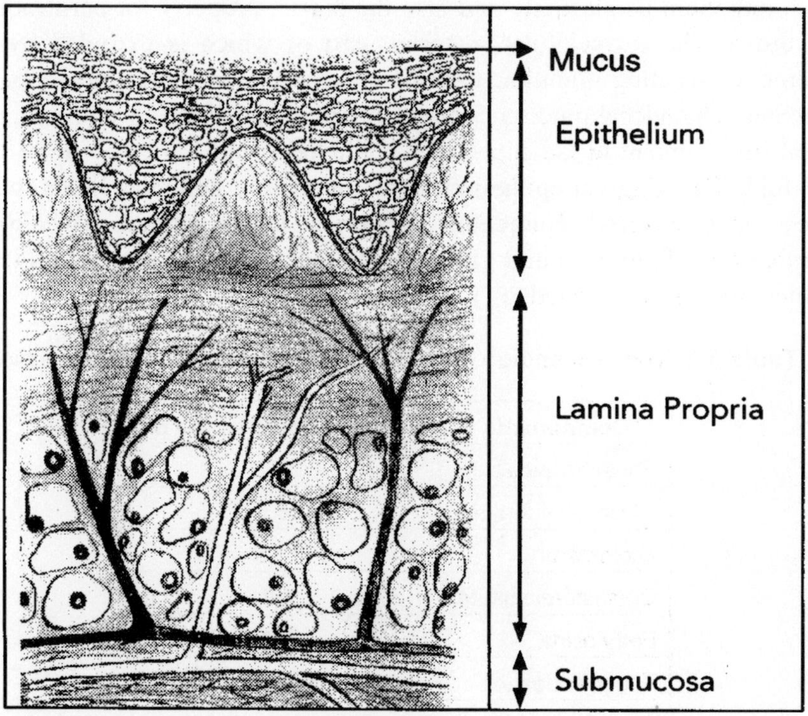

Figure 3.2 Schematic representation of the main tissue components of the oral mucosa

A gel-like secretion known as mucus, which contains mostly water-insoluble glycoproteins, covers the entire oral cavity. Mucus is bound to the apical cell surface and acts as a protective layer to the cells below (9). It is also a visco-elastic hydrogel, and primarily consists of 1–5% of the water insoluble glycoproteins and lipids, mineral salts and free proteins, 95 % water, and several other components in small quantities, such as enzymes, electrolytes, and nucleic acids. Lipid analysis of buccal tissues shows the presence of phospholipids 76.3%, glucosphingolipid 23.0% and ceramide NS at 0.72%. Other lipids such as acyl glucosylated ceramide, and ceramides like Cer AH, CerAP, Cer NH, CerAS, and EOHP/NP are completely absent. This composition can vary based on the origin of the mucus secretion in the body (10).

3.1.2 Biochemistry of the Oral Mucosa

The oral mucosa has large amounts of extracellular material, which gives the epithelium its elasticity and also the barrier property for permeability of drugs. The extracellular material, part of which is extruded by the membrane-coating granules, varies from one oral mucosal membrane to another. The buccal mucosa contains mainly polar lipids (Table 3.2). The total lipid content in the superficial epithelial layers varies by more than 10-fold. The gingival epithelium contains mainly nonpolar lipids, while the non-keratinized epithelium comprise fewer neutral glycolipids, particularly cholesterol and glucosyl ceramides, and more glycoproteins, which are easily hydrated (11)

Table 3.2 The percentage composition of buccal mucosa of the pig.

Components	Buccal mucosa of the pig
Phospholipids[a]	38.2
Glycosylceramides	16.5
cholesterol	13.6
Cholesterol sulfate	7.8
Fatty acids	1.6
Triglycerides	15.7
Cholesteryl esters	5.9

(adopted from Raj B et.al., 2005 (11)) [a]phospholipids includes sphingomyelin, phosphatidylcholine, phosphoethanolamine, serine, and inositol phosphatides

3.1.3 Absorption Pathways

Studies with microscopically visible tracers such as small proteins and dextrans suggest that the major pathway across stratified epithelium of large molecules is via the intercellular spaces and that there is a barrier to penetration as a result of modifications to the intercellular substance in the superficial layers. However, rate of penetration varies depending on the physicochemical properties of the molecule and the type of tissue being traversed. This has led to the suggestion that materials uses one or more of the following routes simultaneously to cross the barrier region in the process of absorption, but one route is predominant over the other depending on the physicochemical properties of the diffusant (12).

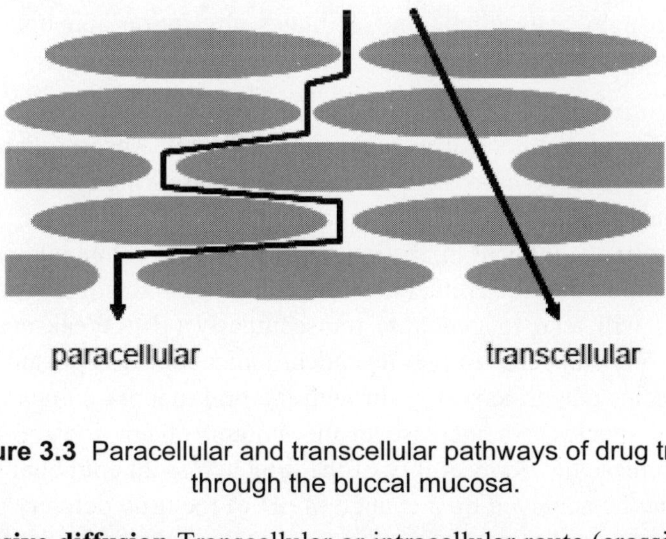

Figure 3.3 Paracellular and transcellular pathways of drug transport through the buccal mucosa.

Passive diffusion Transcellular or intracellular route (crossing the cell membrane and entering the cell) Paracellular or intercellular route (passing between the cells) **Carrier mediated diffusion, active transport, and pinocytosis or endocytosis.** The flux of drug through the membrane under sink condition for paracellular route can be written as Eq. (3.1)

$$Jp = \frac{D_p \varepsilon}{h_p} C_d \qquad .....(3.1)$$

where, D_p is diffusion coefficient of the permeate in the intercellular spaces, h_p is the path length of the paracellular route, ε is the area fraction of the paracellular route and C_d is the donor drug concentration. Similarly, flux of drug through the membrane under sink condition for transcellular route can be written as Eq. (3.2).

$$J_c = \frac{(1-\varepsilon)D_c K_c}{h_c} \quad \text{.....(3.2)}$$

where, K_c is partition coefficient between lipophilic cell membrane and the aqueous phase, D_c is the diffusion coefficient of the drug in the transcellular spaces and h_c is the path length of the transcellular route (13). In very few cases absorption also takes place by the process of endocytosis where the drug molecules were engulfed by the cells. It is unlikely that active transport processes operate within the oral mucosa; however, it is believed that acidic stimulation of the salivary glands, with the accompanying vasodilatation, facilitates absorption and uptake into the circulatory system (14).

The absorption potential of the buccal mucosa is influenced by the lipid solubility and molecular weight of the diffusant. Absorption of some drugs via the buccal mucosa is found to increase when carrier pH is lowered and decreased with an increase of pH (15). However, the pH dependency that is evident in absorption of ionizable compounds reflects their partitioning into the epithelial cell membrane, so it is likely that such compounds will tend to penetrate transcellularly (16). Weak acids and weak bases are subjected to pH-dependent ionization. It is presumed that ionized species penetrate poorly through the oral mucosa compared with non-ionized species. An increase in the amount of non-ionized drug is likely to increase the permeability of the drug across an epithelial barrier, and this may be achieved by a change of pH of the drug delivery system. It has been reported that pH has effect on the buccal permeation of drug through oral mucosa. The diffusion of drugs across buccal mucosa was not related to their degree of ionization as calculated from the Henderson–Hasselbalch equation and thus it is not helpful in the prediction of membrane diffusion of weak acidic and basic drugs (17).

In general, for peptide drugs, permeation across the buccal epithelium is thought to be through paracellular route by passive diffusion. Recently, it was reported that drugs that have a monocarboxylic acid residue could be delivered into systemic circulation from the oral mucosa via its carrier (18). The permeability of oral mucosa and the efficacy of penetration enhancers have been investigated in numerous in vivo and in vitro models. Various kinds of diffusion cells, including continuous flow perfusion chambers, Ussing chambers, Franz diffusion cells and Grass–Sweetana, have been used to determine the permeability of oral mucosa (19). Cultured epithelial cell lines have also been developed as an in vitro model for studying drug transport and metabolism at biological barriers as well as to elucidate the possible mechanisms of action of penetration enhancers (20). Recently, TR146 cell culture model was suggested as a valuable in vitro model of human buccal mucosa for permeability and metabolism studies with enzymatically labile drugs, such as leu-enkefalin, intended for buccal drug delivery (21).

3.1.4 Permeability Barrier of the Oral Mucosa

The permeability barrier property of the oral mucosa is predominantly due to intercellular materials derived from the so-called 'membrane coating granules' (MCGs) (1). MCGs are spherical or oval organelles that are 100–300 nm in diameter and found in both keratinized and non-keratinized epithelia. These organelles have also been referred to as 'small spherically shaped granules', 'keratinosomes', 'transitory dense bodies', and 'cementsomes' (22). MCGs were first named as such because it was believed that they were subject to exocytosis from the cytoplasm of the stratum spinosum of keratinized epithelia following thickening of these cells. Nonetheless, it is actually the contents of MCGs that are subject to exocytosis prior to the onset of membrane thickening.

MCGs are found near the upper, distal, or superficial border of the cells, and a few occur near the opposite border. Several hypotheses have been suggested to describe the functions of MCGs, including a membrane thickening effect, cell adhesion, production of a cell surface coat, cell desquamation, and permeability barrier. Hayward has reviewed the literature related to these functions, and it appears that the permeability barrier is most often attributed to MCGs (22). They discharge their contents into the intercellular space to ensure epithelial cohesion in the superficial layers, and this discharge forms a barrier to the permeability of various compounds. Cultured oral epithelium devoid of MCGs has been shown to be permeable to compounds that do not typically penetrate oral epithelium (23). In addition, permeation studies conducted using tracers of different sizes have demonstrated that these tracer molecules did not penetrate any further than the top 1–3 cell layers. When the same tracer molecules were introduced sub-epithelially, they penetrated through the intercellular spaces. This limit of penetration coincides with the level where MCGs are observed. This same pattern is observed in both keratinized and non keratinized epithelia (1), which indicates that keratinization of the epithelia, in and of itself, is not expected to play a major role as a barrier to permeation (24).

Another barrier to drug permeability across buccal epithelium is enzymatic degradation. Saliva contains no proteases, but does contain moderate levels of esterases, carbohydrases, and phosphatases (25). However, several proteolytic enzymes have been found in the buccal epithelium (26). Walker et al. (27) reported that endopeptidases and carboxypeptidases were not present on the surface of porcine buccal mucosa, whereas aminopeptidases appeared to be the major enzymatic barrier to the buccal delivery of peptide drugs. Aminopeptidase N and A (plasma membrane-bound peptidases) and aminopeptidase B (cytosolic enzyme) have been found in the buccal tissue (28). The use of mucoadhesive polymers as enzyme inhibitor agents has been developed to overcome this obstacle in peptide and protein delivery.

3.1.5 Oral Mucosal Permeation Enhancers

Drug permeation through membrane is the limiting factor in the development of buccal adhesive delivery devices. The epithelium that lines the buccal mucosa is a very effective barrier to the absorption of drugs. Substances that facilitate the permeation through buccal mucosa are referred as permeation enhancers (29). As most of the penetration enhancers were originally designed for purposes other than absorption enhancement, a systemic search for safe and effective penetration enhancers must be a priority in drug delivery. The goal of designing penetration enhancers, with improved efficacy and reduced toxicity profile is possible by understanding the relationship between enhancer structure and the effect induced in the membrane and the mechanism of action. However, the selection of enhancer and its efficacy depends on the physicochemical properties of the drug, nature of the vehicle and other excipients. In some cases usage of enhancers in combination has shown synergistic effect than the individual enhancers. Penetration enhancement to the buccal membrane is drug specific. These permeation enhancers should be safe and non-toxic, pharmacologically and chemically inert, non-irritant, and non-allergenic. One of the major disadvantages associated with buccal drug delivery is the low flux which results in low drug bioavailability (30). Various compounds have been investigated for their use as buccal penetration enhancers in order to increase the flux of drugs through the mucosa are classified in **Table 3.3.**

Table 3.3 Buccal penetration enhancers and its mechanism of action

Classification	Examples	Mechanism of action
Surfactants	Anionic: Sodium lauryl sulfate Cationic: Cetylpyridinium chloride Nonionic: Poloxamer, Brij, Span, Myrj, Tween	Perturbation of intercellular lipids, protein domain integrity
Bile salts	Sodium glycodeoxycholate, Sodium glycocholate, Sodium taurodeoxycholate, Sodium taurocholate, Azone	Perturbation of intercellular lipids, protein domain integrity
Fatty acids	Oleic acid, Caprylic acid Lauric acid, Lauric acid/ Propylene glycol, Methyloleate,Lysophosphati dylcholine, Phosphatidylcholine	Increase fluidity of phospholipid domains
Cyclodextrins	α, β, γ, Cyclodextrin, methylated β –cyclodextrins	Inclusion of membrane Compounds
Chelators	EDTA, Citric acid, Sodium salicylate, Methoxy salicylates	Interfere with Ca Polyacrylates
Positively charged Polymers	Chitosan, Trimethyl chitosan	Ionic interaction with negative charge on the mucosal surface
Cationic Compounds	Poly-L-arginine, L-lysine	Ionic interaction with negative charge on the mucosal surface

Mechanism of action of permeation enhancers

Mechanisms by which penetration enhancers are thought to improve mucosal absorption are as follows.

(i) ***Changing mucus rheology:*** Mucus forms viscoelastic layer of varying thickness that affects drug absorption. Further, saliva covering the mucus layers also hinders the absorption. Some permeation enhancers' act by reducing the viscosity of the mucus and saliva overcomes this barrier.

(ii) ***Increasing the fluidity of lipid bilayer membrane:*** The most accepted mechanism of drug absorption through buccal mucosa is intracellular route. Some enhancers disturb the intracellular lipid packing by interaction with either lipid or protein components.

(iii) ***Acting on the components at tight junctions:*** Some enhancers act on desmosomes, a major component at the tight junctions there by increases drug absorption.

(iv) ***By overcoming the enzymatic barrier:*** These act by inhibiting the various peptidases and proteases present within buccal mucosa, thereby overcoming the enzymatic barrier. In addition, changes in membrane fluidity also alter the enzymatic activity indirectly.

(v) ***Increasing the thermodynamic activity of drugs:*** Some enhancers increase the solubility of drug there by alters the partition coefficient. This leads to increased thermodynamic activity resulting better absorption.

3.2 Mucoadhesion/Bioadhesion

3.2.1 Definition

In 1986, Longer and Robinson defined the term "bioadhesion" as the "attachment of a synthetic or natural macromolecule to mucus and/or an epithelial surface" (31). The general definition of adherence of a polymeric material to biological surfaces (bioadhesives) or to the mucosal tissue (mucoadhesives) still holds.

3.2.2 The Mucoadhesive/Mucosa Interaction

3.2.2.1 Chemical bonds

For adhesion to occur, molecules must bond across the interface. These bonds can arise in the following way (32).

(a) ***Ionic bonds:*** Where two oppositely charged ions attract each other via electrostatic interactions to form a strong bond (e.g. in a salt crystal).

(b) ***Hydrogen bonds:*** Hydrogen bonding is basically an electrostatic interaction that arises when a hydrogen atom bound to an electronegative atom, e.g., nitrogen, oxygen, or fluorine, interacts with another electronegative atom. The result is a dipolar molecule. The hydrogen atom has a partial positive charge and hence can interact with another highly electronegative atom in an adjacent molecule. This results in a stabilizing interaction that binds the two molecules together. The force is short range and highly directional. In a more hydrophobic environment, hydrogen bonds become significant and are essential in the formation of stable structures. Bond energy serves as a measure of strength of bonds. Magnitude of bond energy for hydrogen bond is between 10 and 20 kJ/mol (34). Role of hydrogen bonding in interaction between mucoadhesive and mucin at gastric pH was studied by Tobyn et al. (35). The bonding is stronger and is directional. The directional nature of hydrogen bonding requires the two molecules to adopt a specific relative geometry. The hydrogen can be thought of as being shared, and the bond formed is generally weaker than ionic or covalent bonds.

(c) ***Van-der-Waals bonds:*** The attractive forces included in the DLVO theory are normally termed van der Waal's forces and will arise in a number of ways. These may be further divided into the following three components (36):

(i) *London dispersion forces:* also called as dispersion forces. These originate out of the electronic motions in paired molecules and give rise to attractive interactions. These forces involve the attraction between temporarily induced dipoles in nonpolar molecules (often disappear within a second) (37). This polarization can be induced either by a polar molecule or by the repulsion of negatively charged electron clouds in nonpolar molecules. These results when two atoms belonging to different molecules are brought sufficiently close together. These interactions involve a force of about 0.5–1 K cal/mole. London Dispersion forces exist between all atoms (38).

(ii) *Dipole–dipole interactions:* also called Keesom interactions after Willem Hendrik Keesom who produced the first mathematical description in 1921, are the forces that occur between two molecules with permanent dipoles. Dipole–dipole interactions work in a similar manner to ionic interactions, but are weaker because only partial charges are involved and are due to attraction between polar groups. Dipole–dipole interactions have force of 1–7 K cal/mole. Dipole– dipole interactions also come from partial charges another order of magnitude weaker (39).

(iii) *Debye type forces:* These are the interactions between permanent and induced dipoles. Permanent dipoles can induce a transient electric dipole in non-polar molecules and produce dipole induced dipole interactions. These interactions involve a force of about 1–3 K cal/mole (37).

The non-retarded van der Waal's force is inversely proportional to the square of the distance between two spherical particles, where the proportionality constant is the Hamaker constant, which has the dimension of energy, can be used to describe the strength of the van der Waal's interaction and is dependent on the properties of the involved particles and on the medium where the interaction takes place (40).

(d) ***Hydrophobic bonds:*** Hydrophobic effect is another particularly important phenomenon with respect to bioadhesion related to the presence of water. It is the property that nonpolar molecules like to self-associate in the presence of aqueous solution. It has been assigned to the tendency of water molecules to form ordered structures in proximity to non-polar molecular domains and may give rise to attractive interactions between non-polar residues such as hydrocarbon side chains. The hydrophobic effect is usually described in the context of protein folding, protein–protein interactions, nucleic acid structure, and protein–small molecule interactions. In the case of protein folding, it is used to explain why many proteins have a hydrophobic core which consists of hydrophobic amino acids, such as alanine, valine, leucine, isoleucine, phenylalanine, and methionine grouped together; often coiled-coil structures form around a central hydrophobic axis. The energetics of DNA tertiary structure assembly were determined by Eric Kool to be mostly caused by the hydrophobic effect, as opposed to Watson–Crick base pairing (41).

The hydrophobic effect can be nullified to a certain extent by lowering the temperature of the solution to near zero degrees; at such temperatures, water prefers to be in an ordered structure and the order generated by hydrophobic patches is no longer as energetically unfavorable. This is neatly demonstrated by the increased solubility of benzene in water at temperatures lower than room temperature. On the macroscopic level, long-range attractive forces have been observed between hydrophobic surfaces formed by adsorption or deposition of amphiphilic molecules and are believed to be non-equilibrium force (42). It should be noted that the origin of the long-range attractive forces between hydrophobic surfaces is controversial, but their occurrence has been related to instability of the deposited monolayer. Strength of these interactions is about 0.37 kcal/mol (37).

(e) ***Covalent bonds:*** Covalent bonds are characterized by the electrons that are shared between the engaged atoms. Covalent bonds operate only over short inter atomic distances (1– 2×10^{-1} nm). They tend to decrease in strength with increasing bond-length, and are oriented at well-defined angles. Unless chemical reactions take place, based on the formation or breakup of for example disulphide bridges, covalent bonds are unlikely to be important in bioadhesion processes under physiological conditions.

On the basis of molecular interactions, the interaction between two molecules is composed of attraction and repulsion. Attractive interactions arise from weak forces such as Van der Waal's forces, electrostatic attraction, hydrogen bonding, hydrophobic interactions and/or strong forces, which are covalent in nature. Repulsive interactions occur because of electrostatic and steric repulsion. For muco/bioadhesion to occur, the attractive interactions should be larger than nonspecific repulsion (33).

Following steps are involved in the process of muco/bioadhesion:

(i) Spreading, wetting, swelling and dissolution of bio/mucoadhesive polymer at the interface, initiates intimate molecular contact at the interface between the polymer and the epithelial/mucus layer.

(ii) Interdiffusion and interpenetration between the chains of the adhesive polymer and the mucus/epithelial surface resulting physical cross links or mechanical interlocking.

(iii) *Adsorption:* The orientation of the polymers at the interface so that adhesive bonding across the interface is possible.

(iv) Formation of secondary chemical bonds between the polymer chains and mucin molecules.

3.2.2.2 Theories of bioadhesion

There are six general theories of adhesion, which have been adapted for the investigation of mucoadhesion (43).

(a) ***Electronic theory:*** The electronic theory suggests that electron transfer occurs upon contact of adhering surfaces due to differences in their electronic structure. This is proposed to result in the formation of an electrical double layer at the interface, with subsequent adhesion due to attractive forces.

(b) ***Wetting theory:*** The wetting theory is primarily applied to liquid systems and considers surface and interfacial energies. It involves the ability of a liquid to spread spontaneously onto a surface as a prerequisite for the development of adhesion. The affinity of a liquid for a surface can be found using techniques such as contact angle goniometry to measure the contact angle of the liquid on the surface, with the general rule being that the lower the contact angle, the greater the affinity of the liquid to the solid. The spreading coefficient (S_{AB}) can be calculated from the surface energies of the solid and liquids using the equation 3.3.

$$S_{AB} = \gamma_B - \gamma_A - \gamma_{AB} \qquad \text{.....(3.3)}$$

where γ_A is the surface tension (energy) of the liquid A, γ_B is the surface energy of the solid B and γ_{AB} is the interfacial energy between the solid and liquid. S_{AB} should be positive for the liquid to spread spontaneously over the solid.

The work of adhesion (WA) represents the energy required to separate the two phases, and is given by equation 3.4.

$$WA = \gamma_A + \gamma_B - \gamma_{AB} \qquad \text{.....(3.4)}$$

The greater the individual surface energies of the solid and liquid relative to the interfacial energy, the greater the work of adhesion.

(c) ***Adsorption theory:*** According to the adsorption theory, after an initial contact between two surfaces, the material adheres because of surface forces acting between the atoms in the two surfaces. Two types of chemical bonds resulting from these forces can be distinguished:

(i) Primary chemical bonds of covalent nature, which are undesirable in bioadhesion because their high strength may result in permanent bonds.

(ii) Secondary chemical bonds having many different forces of attraction, including electrostatic forces, van der Waals forces, hydrogen bonds and hydrophobic bonds.

(d) ***Diffusion theory:*** The diffusion theory describes inter diffusion of polymers chains across an adhesive interface. This process is driven by concentration gradients and is affected by the available molecular chain lengths and their mobilities. The depth of interpenetration depends on the diffusion coefficient and the time of contact. Sufficient depth of penetration creates a semi-permanent adhesive bond. The following equation can be used to estimate the depth of penetration equation 3.5.

$$\mathbf{L = (tD_b)^{0.5}} \quad .....(3.5)$$

where t is the contact time and D_b is the diffusion coefficient of the biomaterial in the mucus. The depth of the penetration (L) depends upon the diffusion coefficient, time of contact and other experimental variables. Good mucoadhesion is expected when the depth of chain penetration is equal to the end-to-end distance of the mucoadhesive polymer. Swelling and chain diffusion characteristics of polymers can be manipulated by adhesion promoters. Chemical grafting of long chain polyethylene glycols to polyacrylic acid hydrogels was shown to improve the bioadhesion properties as a result of higher diffusion (131).

(e) ***Mechanical theory:*** The mechanical theory assumes that adhesion arises from an interlocking of a liquid adhesive (on setting) into irregularities on a rough surface. However, rough surfaces also provide an increased surface area available for interaction along with an enhanced viscoelastic and plastic dissipation of energy during joint failure, which are thought to be more important in the adhesion process than a mechanical effect (44).

(f) ***Fracture theory:*** This theory differs a little from the other five in that it relates the adhesive strength to the forces required for the detachment of the two involved surfaces after adhesion. This assumes that the failure of the adhesive bond occurs at the interface. However, failure normally occurs at the weakest component, which is typically a cohesive failure within one of the adhering surfaces. The fracture theory is related to the force required for the separation of the two layers after adhesion. The fracture strength **σ,** which is the adhesive strength, is given by equation 3.6.

$$\boldsymbol{\sigma = (E\varepsilon/l)^{0.5}} \quad .....(3.6)$$

where ε is the fracture energy, E is the Young's modulus and l is the critical crack length.

3.3 Factors Affecting Muco/Bioadhesion in the Oral Cavity

Muco/bioadhesive characteristics are a factor of both the bioadhesive polymer and the medium in which the polymer will reside. Different factors that affecting the mucoadhesive properties of polymer-related, such as molecular weight, flexibility, hydrogen bonding capacity, cross-linking density, charge, concentration, and hydration of a polymer, and environmental factors such as pH, initial contact time, mucin turnover, and disease state are discussed

3.3.1 Polymer-Related Factors

(a) ***Molecular weight:*** Numerous studies have demonstrated that the optimum molecular weight for maximum bioadhesion, depends on the type of bioadhesive polymer used. It is generally understood that the threshold required for successful bioadhesion is at least 100,000 molecular weight. For example, polyethylene glycol (PEG), with a molecular weight of 20,000, has little adhesive character, whereas PEG with 200,000 molecular weight has improved, and a PEG with 400,000 has superior adhesive properties. The fact that bioadhesiveness improves with increasing molecular weight for linear polymers implies two things: (1) interpenetration is more critical for lower molecular weight polymers to be a good bioadhesive, and (2) entanglement is important for higher molecular weight polymers. Adhesiveness of a nonlinear structure, by comparison, follows a quite different trend. The adhesive strength of dextran, with a very high molecular weight of 19,500,000, is similar to that of PEG, with a molecular weight of 200,000. The reason for this similarity may be that the helical conformation of dextran may shield many of the adhesive groups, which are primarily responsible for adhesion, unlike the conformation of PEG (45).

(b) ***Flexibility:*** Bioadhesion starts with the diffusion of the polymer chains in the interfacial region. Therefore, it is important that the polymer chains contain a substantial degree of flexibility in order to achieve the desired entanglement with the mucus. In general, mobility and flexibility of polymers can be related to their viscosities and diffusion coefficients, where higher flexibility of a polymer causes greater diffusion into the mucus network (46). A recent publication demonstrated the use of tethered poly(ethylene glycol)–poly(acrylic acid) hydrogels and their copolymers with improved mucoadhesive properties (47). The increased chain interpenetration was attributed to the increased structural flexibility of the polymer upon incorporation of poly(ethylene glycol).

(c) ***Hydrogen bonding capacity:*** Hydrogen bonding is another important factor in mucoadhesion of a polymer. Park and Robinson found that in order for mucoadhesion to occur, desired polymers must have functional groups that are able to form hydrogen bonds (48). They have also confirmed that flexibility of the polymer is important to improve this hydrogen bonding potential. Polymers such as poly(vinyl alcohol), hydroxylated methacrylate, and poly(methacrylic acid), as well as all their copolymers, are polymers with good hydrogen bonding capacity (49).

(d) ***Cross-linking density:*** The average pore size, the number average molecular weight of the cross-linked polymers, and the density of crosslinking are three important and interrelated structural parameters of a polymer network. Therefore, it seems reasonable that with increasing density of cross-linking, diffusion of water into the polymer network occurs at a lower rate which, in turn, causes an insufficient swelling of the polymer and a decreased rate of interpenetration between polymer and mucin (46). Flory (50) has reported this general property of polymers, in which the degree of swelling at equilibrium has an inverse relationship with the degree of cross-linking of a polymer.

(e) ***Charge:*** Some generalizations about the charge of bioadhesive polymers have been made previously, where nonionic polymers appear to undergo a smaller degree of adhesion compared to anionic polymers. Peppas and Buri have demonstrated that strong anionic charge on the polymer is one of the required characteristics for mucoadhesion (49). It has been shown that some cationic polymers are likely to demonstrate superior mucoadhesive properties, especially in a neutral or slightly alkaline medium. Additionally, some cationic high-molecular-weight polymers, such as chitosan, have shown to possess good adhesive properties (51).

(f) ***Concentration:*** When the concentration of the polymer is too low, the number of penetrating polymer chains per unit volume of the mucus is small, and the interaction between polymer and mucus is unstable (49). In general, the more concentrated polymer would result in a longer penetrating chain length and better adhesion.

However, for each polymer, there is a critical concentration, above which the polymer produces an 'unperturbed' state due to a significantly coiled structure. As a result, the accessibility of the solvent to the polymer decreases, and chain penetration of the polymer is drastically reduced. Therefore, higher concentrations of polymers do not necessarily improve and, in some cases, actually diminish mucoadhesive properties. One of the studies addressing this factor demonstrated that high concentrations of flexible polymeric films based on polyvinylpyrrolidone or poly(vinyl alcohol) as film-forming polymers did not further enhance the mucoadhesive properties of the polymer. On the contrary, it decreased the desired strength of mucoadhesion (52). Thus the importance of this factor lies in the development of a strong adhesive bond with the mucus, and can be explained by the polymer chain length available for penetration into the mucus layer.

(g) ***Hydration (swelling):*** Swelling characteristics are related to the bioadhesive itself and its environment. Swelling depends on the polymer concentration, ionic strength, as well as the presence of water. During the dynamic process of bioadhesion, maximum bioadhesion *in vitro* occurs with an optimum water content. Overhydration results in the formation of a wet slippery mucilage without adhesion. However, a critical degree of hydration of the mucoadhesive polymer exists where optimum swelling and bioadhesion occurs (49).

3.3.2 Environmental Factors

(a) ***pH:*** pH can influence the formal charge on the surface of mucus as well as certain ionizable bioadhesive polymers. Mucus will have a different charge density depending on pH due to differences in dissociation of functional groups on the carbohydrate moiety and the amino acids of the polypeptide backbone. Some studies (53) have shown that the pH of the medium is important for the degree of hydration of cross-linked polyacrylic acid, showing consistently increased hydration from pH 4 through pH 7, and then a decrease as alkalinity and ionic strength increases. For example, polycarbophil does not show a strong bioadhesive property above pH 5 because uncharged, rather than ionized, carboxyl groups react with mucin molecules, presumably through numerous hydrogen bonds. However, at higher pH, the chains are fully extended due to electrostatic repulsion of the carboxylate anions. The polymer chains are also repelled by the negatively charged mucin molecules. It has been also observed the, due to hydrogen bonding between hydroxypropyl cellulose and carbopol 934, interpolymer complexes form pH value below 4.5.

(b) ***Initial Contact Time:*** Contact time between the bioadhesive and mucus layer determines the extent of swelling and interpenetration of the bioadhesive polymer chains. Along with the initial pressure, the initial contact time dramatically effect the performance of a system. Moreover, bioadhesive strength increases as the initial contact time increases (54).

(c) ***Physiological Variables (Mucin Turnover, and Disease States):*** In many routes of administration, surface mucus is encountered by the bioadhesive before it reaches the tissue. The extent of interaction between the polymer and the mucus depends on mucus viscosity, degree of entanglement, and water content. How long the bioadhesive remains at the site depends on whether the polymer is soluble or insoluble in water and the associated turnover rate of mucin. Estimates of mucin turnover vary widely, depending on location and method of measurement. Values ranging from a few hours to a day have been reported. The residence time of dosage forms is limited by the mucin turnover time, which has been calculated to range between 47 and 270 min in rats (55) and 12–24 h in humans (56). Movement of the buccal tissues while eating, drinking, and talking, is another concern which should be considered when designing a dosage form for the oral cavity. Movements within the oral cavity continue even during sleep, and can potentially lead to the detachment of the dosage form. Therefore, a tentative goal of a bioadhesive, whether it sticks to mucin or to an epithelial surface is to remain in place long enough for once daily dosing. It is not known what influence certain disease states have on bioadhesive retention, and this is an area requiring additional investigations.

3.4 Mucoadhesive polymers used in the buccal delivery

3.4.1 Desired characteristics

(i) Polymer and its degradation products should be non-toxic, non-irritant and free from leachable impurities.

(ii) Polymers used in buccal delivery should have good spreadability, wetting, swelling and solubility and biodegradability properties.

(iii) pH should be biocompatible and should possess good viscoelastic properties.

(iv) Should adhere quickly to buccal mucosa and should possess sufficient mechanical strength.

(v) Should possess peel, tensile and shear strengths at the bioadhesive range.

(vi) Polymer must be easily available and its cost should not be high.

(vii) Should show bioadhesive properties in both dry and liquid state.

(viii) Should demonstrate local enzyme inhibition and penetration enhancement properties.

(ix) Should demonstrate acceptable shelf life.

(x) Should have optimum molecular weight.

(xi) Should possess adhesively active groups.

(xii) Should have required spatial conformation.

(xiii) Should be sufficiently cross-linked but not to the degree of suppression of bond forming groups.

(xiv) Should not aid in development of secondary infections such as dental caries.

3.4.2 Classification

In general, adhesive polymers can be classified as obtained from natural source and synthetic source, water-soluble and water insoluble, charged and uncharged polymers. Examples of the recent polymers classified based on the charge are listed in Table 3.4. Natural bioadhesive macromolecules share similar structural properties with the synthetic polymers. They are generally linear polymers with high molecular weight, contain a substantial number of hydrophilic, negatively charged functional groups, and form three-dimensional expanded networks (46). In the class of synthetic polymers, poly(acrylic acid), cellulose ester derivatives, and polymethacrylate derivatives are the current choices.

Chitosan and examples of various gums, such as guar and hakea (from Hakea gibbosa), are classified as semi-natural/natural bioadhesive polymers. Poly(acrylic acid), a linear or random polymer, and polycarbophil, a swellable polymer, represent water-soluble and water-insoluble polymers, respectively. The charged polymers are divided into cationic and anionic polymers, such as chitosan and polycarbophil, respectively, while hydroxypropylcellulose is an example of uncharged bioadhesive polymers (57).

Table 3.4 Classification of bioadhesive polymers.

Anionic	Cationic	Neutral
Carboxymethylcellulose (CMC) Carbopol (CP) Polyacrylic acid Pectin Polycarbophil (PC) Sodium alginate Sodium CMC Carageenan Chitosan-EDTA Xanthan gum	Chitosan Aminodextran Dimethylaminoethyl-dextran Trimethylated chitosan Polylysene	Hydroxyethyl starch Hydroxyl propyl cellulose(HPC) Poly(ethylene oxide) Poly vinyl alcohol(PVA) Poly vinyl pyrrolidone(PVP) Polyethylene glycol Dextran

3.5 Novel Second-generation Mucoadhesives

The major disadvantage in using traditional non-specific mucoadhesive systems (first generation) is that adhesion may occur at sites other than those intended. A scenario that is particularly true for platforms designed to adhere to a distal target such as those hypothesised in targeted mucoadhesion within the GI tract. Unlike first-generation non-specific platforms, certain second-generation polymer platforms are less susceptible to mucus turnover rates, with some species binding directly to mucosal surfaces; more accurately termed ''cytoadhesives''. Furthermore as surface carbohydrate and protein composition at potential target sites vary regionally, more accurate drug delivery may be achievable.

3.5.1 Lectins

Lectins are naturally occurring proteins that play a fundamental role in biological recognition phenomena involving cells and proteins. For example, some bacteria use lectins to attach themselves to the cells of the host organism during infection. Enhancement of mucosal delivery may be obtained through the use of appropriate cytoadhesives that can bind to mucosal surfaces. The most widely investigated of such systems in this respect are lectins. Lectins belong to a group of structurally diverse proteins and glycoproteins that can bind reversibly to specific carbohydrate residues (58). After initial mucosal cell-binding, lectins can either remain on the cell surface or in the case of receptor-mediated adhesion possibly become internalised via a process of endocytosis. Such systems could offer duality of function in that lectin based platforms could not only allow targeted specific attachment but additionally offer a method of controlled drug delivery of macromolecular pharmaceuticals via active cell-mediated drug uptake (59). Whilst lectins offer significant advantages in comparison to first-generation platforms, it is worth noting that such polymers suffer at least in part from premature inactivation by shed off mucus. This phenomenon has been reported to be advantageous, given that the mucus layer provides an initial yet fully reversible binding site followed by distribution of lectin-mediated drug delivery systems to the cell layer (60). Although lectins offer significant advantages in relation to site targeting, many are toxic or immunogenic, and the effects of repeated lectin exposure are largely unknown. It is also feasible that lectin-induced antibodies could block subsequent adhesive interactions between mucosal epithelial cell surfaces and lectin delivery vehicles. Moreover, such antibodies may also render individuals susceptible to systemic anaphylaxis on subsequent exposure (59).

3.5.2 Bacterial Adhesions

Pathogenic bacteria readily adhere to mucosal membranes in the gastrointestinal tract, a phenomenon that has been exploited as a means by which target-specific drug delivery may be achieved. K99-fimbriae, an attachment protein derived from E. coli, has been covalently attached to polyacrylic acid networks (61). The formulated polymer–fimbriae platform exhibited a significant increase in adhesion in vitro in comparison to the control (unmodified polymer).

3.5.3 Thiolated Polymers

Thiolated polymers (thiomers) are a type of second-generation mucoadhesive derived from hydrophilic polymers such as polyacrylates, chitosan or deacetylated gellan gum (62). The presence of thiol groups allows the formation of covalent bonds with cysteine- rich sub domains of the mucus gel layer, leading to increased residence time and improved bioavailability (63). In this respect thiomers mimic the natural mechanism of secreted mucus glycoproteins that are also covalently anchored in the mucus layer by the formation of disulphide bonds (64). Whilst first-generation mucoadhesive platforms are facilitated via non-covalent secondary interactions, the covalent bonding mechanisms involved in second- generation systems lead to interactions that are less susceptible to changes in ionic strength and/or the pH (65). Moreover the presence of disulphide bonds may significantly alter the mechanism of drug release from the delivery system due to increased rigidity and cross-linking. In such platforms a diffusion-controlled drug release mechanism is more typical, whereas in first-generation polymers anomalous transport of API into bulk solution is more common.

3.6 Dosage Forms for Buccal Drug Delivery

Dosage forms such as mouthwashes, erodible/ chewable buccal tablets, and chewing gums allow only a short period of release, and reproducibility of drug absorption is poor. Application of bioadhesive semisolid gels creates considerable technical problems. Bioadhesive buccal films/patches and tablets are the suitable buccal dosage forms. These bioadhesive buccal films/patches and tablets were usually fabricated in different geometry, as shown in Fig. 3.4.

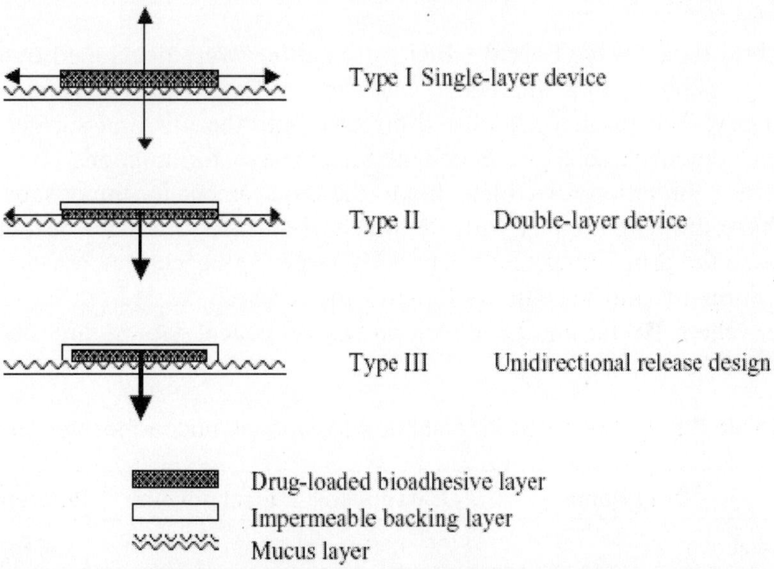

Figure 3.4 Schematic representation of buccal dosage form design.

There are three types of devices can be fabricated for buccal delivery of patches and tablets. Type I is a single-layer device, from which drug can be released multidirectionally. Type II device has a impermeable backing layer on top of the drug-loaded bioadhesive layer, and drug loss into oral cavity can be greatly decreased. Type III is a unidirectional release device, from which drug loss will be avoided and drug can penetrate only via the buccal mucosa (66). The device should be fabricated so that the swelling rate of bioadhesive polymer is optimized to ensure a prolonged period of bioadhesion as well as a controlled or sustained drug release.

Bioadhesive buccal films/patches are commonly manufactured by solvent casting methods using adhesive coating machines, which involve dissolving a drug in a casting solution, casting film, and drying and laminating with a backing layer or a release liner. The processing technology is quite similar to pressure sensitive adhesive-based patch manufacturing. Very recently, a hot-melt extrusion method was reported to fabricate hot-melt extruded films for buccal delivery, which overcomes the disadvantages associated with a solvent casting method such as environmental concerns, long processing times, and high costs. (67).

3.7 Tablets

Several bioadhesive buccal tablet formulations were developed by direct compression method in recent years either for local or systemic drug delivery. Tablets that are placed directly onto the mucosal surface have been demonstrated to be excellent bioadhesive formulations. However, size is a limitation for tablets due to the requirement for the dosage form to have intimate contact with the mucosal surface. They are designed to release the drug either unidirectionally targeting buccal mucosa or multi-directionally into the saliva. Some of the research carried out by various researchers so far in the development of buccal tablets are listed in Table 3.5.

Table 3.5 List of drugs investigated for buccal mucoadhesive tablets.

Drug Name	Bioadhesive buccal polymer	Reference
Acitretin	CP 934P and HPMC	76
Buspirone HCL	CP974, HPMCK4M	77
Calcitonin	Hakea gum	78
Chlorpheneramine maleate	Hakea gum, CP 934, HPMC	78, 125
Clotrimazole	CP974P, HPMC K4M	79
Carvedilol	CP 934 with HPC, HPMC	126
Carbamazepine	HPMC and CP	80
Danazol	PC or HPMC	81
Diltiazem HCl	CP 934, HPMCK4M	127

Contd...

Drug Name	Bioadhesive buccal polymer	Reference
Ergotamine tartrate	Carboxyvinyl polymer and HPC	82
Hydralazine HCL	CP 934P and CMC	83
Hydrocortisone acetate	HPMC, CP 974P, or PC	84
Insulin	CP 934 with HPC or HPMC	85
Ketoprofen	Chitosan and sodium alginate	86
metaclopromide	CP, HPMC, PC, SodiumCMC	87
Metronidazole	HEC, HPC, HPMC, or NaCMC combined with CP940,	88
Miconazole nitrate	Mixtures of HPMC, sodium CMC, CP 934P, and sodium Alginate	89
Nalbuphine	CP 934 and HPC	90
Nifedipine	CMC and CP	91
Nystatin	Carbomer, HPMC	92
Omeprazole	Sodium alginate, HPMC	93
Pindolol	CP 934 and sodium CMC (bioadhesive polymers); HPMC and HPC (matrix-forming polymer)	94
Piroxicam	HPMC and CP 940	95
Propranolol	PAA, HPMC, and HPC	96
Propranolol HCl	HPMC and PC	96
Sodium fluoride	Eudragit® and/or EC	97
Testosterone	Starch-g-PAA copolymers or starch/PAA mixtures	98
Theophylline	Starch–acrylic acid graft Copolymers	99

3.8 Buccal Patches/films

Flexible films/patches were prepared either by solvent casting technique or by hot melt extrusion techniques used to deliver drugs directly to a mucosal membrane. They also offer advantages over creams and ointments in that they provide a measured dose of drug to the site.

3.8.1 Solvent Casting Technique

Weighed quantity of suitable polymer was taken and to this required volume of solvent system (approximately 20-25 mL) was added and vortexed. Sufficient care was taken to prevent the formation of lumps and the boiling tube was set-aside for 6 hours to allow the polymer to swell. After swelling, to this mixture, measured quantity of plasticizer (propylene glycol or glycerin or dibutyl phthalate) was added and vortexed. Finally weighed quantity of drug was dissolved in 5 ml of solvent system and added to the polymer solution and mixed well. It was set aside for some time to remove any entrapped air and transferred into a previously cleaned anumbra petri plate. Drying of these patches was carried out in an oven placed over a horizontal surface, with temperature being maintained at 40^0C. The patches formed were removed carefully and stored in a desiccator till the evaluation tests were performed.

Table 3.6 List of some drug substances processed by solvent casting technique

Drug Name	Polymers Used	Reference
Acyclovir	Copolymers of acrylic acid and PEG monomethylether monomethacrylate	100
Buprenorphine	CP-934, PIB and PIP	101
Carvedilol	HPMC E15, HPC	129
Chlorpheniramine Maleate	HEC	128
Chlorhexidine	Chitosan	102
Felodipine	HPMC E15, Eudragit RL100	142
Isosorbide dinitrate	HPC, HPMC	103
Lidocaine	HPC, CP	104
Miconazole nitrate	SCMC, Chitosan, PVA, HEC and HPMC	105
Nifedipine	Sodium alginate	106
Protirelin (TRH)	HEC, HPC, PVP, or PVA	107
Oxytocin	CP 974P	108
Terbutaline sulfate	CP 934, CP 971, HPMC, HEC, or SCMC	109
Triamcinolone acetonide	CP, poloxamer, and HPMC	110

3.8.2 Hot Melt Extrusion Technique

Hot-melt extrusion (HME) is one of the most widely used processing techniques within the plastics industry. Several research groups have demonstrated HME processes as a viable method to prepare pharmaceutical drug delivery systems, including granules, pellets (68, 69), sustained release tablets (70), transdermal and transmucosal drug delivery systems (71) and implants (72). The HME technique is an attractive alternative to traditional processing methods and it offers many advantages over the other pharmaceutical processing techniques.

(i) Molten polymers during the extrusion process can function as thermal binders and act as drug depots and/or drug release retardants upon cooling and solidification.

(ii) Solvents and water are not necessary thereby reducing the number of processing steps and eliminating time-consuming drying steps.

(iii) A matrix can be massed into a larger unit independent of compression properties. The intense mixing and agitation imposed by the rotating screw cause de-aggregation of suspended particles in the molten polymer resulting in a more uniform dispersion and the process is continuous and efficient.

(iv) Bioavailability of the drug substance may be improved when it is solubilized or dispersed at the molecular level in HME dosage forms.

3.8.2.1 Hot-Melt extrusion equipment

Pharmaceutical Hot-Melt Extrusion processes can be categorized as either ram extrusion or screw extrusion.

Ram extrusion operates with a positive displacement ram capable of generating high pressures to push materials through the die. During ram extrusion, materials are introduced into a heated cylinder. After an induction period to soften the materials, a ram (or a piston) pressurizes the soft materials through the die and transforms them into the desired shape (73). High-pressure is the operating principle of ram extrusion. This technique is well suited for the precision extrusion of highly valuable materials. The ram exerts modest and repeatable pressure as well as a very consistent extrudate diameter. The major drawback of ram extrusion is limited melting capacity that causes poor temperature uniformity in the extrudate and having lower homogeneity, in comparison with extrudates processed by screw extrusion.

Screw extrusion consists of a rotating screw inside a heated barrel and it provides more shear stress and intense mixing. Screw extruder consists of three distinct parts: a conveying system for material transport and mixing, a die system for forming, and downstream auxiliary equipment for cooling, cutting or collecting the finished goods. Individual components within the extruder are the feed hopper, a temperature controlled barrel, a rotating screw, die and heating and cooling systems. Additional systems include mass flow feeders to accurately meter materials into the feed hopper, process analytical technology to measure extrudate properties (near infra red systems and laser systems), liquid and solid side stuffers, vacuum pumps to devolitize extrudates, pelletizers, and calendaring equipment. Standard process control and monitoring devices include zone temperature and screw speed with optional monitoring of torque, drive amperage, and pressure and melt viscosity. Temperatures are normally controlled by electrical heating bands and monitored by thermocouples.

Screw Extruders are of two types (a) Single Screw Extruder and (b) Twin-Screw Extruders

(a) ***Single Screw Extruder:*** The single screw extruder is the most widely used extrusion system in the world. One screw rotates inside the barrel and is used for feeding, melting, devolatilizing, and pumping. Mixing is also accomplished for less demanding applications. Single screw extruders can be either flood or starve fed, depending upon the intended manufacturing process (68, 74). Single screw extruders are continuous, high-pressure pumps for viscous materials that can generate thousands of pounds of pressure while melting and mixing. Most extruder screws are driven from the hopper end. However, once screws are reduced to less than 18 mm, the screw becomes weak and solids transportation is far less reliable. To overcome these shortcomings, a vertical screw, driven from the discharge end, may be used. The discharge of such screws is two to four times stronger increasing solids transport (74).

There are three basic functions of a single screw extruder: solids conveying, melting and pumping. The forwarding of the solid particles in the early portion of the screw is a result of friction between the material and the feed section's bore. After solids conveying the flight depth begins to taper down and the heated barrel causes a melt to form. The energy from the heaters and shearing contribute to melting. Ideally, the melt pool will increase as the solid bed reduces in size until all is molten at the end of the compression zone. Finally, the molten materials are pumped against the die resistance to form the extrudate (74).

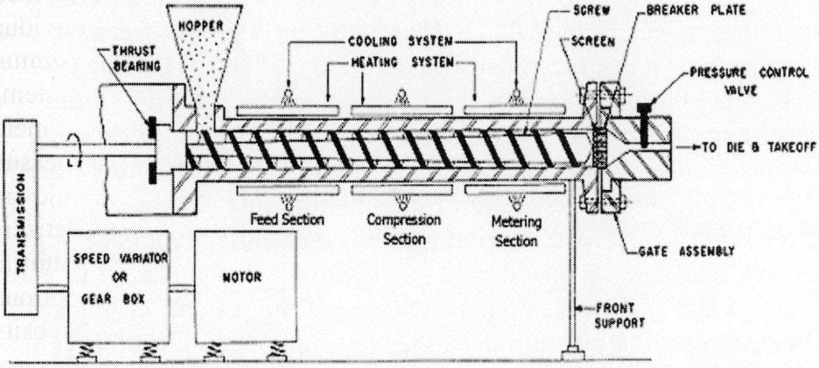

Figure 3.5 Schematic representation of a Single-Screw extruder. (Adopted from Repka et al., 2007 (68)).

(b) ***Twin-Screw Extruders:*** Twin-screw extruders have several advantages over single screw extruders, such as easier material feeding, high kneading, and dispersing capacities, less tendency to over-heat and shorter transit time. The first twin-screw extruders were developed in the late 1930's in Italy, with the concept of combining the machine actions of several available devices into a single unit. As the name implies, twin-screw extruders utilize two screws usually arranged side by side (Figure 3.6). The use of two screws allows a number of different configurations to be obtained and imposes different conditions on all zones of the extruder, from the transfer of material from the hopper to the screw, all the way to the metered pumping zone (75).

Figure 3.6 Twin screw extruder (Adopted from Repka et al., 2007)

In a twin-screw extruder, the screws can either rotate in the same (co-rotating extruder) or the opposite (counter-rotating extruder) direction. The counter-rotating designs are utilized when very high shear regions are needed as they subject materials to very high shear forces as the material is squeezed through the gap between the two screws as they come together. Also, the extruder layout is good for dispersing particles in a blend. Generally, counter-rotating twin-screw extruders suffer from disadvantages of potential air entrapment, high-pressure generation, and low maximum screw speeds and output. Co-rotating twin-screw extruders on the other hand are generally of the intermeshing design, and are thus self-wiping (68). They are industrially the most important type of extruders and can be operated at high screw speeds and achieve high outputs, while maintaining good mixing and conveying characteristics. Unlike counter-rotating extruders, they generally experience lower screw and barrel wear as they do not experience the outward "pushing" effect due to screw rotation. These two primary types can be further classified as non-intermeshing and fully intermeshing. The fully intermeshing type of screw design is the most popular type used for twin-screw extruders (Figure 3.7).

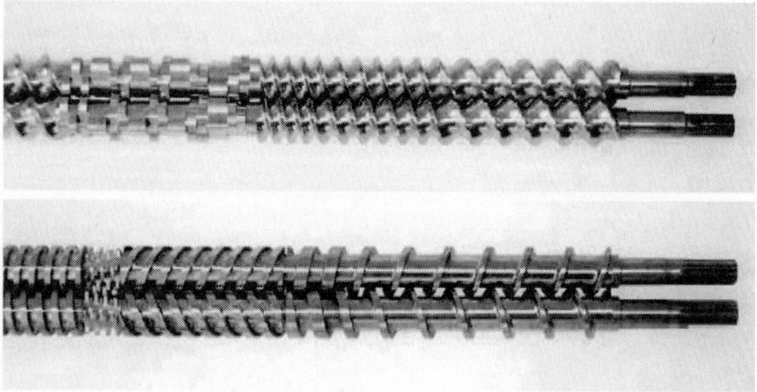

Figure 3.7 Twin screw design examples: intermeshing co-rotating twin-screw (top), and intermeshing counter-rotating twin-screw (bottom). (Adopted from Repka et al., 2007).

This design itself is self-wiping, where it minimizes the non-motion and prevents localized overheating of materials within the extruder. The extruder operates by a first in/first out principle since the material does not rotate along with the screw. Non-intermeshing extruders, on the other hand, are often used for processing when large amounts of volatiles need to be removed and when processing highly viscous materials. Non-intermeshing extruders allow large volume de-volatization via a vent opening since the screws are positioned apart from one another. Non-intermeshing extruders are not susceptible to high torques generated while processing highly viscous materials for the same reasons (75).

Table 3.7 List of drug substances processed by hot melt extrusion techniques.

Name of the drug	Melting temperature	Reference
Acetaminophen	170	111
Acetylsalicylic Acid	135	112
Carbamazepine	192	113
Chlorpheniramine Maleate	135	114
Diclofenac Sodium	284	115
Ethinyl estradiol	144	116
Hydrochlorothiazide	274	117
Hydrocortisone	220	118
Itraconazole	166	119
Ketoconazole	150	120
Ketoprofen	94	121
Lacidipine	185	122
Nifedipine	175	123
Piroxicam	205	124
Tolbutamide	129	124

Table 3.8 Comparison between Solvent Casting Technique and Hot Melt extrusion Technique

Solvent casting technique	Hot melt extrusion technique
Solvents and water are necessary to formulation of films or patches	Solvents and water are not necessary thereby reducing the number of processing steps
Time consuming drying steps are involved	Eliminating time-consuming drying steps.
Suitable solvents are needed to solubilize the drug and polymers to casting the films	Molten polymers during the extrusion process can function as thermal binders and act as drug depots and/or drug release retardants upon cooling and solidification.

Contd...

Solvent casting technique	Hot melt extrusion technique
Solvent casting method involves lots of time consuming process	Shorter and more efficient processing times
Labor intensive	Less labor and equipment demands
Low cost	Favorable cost
Not for continuous	Potential for continuous process

3.9 Evaluation

In addition to the routine evaluation tests such as weight variation, thickness variation, friability, hardness, content uniformity, in vitro dissolution for tablets; tensile strength, film endurance, hygroscopicity etc for films and patches; viscosity, effect of aging etc for gels and ointments; buccal adhesive drug delivery devices are also to be evaluated specifically for their mucoadhesive strength and permeability.

3.9.1 Moisture Absorption Studies for Buccal Patches

The moisture absorption studies give an indication about the relative moisture absorption capacities of polymers and an idea whether the formulations maintain their integrity after absorption of moisture. Moisture absorption studies were performed in 5 % w/v agar in distilled water, which in hot condition was transferred to Petri plates and allowed to solidify. Then six patches from each formulation were selected and weighed. They were placed in desiccator overnight prior to the study to remove moisture if any and laminated on one side with water impermeable backing membrane. They were placed on the surface of the agar plate and incubated at 37° C for 2 hr in incubator. The patches were weighed again and the percentage of moisture absorbed was calculated using the equation 3.7.

$$\%\ \text{Moisture absorbed} = \frac{\text{Final weight} - \text{initial weight}}{\text{Intialweight}} \times 100 \quad \text{.....(3.7)}$$

3.9.2 Swelling Studies for Buccal Tablets

Water uptake of the tablets was determined gravimetrically in phosphate buffer, pH 7.4. The tablets were attached to pre-weighed glass supports using a cyanoacrylate adhesive sealant. The supports with tablets were immersed into the phosphate buffer at 37 °C. At predetermined time intervals, the devices were removed from the media, blotted with tissue paper to remove excess water, and weighed. Water uptake or Swelling index (S.I) was calculated according to the equation 3.8.

$$S.I. = W_s - W_d / W_d \qquad .....(3.8)$$

where W_d and W_s are the weights of dry and swollen devices, respectively. The swelling of the formulations were dependant on both, the type and concentration of the polymer included in the formulations.

3.9.3 Surface pH Study

The bioadhesive buccal tablets were made in contact with 1 mL of distilled water and allowed to swell for 1-2 h at room temperature. The surface pH of the tablets or patches was measured by bringing the pH meter electrode in contact with the surface of the patch or tablet and allowing it to equilibrate for 1 minute.

3.9.4 Measurement of Mechanical Properties

Mechanical properties of the films (patches) were evaluated using a microprocessor based advanced force gauze equipped with a motorized test stand (Ultra Test, Mecmesin, West Sussex, UK), equipped with a 25 kg load cell (Figure 3.8).

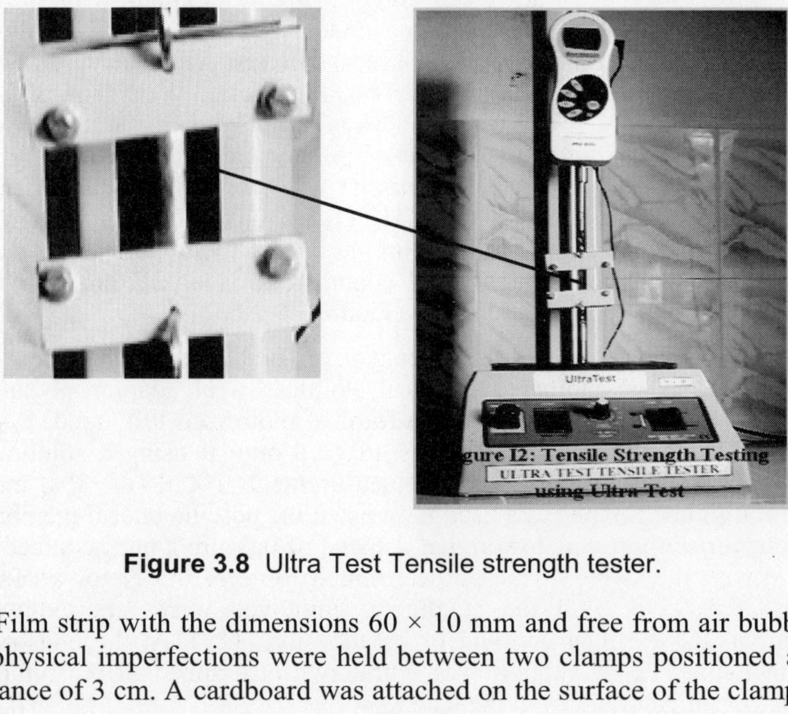

Figure 3.8 Ultra Test Tensile strength tester.

Film strip with the dimensions 60 × 10 mm and free from air bubbles or physical imperfections were held between two clamps positioned at a distance of 3 cm. A cardboard was attached on the surface of the clamp to prevent the film from being cut by the grooves of the clamp. During measurement, the strips were pulled by the top clamp at a rate of 2.0 mm/s to a distance till the film broke. The force and elongation were measured when the films were broken. Results from film samples, which were broken at end and not between the clamps were not included in observations. Measurements were run in six replicates for each formulation. The following equations 3.9, 3.10 were used to calculate the mechanical properties of the films.

$$\text{T.S. } (\text{kg.mm}^{-2}) = \frac{\text{Force at break (kg)}}{\text{Initial cross sec tional area of the sample (mm}^2)} \quad \text{.....(3.9)}$$

$$\text{E/B } (\%.\text{mm}^{-2}) = \frac{\text{Inital in length (mm)}}{\text{Original length}} \times \frac{100}{\text{Cross sections area (mm}^2)} \quad \text{.....(3.10)}$$

3.9.5 *In vitro* Bioadhesion Measurement

In vitro Bioadhesion Measurement method first reported by Wong *et al.*, 1999 to evaluate the adhesive properties of patches using a microprocessor based advanced force gauze equipment with porcine buccal membrane as a model tissue under simulated buccal conditions was followed. A microprocessor based advanced force gauze equipped with a motorized test stand (Ultra Test, Mecmesin, West Sussex, UK), equipped with a 5 kg load cell was employed to determine the bioadhesion using inverted surface of porcine buccal membrane as the model tissue. The porcine buccal tissue was stored in simulated saliva solution (2.38 g Na_2HPO_4, 0.19 g KH_2PO_4 and 8.00 g NaCl in 1000 ml of distilled water at pH 6.75). The porcine buccal membrane was secured tightly to a circular stainless steel adaptor (diameter 2.2 cm) provided with the equipment as an accessory (Figure 3.9).

This was fixed to advanced force gauze. The buccal patch to be tested was placed over another cylindrical stainless steel adaptor of similar diameter and mounted onto the platform of motorized test stand. Buccal patch with backing membrane was adhered onto it using a solution of cyanoacrylate adhesive. During measurement 100µl of 1% mucin solution (crude mucin) was used to moisten the porcine buccal membrane. The upper support was lowered at a speed of 0.5 mm/s until contact was made with the tissue at the predetermined force of 0.5 N for a contact time of 180 sec. All the conditions employed were reported to be optimum for the measurement of bioadhesion (129). At the end of the contact time, upper support was withdrawn at a speed of 0.5 mm/s to detach the membrane from the patch.

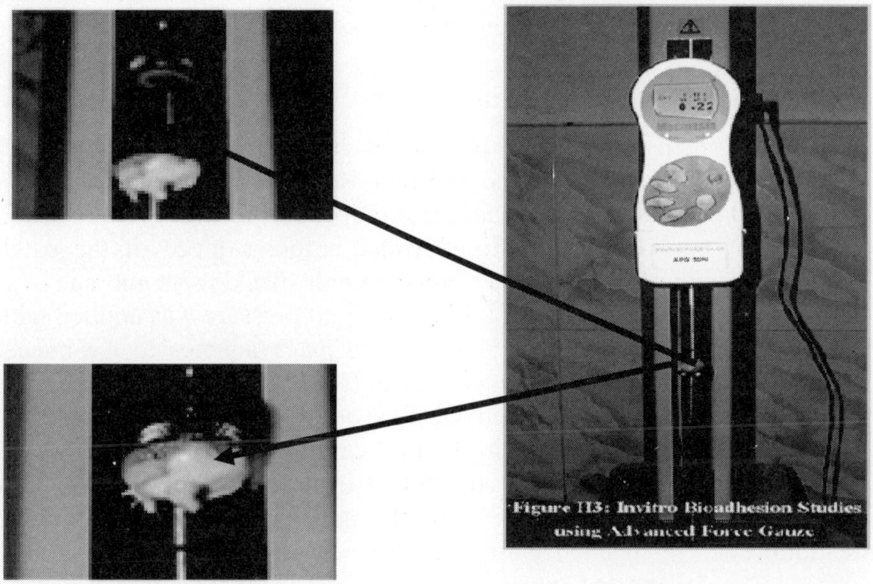

Figure 3.9 In vitro bioadhesion measurement using Ultra Test Mecmesin Advanced Force Gauze

Data collection and calculations were performed using the Data Plot software package of the instrument. Two parameters, namely the work of adhesion and peak detachment force were used to study the buccal adhesiveness of patches. The work of adhesion was determined from the area under force-distance curve while the peak detachment force was the maximum force required to detach the film from the tissue.

3.9.6 Ex Vivo Residence Time

Ex vivo residence tine was determined using a modified USP disintegration apparatus. Nakamura et al. (130) applied this method by taking the disintegration medium composed of 800 ml phosphate buffer pH 6.6 maintained at 37 °C. The porcine buccal tissue was tied to the surface of a glass slab, vertically attached to the apparatus. The buccal tablet was hydrated from one surface using 0.5 mL of pH 6.6 phosphate buffer and then the hydrated surface was brought into contact with the mucosal membrane. The glass slab was vertically fixed to the apparatus and allowed to run in such a way that the tablet completely immersed in the buffer solution at the lowest point and was out at the highest point.

The time taken for complete erosion or detachment of the tablet from the mucosal surface was recorded and considered as ex vivo residence time.

3.9.7 *In vivo* Residence Time

In vivo Residence time was performed in eight healthy adult male volunteers, aged between 22 and 28 years. Prior to the test, the volunteers were elucidated with the procedure and purpose of test. They were asked to rinse their mouth with distilled water before a piece of the patch (12.7 mm diameter) with water impermeable backing membrane was placed on their buccal mucosa. Initially a slight pressure was applied with the help of fingers for one minute till the patch adheres to the buccal mucosa. The volunteers were refrained from food, drinks during the evaluation period. The volunteers were asked to record the residence time of the film on buccal mucosa in the oral cavity, which was taken as the time for the patch to dislodge completely from the buccal mucosa. This was done by continual sensation of the patch as well as the backing membrane. *In vivo* residence time was recorded in each case.

3.9.8 Permeation Studies

During the preformulation studies, buccal absorption/permeation studies must be conducted to determine the feasibility of this route of administration for the drug candidate and to determine the type of enhancer and its concentration required to control the rate of permeation of drugs. These studies involve methods that would examine in vitro, ex vivo and/or in vivo buccal permeation profile and absorption kinetics of the drug.

3.9.8.1 Ex Vivo Permeation

3.9.8.1.a Tissue Preparation (Isolation)

Porcine buccal tissue from domestic pigs was obtained from local slaughterhouse and used within 2 hours of slaughter. The tissue was stored in Krebs buffer at 4° C after collection. The epithelium was separated from the underlying connective tissue with surgical technique and then the remaining buccal mucosa was carefully trimmed with the help of surgical scissors to a uniform thickness of about 500 μm. The thickness of the tissues was measured with a digital micrometer and recorded. This helped to minimize variations between tissue specimens. Finally the membrane was allowed to equilibrate for approximately one hour in receptor buffer to regain the lost elasticity.

3.9.8.2 Ex Vivo Drug Permeation Studies

Most of the ex vivo studies examining drug transport across buccal mucosa uses buccal tissues from animal models (pigs). Immediately after sacrificing the animals the buccal mucosal tissue is surgically removed from the oral cavity. The buccal epithelium was carefully mounted between the two compartments of a Franz diffusion cell with an internal diameter of 2.1 cm (3.46 cm^2 area) with a receptor compartment volume of 25.0 mL. Phosphate buffer saline (PBS) pH 7.4 was placed in the receptor compartment. The donor compartment contained 4 mL solution of PBS pH 7.4 in which drug was dissolved. The donor compartment also contained phenol red a non absorbable marker compound at a concentration of 20 μg mL^{-1}. The entire set up was placed over magnetic stirrer and temperature was maintained at 37° C as shown in Figure 3.10. Samples of 1 mL were collected at predetermined time points from receptor compartment and replaced with an equal volume of fresh solution. Amount permeated can be analyzed using High performance liquid chromatography.

Figure 3.10 In vitro drug permeation studies through porcine buccal membrane set up.

3.9.9 Buccal Absorption Test

Beckett and Triggs (132) developed a method to measure the kinetics of drug absorption.

It is carried out by swirling of a 25 ml sample of the test solution for 15 min by human volunteers followed by the expulsion of the solution. The amount of drug remaining in the expelled volume is then determined to assess the amount of drug absorbed. The drawbacks of this method are inability to localize the drug solution within a specific site of the oral cavity, accidental swallowing of a portion of the sample solution and the salivary dilution of the drug.

3.9.9.1 Modified Beckett's test

Various investigators have subsequently modified Beckett's test. Dearden and Tomlinson (133) in 1971 and further in 1979 added a correction for the production of saliva, due to which the drug concentration in the samples decreased. In 1978, Schurmann and Turner (134) modified the test by adding phenol red as a marker for drug dilution by saliva secretion as well as for accidental swallowing of the drug solution. Tucker (135), 1988 optimized the 'Schurmann and Turner Test' by taking a small sample of the solution in the oral cavity every few minutes, without removing the residual solution. In this way he was able to study absorption kinetics in a single 15-20 minutes test. Advantages of this type of test over the original absorption test are; corrections for saliva secretion, accidental swallowing and changes in pH can be made and that a complete absorption curve can be measured in one single test. Still, the disadvantage is the uncertainty with respect to the amount of drug that actually reaches the systemic circulation.

Volume of Drug solution: Generally volumes of 20 or 25 ml are used. Such small volume solutions are not too large to cause subject discomfort, allow homogenous mixing of contents during a test and provide an adequate volume in which the drug dissolves.

Contact time: Test duration time varies depending on study conditions employed. In general there was an initial fast disappearance of drug upto 5min, followed by a slower progressive loss. This time period depends upon the physicochemical properties of the drug.

Non-absorbable marker compounds: Non-absorbable marker compounds were added to the buffered drug solution and the amount of such compounds is determined before and after buccal absorption test, to determine the non-absorbable losses that might occur due to swallowing of drug solution during buccal absorption test. Inulin, ^{125}I-labelled PVP,

PEG, and phenol red have been used to assess the extent of drug loss arising from non-absorbable sources during buccal absorption test.

Pre-test modifications: Pre-test wash: In order to cleanse mouth and adjust pH prior to buccal absorption test, Mc Elnay and Temple, 1982 (136) performed a 30 sec swirling using buffer solution without drug and Tucker, 1988 (135) followed a similar procedure using a few ml of distilled water.

Equilibration of buffered drug solution to 37ºC: Evered and his co-workers (Manning and Evered, 1976) (137) pre-equilibrated the buffer drug solution to 37ºC immediately prior to buccal absorption test in order to reduce changes with temperature.

Post-test rinsing: Immediately after a buccal absorption test, subjects rinsed their mouth with an aliquot of fresh, drug free buffer or distilled water for a short period of time to remove unabsorbed drug that may still be present in the oral cavity. Post swirling period, should involve small volumes and be restricted to short periods of time. This is particularly important with drugs that can readily return to the oral cavity from the membranes.

Time between repeats: Dearden and Tomlinson (133) found that minimal period between successive tests, for satisfactory reproducible values to be obtained, was about 15 min after a 5 min contact time and 50 min after 30 min contact time. A waiting time of 30 min after a buccal absorption test of 5 or 6 min was considered as sufficient length by a number of authors (138, 139). Other authors preferred waiting times between repeats of 2 hr (140), 5 hr (135) or 24 hr (141). Time period depends on the physicochemical properties of the drug and duration of buccal absorption test. Longer lapses of time must be allowed when longer buccal absorption test contact times are used or when very lipid soluble drugs are investigated.

Limitations of the method: Drug absorption takes place all over the mucosal surfaces; the technique cannot provide information on the relative permeabilities of different regions of the oral cavity. There is a continual and erratic secretion of saliva throughout the duration of the test. The saliva will continually change drug solution pH and volume, and can potentially interact with drug leading to possible interference with assay procedures and problems in the interpretation of buccal absorption test data.

3.10 Research Case Studies for Buccal Drug Delivery

3.10.1 Case study (Patch)

Development of Mucoadhesive Patches for Buccal Administration of Carvedilol

Y. Vamshi Vishnu, K. Chandrasekhar, G. Ramesh and Y. Madhusudan Rao

***Current Drug Delivery*, 2007, *4*, 27-39**

Introduction

Drug Selection Criteria: Carvedilol is a non-selective and β-adrenergic antagonist with no intrinsic sympatomimetic activity and is widely used to treat essential hypertension and angina pectoris. Although it is completely absorbed from the gastrointestinal tract, the systemic availability is approximately 25- 35% because of high first-pass metabolism. Carvedilol was selected as a model drug for the investigation because its low oral dose (6.25-25mg) and bioavailability (25-35%). Carvedilol is metabolized primarily by aromatic ring glucuronidation. The oxidative metabolites are metabolized by conjugation via glucuronidation and sulfation.

Polymer Selection Criteria: A suitable buccal drug delivery system should be flexible and possess good bioadhesive properties, so that it can be retained in the oral cavity for the desired duration. In addition, it should release the drug in a predictable manner to elicit the required therapeutic response. In the present study, flexible buccal patches were developed using water soluble polymer, hydroxypropyl methylcellulose (HPMC E15) and hydroxypropyl cellulose (HPC JF).

Methods

Fabrication of Bioadhesive Films/Patches: Bioadhesive buccal films/patches were prepared using Solvent casting method. The polymeric solution was kept for swelling (4hrs), 15% w/v of propylene glycol was added as plasticizer then drug solution was added after 30 min. These patches were dried at room temperature for 12 hrs. The films were observed and checked for possible imperfections upon their removal from the petri dish. They were stored in a desiccator till the evaluation tests were performed.

Table 3.9 Formulation ingredients of buccal patches.

Formulation	Carvedilol (mg)	HPMC E15 (mg)	HPC JF (mg)	PG (µl)	Solvent Mixture (ml)
AB1	334		1750	225	25
AB2	334		2000	300	25
AB3	334		2250	337.5	25
AB4	334		2500	375	25
AB5	334		2750	412.5	25
AB6	334		3000	450	25
AC1	334	1750		225	25
AC2	334	2000		300	25
AC3	334	2250		337.5	25
AC4	334	2500		375	25
AC5	334	2750		412.5	25
AC6	334	3000		450	25

Two different polymers were used to prepare buccal patches with different concentrations (Table 3.9). Formulated patches were subjected to weight variation, thickness variation, content uniformity tests. Patches with any imperfections, entrapped air, or differing in thickness, weight (or) content uniformity were excluded from further studies.

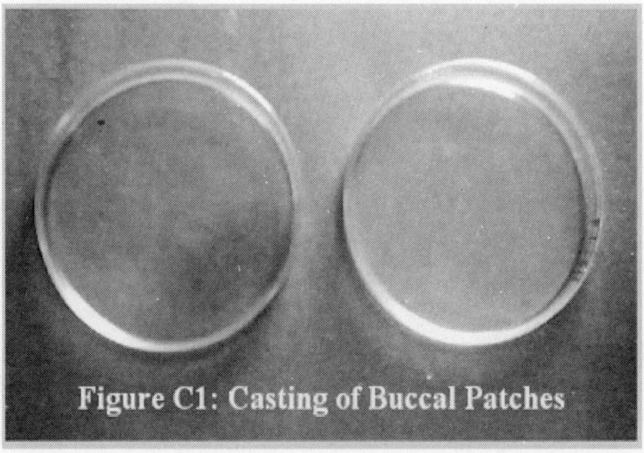

Figure C1: Casting of Buccal Patches

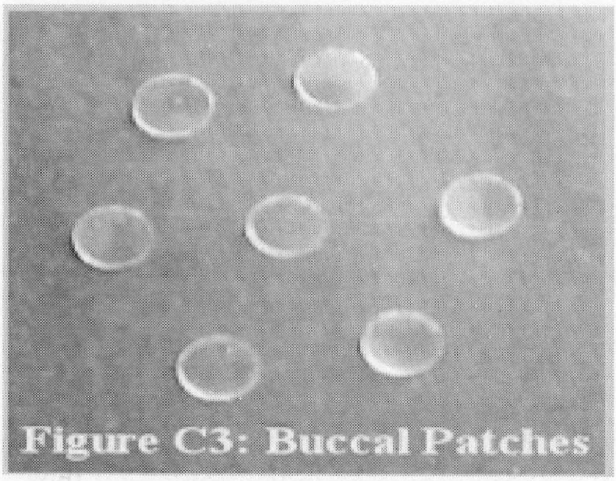

Figure 3.11 Fabrication of buccal patches using solvent casting technique.

In vitro Release Study from Carvedilol Loaded Films

In vitro drug release from the buccal patches was studied by using Lab India dissolution test apparatus Disso 2000 equipped with an auto sampler and fraction collector for the collection and replenishment of sample and dissolution medium respectively. Patches of 1.27cm^2 area were cut as shown in Fig.3.11, patches were meant to release the drug from one side; therefore an impermeable backing membrane was placed on the other side of the patch. The assembly for release studies was prepared by

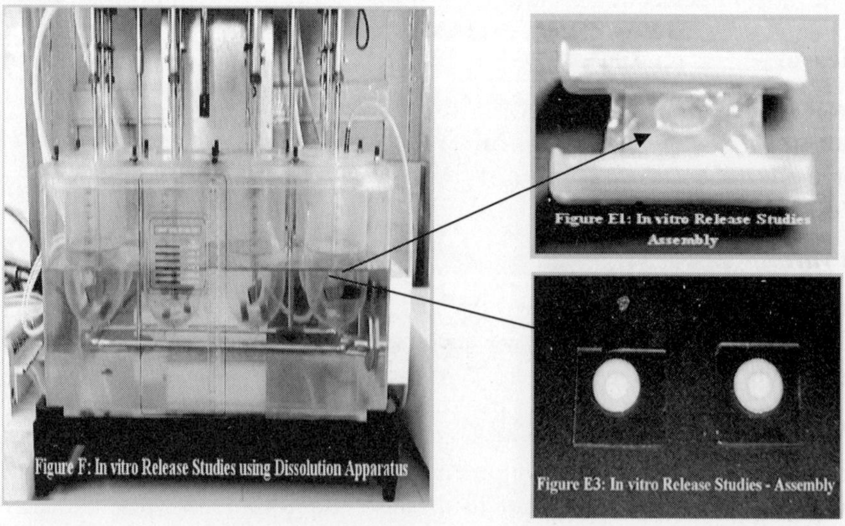

Figure 3.12 *In vitro* release studies using dissolution apparatus.

sandwiching the patch in dialysis membrane (Hi Media molecular weight 5000, Mumbai, India). A piece of glass slide was placed as support to prevent the assembly from floating. The dialysis tubing with patch inside was secured from both ends using closure clips (Hi Media). Then it was placed in dissolution apparatus (Fig. 3.12). The medium was 500ml of 0.5% w/v of sodium lauryl sulphate solution at 50 rpm at a temperature of 37 ± 0.5^0 C. Samples of 5 ml were collected at different time intervals and analysed by using UV-Visible spectrophotometer at 244 nm. The experiment was performed in six replicates.

Moisture Uptake Studies (or) Swelling Index

The moisture uptake studies give an indication about the relative moisture absorption capacities of polymers and an idea whether the formulations maintain their integrity after absorption of moisture. This test was carried out using 5% w/v agar which was dissolved in hot water. It was transferred into Petri plates and allowed to solidify. Six drug free patches from each formulation were selected and weighed. They were placed in vacuum oven overnight prior to the study to remove moisture if any and laminated on one side with water impermeable backing membrane (3M, St.Paul, MI, USA). They were then incubated at 37^0C for one hour, removed and reweighed. The percentage moisture absorption was calculated by using the equation 3.7.

$$\%\ \text{Moisture absorption} = \frac{\text{Final weight} - \text{Initial weight}}{\text{Initial weight}} \times 100$$

Measurement of Mechanical Properties

Mechanical properties (Tensile strength and Elongation at break) of the films/patches were evaluated using a microprocessor based advanced force gauze equipped with a motorized test stand (Ultra Test, Mecmesin, West Sussex, UK).

***In vitro* Bioadhesive Strength**

The bioadhesive strength (Peak detachment force and Work of adhesion) of the buccal patches was determined using an ultra test (Mecmesin, west Sussex UK) equipped with a 5 kg load cell.

***In vivo* Bioadhesion Studies**

Eight healthy male adult volunteers, aged between 22 and 28 years, participated in the study. Prior to the test, the volunteers were educated with the procedure and purpose of test. They were asked to rinse their mouth with distilled water before a piece of the drug free patch with water impermeable backing membrane was placed on their buccal mucosa.

Initially a slight pressure was applied with the help of fingers for one minute till the patch adhered to the buccal mucosa. The volunteers were refrained from food and drinks during the test. The volunteers were asked to record the residence time of the film on buccal mucosa in the oral cavity, which was taken as the time for the patch to dislodge completely from the buccal mucosa. This was done by continual sensation of the patch as well as the backing membrane. The residence time *in vivo* was recorded in each case.

In vitro Permeation of drug through Porcine Buccal Membrane from Buccal Patch

In vitro permeation of carvedilol from buccal patches for the selected formulation (AC5) through porcine buccal membrane was studied. The membrane was mounted over a Franz diffusion cell whose ID is 2.1 cm and a buccal patch was placed over it and which was on a dialysis membrane with molecular weight cut off 5000 so as to secure the patch tightly from getting dislodged from the membrane. The buccal patch was sandwiched between the buccal mucosa and the dialysis membrane.

In vivo Bioavailability Study

Carvedilol a cardiovascular drug may some times produce hypotension in normal patients that may be difficult to control and hence it was suggested to conduct the study in animals. The study was conducted with the approval of the institutional animal ethical committee. Pigs were selected as the animal model, because their buccal membrane closely resembles the human buccal membrane in terms of structure and permeability. Pigs (weighing about 25 kg), from the breeding farm and housed in separate cages. The pigs selected for the study had no medication for two weeks prior to the study. They were restrained from eating and drinking during the study with buccal patch. The formulated test patches (AC5) were laminated on one side with a water impermeable backing layer. The patches were placed in the buccal position of the oral cavity with the polymer side facing the mucosa of buccal cavity. A gentle pressure was applied for one minute. During the first 4hrs of the experiment, the pigs were sedated with an intramuscular injection of ketamine (25mg/kg). The pigs were placed on their side on surgery table and 5ml of blood sample was collected from the tail vein (Fig. 3.13). The blood samples were allowed to coagulate and whole blood samples were centrifuged and serum was separated and stored at –20°C until analysed. The pigs were moved to their cages and blood samplings were collected up to 10hrs. Bioavailability from buccal patch AC5 containing 6.25 mg of carvedilol was compared with that of oral solution containing 6.25 mg of carvedilol in phosphate buffer containing 20 % v/v of ethanol. The drug was not administered in the form of tablet, as it could not be assured whether the pigs would swallow it and it might have required special tubing for the tablet to be administered. It is relatively easy to administer a solution as one could confirm the swallowing by the pig. The bioavailability with an oral solution was compared with that of buccal patch containing the same amount of carvedilol. All the Pharmacokinetic parameters were calculated using KINETICA TM software (Inna Phase Corp., 2000). All the data was statistically analysed by using sigmastat software package (Jandel Corp., California).

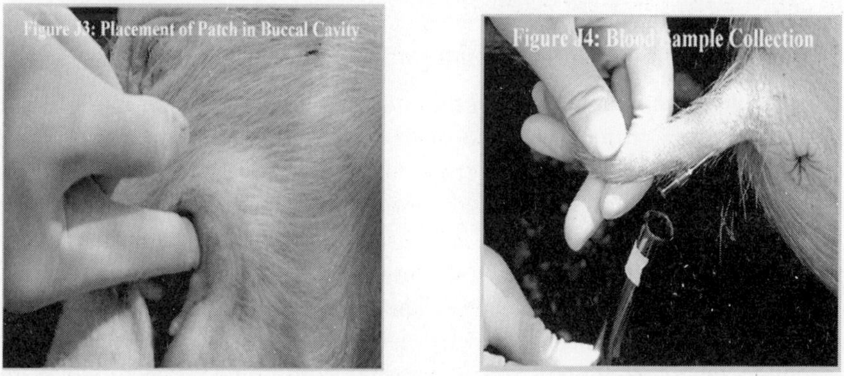

Figure 3.13 Placement of the buccal patch in the oral cavity and blood sample collection.

Analysis of Serum Samples by HPLC Method

The quantitative determination of carvedilol in pig serum was carried out by HPLC method using Shimadzu HPLC system equipped with a LC-10AT pump, shimadzu-RF10AXL Spectrofluorimetric detector and a RP C18 column (250 ± 4.6mm ID, particle size 5μ) at ambient temperature. Serum (300μl) was taken into a glass tube, followed by the addition of 100μl of 500ng/ml solution of flecainide acetate (internal standard). To the above sample 150 μl of 100mM sodium hydroxide was added, vortexed for 3 min followed by the addition of 5 ml of solvent mixture (70:30 of n-hexane, dichloromethane). This was vortexed for 10 min and centrifuged at 5000 rpm. The supernatant was separated and allowed to evaporate in vacuum oven. The residue was reconstituted with 100 μl of mobile phase and 20 µl was injected into HPLC. The eluents were detected at 285 and 380 nm for excitation and emission wavelength, respectively. The sensitivity of detection is medium and the data were acquired, stored and analysed with the software class LC 10 software. The peak area ratio of carvedilol to that of flecainide acetate was determined, and this was used to find the carvedilol from the calibration curve. The calibration curve was constructed by spiking drug-free serum with varying amounts of carvedilol (5-100 ng/ml) and a fixed quantity of internal standard (50ng/100µl of flecainide acetate) and treating the serum samples as described above. A good linear relation was observed between the serum concentration of carvedilol and the peak area ratio of carvedilol to that of internal standard with a high correlation coefficient (0.995) in the range of (5-100ng/ml).

Results and Discussion

Drug penetration studies through porcine buccal membrane

Porcine buccal mucosa has been the most frequently chosen model for *in vitro* permeation studies because of its similarity to human tissue and is available in large quantities from slaughterhouses. The cumulative amount of carvedilol permeated through the buccal epithelium was indicated that the penetration of drug through the porcine buccal epithelium was rapid upto the first 3 hours followed by a low penetration in the next 3 hours (Figure 3.14). The flux was calculated to be 0.116 µg/hr cm^2. The tissue could be isolated successfully because no detectable levels of phenol red, (a marker compound) was found in the receiver compartment, where as carvedilol could penetrate freely.

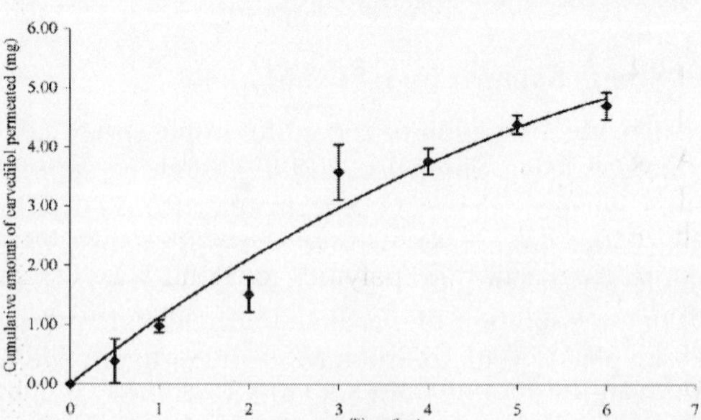

Figure 3.14 *In vitro* permeation of carvedilol through porcine buccal mucosa, values represented mean ± S.D, (n = 3).

Buccal Absorption Study

The results of buccal absorption study revealed that carvedilol could penetrate through the oral cavity. Calculations were performed and results are presented in Fig. 3.15. It was observed that about 60.17% of the drug was absorbed through the buccal membrane in 16 minutes. The drug was absorbed at a rapid rate till the first 2 min and then onwards the drug absorption was of a uniform rate. The total amount of phenol red present in the 8 collected samples was found to be the same when compared to the initial collected samples of phenol red (400 μg) in solution. This indicated that the volunteers did not swallow the solution. The total amount of saliva secreted during the 16-minutes of study was found to be averaging 27.07 ml. The volunteers reported numbness in the mouth for about 15 to 20 minutes after the test.

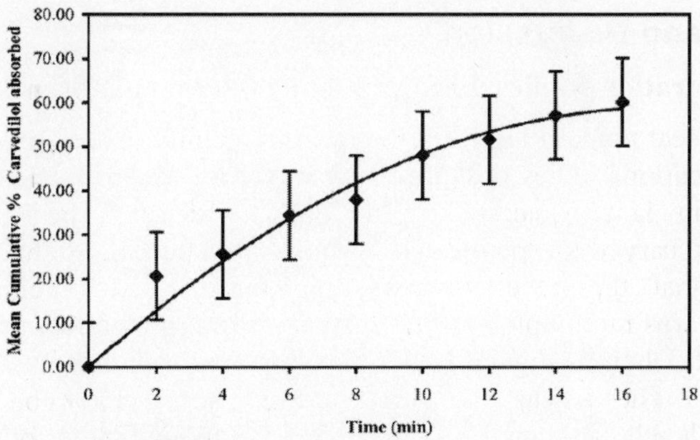

Figure 3.15 Buccal absorption of carvedilol in healthy human volunteers, values represented mean ± S.D (n = 3).

***In vitro* Drug Release Studies**

The drug release profiles of carvedilol from buccal patches are shown in Figs. 3.16 A. & B. It is clear from the plots that formulation containing HPMC E15LV the drug release was governed by polymer content. No lag time was observed as the patch was directly exposed to the dissolution medium. An increase in the polymer content was associated with decrease in the drug release rates. There appeared no significant difference in the final percentage of drug release, which might be due to the fact that in all the formulations the drug dissolved completely in the dissolution medium. In case of HPMC E15 LV (AC5) 84.85% of the drug was released when compared with HPC JF (AB5) from which only 70.25% of drug was released. This can be attributed to lower solubility of

HPC JF to HPMC E15 LV, which is soluble in water at a temperature below 38° C, in addition the drug was hydrophobic in nature. Increasing the amount of the polymer in the patches produced the water-swollen gel like state that could substantially reduce the penetration of the dissolution medium into the patches and so the drug release was retarded.

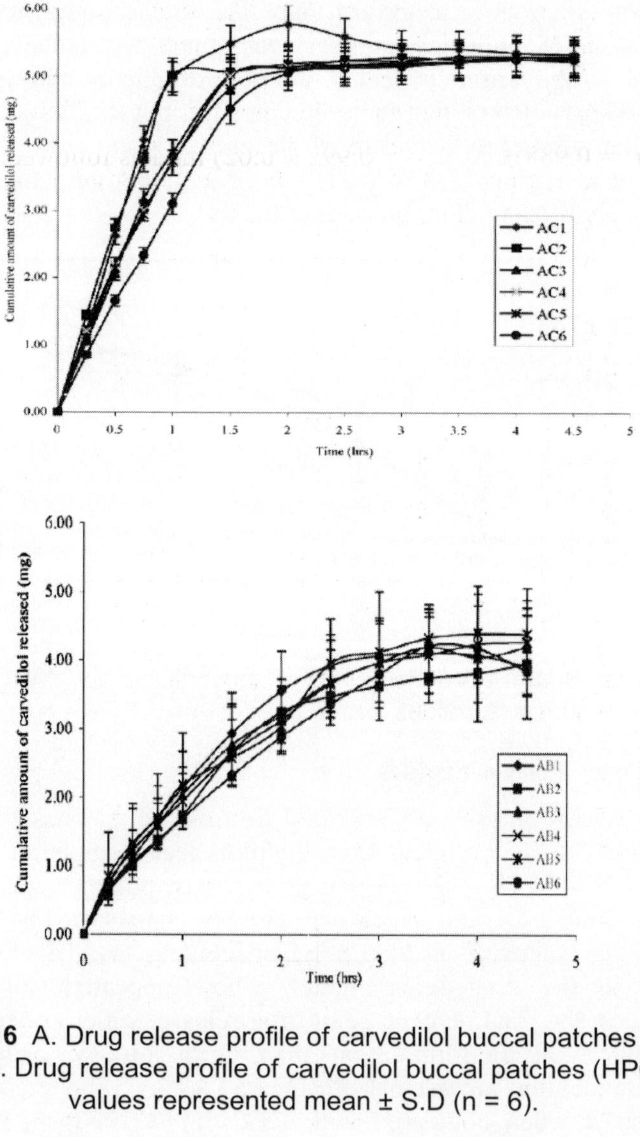

Figure 3.16 A. Drug release profile of carvedilol buccal patches (HPMC E 15), B. Drug release profile of carvedilol buccal patches (HPC JF) values represented mean ± S.D (n = 6).

The formulation (AC5) with a drug to polymer ratio of 1.5 was chosen to evaluate the drug release and bioadhesive properties of the patches. The *in vitro* drug release pattern was interpreted by using 'PCP Disso v2.08' soft ware. Data of the *in vitro* release were fit to different equations and kinetic models to explain the release kinetics of carvedilol from these buccal patches. The kinetic models used were a zero-order equation, first-order equation, Hixson-Crowell equation, Higuchi release and Korsmeyer & Peppas models. The best fit with the highest correlation r and determination r^2 coefficients was shown by both the Higuchi (r^2 = 0.992 ± 0.001, r^2 = 0.988 ± 0.002) and Korsmeyer & Peppas (r^2 = 0.988 ± 0.01, r^2 = 0.992 ± 0.02) models followed by the first order equation and the Hixson-Crowell equation. All the formulations formulated with HPC follow matrix model except AB1 which follows Peppas model. In case of HPMC E15 LV (AC) AC1, AC4, AC6 formulations follow Peppas (non- Fickian) model whereas AC2, AC3, AC5 follow matrix model.

Moisture Uptake Studies

The results of moisture absorption studies are presented in Table 3.10. Results show that there are differences in moisture absorption with HPMC E 15 LV and HPC JF.

Table 3.10 Moisture absorption studies and Mechanical properties of buccal patches.

Formulation	Mean % moisture absorbed	Tensile strength (Kg/mm^2)	Elongation at Break (% mm^2)
AB1	91.66 ± 4.57	2.36 ± 0.43	36.34 ± 4.36
AB2	76.07 ± 4.80	5.73 ± 1.13	30.36 ± 3.82
AB3	65.29 ± 9.37	8.12 ± 2.36	16.72 ± 2.38
AB4	69.65 ± 4.62	10.3 ± 1.96	10.41 ± 2.27
AB5	57.82 ± 4.94	11.8 ± 2.21	12.23 ± 2.82
AB6	56.46 ± 4.03	13.1 ± 2.34	14.02 ± 3.04
AC1	ERODED	4.02 ± 0.81	125.4 ± 11.23
AC2	ERODED	6.93 ± 1.36	90 ± 7.82
AC3	64.16 ± 10.3	10.5 ± 2.14	83 ± 5.36
AC4	68.44 ± 5.82	12.3 ± 1.97	70 ± 5.82
AC5	81.50 ± 17.8	15.3 ± 2.43	54 ± 4.36
AC6	146.2 ± 29.1	18.3 ± 3.42	46.5 ± 5.21

The percentage moisture observed ranged from about 64.44% to 146.21% w/w for various formulation with HPMC E 15 LV and from 56.46% to 91.66% w/w for HPC JF. When the patches were placed without backing membrane complete swelling followed by erosion was observed indicating that the drug release mechanism involves swelling of the polymer initially followed by drug release from the swollen matrix by diffusion. The swelling was slower with HPC JF than HPMC E 15 LV.

Mechanical Properties of Films

The results of the mechanical properties of tensile strength and elongation at break are presented in Table 3.10. Tensile strength increased with increase in the polymer content but elongation at break values decreased with the increase in polymer content. Similar pattern was observed in formulations with both the polymers HPMC E 15 LV and HPC JF. Tensile strength values indicate that there is no statistically significant difference between the next immediate formulations. But statistically significant difference was observed in elongation at break values between the next immediate formulations at $p < 0.05$.

***In vitro* Bioadhesion Studies**

Results of *in vitro* bioadhesion of hydroxypropyl methyl cellulose (AC) and hydroxyl propyl cellulose (AB) patches were obtained. The formulation (AC5) was chosen to evaluate the bioadhesive properties of the patches. The peak detachment force and work of adhesion for the AB5 patch were 0.7500 ± 0.21 N and 0.423 ± 0.23 mJ respectively. For AC5 patch, the work of adhesion and peak detachment force was 4.77 ± 0.86 N and 0.925 ± 0.32 mJ respectively. Buccal patches developed for carvedilol with the above mentioned polymers possess reasonable bioadhesion in terms of peak detachment force and elongation at break values.

***In vivo* Bioadhesion**

In vivo bioadhesion characteristics of the formulated patches, without drug, were evaluated in six healthy male human volunteers. The results exhibited a reasonable adhesion of the device in the oral cavity for about 1 hr for formulation AC1 about 1.5 hrs for AC2 and about 2.5 hours for AC3, AC4 and AC5. No dislocation of the device was observed during this period. The duration of bioadhesion was higher for HPC JF than HPMC E15 LV except in case of the formulation AB1, which dislocated in two hours. All other formulations of HPC remained intact even after three hours. Visual examination of the mucosal tissue after the removal of the patch revealed no signs of damage to the mucosa with either of the polymers. Volunteers reported no irritation during or after the study.

In vitro Permeation of drug through porcine buccal membrane from Buccal Patch

The results of drug permeation from buccal patches of carvedilol through the porcine buccal mucosa reveal that carvedilol was released from the formulations and permeated through the porcine buccal membrane and hence could possibly permeate through the human buccal membrane. The results indicated that the drug permeation was slow and steady and about 38.69% of carvedilol could permeate through the buccal membrane in 4hrs.

In vivo Evaluation of Buccal Patch of Carvedilol

The mean plasma concentration of carvedilol at different time intervals following the application of buccal patch and after oral administration of solution in pigs was shown in Fig. 3.17. The plasma concentration of carvedilol gradually increased and attained maximum after which average steady state level of drug declined gradually upto 12 hr. Thus, the steady state concentration of carvedilol was maintained for 12 hr. The pharmacokinetic parameters such as Cmax, Tmax, AUC total and relative bioavailability were given in Table 3.11. The pharmacokinetic parameters of carvedilol after the application of buccal patch were significantly different from that obtained after oral administration of the solution.

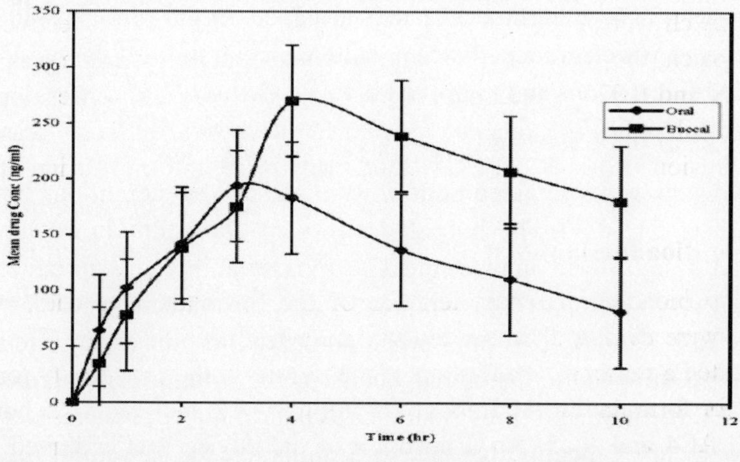

Figure 3.17 Mean serum levels of carvedilol administered by oral and buccal routes in pigs, the values represented mean ± S.D (n = 6)

Table 3.11 pharmacokinetic parameters of carvedilol 6.25mg administered by oral and buccal routes in healthy pigs.

Route	Cmax (ng/mL)[a]	Tmax (h)[a]	AUC total (ng-h /mL)[a]
Oral	207.08 ± 20.6	3.5 ± 0.54	1813.7 ± 42.5
Buccal	314.83 ± 31.4	5.6 ± 1.96	4154.3 ± 80.2

[a]values represented as mean ± SD (n = 6)

Unlike oral dosing the drug concentration after the application of buccal patches was constant over 10 hr. It took about 5.6 h (Tmax) to reach maximum concentration of 314.83 ± 31.4 ng/ml (Cmax). However, on oral administration as solution, the Cmax (207.08 ± 20.64 ng/ml) of carvedilol reached within 3.5 h and declined slowly after 8 hr. The area under the curve (AUC total) of carvedilol with buccal patch was found to be significantly ($P< 0.005$) higher indicating the improved bioavailability due to buccal patch (Table 3.11). It was reported earlier that low bioavailability (25%) is due to the extensive first pass metabolism. However, buccal patch designed in the present study, was found to enhance the bioavailability of carvedilol by 2.29 times with reference to an oral solution of carvedilol. The buccal patch used in the present study provided 57.26% of bioavailability. This increased bioavailability may be due to the elimination of hepatic first-pass metabolism on buccal delivery of the drug. Thus, the buccal patch designed in the present study, was found to provide prolonged steady-state concentration of carvedilol with minimal fluctuations and improved bioavailability.

Remarks of the Case Study

Good results were obtained both *in vitro* and *in vivo* conditions for films of HPMC E15 LV. The buccal delivery of carvedilol in healthy pigs showed a significant improvement in bioavailability of carvedilol from patches when compared to oral route. The bioavailability of carvedilol increased by about 2.29 times when compared to oral route. The results can be extrapolated to the human beings as the structure and permeability of buccal membrane of pigs is similar to that of human beings. Hence the development of bioadhesive buccal formulation for carvedilol may be a promising one as the dose of carvedilol may be decreased and hence side effects may be reduced.

3.10.2 Case study: Buccal tablet

Development and *in vitro* evaluation of buccoadhesive carvedilol tablets

Acta Pharm. 57 (2007) 185–197

Vamshi Vishnu Yamsani, Ramesh Gannu, Chandrasekhar Kolli, M. E. Bhanoji Rao, Madhusudan Rao Yamsani.

Drug Selection Criteria: Carvedilol was reported to be well absorbed following oral administration, but undergoes extensive first pass metabolism; leading to poor bioavailability. From both, physicochemical (low molecular weight 406.5g/mol, low dose 6.25-25 mg) and pharmacokinetic (absolute bioavailability about 25-35 %) views, Carvedilol is considered to be suitable for buccal delivery. A suitable buccal drug delivery system should possess good bioadhesive properties, so that it can be retained in the oral cavity for the desired duration. In addition, it should release the drug in a predictable manner to elicit the required therapeutic response.

Polymer Selection Criteria: In buccal patches of carvedilol released the drug for 5 h. To prolong the drug release and to reduce dosing frequency, a suitable formulation was required with a controlled rate to treat hypertension. In the present study, buccoadhesive tablets were developed using a hydrophilic polymer, hydroxypropyl methylcellulose (HPMC K4M and K15M) and Carbopol 934 to get controlled and zero order release.

Methods

Buccoadhesive tablets preparation: Carvedilol was mixed manually in glass bottles with different ratios of Methocel K4M and K15M, Carbopol 934 as mucoadhesive polymers and Pearlitol SD 200 (mannitol) as diluent (Table 3.12) for 10 min. The blend was lubricated with sodium stearyl fumarate (SSF) for 3–5 min and then compressed into tablets by the direct compression method using 8-mm flat-faced punches (Cadmach Machinery, Mumbai, India). The mass of the tablets were determined using a digital balance (Shimadzu, Japan) and thickness with a digital screw gauge (Mitatyo, Japan).

Table 3.12 Composition of carvedilol buccal tablets.

Formulation	Drug (mg)	HPMC K4M	HPMC K15M	Carbo pol	Perlitol (mg)	SSF (mg)
BC1	6.25	6.25			105.1	2.40
BC2	6.25	12.50			98.85	2.40
BC3	6.25	18.75			92.60	2.40
BC4	6.25	25.00			86.35	2.40
BC5	6.25	31.25			80.10	2.40
BD1	6.25		6.25		105.1	2.40
BD2	6.25		12.50		98.85	2.40
BD3	6.25		18.75		92.60	2.40
BD4	6.25		25.00		86.35	2.40
BD5	6.25		31.25		80.10	2.40
BE1	6.25			1.562		2.40
BE2	6.25			3.125		2.40
BE3	6.25			4.687		2.40
BE4	6.25			6.250		2.40
BE5	6.25			9.375		2.40

SSF – sodium stearyl fumarate

Assay of carvedilol

Twenty tablets were taken and powdered; powder equivalent to one tablet was taken and allowed to dissolve in 100 mL of 0.5% of sodium lauryl sulphate solution on a rotary shaker overnight. The solution was centrifuged and the supernatant was collected. The absorbance was measured using a UV-Vis Spectrophotometer (Elico, India) at 244 nm.

In vitro *release studies*

The drug release rate from buccal tablets was studied using the USP type II dissolution test apparatus. Tablets were supposed to release the drug from one side only; therefore an impermeable backing membrane was placed on the other side of the tablet. The tablet was further fixed to a 2x2 cm glass slide with a solution of cyanoacrylate adhesive. Then it was placed in the dissolution apparatus. The dissolution medium was 500 mL of 0.5% of sodium lauryl sulphate solution at 50 rpm at a temperature of 37 ± 0.5 °C. Samples of 5 mL were collected at different time intervals upto 8h and analyzed spectrophotometrically.

In vitro *bioadhesion studies*

The bioadhesive strength of the tablets was measured using the Ultra test (Mecmesin, UK) equipped with a 5 kg load cell.

Moisture absorption

Moisture absorption test was carried out using 5% w/v agar which was dissolved in hot water. It was transferred into Petri plates and allowed to solidify. Six buccal tablets from each formulation were placed in a vacuum oven overnight prior to the study to remove moisture, if any, and laminated on one side with a water impermeable backing membrane. They were then placed on the surface of the agar and incubated at 37 °C for one hour. Then the tablets were removed and weighed and the percentage of moisture absorption was calculated.

Results and Discussion

Mass, thickness and drug uniformity

The mass and the thickness of the tablets were within the limits of uniformity. The mass ranged from 119.2 to 122.3 mg with RSD values 0.7–1.2%. Thickness ranged between 1.74 and 2.00 mm with RSDs of 0.5 to 1.2%. The drug content ranged from 96.7 ± 0.4% in formulation BC1 to 101.4 ± 0.3% in formulation BC5, 97.0 ± 0.8% in formulation BD1 to 103.2 ± 3.0 in formulation BD5 and 98.5 ± 0.9% in formulation BE1 to 102.4 ± 1.7% in formulation BE4.

In vitro drug release studies

The release of carvedilol from buccoadhesive tablets (Figs. 3.18a–c) varied according to the type and ratio of matrix forming polymers. The drug release was governed by the amount of matrix forming polymer. Burst release was observed in formulations BC1 and BD1. The concentration of HPMC K4M and HPMC K15M was the lowest in BC1 and BD1 among all within the series. The most important factor affecting the rate of release from the buccal tablets is the drug: polymer ratio. An increase in polymer concentration causes an increase in the viscosity of the gel as well as formation of a gel layer with a longer diffusional path. This could cause a decrease in the effective diffusion coefficient of the drug and therefore a reduction in the drug release rate. In the present study, the results followed this predictable behavior (Fig. 3.18a–c). Buccal tablets that contained lower concentrations of either HPMC K4M, HPMC K15M or Carbopol 934 in BC, BD and BE series, respectively, tended to release the drug in shorter time periods, while the release slowed down as the concentration of the gelling polymer increased, thus confirming the dominant role of the swellable hydrophilic polymer in the release of carvedilol from buccal tablets.

Data of the *in vitro* release was fit into different equations and kinetic models to explain the release kinetics of carvedilol from buccal tablets. The kinetic models used were zero-order, first-order, Higuchi and Korsemeyer-Peppas models. Formulations with HPMC K4M (BC1, BC2, BC3) (correlation coefficient between 0.800 and 0.980) and HPMC K15M (BD1, BD2, BD3) (correlation coefficient between 0.844 and 0.985) followed the Higuchi model whereas formulations BC4, BC5 (correlation coefficient 0.985 and 1.000), BD4 and BD5 (correlation coefficient 0.998 and 0.999) followed zero-order release. In the case of Carbopol 934, BE series BE1, BE2 and BE5 (correlation coefficient between 0.961 and 0.981) followed the Higuchi model; BE3 and BE4 (correlation coefficient 0.985 and 0.997), followed zero-order model. The results indicate that as the concentration of each polymer increases in the respective series, Higuchi diffusion mechanism turns to zero-order release profile.

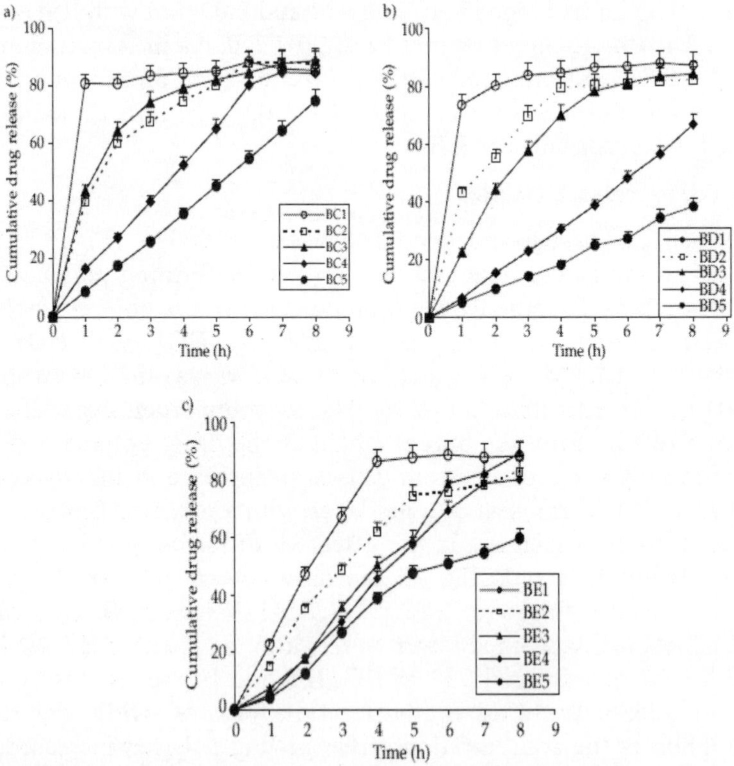

Figure 3.18 Drug release profile of carvedilol buccal tablets formulated with: (a) HPMC K4M, (b) HPMC K15M and (c) Carbopol 934 (mean ± SD, $n = 6$).

Formulation BC3 (88.7 ± 0.4%) composed of 1:3 drug: HPMC K4M ratio; BD3 (84.2 ± 0.3%) 1:3 drug: HPMC K15M ratio and BE3 (89.2 ± 0.3%) 1:0.75 drug: Carbopol 934 ratio showed maximum release among their respective series. Increasing the concentration of the polymer in the formulations showed a sustained effect on carvedilol release. The rapidly hydrating polymer dominated in controlling the release of carvedilol from the buccal tablets, as seen from the dissolution profiles and moisture absorption data.

Release rates slowed down when the concentration of HPMC K4M or HPMC K15M or Carbopol 934 increased from 1:1 to 1:5 ratios and 1:0.25 to 1:1.50 in BC, BD and BE series, respectively. This is because as the proportion of these polymers in thc matrix increased, there was an increase in the amount of water uptake and proportionally greater swelling leading to a thicker gel layer. Zero-order release from swellable hydrophilic matrices occurs as a result of constant diffusional pathlengths. When the thickness of the gelled layer and thus the diffusional pathlengths remain constant, zero-order release can be expected, as seen for formulations BC4, BC5, BD4, BD5, BE3 and BE4.

In vitro bioadhesion studies

The peak detachment force and work of adhesion for the optimized formulation BC3 were 1.62 ± 0.15 N and 0.24 ± 0.11 mJ. In all the formulations, as the polymer concentration increased, both the peak detachment force and work of adhesion increased (Table II). The order of bioadhesion was HPMC K4M < HPMC K15M < Carbopol. Buccal tablets formulated with Carbopol 934 and HPMC K15M showed stronger mucoadhesion than HPMC K4M formulations. Very strong bioadhesion could damage the epithelial lining of the buccal mucosa.

Moisture absorption

The moisture absorption studies give an indication of the relative moisture absorption capacities of polymers and whether the formulations maintain their integrity after moisture absorption. The order of increasing moisture absorption was HPMC K4M < HPMC K15M < Carbopol 934 (Table 3.13). This may be due to the more hydrophilic nature of the polymer Carbopol.

Table 3.13 Peak detachment force, bioadhesion and moisture absorption studies.

Formulation	P.F (N)[a]	W.A (MJ)[b]	M.A (%)[c]
BC1	0.81 ± 0.01	0.09 ± 0.01	31.2 ± 1.0
BC2	1.29 ± 0.09	0.12 ± 0.03	31.3 ± 3.1
BC3	1.62 ± 0.15	0.24 ± 0.11	31.7 ± 1.4
BC4	2.08 ± 0.18	0.33 ± 0.05	33.1 ± 1.2
BC5	2.50 ± 0.20	0.43 ± 0.12	45.5 ± 2.9
Formulation	P.F (N)[a]	W.A (MJ)[b]	M.A (%)[c]
BD1	0.92 ± 0.10	0.23 ± 0.11	31.5 ± 0.9
BD2	1.64 ± 0.21	0.35 ± 0.09	32.7 ± 2.6
BD3	2.03 ± 0.30	0.47 ± 0.05	36.8 ± 1.6
BD4	3.00 ± 0.42	0.98 ± 0.10	44.3 ± 3.9
BD5	3.85 ± 0.17	1.02 ± 0.13	48.4 ± 2.8
BE1	2.00 ± 0.11	1.00 ± 0.13	36.2 ± 1.5
BE2	2.50 ± 0.35	1.85 ± 0.35	53.3 ± 5.0
BE3	3.00 ± 0.41	2.25 ± 0.43	68.2 ± 1.3
BE4	4.25 ± 0.40	3.50 ± 0.51	68.6 ± 3.2
BE5	6.00 ± 0.47	4.25 ± 0.40	70.7 ± 3.3

[a]Peak detachment force
[b]work of adhesion
[c]miosture absorbed

Remarks about the Case Study

HPMC K4M shows satisfactory buccoadhesive properties. Formulation BC3 using this polymer in a drug: polymer (1:3) ratio showed significant bioadhesive properties with an optimum release profile and could be useful for buccal administration of carvedilol. Further work is recommended to support its efficacy claims by long term pharmacokinetic and pharmacodynamic studies in human beings.

3.11 Conclusions

The oral transmucosal route is becoming more and more popular for systemic drug delivery because it does have significant advantages compared to the peroral route. Oral transmucosal technology offers an alternative means of administering drugs. It allows more rapid absorption into the bloodstream than is possible with oral administration to the gastrointestinal tract. Oral transmucosal administration is non-invasive, no technical and convenient for patients and the pain factor associated with parenteral routes of drug administration can be totally eliminated. Buccal adhesive systems offer innumerable advantages in terms of accessibility, administration and withdrawal, retentivity, low enzymatic activity, economy and high patient compliance.

Adhesion of buccal adhesive drug delivery devices to mucosal membranes leads to an increased drug concentration gradient at the absorption site and therefore improved bioavailability of systemically delivered drugs. In addition, buccal adhesive dosage forms have been used to target local disorders at the mucosal surface (e.g., mouth ulcers) to reduce the overall dosage required and minimize side effects that may be caused by systemic administration of drugs.

At present global scenario, scientists are finding different ways to develop buccal adhesive systems through various approaches to improve the bioavailability of orally less/inefficient drugs by modifying the formulation strategies like inclusion of pH modifiers, inclusion complexing agents, enzyme inhibitors, permeation enhances etc. Novel buccal adhesive delivery system, where the drug delivery is directed towards buccal mucosa by protecting the local environment is also gaining interest. Currently solid dosage forms, liquids and gels applied to oral cavity are commercially successful. The future direction of buccal adhesive drug delivery lies in vaccine formulations and delivery of small proteins/peptides. Microparticulate bioadhesive systems are particularly interesting as they offer protection to therapeutic entities as well as the enhanced absorption that result from increased contact time provided by the bioadhesive component. Exciting challenges remain to influence the bioavailability of drugs across the buccal mucosa. Many issues are yet to be resolved before the safe and effective delivery through buccal mucosa. Successfully developing these novel formulations requires assimilation of a great deal of emerging information about the chemical nature and physical structure of these new materials.

References

1. Gandhi RB and Robinson JR. Oral cavity as a site for bioadhesive drug delivery. Adv Drug Deliv Rev, 13:43–74 (1994).
2. Shojaei AH., Chang RK and Guo X et al. Systemic drug delivery via the buccal mucosal route. J Pharm Technol, 25(6):70-81 (2001).

3. Shojaei AH, Zhou SL and Li X. Transbuccal delivery of acyclovir: II. Feasibility, system design and in vitro permeation studies. J Pharm Sci, 1(2):66-73 (1998).
4. Zhang H, Zhang J and Streisand JB. Oral mucosal drug delivery: clinical pharmacokinetics and therapeutic applications. Clin Pharmacokinet, 41(9):661-80 (2002).
5. Chen SY and Squier CA. The ultrastructure of the oral epithelium, in: The Structure and Function of Oral Mucosa, Pergamon Press, Oxford, pp.7 –30 (1984).
6. Harris D and Robinson JR. Drug delivery via the mucous membranes of the oral cavity. J Pharm Sci, 81:1–10 (1992).
7. Meyer J and Gerson SJ. A comparison of human palatal and buccal mucosa. Periodontics, 2: 284–291 (1964).
8. Stablein MJ and Meyer J. The vascular system and blood supply, in: J Meyer., CA Squier., SJ Gerson (Eds.), The Structure and Function of Oral Mucosa, Pergamon Press, Oxford, pp. 237– 256 (1984).
9. Allen A, Bell A and McQueen S. Mucus and mucosal protection, in: Allen A, Flemstrom G, Garner A, Silen W and Turnberg LA (Eds.). Mechanisms of Mucosal Protection in the Upper Gastrointestinal Tract, Raven Press, New York, pp.195-202 (1984).
10. Haas J and Lehr CM. Developments in the area of bioadhesive drug delivery systems. Expert Opin. Biol. Ther, 2: 287–298 (2002).
11. Raj B, Ravichandran M, Xiaoling L and Bhaskara RJ. Advances in buccal drug delivery. Critical Reviews in Therapeutic Drug Carrier Systems, 22(3): 295-330 (2005).
12. Sudhakar Y, Kuotsu K and Bandyopadhya AK. Buccal bioadhesive drug delivery- A promising option for orally less efficient drugs. J Control Rel, 114:15–40 (2006).
13. Shojaei AH and Li X. Determination of transport route of acyclovir across buccal mucosa. Proc Int Symp Control Rel Bioact Mater, 24: 427–428 (1997).
14. Chen L, Hui-Nan X and Xiao-Ling L. In vitro permeation of tetramethylpyrazine across porcine buccal mucosa. Acta Pharmacol Sin, 23: 792–796 (2002).
15. Nielsen HM and Rassing MR. TR146 cells grown on filters as a model of human buccal epithelium. III. Permeability enhancement

by different pH value, different osmolarity value and bile salts. Int J Pharm, 185, 215–225 (1999).

16. Zhang H and Robinson JR. In vitro methods for measuring permeability of the oral mucosa, in: J. Swarbrick, JC. Boylan (Eds.), Oral Mucosal Drug Delivery, 1st edition, vol. 74, Marcel Dekker, INC, New York, pp. 85–100 (1996).
17. Randhawa MA, Malik SA and Javed M. Buccal absorption of weak acidic drugs is not related to their degree of ionization as estimated from the Henderson–Hasselbalch equation. Pak. J. Med. Res, 42 (2):145-159 (2003).
18. Utoguchi N, Watanabe Y, Suzuki T, Maeharai J, Matsumoto Y and Matsumoto M. Carrier mediated transport of monocarboxylic acids in primary cultured epithelial cells from rabbit oral mucosa. Pharm. Res, 14:320–324 (1997).
19. Squier CA, Kremer MJ and Wertz PW. Continuous flow mucosal cells for measuring in vitro permeability of small tissue samples. J. Pharm. Sci, 86:82–84 (1997).
20. Brun PPHL, Fox PLA, Vries MED and Bodde HE. In vitro penetration of some β-adrenoreceptor blocking drugs through porcine buccal mucosa. Int J Pharm, 49:141–145 (1989).
21. Nielsen HM and Rassing MR. TR146 cells grown on filters as a model of human buccal epithelium. III. Permeability enhancement by different pH value, different osmolarity value, and bile salts. Int J Pharm, 185:215–225 (1999).
22. Hayward AF. Membrane-Coating Granules. Int. Rev. Cyt, 59:97–127 (1979).
23. Squier CA, Eady RA and Hopps RM. The permeability of epidermis lacking normal membrane-coating granules:an ultrastructural tracer study of Kyrle–Flegel disease. J Invest Dermatol, 70:361–364 (1978).
24. Squier CA and Hall BK.The permeability of mammalian nonkeratinized oral epithelia to horseradish peroxidase applied in vivo and in vitro. Arch Oral Biol, 29:45–50 (1984).
25. Robinson JR and Yang X. Absorption enhancers, J. Swarbrick, J.C. Boylan (Eds.), Encyclopedia of Pharmaceutical Technology, vol. 18, Marcel Dekker, Inc., New York, pp. 1– 27 (2001).

26. Veuillez F, Kalia YN, Jacques Y, Deshusses J and Buri P. Factors and strategies for improving buccal absorption of peptides. Eur J Pharm Biopharm, 51:93–109 (2001).

27. Walker GF, Langoth N and Bernkop-Schnurch A. Peptidase activity on the surface of the porcine buccal mucosa. Int J Pharm, 233:141–147 (2002).

28. Kashi SD and Lee VHL. Enkephalin hydrolysis in homogenates of various absorptive mucosae of the albino rabbit:similarities in rates and involvement of aminopeptidases. Life Sci, 38:2019–2028 (1986).

29. Chattarajee SC and Walker RB. Penetration enhancer classification, in: EW Smith, HI Maibach (Eds.), Percutaneous Penetration Enhancement, CRC Press, Boca Raton, FL, pp. 1–4 (1995).

30. Smart JD. The basics and underlying mechanisms of mucoadhesion. Adv Drug Del Rev, 57:1556–1568 (2005).

31. Longer MA and Robinson JR. Fundamental aspects of bioadhesion. Pharm. Int, 7:114–117 (1986).

32. Laidler KJ, Meiser JH and Sanctuary BC. Physical Chemistry, Fourth edition, Houghton Mifflin Company, Boston (2003).

33. Kamath KR and Park K. Mucosal adhesive preparations, in: J. Swabrick, JC. Boylan (Eds.), Encyclopedia of Pharmaceutical Technology, vol. 10, Marcel Dekker, New York, pp. 133–163 (1994).

34. Larsson K and Glantz PO. Microbial adhesion to surfaces with different charges. Acta Odontol Scand, 39:79–82 (1981).

35. Tobyn MJ, Johnson JR and Gibson SAW. an invitro assessment of mucus/ mucoadhesive interactions. J Pharm Pharmacol, 44:1048 (suppl) (1992).

36. Hiemenz PC. Principles of Colloid and Surface Chemistry. Marcel Dekker, New York (1986).

37. Martin A, Bustamante P and Chun P(eds). BI Waverly Pvt ltd. New Delhi, 4th edition (1986).

38. Whitesides GM, Mathias JP and Seto CT. Molecular self-assembly and nanochemistry: a chemical strategy for the synthesis of nanostructures. Science, 254:1312-1319 (1991).

39. Rawlins EA(Ed.), Bentley's Textbook of Pharmaceutics, 8th edition, ELBS publishers. pp. 32–33 (1984)

40. Ninham BW and Yaminsky V. Ion binding and ion specificity: the Hofmeister effect, Onsager and Lifshitz theories. Langmuir, 13:2097–2108 (1990).

41. Dill KA. The meaning of hydophobicity. Science, 250: 297-298 (1990).

42. Claesson PM and Christensson HK. Very long-range attractive forces between uncharged hydrocarbon and fluorocarbon surface in water. J Phys Chem, 92:1650–1655 (1988).

43. Ahuja RP and Khar JA. Mucoadhesive drug delivery systems, Drug Dev. Ind Pharm. 23:489–515 (1997).

44. Peppas NA and Sahlin JJ. Hydrogels as mucoadhesive and bioadhesive materials: a review. Biomaterials, 17:1553–1561 (1996).

45. Duchene D, Touchard F and Peppas NA. Pharmaceutical and medical aspects of bioadhesive systems for drug administration. Drug Dev Ind Pharm, 14:283–318 (1988).

46. Gu JM, Robinson JR and Leung SHS. Binding of acrylic polymers to mucin/epithelial surfaces:structure–property relationships. Crit Rev Ther Drug Carr Syst, 5:21–67 (1998).

47. Huang Y, Leobandung W, Foss A and Peppas NA. Molecular aspects of muco-and bioadhesion: tethered structures and site-specific surfaces. J Control Rel, 65:63–71 (2000).

48. Park H and Robinson JR. Mechanisms of mucoadhesion of poly (acrylic acid) hydrogels. Pharm Res, 4: 457–464 (1987).

49. Peppas NA and Buri PA. Surface, interfacial and molecular aspects of polymer bioadhesion on soft tissues. J Control Rel, 2:257–275 (1985).

50. Flory PJ. Principle of Polymer Chemistry, Cornell University Press, Ithaca, New York, p. 541 (1953).

51. Lehr CM, Bouwstra JA, Schacht EH and Junginger HE. In vitro evaluation of mucoadhesive properties of chitosan and some other natural polymers. Int J Pharm, 78:43–48 (1992).

52. Solomonidou D, Cremer K, Krumme M and Kreuter J. Effect of carbomer concentration and degree of neutralization on the

mucoadhesive properties of polymer films. J Biomater Sci Polym Ed, 12:1191–1205 (2001).

53. Park H and Robinson JR. Physico-chemical properties of water insoluble polymers important to mucin epithelial adhesion. J Control Rel, 2:47–57 (1985).

54. Kamath KR and Park K. Mucosal adhesive preparations. In: Swarbrick J, Boylan JC, editors. Encyclopedia of Pharmaceutical Technology. New York: Marcel Dekker. p133–163 (1994).

55. Lehr CM, Poelma FGJ, Junginger HE and Tukker JJ. An estimate of turnover time of intestinal mucus gel layer in the rat in situ loop, Int J Pharm, 70:235–240 (1991).

56. Forstner JF. Intestinal mucins in health and disease. Digestion, 17:234–263 (1978).

57. Lee JW, Park JH and Robinson JR. Bioadhesive-based dosage forms:the next generation. J Pharm Sci, 89:850–866 (2000).

58. Clark MA, Hirst B and Jepson M. Lectin-mediated mucosal delivery of drugs and microparticles. Adv Drug Deliv Rev, 43:207–223 (2000).

59. Lehr C Lectin-mediated drug delivery: the second generation of bioadhesives. J Control Rel, 65:19–29 (2000).

60. Wirth M, Gerhardt K, Wurm C and Gabor F. Lectin-mediated drug delivery: influence of mucin on cytoadhesion of plant lectins in vitro. J Control Rel, 79:183–191 (2002).

61. Bernkop-Schnürch A, Gabor F, Szostak M and Lubitz W. An adhesive drug delivery system based on K99-fimbriae. Eur J Pharm Sci, 3:293–299 (1995).

62. Leitner V and Walker GA. Bernkop-Schnürch, Thiolated polymers: evidence for the formation of disulphide bonds with mucus glycoproteins. Eur J Pharm Biopharm, 56:207–214 (2003).

63. Albrecht K, Greindl M, Kremserm C, Wolf C, Debbage P and Bernkop-Schnürch A. Comparative in vivo mucoadhesion studies of thiomer formulations using magnetic resonance imaging and fluorescence detection. J Control Rel, 115:78–84 (2006).

64. Bernkop-Schnürch A. Thiomers: a new generation of mucoadhesive polymers. Adv Drug Deliv Rev, 57:1569–1582 (2005).

65. Roldo M, Hornof M, Caliceti P and Bernkop-Schnürch A. Mucoadhesive thiolated chitosans as platforms for oral controlled drug delivery: synthesis and in vitro evaluation. Eur J Pharm Biopharm, 57:115–121 (2004).

66. Jinsong Hao and Paul WSH. Buccal Delivery Systems. Drug dev Ind Pharm, 29:821–832 (2003).

67. Repka MA, Repka SL and McGinity JM. Bioadhesive Hot-Melt Extruded Film for Topical and Mucosal Adhesion Applications and Drug Delivery and Process for Preparation Thereof. US Patent 6,375,963: April 23, 2002.

68. Repka MA, Sridhar T, Sampada BU, Sunil kumar B, McGinity JW and Martin C. Pharmaceutical applications of Hot-Melt Extrusion:Part I. Drug Dev Ind Pharm, 33:909-926 (2007).

69. Follonier N, Doelker E and Cole ET. Evaluation of hot-melt extrusion as a new technique for the production of polymer-based pellets for sustained release capsules containing high loadings of freely soluble drugs. Drug Dev Ind Pharm, 20(8):1323–1339 (1994).

70. Crowley MM, Schroeder B, Fredersdorf A, Obara S, Talarico M and Kucera S. et al. Physicochemical properties and mechanism of drug release from ethyl cellulose matrix tablets prepared by direct compression and hot-melt extrusion, Int J Pharm, 271(1–2):77–84 (2004b).

71. Prodduturi S, Manek RV, Kolling WM, Stodghill SP and Repka MA. Solid-state stability and characterization of hot-melt extruded poly(ethylene oxide) films. J Pharm Sci, 94(10):2232–2245 (2005).

72. Rothen-Weinhold SA, Oudry N, Schwach-Abdellaoui K, Frutiger-Hughes S, Hughes GJ and Jeannerat D. Formation of peptide impurities in polyester matrices during implant manufacturing. Eur J Pharm Biopharm, 49(3):253–257 (2000).

73. Perdikoulias J and Dobbie T. Die Design. In I. Ghebre-Sellassie & C. Martin(Eds.), Pharmaceutical extrusion technology. Drugs and the pharmaceutical sciences (Volume 133: pp. 99–110(2003).). New York: Marcel Dekkr, Inc.

74. Luker K. Single-screw extrusion and screw design. In I. Ghebre-Sellassie & C. Martin(Eds.), Pharmaceutical extrusion technology.Drugs and the pharmaceutical sciences (Volume 133: pp. 39–68 (2003). Newyork: Marcel Dekker.

75. Mollan M. Historical overview. In I.Ghebre-Sellassie & C. Martin (Eds.), Pharmaceutical extrusion technology (Volume 133, pp. 1–18 (2003).NewYork, Ny: Marcel Dekker, Inc.

76. Gaeta GM, Gombos F, Femiano F, Battista C, Minghetti P, Montanari L, Satriano RA and Argenziano G. Acitretin and treatment of the oral leucoplakias. A model to have an active molecules release. J Eur Acad Dermatol. Venereol, 14:473–478 (2000).

77. Du Q, Ping QN and Liu GJ. Preparation of Buspirone hydrochloride buccal adhesive tablet and study on its drug release mechanism. Yao Xue Xue Bao, 37(8):653–656 (2002).

78. Alur HH, Pather SI, Mitra AK and Johnston TP. Transmucosal sustained-delivery of chlorpheniramine maleate in rabbits using a novel natural mucoadhesive gum as an excipient in buccal tablets. Int J Pharm, 188:1–10 (1999).

79. Rajesh K, Agarwal SP and Ahuja A. Buccoadhesive erodible carriers for local drug delivery: design and standardization. Int J Pharm, 138:68–73 (1996).

80. Ikinci G, Capan Y, Senel S, Alaaddinoglu E, Dalkara T and Hincal AA. In vitro/in vivo studies on a buccal bioadhesive tablet formulation of Carbamazepine. Pharmazie, 55:762–765 (2000).

81. Jain AC, Aungst BJ and Adeyeye MC. Development and in vivo evaluation of buccal tablets prepared using danazol-sulfobutylether-7h-cyclodextrin (SBE7) complexes. J Pharm Sci, 91:1659–1668 (2002).

82. Ahuja A, Khar RK and Ali J. Mucoadhesive drug delivery systems. Drug Dev Ind Pharm, 23(5):489–517 (1997).

83. Ceschel GC, Maffei P, Borgia SL and Ronchi C. Design and evaluation of buccal adhesive hydrocortisone acetate tablets. Int J Pharm, 238:161–170 (2002).

84. Ceschel GC, Maffei P, Lombardi Biogia S and Ronchi C. Design and evaluation of buccal adhesive hydrocortisone acetate (HCA) tablets. Drug Deliv, 8:161–171 (2001).

85. Hosny EA, Elkheshen SA and Saleh SI. Buccoadhesive tablets for insulin delivery: in-vitro and in-vivo studies. Boll Chim Farm, 141:210–217 (2002).

86. Miyazaki S, Nakayama A, Oda M, Takada M and Attwood D. Chitosan and sodium alginate based bioadhesive tablets for intraoral drug delivery. Biol Pharm Bull, 17:745–747 (1994).

87. Taylan B, Capan Y, Guven O, Kes S and Hincal AA. Design and evaluation of sustained release and buccal adhesive propronolol hydrochloride tablets. J Control Rel, 38(1):11–20 (1996).

88. Perioli L, Ambrogi V, Rubini D, Giovagnoli S, Ricci M, Blasi Pand Rossi C. Novel mucoadhesive buccal formulation containing metronidazole for the treatment of periodontal disease. J Control Rel, 95:521–533 (2004).

89. Mohammed FA and Khedr H. Preparation and in vitro/in vivo evaluation of the buccal bioadhesive properties of slow release tablets containing miconazole nitrate. Drug Dev Ind Pharm, 29:321–337 (2003).

90. Han RY, Fang JY, Sung KC and Hu OYP. Mucoadhesive buccal disks for novel nalbuphine prodrug controlled delivery: effect of formulation variables on drug release and mucoadhesive performance. Int J Pharm, 177:201–209 (1999).

91. Varshosaz J and Dehghan Z. Development and characterization of buccoadhesive nifedipine tablets. Eur J Pharm Biopharm, 54:135–141 (2002).

92. Labot JM, Manzo RH and Allemandi A. Double layered mucoadhesive tablets containing nystatin. AAPS Pharm SciTech, 3(3): 22-31 (2002).

93. Yong CS, Jung JH, Rhee JD, Kim CK and Choi HG. Physicochemical characterization and evaluation of buccal adhesive tablets containing omeprazole. Drug Dev Ind Pharm, 27:447–455 (2001).

94. Dortunc B, Ozer L, Uyanik N. Development and in vitro evaluation of a buccoadhesive pindolol tablet formulation. Drug Dev Ind Pharm, 24:281–288 (1998).

95. Jug M and Becirevic-Lacan M. Influence of hydroxypropyl-B-cyclodextrin complexation on piroxicam release from buccoadhesive tablets. Eur J Pharm Sci, 21:251–260 (2004).

96. Celebi N and Kislal O. Development and evaluation of a buccoadhesive propranolol tablet formulation. Pharmazie, 50:470–472 (1995).

97. Diarra M, Pourroy G, Boymond C and Muster D. Fluoride controlled release tablets for intra buccal use. Biomaterials, 24:1293–1300 (2003).

98. Ameye D, Voorspoels J, Foreman P, Tsai J, Richardson P, Geresh S and Remon JP. Ex vivo bioadhesion and in vivo testosterone bioavailability study of different bioadhesive formulations based on starch-g-poly (acrylic acid) copolymers and starch/poly(acrylic acid) mixtures. J Control Rel, 79:173–182 (2002).

99. Geresh S, Gdalevsky GY, Gilbo I, Voorspoels J, Remon JP and Kost J. Bioadhesive grafted starch copolymers as platforms for peroral drug delivery: a study of theophylline release. J Control Rel, 94:391–399 (2004).

100. Shojaei AH, Zhuo SL and Li X. Transbuccal delivery of acyclovir (II): feasibility, system design, and in vitro permeation studies. J Pharm Pharm Sci, 1:66–73 (1998).

101. 101.Guo JH. Investigating the surface properties and bioadhesion of buccal patches. J Pharm Pharmacol, 46:647–650 (1994).

102. Senel S, Ikinci G, Kas S, Yousefi-Rad A, Sargon MF and Hıncal AA. Chitosan films and hydrogels of chlorhexidine gluconate for oral mucosal delivery. Int J Pharm, 193:197–203 (2000).

103. Danjo K, Kato H, Otsuka A and Ushimaru K. Fundamental study on the evaluation of strength of granular particles. Chem Pharm Bull, 42:2598–2603 (1994).

104. Ishida M, Nambu N, Nagai T. Mucosal dosage form of lidocaine for toothache using hydroxypropyl cellulose and Carbopol. Chem Pharm Bull, 30:980–984 (1982).

105. Nafee NA, Ismail FA, Nabila V and Boraie LM. Mucoadhesive buccal patches of miconazole nitrate: in vitro/in vivo performance and effect of ageing. Int J Pharm, 264:1–14 (2003).

106. Save T, Shah UM, Ghamande AR, Venkatachalam P. Comparative study of buccoadhesive formulations and sublingual capsules of nifedipine. J Pharm Pharmacol 46(3):192-195 (1994).

107. Anders R and Merkle HP. Evaluation of laminated muco-adhesive patches for buccal drug delivery. Int J Pharm, 49:231–240 (1989).

108. Li C, Bhatt PP and Johnston TP. Transmucosal delivery of oxytocin to rabbits using a mucoadhesive buccal patch. Pharm Dev Technol, 2:265–274 (1997).

109. Mohamed MI and Mortada ND. Development and characterization of a buccoadhesive dosage form of terbutaline sulphate. Mansoura. J Pharm Sci, 16:69–81 (2000).

110. Chun MK, Kwak BT and Choi HK. Preparation of buccal patch composed of carbopol, poloxamer and hydroxypropyl methylcellulose. Arch Pharm Res 26:973–978 (2003).

111. Ndindayino F, Henrist D, Kiekens F, Van Den Mooter G, Vervaet C and Remon JP. Direct compression properties of melt-extruded isomalt. Int J Pharm, 235(1–2):149–157 (2002a).

112. Stepto RFT. Thermoplastic Starch. Paper Presented At The Macromol. Symp(2000).

113. Perissutti B, Newton JM, Podczeck F and Rubessa F. Preparation of extruded carbamazepine and PEG 4000 as a potential rapid release dosage form. Eur J Pharm Biopharm, 53(1):125–132 (2002).

114. Fukuda M, Peppas NA and McGinity JW. Floating hot-melt extruded tablets for gastroretentive controlled drug release system. J Cont Rel, 115(2):121–129 (2006a).

115. Lyons JG, Kennedy JE, Geever LM, O'sullivan P and Higginbotham CL. The use of agar as a novel filler for monolithic matrices produced using hot melt extrusion. Eur J of Pharm and Biopharm, 64(1):75–81 (2006).

116. Van Laarhoven JAH, Kruft MAB and Vromans H. In vitro release properties of etonogestrel and ethinyl estradiol from a contraceptive vaginal ring. Int J Pharm, 232(1-2):163–173 (2002).

117. Ndindayino F, Vervaet C, Van Den Mooter G and Remon JP. Direct compression and moulding properties of co-extruded isomalt/drug mixtures. Int J Pharm, 235(1–2):159–168 (2002c).

118. Repka MA, Gerding TG, Repka SL and McGinity JW. Influence of plasticizers and drugs on the physical-mechanical properties of hydroxypropylcellulose films prepared by hot-melt extrusion. Drug Dev Ind Pharm, 25(5):625–633 (1999a).

119. Miller D, Yang W, Williams RO and McGinity JW. Hot-melt extrusion for enhanced delivery of drug particles. J Pharm Sci, 96:361–376 (2006).

120. Mididoddi PK, Prodduturi S and Repka MA. Influence of tartaric acid on the bioadhesion and mechanical properties of hot-melt extruded hydroxypropyl cellulose films for the human nail. Drug Dev Ind Pharm, 32:1059–1066 (2006).

121. Crowley MM, Fredersdorf A, Schroeder B, Ucera S, Prodduturi S and Repka MA. The influence of guaifenesin and ketoprofen on the properties of hot-melt extruded polyethylene oxide films. Eur J Pharm Sci, 22(5):409–418 (2004a).

122. Forster A, Hempenstall J, Tucker I and Rades T. Selection of excipients for melt extrusion with two poorly water-soluble drugs by solubility parameter calculation and thermal analysis. Int J Pharm, 226:147–161 (2001c).

123. Nakamichi K, Nakano T, Yasuura H, Izumi S and Kawashima Y. The role of kneading paddle and the effects of screw revolution speed and water content on the preparation of solid dispersions using a twin-screw extruder. Int J Pharm, 241:203–211 (2002).

124. Forster A, Hempenstall J, Tucker I and Rades T. The potential of small-scale fusion experiments and the gordon-taylor equation to predict the suitability of drug/polymer blends for melt extrusion. Drug Dev Ind Pharm, 27(6):549–560 (2001b).

125. Madhusudan Rao Y, Chandra Sekhar K, Vamshi Vishnu Y, Ramesh G and Naidu KVS. Novel buccoadhesive formulation of chlorpheniramine maleate, in vitro and in vivo characterization. Asian J Pharm, 1(1):33-40 (2006).

126. Vamshi Vishnu Y, Ramesh G, Chandra Sekhar K, Bhanoji Rao ME and Madhusudan Rao Y. Development and in vitro evaluation of buccoadhesive carvedilol tablets. Acta Pharm, 57:185–197 (2007).

127. Shayeda, Ramesh G, Chinna Reddy P and Madhusudan Rao Y. Development of Novel Bioadhesive Buccal Formulation of Diltiazem: In vitro and In vivo Characterization. PDA J Pharm Sci Tech, 63(4):1-9 (2009).

128. Chandra Sekhar K, Naidu KVS, Vamshi Vishnu Y, Ramesh G, Kishan V and Madhusudan Rao Y. Transbuccal Delivery of Chlorpheniramine Maleate from Mucoadhesive Buccal Patches. Drug Delivery, 15:185–191 (2008).

129. Vamshi Vishnu Y, Chandrasekhar K, Ramesh G and Madhusudan Rao Y. Development of Mucoadhesive Patches for Buccal Administration of Carvedilol. Current Drug Delivery, 4:27-39 (2007).

130. Nakamura F, Ohta R, Machida Y and Nagai T. In vitro and in vivo nasal mucoadhesion of water soluble polymers. Int J Pharm, 134:173–81 (1996).

131. Huang Y, Leobandung W, Foss A and Peppas NA. Molecular aspects of muco-and bioadhesion: tethered structures and site-specic surfaces. J Control Rel, 65(1-2):63-71 (2000).

132. Beckett AH and Triggs EJ. Buccal absorption of basic drugs and its application as an in vivo model of passive drug transfer through lipid membranes. J Pharm Pharmacol, 19:31S–41S (1967).

133. Dearden JC and Tomlinson E. Correction of effect of dilution on diffusion through a membrane. J Pharm Sci, 60:1278-1279 (1971).

134. Schurmann W and Turner P. A membrane model of human oral mucosa as derived from buccal absorption and physicochemical properties of beta blocking drugs atenolol and propranolol. J Pharm Pharmcol, 30:137-147 (1978).

135. Tucker IG. A method to study the kinetics of oral mucosa of drug absorption from solutions. J Pharm Pharmcol, 40:679-683 (1988).

136. Mc Elnay JC and Temple DJ. The use of buccal partitioning as a model to examine the effects of aluminium hydroxide gel on the absorption of propranolol. Br J Clin Pharmacol, 13:399-403 (1982).

137. Manning AS and Evered DF. The absorption of sugars from the human oral cavity. Clin Sci Mol Med, 51:127-132 (1976).

138. Beckett AH and Moffat AC. The influence of alkyl substitution in acids on their performance in the buccal absorption test. J Pharm Pharmacol, 20:139S-247S (1968).

139. Arbab AG and Turner P. Influence of pH on absorption of thymoxamine through buccal mucosa in man. Br J Clin Pharmacol, 43:479-480 (1971).

140. Edwards G, Breckenridge AM, Adjenpon YK, Orme MLE and Ward SA. The effect of variations in urinary pH on the pharmacokinetics of diethylcarbamazine. Br J Clin Pharmacol, 12:807-812 (1981).

141. Randhawa MA and Turner P. Buccal absorption of drugs: an *in vivo* measurement of their innate lipophilicity. Int J Clin Pharm Res, 8:1-4 (1988).

142. Chinna Reddy P, Ramesh G, Vamshi Vishnu Y, Shravan Kumar Y and Madhusudan Rao Y. Development of bilayered mucoadhesive patches for buccal delivery of felodipine: in vitro and ex vivo characterization. Curr Trends Biotech Pharm, 4(2):673-683 (2010).

4
Multiple Emulsions

B. Chandrasekhar Reddy and A. V. Jithan

Vaagdevi College of Pharmacy, Warangal, India

4.1 Introduction

Drug research is functionally divided into two phases, drug discovery or design mainly focusing on synthesis and screening of new molecules for *in vivo* and *in vitro* biological activity and drug development focusing on design of novel drug delivery system (NDDS). The idea of drug carrier system with targeted specificity has fascinated scientists. Drug delivery systems such as nanospheres, microspheres, microemulsions, emulsion particles, self emulsifying drug delivery system (SEDDS), liposomes and mixed micelles have been investigated as potential carrier systems for the delivery and/or targeting of the drugs to specific sites in the body. Among these carrier systems, emulsions have always invited particular interest as a carrier for lipophilic drugs due to their biocompatibility, satisfactory stability and ease of manufacture on an industrial scale (1, 2). A simple emulsion is a heterogeneous system of one immiscible liquid dispersed in another in the form of droplets. Simple emulsions are classified according to the nature of their continuous or dispersed phase as, water-in-oil (w/o) or oil-in-water (o/w). An emulsifier is present to stabilize the system.

Multiple emulsions are more complex systems, as the drops of the dispersed phase themselves contain even smaller dispersed droplets, in most cases identical with the continuous phase, but separated physically from the continuous phase. They are called emulsions of emulsions (3). Based on the nature of dispersed medium even multiple emulsions are of two types viz., systems in which oil is continuous phase or water. The simplest multiple systems representing the two classes are o/w/o and w/o/w respectively. Among water-in-oil-in-water (w/o/w) and oil-in-water-in-oil (o/w/o) type multiple emulsions, the former has wider areas of application and hence have been studied in great detail. Since water-soluble drugs are enclosed in an inner water phase of w/o/w emulsion, these emulsions can be used as carriers of water soluble drug for sustained release delivery. Additionally, w/o/w emulsions possess many advantages over w/o emulsions and have low viscosity due to the external aqueous phase, which makes them convenient to handle and use, especially in parenteral formulation as targeting drug delivery system (3).

Multiple emulsions have a long history. Seifriz (4) observed the oil globules in which water globules are present. He also observed the oil globules inside the water globules of a w/o emulsion. He named them as bi-multiple emulsions. He found that more complex trimultiple, quatremultiple and quinquemultiple systems existed. The potential of these systems for application in pharmacy and separation technology has generated increased attention and comprehensive reviews have appeared dealing with their preparation, stability and pharmaceutical uses (5,6,7,8,9,10). The basic rationale for the use of w/o/w and o/w/o type multiple emulsions as a means of prolonged delivery of drugs is that the drug contained in the innermost phase is forced to partition itself through several phases prior to release at the absorption site. Thus, the partition and diffusion coefficient of the drug and the strength of the middle membrane phase, which is a multimolecular layer of oil, water and emulsifier molecules at both the interfaces of multiple emulsion system, controls the drug release from these systems (11). Multiple emulsions have also been termed as liquid membrane system as the liquid film which separates the other liquid phases acts as a thin semi permeable film through which solute must diffuse moving from one phase to another (3,11). Liquid membrane emulsions of the o/w/o type have been used to separate hydrocarbons where the aqueous phase serves as a membrane and the solvent as the external phase (12). The prominence to multiple emulsions also has increased because of the introduction of biotechnology based drugs. The oral route for delivering therapeutic peptides to the body remains one of the major challenges of pharmaceutical technology. Considerable attempts have been made to develop an efficient drug delivery system for the oral administration of these peptides. Trials for the oral route have used mucoadhesive nanoparticles, W/O/W multiple emulsions, mucoadhesive submicron emulsions and surface- coated liposomes. Many researchers have reported that mucoadhesive polymers and multiple emulsions improved the bioavailability of peptides with poor absorption characteristics. The release mechanism of all multiple emulsions was the swelling–

breakdown phenomenon after dilution of the emulsions under hypo-osmotic conditions. Similarly, several applications to multiple emulsions have been reported. This chapter reviews formulation, stability, and potential applications of multiple emulsions.

4.2 Nomenclature (13)

Due to the complexity of multiple emulsions, it is useful to have an unambiguous system of nomenclature. In a w/o/w emulsion, the continuous oil phase of a primary (w/o) emulsion used to prepare the multiple emulsion, becomes the disperse oil phase of the multiple emulsion. The surfactant or surfactant mixture and the volume fraction of water in the w/o emulsion becomes the internal aqueous phase volume of the w/o/w emulsion. A useful subscript notation for complex emulsions was suggested by Sheppard and Tcheurek to define the individual components, and their sources in order to avoid ambiguity. For example, to describe a w/o/w emulsion, the aqueous dispersed phase w, of the primary w/o emulsion is dispersed in the oil phase 'o'. The w_1/o emulsion is redispered in the external, continuous phase of the multiple emulsion w_2. Thus the system can be notated as $w_1/o/w_2$. In most cases the two aqueous phases will be identical, and thus the system can be represented as $w_1/o/w_1$. Conversely an oil/water/oil system can be written as $o_1/w/o_2$.

Sheppard and Tcheurek suggested the use of order terminology, which may be of use in describing more complex systems. A $w_1/o/w_2$ or $o_1/w/o_2$ system would be described as a second order emulsion (two liquid/liquid inter faces). More complex systems could be described as 3rd, 4th etc order emulsions. Thus simple emulsions would be first order emulsions. A $w_1/o/w_1$ emulsion is a two component second order emulsion and an $o_1/w/o_2$ is a three component second order emulsion and so on.

Mulley and Marland have studied multiple emulsion drop formation in systems containing three different immiscible liquid phases. They discussed many possible forms, such systems could take depending on conditions. For example, $o_1/w/o_2$, $o_2/w/o_1$, $o_1/w/o_1$, $o_2/w/o_2$, $w/o_1/w$, $w/o_2/w$, $o_1/o_2/w$, $o_2/o_1/w$, $w/o_2/o_1$, $w/o_1/o_2$. The authors point out that some of these combinations may exist in a particular system. Even more complex situation can be envisaged.

4.3 Advantages

Multiple emulsions have a number of potential benefits over traditional emulsions, such as controlled or triggered release, reduction of fat content, and protection of labile ingredients (14,15,16). Nevertheless, there have been many difficulties associated with preparing this type of multiple emulsion with sufficient stability for utilization within the food industry, e.g. due to droplet coalescence or due to diffusion of water molecules from the internal aqueous phase to the bulk aqueous phase (14,15,16).

Multiple emulsions have shown promise, particularly in cosmetic, pharmaceutical, and separation sciences. Potential pharmaceutical applications, that take advantage of the presence of a reservoir phase inside droplets of another phase, include adjuvant vaccines, prolonged drug delivery systems, sorbent reservoirs in drug overdose treatment, and immobilization of enzymes. Multiple emulsions have been used as the basis of liposome-like lipid vesicles and as a means for the preparation of Lupron Depot and other microsphere injectables (17, 18). Despite their potential usefulness, applications of multiple emulsions have been limited. Currently, there are no pharmaceutical products formulated as multiple emulsions available on the United States market. This is probably due to the inherent thermodynamic instability associated with these emulsions and the complexity of their structures. Emulsion instability arises from the immiscibility of the dispersed and continuous phases and is manifested as interfacial tension. When the dispersed phase is broken into droplets, the surface free energy is increased. The increase in interfacial free energy causes thermodynamic instability of the dispersed phase, leading to droplet coalescence.

Though, multiple emulsions are relatively complex to formulate, bulky and prone to various routes of physical degradation, still there are certain advantages of multiple emulsions which make them a challenging drug delivery system. Multiple emulsions have excellent and exciting opportunity for slow or controlled release of active entrapped compound (19).

The advantages include:

1. A remarkable degree of biocompatibility.
2. Complete biodegradability and the lack of toxic products resulting from carrier degradation.
3. Avoidances of any undesired immune responses against the encapsulated drug.
4. Hydrophilic as well as hydrophobic drugs can be entrapped within the multiple emulsions.

5. Protection of the loaded compound from inactivation by the endogenous factors.
6. Remarkable decrease in fluctuation of steady state concentration as compared to the conventional method of drug administration which is a common advantage for most of the new drug delivery systems.
7. Considerable increase in drug dosing intervals with drug concentration in therapeutic window for a relatively longer time.
8. Possibility of drug targeting to reticuloendothelial system (RES) and other organs.
9. Can be used in taste masking of bitter drug.

4.4 Preparation Techniques

In principle, multiple emulsions can be prepared by any of the numerous methods for the preparation of conventional emulsion systems, including sonication, agitation, and phase inversion (13). Great care must be exercised in the preparation of the final system, however, because vigorous treatments normally employed for the preparation of primary emulsions will often break the primary emulsion, resulting in loss of phase identity. Multiple emulsions have reportedly been prepared conveniently by the phase inversion technique. However, such systems have generally been found to have limited stability. It generally requires a very judicious choice of surfactant or surfactant combinations to produce a multiple emulsion system that has useful characteristics of formation and stability. Here in preparing multiple emulsion system, two different surfactants of opposite nature is used. One surfactant stabilizes the w/o (lipophilic) emulsion while the other stabilizes the o/w (hydrophilic) emulsion (3, 20). Emulsifiers get adsorbed at the surface of droplets during emulsion formation and prevent them from drawing close enough to aggregate. Phase diagrams can help in producing multiple emulsions by using only one surfactant to stabilize both the primary and secondary emulsion. The presence of a liquid crystalline phase in the ternary system consisting of water-emulsifier-oil, where the aqueous micellar solution of the emulsifier and the oil rich solution of the emulsifier are in equilibrium with each other, has been shown to greatly improve the stability of emulsions (21).

4.4.1 Two Step Emulsification Method (13)

It is the most common method for preparing multiple emulsions because it is very easy and gives high yield with reproducibility. Multiple emulsions may be prepared in the laboratory by the re-emulsification of a primary emulsion. A two stage procedure is therefore necessary. The first state involves the preparation of the primary emulsion either w/o or o/w type which is then re-emulsified with an excess of aqueous phase or oil phase in presence of second emulsifier to get the multiple emulsions of w/o/w or o/w/o type respectively. A general procedure for the preparation of a W/O/W multiple emulsion may involve the formation of a primary emulsion of water in oil using a lipophilic surfactant with a low HLB (2–8) suitable for the stabilization of such W/O systems. The primary emulsion will then be emulsified in a second aqueous solution containing a second hydrophilic surfactant (HLB 6–16) to promote o/w emulsification. Because of the possible instability of the primary emulsion, great care must be taken in the choice of the secondary dispersion method. The second emulsification step is carried out in a low shear device so as to avoid expulsion of internal droplets to the external continuous phase that could result in gross coalescence of the primary emulsion and the production of essentially ''empty'' oil droplets. Whatever method is employed, it is unavoidable that some of the internal aqueous phase is undoubtedly lost into the external aqueous phase. The evaluation of the yield of filled secondary emulsion drops, therefore, is very important in assessing the value of different preparation methods and surfactant combinations. The nature of the droplets in a multiple emulsion will depend on the size and stability of the primary emulsion.

In recent studies, a modified two-step emulsification technique for the preparation of multiple emulsions is being used whereby each step is being further divided in two sub-steps (pre-emulsification by sonication and than stirring) leading to a 2 x 2, two-step method of emulsification. This modified method gives high yield and stable emulsion (22, 23). Second step involving temperature and pressure as production parameters influences the size of oil droplets in w/o/w multiple emulsions. Fine w/o emulsions used for second step emulsification produced high encapsulation efficiency (24).

4.4.2 Phase Inversion Method (one step method)

There have been several reports (25, 26, 27, 28) on the one-step emulsification for the preparation of w/o/w double emulsions which included strong mechanical agitation of the water phase containing an hydrophilic emulsion and an oil phase containing large amount of hydrophobic surfactant. A w/o emulsion is formed, but it tends to invert and form a w/o/w double emulsion. In addition, double emulsions can be prepared by forming w/o emulsion with a large excess of relatively hydrophobic emulsifier and small amount of hydrophilic emulsifier followed by heat treating the emulsion until, atleast in part, it will invert. At a proper temperature, and with the right HLB of the emulsifiers, w/o/w emulsion can be found in the system. Phase inversion technique has a wide globule size distribution and indefinite number of encapsulated compartments. This is why this technique is not being used for preparing multiple emulsion of uniform size (29). A single step emulsification process for the preparation of multiple emulsions by introduction of two surfactants and two polymers has been recently reported. This method was more efficient in comparison to two-step emulsification, by using two polymers, which improves stability of emulsion (30).

4.4.3 Membrane Emulsification Technique

This is a new, convenient and more useful technique where a microporous glass membrane with a defined pore size can be used as emulsifying tool. It is based on the principle of dispersing one immiscible phase (dispersing phase) into other phase (continuous phase) by applying pressure. Other methods of preparation give high variability in the globule size but technique gives emulsion of desired size by passing through required membrane size (31, 32). The particle size of the w/o/w emulsion can be controlled with the proper selection of porous glass membrane and the relation between membrane pore size and particle size of emulsion exhibits good correlation as described by the formula:

$$Y = 5.03\ X + 0.19$$

Where X is the pore size and Y is the mean particle size of the multiple emulsions prepared by using membrane emulsification technique (33, 34).

A microporous glass membrane with narrow pore size range was used successfully for preparing stable simple (o/w) and water-oil-water (w/o/w) type emulsions. Narrow range of pore size was used for controlling size of emulsion and egg yolk phospholipids and soybean oil was employed as oil phase. Both emulsions were stable for 6 weeks when stored at 5°C (35). A homogenous w/o/w emulsions encapsulating cytarabine, doxorubicin and vancomycin were prepared by membrane emulsification. These emulsions were having effective encapsulation and controlled release delivery (32).

4.4.4 Microchannel emulsification

Microchannel emulsification (MC) is a novel technique devised for preparing monodispersed emulsions. A two-step emulsification process employing MC emulsification as the second step was used for preparing w/o/w emulsions. The behavior of internal water droplets penetrating the MC was investigated using ethyl oleate, and medium-chain triglyceride (MCT) as oil phases. They observed successful MC emulsification and prepared monodispersed oil droplets that contained small water droplets. The w/o/w emulsion entrapment yield was measured fluorometrically and was found to be 91% (36).

4.5 Formation of Multiple Emulsions

Recent advances in the formulation of multiple emulsions, and their various uses have attracted many workers. But formation of a stable emulsion with potential entrapment is still a task to the formulators. The presence of at least two surfactants, and stabilization of both w/o, and o/w emulsions in the same systems rendered it further difficult. Although a number of factors are expected to alter the quantity of water entrapped in, and the stability of multiple emulsions, no systematic study has so far been reported.

Frenkel et al (37, 38) have studied recently the inversion of w/o/w emulsions, to o/w systems as a function of H.L.B of external emulsifier, its concentration and emulsion drop size. Factors effecting the stability and drug release have been reported recently. Regarding the methods of preparation, a number of reports are available. But in almost all reports, simple preparation procedures are chalked out, without considering in detail, the different factors responsible for better entrapment and stability.

Hydrophilic – lipophilic balance of the emulgents, and oil phases are the foremost conditions that determine the emulgent interaction and orientation at the interphase of a dispersed system, which inturn determines its stability. A lot of work has been done in this field, in order to determine the critical H.L.B. that inherently provides stability to the system. When both sides of an oil globule meet at interface, the significance of emulgent interaction and orientation is increased further. In multiple emulsion systems the stabilizing factors are two important H.L.B. values, i.e., one for w/o and the other for o/w emulsion. This is evident because, the multiple emulsion consists of two interphases in the same system. Therefore, the stable entrapment, and the inversion conditions is greatly dependent on this H.L.B. of the multiple system.

Studies with a range of H.L.B. revealed, that stable multiple emulsion was obtained from stable w/o emulsion, at the critical weighted H.L.B. Below that point, inversion in the second step was quite difficult. On the other hand, at higher H.L.B during inversion, leaking out of aqueous droplets occurred, and a negligible entrapment remained. This study has indicated that the outer apparent H.L.B. might have an influence over the internal droplet stability.

The interfacial tension, or the surface free energy acts from two angles in the multiple emulsion system. Internal aqueous droplets coalesce, and oil droplets coalesce to ultimate phase separation. Thus emulsification process requires reduction of this free energy to sufficiently small value to obtain a stabilized system. By increasing the concentration of oil phase, the concentration of globules increases. This makes the globule more susceptible to coalescence, and renders the system less stable. At controlled conditions, with optimum emulgent concentration, entrapment is more, even at high phase volume ratio.

Entrapment of water in a multiple system is reduced due to high speed and prolonged agitation. Organic liquids have very low viscosity, and the density difference facilitates coalescence due to gravity, which ultimately ease out the leaking out of the entrapped globules. This renders the organic liquids unsuitable for multiple emulsions. Fixed or mineral oils are advantageous in formulation of multiple emulsions. Both formulation, and process variables, greatly influences the extent of water entrapment.

Oil

The nature of the oil can markedly affect the behaviour of the system. The oils used to prepare multiple emulsions include liquid paraffin, vegetable oils such as sesame oil, olive oil, arachis oil, isopropyl myristate and others. Mixtures of oils can also be used to minimize the differences in specific gravity between the oil and aqueous phase of emulsion (39) and to vary the viscosity of the oil phase in order to control the movement of solute across the oil membrane (40). In case of o/w/o emulsion system, the two oil phases can be same or different (41, 42, 43). The most widely used oils in oral preparations are nonbiodegradable mineral and castor oils that provide a local laxative effect, and fish liver oils or various fixed oils of vegetable origin as nutritional supplements. A novel o/w/o emulsion containing castor oil as the internal phase and a fluorocarbon as the external oil phase has been described for pulmonary delivery of the drug (42).

Selection of oil phase can affect various emulsion parameters like yield, release profile, particle size and emulsion stability. The mineral oils give much higher yield than the vegetable oils (3, 11, 44). The release of drug from multiple emulsions is affected by the nature of the oil phase due to difference in partition coefficient of different oil phases for the drug (45, 46, 47). W/o/w emulsions prepared with high viscosity oils tend to have larger particle sizes (48). Viscous oils produce w/o/w emulsions which are more stable in terms of percentage breakdown (49).

The two aqueous phases of w/o/w emulsions can be simple aqueous solutions of drugs, buffered solutions, aqueous suspensions of the drug, gelled aqueous phases and aqueous phases containing viscosity enhancers. Release rate can be modified by changing the p^H of the two aqueous phases (50). Increase in p^H difference between two aqueous phases destabilizes the w/o/w emulsions (51).

Surfactants

The selection of surfactant depends upon the use of the multiple emulsions. For the internal pharmaceutical use and as cosmetics, the toxicity of the surfactant is to be considered. Non–ionic emulgents are preferred due to their low toxicity, and they do not react with ionic compounds. Matsumoto et al (3, 20, 22) have found that non-ionic surfactants gave better yields of multiple drops, than the ionic surfactants.

It is necessary to use two surfactants for the preparation of multiple emulsions, one for the primary emulsion, and the other for the dispersion of the primary emulsion. The optimum concentration of the surfactant, to emulsify given oil can be determined, by the use of the hydrophilic lipophilic balance (HLB). In general in w/o/w emulsions, the optimal HLB value of the primary surfactant will be in the range of 2-7 and for the secondary surfactant 6-16. Equilibration of the system after mixing, will result in the transfer of surfactant between the aqueous and non aqueous components.

Concentration of surfactant also affects the emulsion yield. The use of very low or very high concentrations may not be able to stabilize the emulsions while use of very high concentration may cause toxicity (3). Matsumoto et al (25) suggested that the concentration of lipophilic surfactant span 80 required for 90% and more yield was more than 30% w/w of the oil phase but high concentration of hydrophilic surfactant in the external aqueous phase of w/o/w emulsion decreased the yield. Also, an excess of lipophilic surfactant can cause the inversion of w/o/w emulsion to simple o/w emulsion. For preparation of w/o/w emulsions,

Jager- Lezer et al (52) calculated the minimal lipophilic surfactant concentration needed to saturate the primary w/o interface by taking into account the primary emulsion composition, molecular mass of lipophilic surfactant, average diameter of the internal aqueous droplets and the molecular area occupied by the lipophilic surfactant on the saturated primary interface (determined by interfacial measurements). It was assumed that the oil phase formed a lipophilic surfactant reservoir when surfactant concentration in it was more than the above calculated value. Thus by these calculations optimal surfactant concentration can be found.

Phase volume

Matsumoto found, that internal phase volume of w/o has no significant effect on the yield of w/o/w emulsion. W/o/w can be prepared by using a wide range of internal phase volumes. An optimal (25-50%) internal phase volume can be utilized for the emulsion formulation. The secondary phase volume (w/o/w) influences the yield of multiple drops over a range of low volume fractions only (9, 25). Very high phase volume ratio (70-90%) had also been reported to produce a stable multiple emulsion (53, 54). It is very important to have proper order of phase addition while formulation and dispersed phase should be added slowly into continuous phase for the formulation of a stable multiple emulsion.

Nature of the entrapped material

The nature of the entrapped materials may have a bearing on the stability of the system. Due to the nature of the multiple emulsion, the middle phase may act as a membrane and osmotic effects may become significant. The entrapped solutions may interact with the surfactant or the surface active drugs may be adsorbed at the interphase, resulting in decreased stability.

Shear/agitation

High shear disrupts the large percentage of multiple oil drops and hence results in the instability of system due to tremendous increase in effective surface area. Therefore, with increased homogenization time, the yield of the system falls rapidly. Generally, high agitation speed is used for primary and low speed is used for secondary emulsification for the preparation of multiple emulsions. Very high and low shear rate drastically affect stability of emulsion system hence shearing/agitation time should be optimized. Higher shear stress causes incorporation of air and excessive frothing resulting in loss of surfactant at water-air interface while low shear does not reduce size of globules. Thus high shear combined with air lead to instability of multiple emulsions (3,22,25,49).

Temperature

Temperature has only an indirect effect on emulsification that is attributed to its effect on viscosity, surfactant adsorption and interfacial tension. An increase in the density difference between the oil and water phases result in a decrease in droplet size owing to the different velocities imparting to the two phases during emulsification. Rise in temperature augments the lipophilic character of the hydrophilic emulsifier, as it tends to precipitate. Generally, for the primary emulsion formulation temperature is kept at 70^{0}C, whereas for multiple emulsion preparation it is kept at 10^{0}C (55,56). Large temperature variations during manufacturing, storage, transport and use leads to drastic modifications within emulsions (57).

Rheology

The rheological properties of emulsions are influenced by a number of factors, including the nature of the continuous phase, the phase volume ratio, and to a lesser extent by particle size distribution. A variety of products ranging from mobile liquids to thick semisolids can be formulated by altering the dispersed phase volume and/or the nature and concentration of the emulsifiers. For low internal phase volume emulsions, the consistency of the emulsion is generally similar to that of the continuous phase; thus, o/w/o emulsions are generally thicker than w/o/w emulsions, and the consistency of an w/o/w system can be increased by the addition of gums, clays, and other thickening agents that impart plastic or pseudoplastic flow properties (58).

4.6 Stability of Multiple Emulsions

The complex nature of the multiple emulsions has hindered the study of their stability. A stable multiple emulsion should desire sufficient guarantee, that the dispersed phase will not undergo coalescence, or other physical deteriorations and at the same time, will maintain core droplets to remain inside the globules without inner coalescence, or leaking out (3).

The assessment of stability of the multiple emulsion is confined mainly to the tracer techniques, and to some extent rheological and other physico-chemical methods are used. Conductometric method was followed by estimating the concentration of ions, which migrate from the inner aqueous phase to the aqueous suspending medium, and claimed suitable, for control and stability studies. Permeation of inner aqueous phase has also been studied ensuing markers (59,60) like methylene blue or glucose, in core droplets by dialysis techniques.

Measurement of the number and size of multiple drops over a period of time gives a good indication of stability. The method of photo micrography is used, which is very tedious and involves the observations of random samples under the microscope and comparing the size of the observed multiple, internal and simple drops with the standard size. Apart from the above studies, an attempt was made to establish the method for estimating the stability of w/o/w multiple emulsions, with correlation between the viscosity and the dispersed phase concentration, using the theory postulated by Mooney (61). Later it was observed (62) that the rate of change of viscosity in an initial stage of aging was proportional to the osmotic pressure gradient between the two aqueous phases. Attempts (63) were also made, to analyse the data by swelling or shrinkage of the inner aqueous phase due to migration of water through the oil layer, and there by establishing the stability of multiple phase emulsions, Matsmoto et al (63) studied the water permeability and stability of w/o/w multiple emulsions microscopically, by measuring the rate of change in the size of dispersed globules.

However, it is apparent that conductometric method by tracer technique provides good prediction of stability of a multiple emulsion. Conductivity is related to loss of conductivity and dielectric constant is an important parameter, which could provide quantitative relation to core water permeation through oil layer, and the stability of the preparation. These two parameters, dielectric constant and loss conductivity have been shown to determine quantitatively the stability of simple emulsions and water entrapment in w/o/w emulsions.

The conventional stability testing methods for simple emulsions like freeze –thawing, centrifugation, temperature cycling, measurement of electrical properties and rheological behaviour may also be applicable to multiple emulsions.

A large number of factors affect the stability and drug release characteristics of emulsions. Factors such as method of preparation, viscosity, dispersed phase volume, ionic strength of the aqueous phase etc play important role in stability. As mentioned earlier, the situation becomes very complex in the study of multiple emulsions because at least a couple of surfactants are used as both w/o and o/w emulsions must be stabilized. There are two dispersed phases, two phase volumes which adds to the complexity, and consequently stability of the system. In fact stability of multiple emulsions and drug release is controlled by all factors controlling the stability of multiple emulsions, and also by a number of other factors specific to multiple emulsions.

The storage of emulsion samples at higher temperature slowly degrades the emulgent molecules. As the concentration of the surfactant is decreased due to degradation, interfacially absorbed emulgent molecules are desorbed into the bulk phase, which also degrade on aging at higher temperature. Higher amount of emulgent molecules remain adsorbed at the oil interface but successive desorption process denudes the interface. This ultimately leads to coalescence. The thermal motion of globules further accelerates the coalescence process, and finally oil phase separates out. The temperature effect appears to be more complex in multiple emulsions. The total interfacial area is more in multiple emulsions, since a large number of inner aqueous droplets create internal interface. The presence of opposite type of emulgent combination rather complicates the emulgent orientation at the interface. In multiple emulsions, in addition to desorption process, coalescence of oil globules depends upon the breaking out of aqueous core globules. The process renders the interfacial film sensitivity, and disrupt during leaking of inner globules, thus resulting in easy coalescence. Microscopic observation revealed that, small aqueous droplets also collide inside the oil globules come very close to inside of oil drops due to the density difference and ultimately oil layer ruptures to expel the water droplets. This phenomenon is somewhat analogous to the process of expulsion of water products in amoeba. The inherent force that seems to be the gravitational is based on difference of specific gravity of aqueous and oil phases. The changes in dielectric constant and loss of conductivity of simple and multiple emulsions revealed that, simple emulsions always exhibit uneven change indicating their dependence on rate of leaking out of inner aqueous phase. Even after long storage, some finely entrapped aqueous droplets remain in oil globules. The changes in dielectric constant and loss conductivity indicate that the initial rate of leaking was faster which decreased gradually.

The complex nature of multiple emulsions systems has hindered the study of their stability. Measurement of the number and size of the multiple drops over a period of time, gives a good indication of stability. This is achieved by photomicrography. This is a tedious process which involves the observation of random samples under the micro scope and comparing the sizes of the observed multiple, internal and simple drops with standard size circles on an eyepiece graticule.

As three different size distributions exist, it may difficult to determine, if the drops contain internal droplets, particularly if they are very small (<1, μm) or if they take up a large proportion of multiple drop, and the internal droplet interfaces therefore become indistinct. Small simple drops may pass below large simple drops, giving a false impression of multiple drops. Matsumoto et al (63) reported the preparation of w/o/w emulsions in which the internal droplets could not be seen under the microscope, because of reflected light from the surface of the oil drops. Another problem is possible increase in droplet diameter due to coalescence may be offset by decrease in droplet size due to shrinkage or loss of internal droplets from multiple drops. The problem with photomicrography is freezing the movement of droplets due to Brownian motion, particularly at higher magnifications. This method provides useful information on stability. This method is used by Davis and Burbage (64) and Florence and Whitehill (65) in the study of breakdown mechanisms. Davis and Burbage(66) avoided the problem, with very small internal droplets, by using a Freeze-etching electron microscope technique. From the electron micrographs the internal water droplets were clearly seen and could be seized.

A method of analysis based on the semi permeable nature of the oil film between the two aqueous phases of a w/o/w was developed by Davis et al (64). The emulsion was exposed to an osmotic gradient provided by electrolyte in the external phase. Droplet shrinkage resulted from the transport of water from the internal to external aqueous phase, as a consequence of osmotic gradient. The rate and the amount of shrinkage, related to the surface area and the volume of the internal aqueous phase was measured by coulter counter.

In 1978, using a graphical inflection technique, Davis and Burbage (64) found that initially during storage, mean diameter of the multiple drops fell as internal water was lost, but later increased due to droplet coalescence.

4.7 Factors Controlling the Stability of Multiple Emulsions

Electrolyte effect

Electrolyte presence appeared to be one of the most important factors in determining the stability and release of materials from the internal droplets (the effect of electrolytes is two fold (a) Osmotic and (b) Interfacial: The former being to multiple emulsions).

(a) Osmotic effects

Collings (9) found that w/o/w emulsion broke down rapidly in vivo, close to the site of injection, with the consequence that no significant delay in response to the entrapped drug was obtained compared to aqueous solutions of the drug. It was found that the premature breakdown of the emulsions in vivo was due to the fact that the middle oil phase was acting as a semi permeable membrane, between the two aqueous phases. The osmotic pressure in the external environment (body fluids) was higher than the internal phase, leading to shrinkage of the internal aqueous droplets and/or rupture of the oil layer.

Thus, water molecules may pass from one aqueous phase to the other. If the osmotic pressure is higher in the internal aqueous phase, water may pass to this phase resulting in swelling of the internal droplets which eventually burst, releasing the contents. The reverse is true, if the osmotic pressure is higher in the external aqueous phase, causing shrinkage of the internal droplets. If the osmotic pressure difference across the oil layer is extreme, then passage of water is so rapid that almost immediate rupture of the oil drops occurs with expulsion of the internal droplets. This appears to occur frequently where the oil layer is thin. Materials other than electrolytes, such as protein and sugars and coarse drugs, in either aqueous phase can also exert this effect. Collings (9) partially solved this problem by the addition of small amounts of sodium chloride to the internal aqueous phase, so that this phase was isotonic with the final external phase, but this approach can lead to inequality of pressure on storage which is unsatisfactory. The osmolarity may also be adjusted by the addition of other materials such as glucose or glycerol.

When the oil layer ruptures the inner aqueous phase in the droplet disappears instantaneously followed by mixing of the internal aqueous phase with the external aqueous medium leaving a simple oil drop.

Davis and Burbage (64) investigating the effects of sodium chloride on the size of the multiple drops found that a threshold concentration of electrolyte exists below which little or no change in size occurs. This appears to be related to the drop diameter – larger multiple drops required a higher concentration of sodium chloride to effect shrinkage. Electrolytes then may be used to control the rate and amount of materials transferred from one phase to the other.

Matsumoto and Khoda (68) measured the water permeation coefficient of the oil layer indirectly by measuring the change in viscosity of the w/o/w emulsion to determine change in globule diameter.

$$\phi_w = -PaA(g_2c_2 - g_1c_1) \qquad .....(4.1)$$

Where Pa is the osmotic permeability coefficient of the oil membrane ϕw is the translayer flux of water in moles per unit time. 'A' is the area of the membrane. 'g' is the osmotic coefficient and 'c' is the solute concentration. The trans-layer flux of water can be replaced by (d ϕ_w/dt) concent (cm^3 s^{-1}), because ϕw is the volume of the inner aqueous phase in unit volume of the system, while the area can be replaced by the total surface area of the dispersed drops, in unit volume of freshly prepared emulsion. The osmotic permeability coefficient is used to calculate the diffusion coefficient for water in the oil layer and so is useful in determining the mode of water transport under an osmotic gradient.

(b) Interfacial effects

Brodin et al (69) observed that factors other than osmotic gradients affecting passage of the drug. They suggested that sodium chloride competes for surfactant, for water molecules at the inner w/o interface, which would result in a rigid interfacial layer, which would be a more effective mechanical barrier to drug transfer.

The mechanisms of instability in multiple emulsions are complex and difficult to study. W/o/w multiple drops may coalesce with other oil drops and may lose their internal droplets by rupture of the oil layer, on the surface of the internal droplets, leaving simple oil drops. Under the influence of an osmotic gradient, the middle oil phase acts as a semi permeable membrane, resulting in the passage of water across the oil phase. This leads to either swelling or shrinkage of the internal droplets, depending on the direction of the osmotic gradient. Another possible breakdown mechanism may be coalescence of the internal aqueous droplets within the oil phase. A combination of these mechanisms may take place and the exact breakdown mechanisms remain unclear. The likelihood of events taking place may be predicted by analysis of the Vander Walls attractive forces and free energy changes in these systems (3).

4.8 Mechanisms of Instability (19)

A stable emulsion is considered to be one in which the dispersed droplets retain their initial character and remain uniformly distributed throughout the continuous phase during the desired shelf life. There should be no phase change or microbial contamination on storage, and the emulsion should maintain elegance with respect to odour, colour, and consistency. Instabilities of both chemical and physical origins can occur in emulsion formulations. Emulsion stability is a phenomenon, which depends upon the equilibrium between water, oil and surfactants.

Main mechanisms by which instability occurs in multiple emulsions of w/o/w type have been studied thoroughly and can be summarized to following possible mechanisms (70,71,72).

(i) Aggregation of the internal and multiple emulsion droplets,

(ii) Breakage or disruption of oil layer on the surface of internal droplets

(iii) Seepage of the contents from inner droplet in external phase.

(iv) Shrinkage and swelling of the internal drops due to osmotic gradient across the oil membrane.

(v) Phase separation

4.9 Stabilization of Multiple Emulsion

A number of approaches had been tried out to improve the stability of multiple emulsion and many have been suggested for further investigations. Several reviews had been reported on the stabilization and mechanisms of drug release that concentrates on different mechanism and methods of stabilization (70,71,72).

4.9.1 Stabilization by Forming Polymeric Gel (19)

Gelation of either internal or external phase or the oily phase (membrane) has been employed in various studies to stabilize the system. Stable w/o/w emulsion was formulated by polymeric gel in the internal or external aqueous phase either by cross linking of surfactant molecules through high energy γ-radiation or by polymerization of monomers dissolved in aqueous phase. These methods are based on viscosity enhancement of aqueous phase thereby improving stability of emulsion droplets (73,74).

Stable multiple emulsion were achieved by incorporating poly (acrylamide) (3), gelatin or aminoacid (75). Synthetic polymer (Carbopol 974P) and chemically modified cellulose (Hydroxypropylcellulose) were used as gelling agent for multiple emulsion preparation (76). A stable multiple emulsion was developed by incorporating a co-emulsifier into the external water phase. Incorporation of the co-emulsifier increases the compatibility, elasticity and temperature stability of multiple emulsions (77).

4.9.2 Stabilization by Increased Viscosity

Restricting the mobility of the active matter in the different compartments of the double emulsion will slow down coalescence and creaming, as well as decrease the transport rates of the drug or the marker from the water phase through the oil membrane. Attempts were made: (1) to increase the viscosity of the internal aqueous phase by adding gums/hydrocolloids to the inner water phase. Such a thickener may affect also the external continuous phase since the entrapment is not quantitative and the yields of entrapment are limited and emulsifier-dependent; (2) increase the viscosity of the oil phase (fatty acids salts); and (3) thickening or gelling the external water by gums is limited only to cosmetic or similar applications in which semisolid emulsions are directly applied. Some of these examples are topical skin-care products, creams, and body lotions (78,79). Double emulsions that were solidified after preparation may suffer from destabilization effects. This phenomenon is scarcely considered but in practice it can occur very often. The solidification occurs because of temperature changes during transport or storage. Clausse and coworkers (80,81) studied the phenomenon in W/O/W emulsions by microcalorimetric (DSC) techniques. It was concluded that, out of thermodynamic equilibrium, double emulsions may suffer from water transfer during the solidification. This phenomenon occurs even if partial solidification takes place. In addition, a change in the size distribution of emulsion droplets is observed. The mean diameter of the droplets in the W/O emulsion may shift toward O/W emulsion and the double emulsion can invert. Therefore, it is not always obvious that increasing the viscosity, gelation, or partial solidification improves emulsion stability.

Additives in internal aqueous phase

Electrolytes have been used for stabilizing w/o/w emulsion. These electrolytes include ascorbic acid, acetic acid, sodium chloride and sodium citrate. The addition of lysozyme to the aqueous phase or stearylamine and oleic acid to the oil phase effectively improves the stability of the oil layer. The influence of various additives, e.g. glucose, sucrose, acetic acid, citric acid, ascorbic acid, sodium chloride and sodium acetate on the stability of water-in-olive oilin- water multiple emulsion had been studied and it was observed that glucose and sucrose could increase the viscosity and as a result improves stability of emulsion system. Additives in the internal aqueous phase make it hypertonic and stabilize the system. Microscopic study conducted on such emulsion indicated that as the dextrose concentration (0-2.5%), in internal aqueous phase increased the stability recorded in terms of degree of coalescence of internal droplets and rupture rate of interfacial oily layer, was also increased(22,82).

Addition of sodium chloride and hydrophilic surfactant into the internal aqueous compartment of the multiple emulsions also influences the breakdown of the suspending vesicular globules (83). The stabilization of w/o/w emulsion by making inner aqueous phase hypertonic, addition of chitosan in the inner aqueous phase and phase inversion with porous membrane and its application to transcatheter arterial emobilzation (TAE) therapy was reviewed (72). For long term stability of the double emulsion, it should have a balance between the Laplace and osmotic pressures within emulsion (84). Recently, a substituted cyclodextrin (Hydroxypropyl-cyclodextrin) in internal phase effectively stabilizes antioxidant (Kaojic acid) in w/o/w multiple emulsions (85,86).

4.9.3 Modulating Surfactant Concentration

Several attempts have been made for stabilizing multiple emulsions by modulating emulsifiers systems (87,88,89,90). A suitable surfactant with required HLB is used for preparing stable emulsion. Emulsifiers stabilize emulsions through mechanical or electrical interfacial barrier formation or by reduction of the interfacial tension or both (91). For long term emulsion stability, the strength of the mechanical barrier is more important than interfacial tension (92). Therefore, an optimal HLB for a particular multiple emulsion system depends on the concentration of the primary and secondary emulsifiers (93). Type and concentration of surfactant are the primary factors which influence the stability, while pH and ionic strength insignificantly influences stability of emulsions.

Mixture of emulsifiers (spans and tween) was used to improve stability of the multiple emulsions. Mixed emulsifier gives higher film strength in comparison to single emulsions. This hypothesis was confirmed because addition of bovine serum albumin (BSA) in internal aqueous phase did not improve stability of emulsion (94). The ratio of hydrophilic and hydrophobic surfactants is very important in optimizing multiple emulsions. Long-term stable water-in-oil-in-water multiple emulsions was developed by modulating concentrations of span 83 and tween 80 (95).

4.9.4 Formation of Interfacial Complex Films

Polymerized interfacial membrane has been found to be more effective than previous methods in stabilizing multiple emulsions. It is because resultant interfacial membrane restricts surfactant migration from one interface to other, which is one of the causes of instability in w/o/w emulsion (90). This mode of stabilization can be executed either through *in situ* polymerization at the interface or by interfacial interactions between a polymer and a surfactant. During the search for stable w/o/w emulsion as vehicle for sustained release preparation, numerous researchers have employed interfacial complexation between non-ionic surfactants and macromolecules (3). Another approach to obtain stable w/o/w multiple emulsions is based on the interfacial interactions between a macromolecule, e.g. albumin (96) or polyacrylic acid or bovine serum albumin (97) in internal aqueous phase and lipophilic nonionic surfactant poloxamer 331 in middle oily phase, resulting into formation of stable w/o emulsion. Hydrophilic poloxamer surfactant of high molecular weight, e.g. poloxamer 403 enhanced stability of w/o/w emulsion of the above-mentioned primary emulsion. Interfacial membrane was so strong that it withstood the thinning of oil phase caused by swelling of internal aqueous droplets because of osmotic pressure of influxed water (90, 98). Interfacial complex films between polyvinyl alcohol, acacia, polyvinylpyrrolidone and sorbitan monooleate were used for preparing long-term stability of the emulsions bearing rifampicin (45). Stable w/o/w emulsions were prepared by interfacial films between various pluronic F127: polyacrylacid complexes in the internal aqueous phase and the lipophillic surfactant in oil phase. These emulsions were capable for rapid or sustained drug delivery depending on the pluronic F127: polyacrylacid complexes and type of lipophilic surfactant used (99). The significance of the inner and outer phase pressure, as well as interfacial film strength on w/o/w multiple emulsions stability was studied by using microscopy and long term stability tests. Stability was correlated with the interfacial film strength (measured by interfacial elasticity) of the hydrophobic surfactants at the mineral oil/external continuous aqueous phase interphase. Authors concluded that metastable dimpled structure and long term stability of multiple emulsion was dependent on the osmotic pressure of the inner droplets (100).

Steric stabilization

Steric stabilization had been found to stabilize emulsions. It is due to the ability of steric cloud to prevent their instability in emulsions by surrounding environmental challenges. Steric stabilization can be achieved by using either adsorption or grafting of natural polymers or by using polymeric surfactant or lipid grafted polymers. Multiple emulsions can be stabilized by coating them with hydrophobized polysaccharides (101,102).

4.10 Characterization

Emulsions are mostly characterized by the size distribution of the droplet and other physical proerties such as dielectric properties, optical properties, thermal behaviour, rheological properties, and other microscopic and macroscopic observation.

Macroscopic examination: Primary observations like color, consistency and homogeneity are frequently used to ensure formation of an emulsion. Type of multiple emulsions formed (w/o/w or o/w/o) can be validated by dilution with the external phase (3).

Microscopic examination: The optical microscopy method, calibrated ocular and stage micrometer can be used for globule size determinations of both multiple emulsion droplets as well as droplets of internal dispersed phase(3). This method is simple and provides useful information on the character of multiple emulsions and allows determination of droplet size and size distribution. A suitable magnification may be used for the purpose. However, this method has two main drawbacks; (1) simple small drops passes to form large simple drop which gives a false impression of multiple nature (2) if the internal droplets are very small in size these cannot be viewed due to reflection of light from the oil droplets surface (104). Various other techniques used to characterize emulsions like Coulter-counter, freeze- fracture electron microscopy and scanning electron microscopy are also used to determine average globule size and size distribution of multiple emulsion droplets.

Number of globules: Number of globules per cubic millimeter can be measured using a hemocytometer cell after appropriate dilution of the multiple emulsions. The globules in five groups of 16 small squares (total

80 squares) can be counted and the total number of globules per cubic mm can be calculated using the formula (105);

$$\text{No. of Globules / mm}^3 = \frac{\text{No. of Globules} \times \text{Dilution} \times 4000}{\text{No. of small squares counted}} \quad .....(4.2)$$

Entrapment efficiency/percentage drug entrapment/emulsion yield

Two methods, size analysis and internal tracer technique have been used for evaluating yield of multiple emulsion droplets (106). Particle size distribution analysis of simple and multiple droplets of total system can be estimated for emulsion yield (64,107). Several techniques (photomicrography, coulter counter, electron microscopy) can be used for evaluating particle size analysis. Second method is the use of the internal tracer to establish entrapment efficiency of an impermeable marker molecule in the w1 phase of emulsion. The method basically involves determination of unentrapped marker or drug. Emulsion yield can be measured by dissolving a marker compound in the inner aqueous phase of emulsion and measuring concentration of marker in the outer aqueous phase. Yield can be determined by subtracting the remaining amount of marker from the original amount (93,108). Dialysis (22,109), centrifugation, filtration and conductivity measurement (110) are the four methods generally used for determination of entrapment efficiency by this technique. Several markers have been used like glucose (20), hydrogen ions (111,112)], electrolytes (113,114), new coccine (115,116), ionic drugs (Ephedrine HCl), dyes (sulphane blue, polytartarazine), radioactive tracer (tritiated water), etc (117)

The primary multiple emulsion was prepared, containing a small amount of the market material (glucose) in aqueous phase. Immediately after preparation, the w/o/w emulsion was dialysed against a certain volume of distilled water, and the quantity which has migrated to the external aqueous phase, was determined by microanalysis. The yield of multiple drops was obtained as follows:

$$\text{Yield \%} = 100 - 100\ a/\ (C/V_1 + V_2 + V_3) \quad .....(4.3)$$

Where 'a' is the quantity of migrated glucose in g/ml of distilled water, C is the original weight of glucose in the inner aqueous phase, and V_1, V_2, V_3 are the volume of the inner aqueous phase, external aqueous phase and distilled water respectively. During the 20 hours period of dialysis some deterioration in the system may take place.

Kita et al. (118) reported a similar method involving the measurement of the passage of ions into the external aqueous phase due to rupture of the oil phase. The w/o/w emulsion containing a small amount of sodium chloride in the internal aqueous phase was placed behind a dialysis membrane and ion-sensitive electrodes used to detect the quantity of migrated ions. The multiple drop yield was calculated as:

$$\text{Yield } \% = 100 - 100\text{ x }/V_1 \quad \text{.....(4.4)}$$

$$\text{Where X (ml)} = A\,(V_o + V_d)\,/\,(B\text{-}A) \quad \text{.....(4.5)}$$

and A & B are the concentrations of the particular ion in distilled water and aqueous phase respectively, X(ml) is the volume of the leaked out internal aqueous phase and V_1, V_0 and V_d are the volume of the internal aqueous phase, external phase and distilled water respectively. As water is lost from the internal to external aqueous phase in a w/o/w emulsion, the viscosity of the external phase should decrease somewhat and as the overall viscosity of the emulsion is dependent on continuous phase viscosity. This should be reflected in the rheological properties of the w/o/w system. In view of this Kita et al., attempted to estimate the stability of w/o/w emulsions by following viscosity change with time.

Yield of w/o/w multiple emulsions was improved by addition of various additives like sodium alkylsulfonate, sodium alkylcarbonate, tweens, sodium carbonate, sorbitol in the internal aqueous phase. This increase has been attributed due to decrease in interfacial tension between the internal aqueous phase and oily phase of emulsion (116). Similar results had been reported by the addition of sodium chloride in inner aqueous phase on drug entrapment in w/o/w emulsions containing tryptophan as model drug. Drug entrapment was decreased when sodium chloride was added in the outer aqueous phase (119).

Rheological analysis: Viscosity measurement is an important parameter for the characterization and stability studies of multiple emulsions. Viscosity changes over time show the volume fraction instability of the dispersed globules and therefore multiple emulsion instability. In case of an unstable emulsion system, oil layer ruptures and inner aqueous globules merge with the continuous aqueous phase and disappear instantly. Volume fraction of the dispersed phase decreases gradually in emulsion with increasing oil globules rupture lead to decrease in viscosity of emulsion system (120,121). Solutes and water permeate through oil layer of vesicles causing change in viscosity (122). Newtonian viscosity of the diluted w/o/w emulsion could be influenced by osmotic pressure gradient between the inner and external aqueous phases. This osmotic drift can be on either side of the emulsion depending on viscosity of the aqueous phases and can be used to estimate water permeation coefficient as well (123, 124). Rheological analysis of multiple emulsions is very important as it determines the *in vitro* emulsion stability. Brookfield, rotational or cone and plate viscometers are widely used for viscosity determination of multiple emulsions (22, 110). Elucidation of drug release kinetics can be done by using rheological analysis. Viscosity measurements, can give information about release mechanism from w/o/w emulsions. W/o/w emulsions with sodium lactate in the internal aqueous phase were prepared and diluted with various concentrations of glucose solutions. The diluted emulsions were subjected to rheological and conductometric analysis. Emulsions diluted with glucose solution of certain concentration showed no variation in the apparent viscosity (indicating the stability of multiple droplets) but showed an increase in conductivity value, revealing release of electrolyte through oil membrane and not due to the rupture of oil film. When the w/o/w emulsion was diluted with distilled water, the rheogram showed an increase in apparent viscosity followed by a decrease. This was attributed to initial movement of water from external to internal aqueous phase, leading to swelling of multiple droplets, followed by their rupture.

Rheological behaviour of w/o/w emulsions was studied by kawashima et al using cone and plate type viscometer. A negative thixotropic behaviour became more pronounced and the apparent viscosity increased upon increasing the shear rate, prolonging the shear time, or repeating the shear. Further shearing caused a rapid increase in the shear stress of emulsion and induced phase inversion. This phase inverted emulsion of w/o type and in a semisolid state. This type of rheological behaviour was attributed to the increase in the volume fraction of the oil droplets and by coalescence of the oil droplets upon shearing.

Zeta potential: The electrophoretic mobility of dispersed globules under the influence of applied voltage in zeta-potentiometer can be used to measure the zeta potential and surface charge of multiple emulsions. The apparatus consists of cylindrically bored microelectrophoresis cell equipped with platinum-iridium electrodes to measure the electrophoretic mobility of the diluted w/o/w emulsion (125). It can be calculated by using following equation

$$\zeta = \frac{4\pi\eta u}{\varepsilon E}$$

Where,

ζ = Zeta potential (mv)

η = Viscosity of the dispersion medium (poise)

μ = Migration velocity (cm/sec)

ε = Dielectric constant of the dispersion medium

E = Potential gradient (Voltage applied / distance between electrodes)

Percentage phase separation: Phase separation is a phenomenon by which one phase of emulsion gets separated due to coalescence. Percentage phase separation is the volume of phase in percentage separated from the total volume of emulsion after storage. 20 ml of freshly prepared w/o/w emulsion is kept in 25 ml graduated glass cylinder and allowed to stand for defined period at 40°C. The volume of separated aqueous phase (Vsep) is observed periodically at regular intervals and percentage phase separation (B) is calculated as follows (126);

$$B\ (\%) = 100\ (Vsep/20)\ /\ [(V_1+V_2)\ /\ (V_1+V_2+V_0)]$$

Where V_1, V_2, V_0 represents volumes of inner aqueous phase, dispersion phase and middle oil phase.

In-vitro drug release: The drug release from w/o/w emulsion is generally determined by dialysis method using cellophane tubing. Typically, 5 ml of (w/o/w) multiple emulsion is placed in the dialysis tube which is then tied at both ends by thread and placed in basket (usually 100 rpm) and dialyzed against specified dissolution media (usually 200 ml) at 37 ± 1°C. A sink condition is maintained and samples are withdrawn at different time intervals and replaced by fresh dissolution media (127,128,129).

4.11 Drug Delivery Mechanisms

The release from W/O/W multiple emulsion occurs either by transport through the oily membrane or by its breakdown. In the first case, and depending on the affinity of the molecule for the oily phase, the transport is due to molecular diffusion (Fick diffusion) or diffusion facilitated by certain surfactants that take on the role of carriers.

Release by breakup after swelling

The swelling/breakdown process occurs only if there is a concentration gradient between the internal and the external aqueous phases. The resulting osmotic water flow is observed to travel from the external phase to the internal one, thereby reducing the concentration gradient. This water flow causes the drops to swell until a critical size is reached. Beyond the critical size, a breakup of the oily membrane occurs. The development of such osmotic swelling has been considered by various authors (130,131) not for inducing a controlled release but for estimating the stability of multiple emulsions. In order to prevent the swelling/breakdown kinetics from the beginning, immediately after making the multiple emulsion, which is as soon as a concentration gradient is established, very concentrated multiple emulsions are prepared (typically the volume fraction is higher than 0.75). At such a high concentration the dispersed drops are close packed, so there is not enough room for the drops to swell. When the multiple emulsions are very concentrated, they remain stable with respect to the swelling/breakdown mechanism. When the multiple emulsions are diluted, the swelling/breakdown mechanism can occur.

The swelling/breakdown kinetics depends on various parameters: the concentration gradient, the nature, and the concentration of the two surfactants, and the properties of the oil (132) found that maximum swelling was obtained for the highest concentration gradient. Jager - Lezer et al., (133,134) studied influence of the lipophilic surfactant concentration and found that the maximum swelling increases with the surfactant concentration and the rate of release decreases with the surfactant concentration. The same results were found using lipophilic surfactants of different properties, both polymeric and not. The higher swelling, when the release is lower can be explained as follows: the excess of the lipophilic surfactant, initially located in the oily membrane, migrates at the external interface and fills up the free spaces. As a result breakdown can be delayed and even prevented.

Release by breakup under shear

The release under shear of an active molecule that is initially encapsulated in the aqueous phase of a W/O/W multiple emulsion is a very promising phenomenon for applications in cosmetics or pharmaceuticals. Taylor (135,136) was first to study the deformation of molecules under shear and their bursting in a simple, dilute emulsion. He considered that breakup occurred when shear stress exceeds cohesion stress. He defined this breakup by way of a capillary number. Bursting occurs when the capillary number exceeds a critical value close to unity. The relation shows that the thickening of the continuous phase induces the globules to burst.

4.12 Bioavailability

The administration of emulsions intravenously, presents some interesting relationships between physico-chemical properties and physiological response. The clearance of an emulsion from the blood is determined largely by its interaction with the reticuloendothelial system. It can be stated that:

1. Fine particle size emulsions are cleared more slowly than course particle size emulsions.
2. Negatively charged and positively charged particles are cleared more quickly than neutral particles.
3. Emulsions stabilized by low molecular weight emulsifiers are cleared more rapidly than those stabilized by high molecular emulsifiers.

All particles in the blood carry a negative charge, thus a so called positively charged particle, will become negative on injection through the adsorption of blood components, which could dictate subsequent fate of the particles. The same applies to other colloidal particles, liposomes and their phagocytosis is pertinent to the emulsion situations. The inter relation between surface characteristics and take up by phagocytic cells can be determined by using experimental animals or in vitro turbidimetric methods using polynuclear macro phases.

Using mitomycin – C and bleomycin, Davis et al. noted, that water in oil emulsion produced a much greater delivery of drug into the lymph system, than an o/w emulsion. Sezeki et al therefore suggested that encapsulating a water insoluble drug in oil was highly advantageous for the transport of drug from the intestinal spaces to the lymphatic capillaries. They also noted that the concentration of oil was much higher in the lymph nodes than the thoracic lymph, and that it rose to a maximum quickly and decreased in the former, where it was relatively constant in the later.

Microscopical observations made of the muscle fibers and the regional lymph nodes showed that a multiple w/o in oil internal fluid emulsion was formed in the muscle tissues upon injection. With microsphere in oil emulsion, microsphere (1-2 diameter) were observed dispersed in the oil droplets. The release of the drug is primarily due to the breakdown of the emulsion. The lymph nodes also contained oil droplets smaller in size than in muscle tissues, with aqueous droplets dispersed inside the oil, but also some aqueous droplets escaped from the oil.

Theoretical treatments have been considered to investigate the role of interfacial transfer when mass transfer occurs through polyphase media.

Four type of examples are discussed

1. Transfer across the interface is so small that the oil phase droplets act as impermeable shears and only a little solute enters the internal phase.
2. The transfer rate is still relatively slow, but the mass transfer across the polyphase media is influenced.
3. The solute distribution is relatively rapid and equilibrium between the two phases is essentially maintained.
4. When transfer kinetics are very rapid, the oil droplets are fully equilibrated, and both phases contribute to the effective diffusion coefficient of the heterogeneous system.

4.13 Applications of Multiple Emulsions

From the last 25 years, multiple emulsions had been investigated vigorously with revolution in emulsion technology. Water-in-oil-in-water (w/o/w) multiple emulsions have several potential applications in pharmaceuticals, food technology, separation sciences and in cosmetics. The various pharmaceutical applications include immobilization of enzymes, red blood cell substitute, transdermal delivery, bioavailability enhancement, taste masking, drug targeting, prolonged delivery of drugs etc.

Prolonged/controlled drug delivery

The rationale behind use of multiple emulsions as prolonged and controlled drug delivery systems is that the drug present in the innermost phase has to cross several phases before it is available for absorption for the system. The potential of using controlled release multiple emulsion has already been reported by several workers and a number of research papers are available, these have been summarized in Table 4.1. Different categories of drugs have been tried out for the sustained delivery having very short half-lives. Both oral and parenteral prolonged delivery had been investigated. In parenteral delivery, multiple emulsions act as reservoirs of the drug in blood and releases their content in controlled fashion. W/o/w emulsions for parenteral delivery are more convenient to handle, use, and inject due to lower viscosity of these systems.

Table 4.1 Multiple Emulsions Used for Prolonged/Controlled Drug Delivery.

Category	Drug investigated
Analgesic and antipyretic agents	Indomethacin
	Diclofenac sodium
	Paracetamol
Anticancer agents	Cytarabine
	Doxorubicin
	Etoposide
	5-fluorouracil
	Methotexate
	Tegafur
Antimalarials	Chloroquine
Antitubercular agents	Rifampicin
	Isoniazid
Antiasthmatic agents	Salbutamol
	Theophylline
Antibiotics	Cefadroxil, Cephradine
	Nitrofunrantoin
Antihistaminic	Chlorpheniramine
Peptides	Vancomycin
	Insulin
Miscellaneous	Salmon calcitonin
	Prednisolone
	Antipyrine, 4-Aminoantipyrine
	Pilocarpine
	Naltrexone
	Phenylephrine
	Pentazocine

As oxygen substitute

Multiple emulsions had been tried out as a substitute to blood where hemoglobin has been incorporated in the inner phase of emulsion. A stable hemoglobin (Hb)-in-oil-in-water (Hb/o/w) multiple emulsion was prepared to simulate red blood cell (RBC) properties. Multiple emulsions serve as the liquid membrane across which gases (O_2 and CO_2) are exchanged with hemoglobin incorporated in the inner aqueous phase. But from the last one decade there is no report for multiple emulsion investigated for this purpose.

As bioavailability enhancer

Multiple emulsions have received great interest in facilitating the absorption of water soluble compounds through gastrointestinal tract which do not normally gets absorbed due to instability in the presence of certain physiological, ionic/enzymatic environments in the gastrointestinal tract like proteins, peptides etc. For highly polar drugs also, multiple emulsion system has found a great role in enhancing absorption due to lipophillic characteristics. It is well-known that lipid material like oils, fats, and lipophillic drugs are being absorbed by lymphatic rather than portal system by- passing the hepatic first pass metabolism and pouring the drugs directly to the systemic circulation. Multiple emulsions have also been used to improve bioavailability of those drugs, which have high first pass metabolism. Multiple emulsion increases bioavailability of drugs either by protecting these in physiological, ionic /enzymatic environment in the GIT where otherwise these gets degraded like proteins, peptides or bypassing the hepatic first pass metabolism. Table 4.2 shows a list of drugs investigated for this purpose.

Table 4.2 Multiple Emulsions Used in Bioavailability Improvement/Enhancement.

Category	Drug investigated
Antidiabetic agents	Insulin
Anti-fungal agent	Griseofulvin
Antitubercular agents	Isoniazid
Antibiotics	Vancomycin
	Nitrofurantoin
Miscellaneous	Pyrenetetrasulonic acid tetrasodium salt (PTSA)

Targeted drug delivery system

Site specificity is very important prerequisite for any pharmacotherapy. Ideal approach for a drug delivery system is to deliver drug only at the diseased tissue/organ and not affecting other undiseased tissue (Drug targeting). Drug targeting is an important concept, which minimizes the adverse effects of drugs by specifically concentrating drug in diseased tissue. Several micro particulate systems have been used for drug targeting. It is of particular interest for the cytotoxic drugs (Anti-cancer agent) because of high toxicity for non-diseased tissue. Multiple emulsions system had also been prepared for several cytotoxic agents, targeting different tumors. These can be used as lymphotropic carriers for drug targeting. *In vivo* fate of multiple emulsion involve its uptake by reticuloendothelial system (RES) thereby concentrating these oily vessels in lymphatic system and lymph node. These are of immense potential for the lymphatic targeting; other organs (Liver, brain etc.) have also been targeted. Table 4.3 gives an overview of the potential of multiple emulsions in therefore mentioned areas.

Table 4.3 Multiple Emulsions Used for Drug Targeting.

Target (Tissue /organ)	Drug investigated
Lymphatic system	5-fluorouracil
	Iodohippuric acid
	Bleomycin
	Isoniazid
Tumor	Bleomycin
Brain	Rifampicin
Liver	5-fluorouracil
	Epirubicin
Lungs	Rifampicin
nflammatory tissue	Diclofenac sodium

Taste masking

Multiple emulsions had been investigated for the taste masking potential of the bitter drugs. Water-soluble bitter drugs had been incorporated in the inner aqueous phase of w/o/w multiple emulsion, which is being surrounded by oil layer and leading to masking the taste of the bitter drugs. Several biocompatible/edible oils had been used for masking taste and improving the masking potential. Taste masking of the chlorpromazine was carried out by multiple emulsions. Chloroquine, antimalarial agent is a known bitter drug and can be used as model drug for taste masking. Multiple emulsions have masked bitter taste of chloroquine significantly.

Enzyme immobilization

Enzyme immobilization technique involves the entrapment of enzymes, which catalyze several reactions. Reports of using multiple emulsions for enzyme immobilization goes back to 1972 where hydrocarbon based multiple emulsion were used to entrap urease enzyme for kidney diseases. Later same worker immobilized enzymes and whole cell from *M. denitrificans* for waste water treatment. Presently this technique has been established as a tool for immobilization of various enzymes, proteins, amino acids etc in biotechnology. Immobilized alcohol dehydrogenase has been utilized for the conversion of alcohol to acetaldehyde for industrial use. Like wise Makryalese and coworkers carried out conversion of ketoisocaproate to L-Leucine by L-Leucine dehydrogenase. Enzymatic production of aminoacid (L-Phenyl alanine) from immobilized chymotrypsin was previously reported. Similarly scientists used immobilized lipase enzyme for the hydrolysis of fatty acids. Enzymatic conversion of highly lipophillic, water insoluble substrates like steroids have also been carried out

Drug over dosage treatment/detoxification

Multiple emulsions has not been much investigated for the overdosage treatment. For the first time, Liquid membrane was proposed for the drug overdosage treatment in 1978. This system could be utilized for the overdosage treatment by utilizing the difference in the pH in different compartments of multiple emulsions, which affects the ionization behavior of the drug. These have been utilized for acidic drug overdosage treatment like barbiturates. When emulsion is administered orally, acidic pH of stomach becomes external aqueous phase where barbiturate exists mainly in unionize form which would easily transfer through the oil phase into the inner aqueous phase across the concentration gradient. Basic buffer in inner phase ionizes the barbiturate and prevents its backward transport. Thus, entrapping excess drug in multiple emulsions treats over dosage. Overdosage treatment of quinine sulfate has been reported. Multiple emulsions have been used for detoxification of blood also.

Topical applications

Only few studies have been reported on topical administration of multiple emulsions. First published study shows that multiple emulsions release their contents more slowly than solutions. A typical three-phase emulsion containing model compounds testosterone, caffeine and tritiated water was formulated and the effect of perfluropolymethylisopropylether

(Fomblin HCl) on percutaneous absorption was investigated. Fomblin decreased the percutaneous absorption of testosterone but increased water permeation while in case of caffeine the flux through skin was not affected. A topical w/o/w emulsion containing active substance in each phase: sodium lactate (moisturizing agent) in inner phase; spironolactone (anti acne agent) in intermediate oily phase; and chlorhexidine digluconate (antibacterial) in outer phase was prepared to understand formation mechanisms. The compatibility of these ingredients in the system was evaluated and release of sodium lactate was observed. Ferreira and coworkers developed w/o/w multiple emulsions carrying metronidazole and glucose and compared their percutaneous release with w/o, o/w emulsions. Absorption of metronidazole was similar from w/o/w and o/w emulsions and was lower from w/o emulsion whereas in case of glucose absorption was in following order o/w>w/o/w>w/o. Stability of ascorbic acid was enhanced by incorporating in multiple emulsions. The other drugs that were studied included metronidazole, glucose dihydralazine, hydrocortisone, and benzalkonium chloride.

As a technique for microsphere/microcapsule preparation

Now-a-days multiple emulsion system has been recognize an intermediate step in the formulation of microspheres. Basic technique for microencapsulation involves two-step emulsification method for microsphere preparation. Finally, microspheres can be prepared by gelation of external phase, utilizing one of the following method (1) crosslinking monomer in external aqueous phase by γ-radiations (2) heat denaturation of proteins (3) non-solvent addition following cross linking (4) salt or alcohol addition. A novel technique of w/o/o double emulsion solvent diffusion method was used for encapsulating hydrophilic drugs. Recently, uniform size microcapsules were successfully developed by combining glass membrane and multiple emulsion-solvent evaporation method. Microspheres of several drugs has been prepared and different polymer (ethylcellulose, eudragit, polylactic acid, polyglycolic acid etc.) has been utilize based on the nature of the drug to be encapsulated (Hydrophili or lipophilic). The following is the list of the drugs microencapsulated using multiple emulsions: sulfadiazine, leuprolide acetate, pseudoephedrine, salbutamol sulphate, theophylline, propranolol, retinol, diclofenac sodium, lysozyme, etc.

Miscellaneous applications

Herbert was the first to describe the use of multiple emulsion system for medical application. He has prepared w/o/w emulsions for vaccines.

Herbert re-emulsified the w/o emulsion to produce a less viscous w/o/w emulsion system. The system was not only easier to inject but also given rise to a further improvement in antibody titer. Later scientists have also used w/o/w emulsions as adjuvants of vaccines. A comparison study with an aqueous influenza vaccine comprising a simple and multiple emulsions demonstrated that multiple emulsion was less viscous than simple emulsions and gave a, antibody response greater than that of simple emulsions and considerably greater than aqueous vaccines. Many other workers have used w/o/w emulsion systems as vaccine adjuvant.

The use of w/o/w emulsions to deliver certain anticancer agents is known. A number of other uses of multiple emulsions, as parenteral delivery systems have been found. The multiple emulsions, usually as a w/o/w system can give rise to a sustained release effect. As a consequence improved response and/or reduced dosage may be achieved. The drug dissolved in internal aqueous droplets is entrapped within a lipid environment and drug release should occur slowly, providing a sustained release with reduced dosage. A single dose multiple emulsion of methotrexate (3mg/kg) gives a much greater prolongation of survival time of leukemic mice than did a single dose of 30mg/kg of methotrexate in aqueous solution; and a slightly greater therapeutic effect than five daily doses of (3mg/kg) of aqueous solution. The in vitro release of naltrexone hydrochloride from w/o/w emulsions and naltrexone base from w/o/w emulsions has been investigated. A prolonged parenteral dosage form for narcotic antagonism was observed.

The use of multiple emulsions in prolonged drug release is studied. A slow release w/o/w multiple emulsion of mebeverine, pentazocine were prepared by using various thickening and emulsifying agents and found that bioavailability of these drugs were increased. A sustained release multiple emulsion of lidocaine were prepared and stabilized by using microcrystalline cellulose. Invivo release was investigated and a comparision was made with that of aqueous solution. Anesthetic effects such as duration and tolerability was compared. Multiple emulsion showed a longer duration of action with less eye irritation and improved efficacy than aqueous solution. Multiple emulsions / double emulsions (w/o/w) of various drugs such as ibuprofen, indomethacin, diclofenac sodium, ephedrine HCl, terbutaline sulphate, nitrofurazone and vancomycin (a peptide drug), rifampicin were prepared. A slow and sustained release of drug was observed.

Multiple emulsions of various corticosteroids such as prednisolone, hydrocortsisone were prepared and studied for the effect of phase volume, effect of various HLB emulsifiers on the bioavailability of drugs.

Increase in bioavailability with maximum response was observed with multiple emulsions compared with that of aqueous solution.

4.14 Routes of Administration

The emulsion systems are delivered by three routes 1. parenteral route, 2. topical route and 3. oral route.

Parenteral route

Emulsions are employed as injectables for variety of reasons. These range from their use as vehicles for lipid soluble materials, to the more challenging areas of controlled drug release, targeting of drugs to specific sites in the body. Both o/w and w/o and multiple emulsions are given in the form of injection. The type of emulsion employed is usually determined by the role that the vehicle will play and the intended route of administration. Multiple or double emulsions are used as depot systems. Their potential advantage in drug delivery can be counter balanced by an increased complexity of the dosage form.

Although the scientific literature contains many details about the stable systems produced with variety of oils and emulsifying agents, very few of them can be used for parenteral use. For intravenous administration in humans, the range of oils and surfactants is limited. Purified paraffin oils and vegetable oils and nonionic materials (span, tween etc.) can be used. The administration of essential nutrients via parenteral routes has been known for more than 100 years. Usually a lipophilic solution contain carbohydrates, aminoacids, electrolytes and vitamins. The lipophilic nature of the solution needs its penetration through a large central vein. Fat emulsions have an important role in parenteral nutrition.

Problems may arise in the parenteral administration of drugs that have low water solubility . The use of soyabean oil emulsions as carriers for lipid soluble drugs has been pioneered by Jeppson and Ljunberg in Sweden. The so called slow release emulsions of narcotic antagonistic have been used by Laska and Fennessy. Scientists also explored the possible use of lipid emulsions for administration of anticancer agents.

Multiple emulsion system is prepared by redispersing a w/o emulsion containing the antigen solution in the aqueous phase which contains an emulsifier that promotes an o/w emulsions. Multiple emulsions were first investigated as vaccine adjuvants by Herbert. He also reported that multiple emulsion vaccines had good invitro stability, did not produce persistant subcutaneous depots and much lower viscosity and therefore easy to inject. Multiple emulsions vaccines may be suitable for administration from multidose containers with automatic syringe. Scientists further confirmed that the multiple emulsion was less viscous. And it is less likely to produce local reactions. Despite their early promise, the multiple emulsion system has not been widely used. Their complexity creates formulation difficulties. Multiple emulsions produced from vegetable oils are particularly difficult to make if a high yield of multiple droplets and good stability are required.

Oral route

Orally administered emulsions are almost exclusively o/w emulsions, although the use of multiple emulsions of the o/w/o type has been reported. Scientists used multiple emulsions (w/o/w) for oral administration of insulin and concluded that, this was a means of effecting intestinal absorption of the drug. W/o/w insulin micelles given internally were found to be more effective in lowering blood and urinary glucose levels than the emulsion and may be worth further consideration.

4.15 O/W/O Double Emulsions

Oil-in-water double emulsions were considered to have less potential applications and therefore were less extensively studied. However, in more recent years several new applications have been reported for O/W/O double emulsions, which sound interesting and are worth being mentioned. The modulated release of triterpenic compounds from an O/W/O multiple emulsion formulated with dimethicones, studied with infrared spectrophoto-metric and differential calorimetric approaches, is one of these examples (99). The authors explored the advantages in the release of triterpenic compounds from O/W/O emulsions. They found two principal advantages: (1) the use of low molecular weight sili-cones decreased the oily touch of the final preparation; and (2) owing to the large range of viscosity, these excipients influenced the skin distribution of the active matter after the topical application. The effects of different dimethicones incorporated within multiple emulsions were studied, through in-vitro penetration results. The residual film on the skin was also evaluated. Correlations were established between the sili-cones structure and the distribution of drugs at different skin levels or between the silicone structure and the percutaneous penetration. The in-corporation of silicones within O/W/O multiple emulsions seems to be an efficient means of modulating the penetration and the distribution of drugs in the skin.

In another study the stability of retinol (vitamin A alcohol) was compared in three different emulsions: O/W,W/O, and O/W/O (15). The stability in the O/W/O emulsion was the highest among the three types of emulsions. The remaining percentages, at 50°C after 4 weeks, were of 56.9, 45.7, and 32.3, in the O/W/O, W/O, and O/W emulsions, respectively. However, it was also reported that with increasing peroxide value of O/W and W/O emulsifiers, the remaining percentage of vitaminApalmitate and retinol in the emulsions increased significantly, indicating that peroxides in the formulas accelerate the decomposition of vitamin A. Organophilic clay mineral tan-oil gelling agent and a W/O emulsifier also affected the stability of retinol. The stability of retinol in the O/W/O emulsion increased with increasing inner oil phase ratio, whereas in O/W it was unaffected by the oil fraction. The encapsulation percentage of retinol in the O/W/O emulsion, and the ratio of retinol in the inner oil phase to the total amount in the emulsion, increased with increasing the oil fraction. The remaining percentage of retinol in the O/W/O emulsion was in excellent agreement with the encapsulation percentage, suggesting that retinol in the inner oil phase is more stable than that in the outer oil phase. Addition of antioxidants (*tert*-butylhydroxytoluene, sodium ascorbate, and EDTA) to the O/W/O emulsion improve the stability of retinol up to 77.1% at 50°C after 4 weeks. The authors concluded that the O/W/O emulsion is a useful formula to stabilize vitamin A.

Conclusions

Multiple emulsions have been known for many years, but are not widely used in pharma industry because of a poor understanding of their stability and properties. This chapter addresses some of these problems. It covers the fundamentals of multiple emulsions, including definition and properties; critical aspects of multiple emulsions and their limiting characteristic and stability; and an in-depth discussion on pharmaceutical applications.

References

1. Takino, T.; Konishi, K.; Takakura, Y.; Hashida, M. *Biol. Pharm. Bull.,* **1994**, *17*(1), 121.
2. Nakano, M. *Adv. Drug Del. Rev.,* **2000**, *45,* 1.
3. A.T. Florence and D. Whitehill: "The Formulation and Stability of Multiple Emulsions." Int. J. Pharm. **11,** 277 (1982).

4. Seifriz W. 1925. Studies in emulsions . *J Phys Chem* 29 : 738.
5. Omotosho JA, Florence AT, Whateley TL. 1990. Absorption and lymphatic uptake of 5 - fl uorouracil in the rate following oral administration of W/O/W multiple emulsions . *Int J Pharm* 61 : 51 – 56.
6. Omotosho JA. 1990. The effect of acacia, gelatin and polyvinylpyrrolidone on chloroquine transport from multiple W/O/W emulsions. *Int J Pharm* 62 : 81 – 84.
7. Morimoto Y, Subigayashi K, Yamaguchi Y, Kato Y. 1979. Detoxication capacity of A multiple (W/O/W) emulsion for the treatment of drug overdose: Drug extraction into the emulsion in the gastro - intestinal tract of rabbits. *Chem Pharm Bull* 27 3188 – 3192.
8. Elson LA, Mitchley BCV, Collings AJ, Schneider R. 1970. Chemotherapeutic effect of a water - in - oil - in - water of methotrexate on the mouse L1210 leukaemia , *Rev Europ Etudes Clin Biol* 15 : 87 – 90.
9. Collings AJ. 1971. Improvements in or relating to sustained release preparations, British Patent 1,235,667.
10. Brodin AF, Kavaliunas DR, Frank SG. 1978. Prolonged drug release from multiple emulsions. *Acta Pharm Suec* 15 : 1 – 10.
11. Sinha, V.R.; Kumar, A. *Ind. J. Pharm. Sci.,* **2002**, *64,* 191
12. Li NN, Shrier AL. 1972. Liquid membrane water treating. In: *Recent Development in Separation Science*, Vol. 1, Cleveland : Chem. Rubber Co. , pp 163 – 174.
13. Drew Myers. Surfactant science and technology, 3rd edition (Wiley-Interscience, New jersey), pp. 316 (2006).
14. Garti N, Benichou A. 2004. Recent developments in double emulsions for food applications.In: *Food Emulsions.* Eds. Friberg SE, Larsson K, Sj ö blom J. New York : Marcel Dekker Inc . pp 353 – 412.
15. Garti N, Benichou A. 2001. Double emulsions for controlled - release applications - Progress and trends. In: *Encyclopedic Handbook of Emulsion Technology* , Sj ö blom J , eds. New York : Marcel Dekker , pp 377 – 407.
16. Garti N, Benichou A. 2003. Recent Developments in Double Emulsions for Food Applications. In: *Food Emulsions* , 4th ed. ,

Friberg SE , Larsson K , Sj ö blom J, eds. Food sci technol ser, Vol. 132. New York : Marcel Dekker , pp 281 – 340.

17. Yazan Y, Seiller M, Puisieux F. 1993. Multiple emulsions. Boll Chim. *Farmaceutico* 132 : 187 – 196.

18. Fox C. 1986. Introduction to multiple emulsions. *Cosmet Toilet* 101 : 101 – 106 .

19. Azhar Yaqoob Khan*, Sushama Talegaonkar, Zeenat Iqbal, Farhan Jalees Ahmed and Roop Krishan Khar.multiple emulsions: an overview. Current Drug Delivery, 3: 429-443 (2006).

13. Drew Myers. Surfactant science and technology, 3rd edition (Wiley-Interscience, New jersey), pp. 317 (2006).

20. Matsumoto, S.; Kita, Y.; Yonezawa, D. *J. Col. Inter. Sci.,* **1976**, *57,* 353.

21. Kavaliunas, D.R.; Frank, S.G. *J. Col. Inter. Sci.,* **1978**, *66 (3),* 586.

22. Okochi, H.; Nakano, M. *Chem. Pharm. Bull.,* **1996**, *44(1),* 180.

23. Okochi, H.; Nakano, M. *Adv. Drug Deliv. Rev.,* **2000**, *45,* 5.

24. Lindenstruth, K.; Muller, B.W. *Eur. J. Pharm. Biopharm.,* **2004**, *58,* 621.

25. Matsumoto, S.; Kita, Y.; Yonezawa, D. J Colloid Interface Sci 57: 353-361, 1976.

26. S Matsumoto,WWKang. J Dispers Sci Technol 10: 455— 482, 1989.

27. MFrenkel, R Schwartz, N Garti. J Colloid Interface Sci 94: 174—178, 1983.

28. N Garti, M Frenkel, R Schwartz. J Dispers Sci Technol 4: 237—252, 1983.

29. Hou, W.; Papadopoulos, K.Y. *Colloids Surf. A,* **1997**, *125,* 181.

30. Oh, C.; Shin, S.; Park, J.H.; Oh, S.G. *J. Disp. Sci. Tech.,* **2004**, *25*(1), 53.

31. Nakashima, T.; Shimizu, M.; Kukizaki, M. *Membrane emulsification: operation Manual,* First Edition,:, Ind. Res. Inst. Miyazaki Prefecture, Japan, **1991**.

32. Okochi, H.; Nakano, M. *Chem. Pharm. Bull.,* **1997**, *45*(8), 1323.

33. Higashi, H.; Shimizu, M.; Nakashima, T.; Iwata, K.; Uchiyama, F.; Tateno, S.; Setoguchi, T.; *Cancer,* **1995**, *75,* 1245.

34. Higashi, H.; Shimizu, M.; Nakashima, T.; Tabata, N.; Kondo, K.H.; Maeda, Y. *J. Pharmacol. Exp. Ther.,* **1999**, *289,* 816.

35. Mine, Y.; Shimizu, M.; Nakashima, T. *Colloids Surf. B,* **1996**, *6,* 261. 32. Okochi, H.; Nakano, M. *Chem. Pharm. Bull.,* **1997**, *45*(8), 1323.

36. Sugiura, S.; Nakajima, M.; Yamamoto, K.; Iwamoto, S.; Oda, T.; Satake, M.; Seki, M. J. *Col. Inter. Sci.,* **2004**, *270,* 221.

37. Frenkel, M.; Shwartz, R.; Garti, N. *J. Col. Inter. Sci.,* **1983**, *94,* 174..

38. Magdassi, S.; Frenkel, M.; Garti, N.; Kasan, R. *J. Col. Inter. Sci.,* **1984**, *97,* 374.

39. Okochi, H, and Nakano, M., Chem. Pharm.Bull., 1996,44,180.

40. Frakenfeld, j.w., fuller, g.c. and Rhodes, c.t., drug develop.Commun., 1976,2,405.

41. Baillet, a ., pirishi, f., vaution, c., grossiord, j.l ,and seiller, m., Int .j.cosmet.sci., 1994,16,1.

42. Kraft, P., Riess, J.G. and Zarif, L., PCT Int.Appl.WO98, 05,301 (C1.A61K9/113),12 Feb., 1998,FR APPL.96/10 140, 7 Aug 1996.

43. Laugel , C. Rafidison, P., Potard, G., Aguadisch, L.and Baillet, A., J.Control.Release,2000,63,7.3. A.T. Florence and D. Whitehill: "The Formulation and Stability of Multiple Emulsions." Int. J. Pharm. **11,** 277 (1982). 11. Sinha, V.R.; Kumar, A. *Ind. J. Pharm. Sci.,* **2002**, *64,* 191

44. Davis, S.S.; Walker, I. *J. Col. Inter. Sci.,* **1983**, *17,* 203.

45. Nakhare, S., Juni, K. and Vyas, S.P., Indian j. Pharm.Sci., 1995.57,71.

46. Omotosho, J.A.; Florence, A.T.; Whateley, T.L.; Law, T.K. *J. Pharm. Pharmacol.,* **1986**, *38,* 865.

47. Omotosho JA , Florence AT , Whateley TL . 1990 . Absorption and lymphatic uptake of 5 - fl uorouracil in the rate following oral administration of W/O/W multiple emulsions. *Int J Pharm* 61 : 51 – 56 .

48. Okochi, H.; Nakano, M. *Chem. Pharm. Bull.,* **1996**, *44(1),* 180.

49. Nianxi, Y.; Mingzu, Z.; Peihong, N.I. *J. Microencap.,* **1992**, *9,* 143.

50. Fukushima, S.; Nishida, M.; Nakano, M. *Chem. Pharm. Bull.,* **1983**, *31,* 4048.

51. Nianxi, Y.; Mingzu, Z.; Peihong, N.I. *J. Microencap.,* **1992**, *9,* 143.

52. Jager - Lezer N , Terrisse L , Bruneau F , Tokgoz S , Ferreira L , Clausse D, Seiller M , Grossiord J - L. 1997 . Infl uence of lipophilic surfactant on the release kinetics of water - soluble molecules entrapped in a W/O/W multiple emulsion . *J Controlled Release* 45 : 1 – 13 .

53. Luca, M.; Grossiord, J.L.; Medard, J.M.; Vaution, C. *Cosmetics Toiletries,* **1990**, *15,* 65.

54. Seiller, M.V.; Vaution, C.; Grossiord, J.L.; Rabaron, A. *Actualities Pharmaceutiques,* **1991**, *41,* 55.

55. Geiger, S.; Tokgoz, S.; Fructus, A.; Jager-lezer, N.; Seiller, M. *J. Control. Release,* **1998**, *52,* 99.

56. Khopade, A.J.; Jain, N.K. *Pharmazie,* **1997**, *52,* 562.

57. Clausse, D.; Perzon, I.; Komunjer, L. *Colloids Surf. A,* **1999**, *152*(1-2), 23.

58. Singh, B.N. In *Encyclopedia of Pharmaceutical Technology,* Marcel Dekker Inc., New York, **2002**, pp. 886-909.

59. Y. Kita, S. Matsumoto, and D. Yonezawa: "Permeation of Water Through the Oil Layer in W/O/W-Type Multiple-Phase Emulsions." Nippon Kagaku Kaishi **1,** 11 (1978).

60. E Pilman, K Larsson, E Torenberg. J Dispers Sci Technol 1: 267—281, 1980.

61. Mooney M. The viscosity of a concentrated suspension of spherical particles. J Colloid Sci. 1951;6:162-170.

62. Kita Y, Matsumoto S, Yonezawa D. 1977. Viscometric method for estimating the stability of W/O/W type multiple - phase emulsions. *J Colloid Interface Sci* 62 : 87 – 94 .

63. Matsumoto S, Inoue T, Kohda M, Ikura K. 1980. Water permeability of oil layers in W/O/W emulsions under osmotic pressure gradients . *J Colloid Interface Sci* 77 : 555 – 563 .

64. Davis SS, Burbage AS. 1978. The particle size analysis of multiple emulsions (water - in - oil - in - water). In: *Particle Size Analysis*, Groves MJ, eds. London : Heyden , pp 395 – 410.

65. Florence AT, Whitehill D. 1985. Stability and stabilization of water - in - oil - in - water multiple emulsions. *ACS Symp Ser* 272 (*Macro - and Microemulsions*): 359 – 380.

66. Davis SS, Burbage AS . 1977. Electron micrography of water - in - oil - in - water emulsions. *J Colloid Interface Sci* 62 (2): 361 – 363.

67. Davis SS, Walker I. 1983. Measurement of the yield of multiple droplets by a fluorescent tracer technique. *Int J Pharm* 17 : 203 – 213.

68. Matsumoto S, Kohda M. 1980. The viscosity of W/O/W emulsions: An attempt to estimate the water permeation coefficient of the oil layer from the viscosity changes in diluted system on aging under osmotic pressure gradient . *J Colloid Interface Sci* 73 (1): 13 – 20.

69. Brodin AF, Kavaliunas DR, Frank SG. 1978. Prolonged drug release from multiple emulsions. *Acta Pharm Suec* 15 : 1 – 10 .

70. Garti, N. *Lebensm-Wiss. U.-Technol.,* **1997**, *30,* 222.

71. Garti, N.; Aserin, A. *Adv. Col. Inter. Sci.,* **1996**, *65,* 37.

72. Hino, T.; Kawashima, Y.; Shimabayashi, S. *Adv. Drug. Deliv. Rev.,* **2000**, *45,* 27.

73. Florence, A.T.; Whitehill, D. *J. Pharm. Pharmacol.,* **1980**, *32,* 64P.

74. Florence, A.T.; Whitehill, D. *J. Pharm. Pharmacol.,* **1982**, *34,* 687.

75. Zhang, W.; Miyakawa, T.; Uchida, T.; Goto, S. *YakugakuZasshi,* **1992**, *112*(1), 73.

76. Muguet, V.; Ozer, O.; Barratt, G.; Marty, J.P.; Grossiord, J.L.; Seiller, M. *J. Control. Release,* **2001**, *70,* 37.

77. Terrisse, I.; Seiller, M.; Magnet, A.; Grossiord, J.L.; Hen-Ferrenbach, C.L. *Colloids Surf. A,* **1994**, *91,* 121.

78. A Vaziri, B Warburton. J Microencapsulation 12: 1—5, 1995.

79. S Susuki, JK Lim. J Microencapsulation 11: 197—203, 1994.

80. D Clausse. J Therm Anal Calorim 51: 191—201, 1998.

81. D Clausse, I Pezron, L Komunjer. Colloids SurfacesA152: 23—29, 1999.

82. Kawashima, Y.; Hino, T.; Takeuchi, H.; Niwa, T. *Chem. Pharm. Bull.,* **1992**, *40,* 1240.

83. Ohwaki, T.; Machida, R.; Ozawa, H.; Kawashima, Y.; Hino, T.; Takeuchi, H.; Niwa, T. *Int. J. Pharm.,* **1993**, *93,* 61.

84. Kanouni, M.; Rosano, H.L.; Naouli, N. *Adv. Col. Inter. Sci.,* **2002**, *99,* 229.

85. Kim, J.W.; Kang, H.H.; Suh, K.D.; Oh, S.G. *J. Disp. Sci. Tech.,* **2003**, *24,* 833.

86. Yu, S.C.; Bochot, A.; Bas, G.L.; Cheron, M.; Mahutean, J.; Grossiord, J.L.; Seiller, M.; Duchene, D. *Int. J. Pharm.,* **2003**, *261,* 1.

87. Hou, W.; Papadopoulos, K.Y. *Colloids Surf. A,* **1997**, *125,* 181.

88. Luca, M.; Grossiord, J.L.; Medard, J.M.; Vaution, C. *Cosmetics Toiletries,* **1990**, *15,* 65.

89. Sela, Y.; Magdassi, S.; Garti, N. *J. Control. Release,* **1995**, *33,* 1.

90. Law, T.K.; Whateley, T.L.; Florence, A.T. *J. Control. Release,* **1986**, *3,* 279.

91. Florence, A.T.; Rogers, J.A. *J. Pharm. Pharmacol.,* **1971**, *23,* 153-169.

92. Myers, D. *Surfactant Science and technology.,* VCH, New York, **1988**, pp. 209-253.

93. Magdassi, S.; Frenkel, M.; Garti, N.; Kasan, R. *J. Col. Inter. Sci.,* **1984**, *97,* 374

94. Opawale, F.O.; Burgess, D.J. *J. Pharm. Pharmacol.,* **1998**, *50,* 965.

95. Jiao, J.; Bugress, D.J. *AAPS Pharm. Sci.,* **2003**, *5*(1), E7. Omotosho, J.A.; Whateley, T.L.; Florence, A.T. *J. Microencap.,* **1989**, *6*(2), 183.

96. Garti, N.; Aserin, A.; Cohen, Y. *J. Control. Release,* **1994**, *29,* 41.

97. Florence, A.T.; Law, T.K.; Whateley, T.L. *J. Col. Inter. Sci.* **1985**,*107,* 584.

98. Cole, M.L.; Whateley, T.L. *J. Control. Release,* **1997**, *49,* 51.

99. Jiao, J.; Rhodes, D.G.; Burgess, D.J. *J. Col. Inter. Sci.,* **2002**,*250*(2), 444.

100. Iwamoto, K.; Kato, T.; Kawahara, M.; Koyama, N.; Watanabe, S.; Miayake, Y.; Sunamoto, J. *J. Pharm. Sci.,* **1991**, *80,* 219.

101. Vyas, S.P.; Khar, R.K. *In Targeted and Controlled Drug delivery: Novel Carrier Systems.* First Edition:, CBS Publishers & Distributors, New Delhi, **2002**, pp. 303-330.

102. Khopade, A.J.; Jain, N.K. In *Advances in controlled and novel drug delivery*, Jain, N.K. Ed.; India, **2001**, pp. 381.

103. Chatterjee, C.C. In *Human Physiology,* Eleventh Edition, Medical Allied Agency, Calcutta, **1985**, Vol. *1,* 174.

104. Davis, S.S.; Walker, I. *J. Col. Inter. Sci.,* **1983**, *17,* 203.

105. Burbage, A.S. The stability and drug release characteristics of multiple emulsions. Ph.D Thesis, University of Aston, Birmingham, **1979.**

106. Matsumoto, S.; Ueda, Y.; Kita, Y.; Yonezawa, D. *Agr. Biol. Chem.,***1978**, *42,* 739.

107. Fukushima, S.; Nishida, M.; Nakano, M. *Chem. Pharm. Bull.,* **1983**, *31,* 4048

108. Ozer, O.; Muguet, V.; Roy, E.; Grossiord, J.L.; Seiller, M. *Drug Dev. Ind. Pharm.,* **2000**, *26,* 1185.

109. Martin, T.P.; Davis, G.A. H*ydrometallurgy,* **1976**, *2,* 313.

110. Teramoto, M.; Takihana, H.; Shibutani, M.; Yuosa, J.; Miyake, Y.; Teranishi, H. *J. Chem. Eng. Jap.,* **1981**, *141,* 122.

111. Lee, K.H.; Evans, D.F.; Cussler, F.L. *A.I.Ch. E.J.,* **1978**, *24,* 860.

112. Takahashi K., Ohtsuto, F., Takenchi, H. *Chem. Eng. Jap.,* **1981**, *14,* 416.

113. Ohwaki, T.; Machida, R.; Ozawa, H.; Kawashima, Y.; Hino, T.; Takeuchi, H.; Niwa, T. *Int. J. Pharm.,* **1993**, *93,* 61.

114. Ohwaki, T.; Nitta, K.; Ozawa, H.; Kawashima, Y.; Hino, T.; Takeuchi, H.; Niwa, T. *Int. J. Pharm.,* **1992**, *85,* 19.

115. Sela, Y.; Magdassi, S.; Garti, N. *J. Control. Release,* **1995**, *33,* 1.

116. Kita, Y.; Matsumoto, S.; Yonezawa, D. *J. Col. Inter. Sci.,* **1977**, *62,* 87.

117. Hino, T.; Shimabayashi, S.; Tanaka, M.; Nakano, M.; Okochi, H. *J Microencapsulation.,* **2001**, *18*(1), 19.

118. Rutgers, I.R. *Rheol. Acta,* **1962**, *2,* 202.

119. Matsumoto, S.; Inoue, T.; Kohda, M.; Ohta, T. *J. Col. Inter. Sci.,* **1980**, *77,* 564.

120. Tomita, M.; Abe, Y.; Kondo, T. *J. Pharm. Sci.,* **1982**, *71,* 332.

121. Stroeve, P.; Varansi, P.P. *J. Col. Inter. Sci.,* **1984**, *99,* 360.

122. Matsumoto, S.; Kohdra, M. *J. Col. Inter. Sci.,* **1980**, *73,* 13.

123. Nakhare, S.; Vyas, S.P. *Pharmazie,* **1997**, *52,* 224.

12.4 akhare, S.; Vyas, S.P. *J. Microencap.,* **1996**, *13*(3), 281.

125. Khopade, A.J.; Jain, N.K. *Drug Dev. Ind. Pharm.,* **1998**, *24,* 289.

126. Khopade, A.J.; Jain, N.K. *Drug Dev. Ind. Pharm.,* **1998**, *24,* 677.

127. Roy, S.; Gupta, B.K. *Drug Dev. Ind. Pharm.,* **1993**, *19,* 1965.

128. Matsumoto S , Inoue T , Kohda M , Ikura K . 1980 . Water permeability of oil layers in W/O/W emulsions under osmotic pressure gradients . *J Colloid Interface Sci* 77 : 555 – 563 .

129. Tomita M, Abe Y , Kondo T . 1982 . Viscosity change after dilution with solutions of water/oil/water emulsions and solute permeability through the oil layer . *J Pharm Sci* 71 : 332 – 334 .

130. Grossiord JL, Seiller M. 1998. Rheology of W/O/W multiple emulsions: Formulation, characterization and breakup mechanisms. In: *Multiple Emulsions, Structure, Properties and Applications* , Grossiord JL , Seiller M. eds. Paris : Editions de Sant é , pp 169 – 192 .

131. Geiger S , Tokgoz S , Fructus A , Jager - Lezer N , Seiller M , Lacombe N , Grossiord JL . 1998 . Kinetics of swelling - breakdown of a W/O/W multiple emulsions: Possible mechanisms for the lipophilic surfactant effect . *J Controlled Release* 52 : 99 – 104 .

132. Geiger S , Jager - Lezer N , Tokgoz S , Seiller M , Grossiord JL . 1999. Characterization of the mechanical properties of a W/O/W multiple emulsion oily membrane by a micropipette aspiration technique . *Colloids Surf A* 157 (1 – 3): 325 – 332 .

133. Taylor GI . 1932 . The viscosity of a fluid containing small drops of another fluid. *Proc R Soc London A* 138: 41 – 48.

134. Taylor GI . 1934 . The formation of emulsions in definable fields of flow. *Proc R Soc London A* 146 : 501 – 523.

5 Nanoemulsions

Abheri Das Sarma and Kalyan Kumar Sen
Gupta College of Technological Sciences, Asansol,West Bengal, India

5.1 Introduction

Emulsions are dispersions of droplets of a liquid phase in a different immiscible liquid (1–4). Oil droplets dispersed in water are known as "direct emulsions", whereas water droplets dispersed in oil are called "inverse emulsions". Without shear, emulsions would never form, since the thermo-dynamic lowest energy state of two immiscible liquids is simply a layer of the liquid having lower density on top of a layer of the liquid having higher density. In order to create emulsions, shear must be applied in a way that causes the droplets of one phase to stretch and rupture, through a capillary instability, into smaller droplets. Stretching and rupturing the droplets is possible if the applied shear stress is greater than the characteristic Laplace pressure scale, σ/a, where σ is the liquid-liquid surface tension and a is the droplet's radius. A surfactant, usually soluble in the continuous phase, coats the surfaces of the newly formed droplets and provides a stabilizing repulsion between droplet interfaces that strongly inhibits subsequent recombination, or coalescence, of the droplets. Microscale emulsions can be commonly made in the kitchen using shear stresses generated by whisks, spatulas, and blenders; mayonnaise is a common food emulsion of oil droplets in water stabilized by egg protein. However, obtaining nanoscale emulsions having droplet diameters less than 100 nm with these methods is generally not possible since the applied shear stresses are too low. In order to go beyond conventional emulsification methods into the realm of nanoemulsions, it is necessary to apply extremely high shear stresses to cause violent stretching and rupturing of the droplets. By equating the viscous stress of the driving shear with the Laplace pressure of droplets that no longer rupture in the shear flow, one can obtain the classic Taylor equation for estimating the radius of the emulsion (4–5): $a \approx \sigma /(\eta_c\gamma)$, where γ is the shear rate and η_c is the viscosity of the continuous liquid phase in which the droplets are dispersed. This formula neglects the dissipative effects of viscosity, η_d, of the liquid inside the droplets. Since the shear stress depends on a combination of the shear rate and the emulsion's rheological properties, it is possible to make nanoscale emulsions of strongly immiscible liquids without relying upon thermodynamic self-assembly of nanoscale droplets in microemulsion phases (6,7). Microfluidic and ultrasonic techniques both produce the kind of extreme shear that is required to break emulsions down to diameters below 100 nm (8-10). Sub-micron emulsions are sometimes called "mini-emulsions", and we define "nanoemulsions" to be emulsions having diameters around 100 nm or less (11). Because the sub- microscopic droplet sizes make it difficult to observe the droplets and because very strong shear is required to create them, nanoemulsions have been studied far less than conventional emulsions. Moreover, since it is easier to visualize the rupturing of isolated droplets, most experiments have avoided the complexity inherent in emulsification of concentrated emulsions at a droplet volume fraction φ where droplet interfaces strongly interact. Nanomemulsions are often mentioned as being translucent or transparent, rather than the characteristic opaque, milky white of traditional emulsions (12, 13). While emulsion droplets do become translucent or transparent at sizes smaller than the optical wavelengths of visible light, the transition does not happen precisely at 100 nm. Furthermore, optical properties of an emulsion are also dependent on the volume fraction of the dispersed phase. A milky-white emulsion can become translucent upon dilution, though there may be no reduction in the particle size. Since the properties of the droplet do not change upon dilution, neither should the nomenclature be changed. It has also been stated that nanoemulsions are stable to creaming or sedimentation (13). The stated justification is that at smaller particle sizes the Brownian motion of the droplets is great enough to overcome the effects of gravity (12). However, creaming is dependent not only on size but also on the relative densities of the two phases. An o/w emulsion with an oil density of 0.97 g/mL will be much more resistant to creaming than an o/w emulsion with an oil density of 1.8, even if both emulsions have droplets of the same size. Though it is possible to have nanoemulsions in which creaming does not occur, as with optical properties there is no specific cutoff at 100 nm. Others have used the term nanoemulsions to mean emulsions that have a diameter in the nanometer range, i.e., below

1000 nm. This definition offers the most clarity and precision of meaning, by clearly stating that nanoemulsions are emulsions of a certain size, which is indicated by the prefix. A nanoemulsion should then be defined as a heterogeneous system composed of one immiscible liquid dispersed as droplets within another liquid, where the average droplet diameter is below 1000 nm. However, size is one of the most important variables that can define an emulsion. The name nanoemulsion alone does not eliminate the necessity of using size as a primary defining characteristic. There will be more difference between a nanoemulsion with diameter of 50 nm and one with a diameter of 500 nm, than between a nanoemulsion with diameter of 900 nm and an emulsion with diameter of 1100 nm, even if the names are the same in the former comparison and different in the latter (14).

Unlike microemulsions (which are also transparent or translucent and thermodynamically stable, nanoemulsions are only kinetically stable. However, their long-term physical stability (with no apparent flocculation or coalescence) makes them unique and they are sometimes referred to as ''Approaching Thermodynamic Stability''. The inherently high colloid stability of nano-emulsions can be well understood from a consideration of their steric stabilization (when using nonionic surfactants and/or polymers) and how this is affected by the ratio of the adsorbed layer thickness to droplet radius, as will be discussed below. Unless adequately prepared (to control the droplet size distribution) and stabilized against Ostwald ripening (which occurs when the oil has some finite solubility in the continuous medium), nano-emulsions may lose their transparency with time as a result of increasing droplet size. Nano-emulsions are attractive for application in personal care and cosmetics as well as in health care due to the following advantages: 1.The very small droplet size causes a large reduction in the gravity force and Brownian motion may be sufficient to overcome gravity. This means that no creaming or sedimentation occurs on storage. 2. The small droplet size also prevents any flocculation of the droplets. Weak flocculation is prevented and this enables the system to remain dispersed with no separation. 3. The small droplets size also prevents their coalescence, since these droplets are non-deformable and hence surface fluctuations are prevented. In addition, the significant surfactant film thickness (relative to droplet radius) prevents any thinning or disruption of the liquid film between the droplets. 4. Nanoemulsions are suitable for efficient delivery of active ingredients through the skin. The large surface area of the emulsion system allows rapid penetration of actives. 5. Due to their small size, nanoemulsions can penetrate through the ''rough'' skin surface and this enhances penetration of actives. 6. The transparent nature of the system, their fluidity (at reasonable oil concentrations) as well as the absence of any thickeners may give them a pleasant aesthetic character and skin feel.

7. Unlike microemulsions (which require a high surfactant concentration, usually in the region of 20% and higher), nanoemulsions can be prepared using reasonable surfactant concentrations. For a 20% o/w nanoemulsion, a surfactant concentration in the region of 5–10% may be sufficient. 8. The small size of the droplets allow them to deposit uniformly on substrates – wetting, spreading and penetration may be also be enhanced because of the low surface tension of the whole system and the low interfacial tension of the o/w droplets. 9. Nanoemulsions can be applied for delivery of fragrant, which may be incorporated in many personal care products. This could also be applied in perfumes, which are desirable to be formulated alcohol free. 10. Nanoemulsions may be applied as a substitute for liposomes and vesicles (which are much less stable) and it is possible in some cases to build lamellar liquid crystalline phases around the nanoemulsion droplets. Despite the above advantages, nanoemulsions have only attracted interest in recent years because .1. Their preparation requires, in many cases, special application techniques such as the use of high-pressure homogenizers as well as ultrasonics. Such equipment (such as the Microfluidiser) has became available only in recent years. 2. There is a perception in the Personal Care and Cosmetic Industry that nanoemulsions are expensive to produce. Expensive equipment is required as well as the use of high concentrations of emulsifiers. 3. Lack of understanding of the mechanism of production of submicron droplets and the role of surfactants and co-surfactants. 4. Lack of demonstration of the benefits that can be obtained from using nanoemulsions when compared with classical macroemulsion systems. 5. Lack of understanding of the interfacial chemistry involved in production of nanoemulsions. For example, few formulation chemists are aware of the use of the phase inversion temperature (PIT) concept and how this can be usefully applied for the production of small emulsion droplets. 6. Lack of knowledge on the mechanism of Ostwald ripening, which is perhaps the most serious instability problem with nanoemulsions. 7. Lack of knowledge of the ingredients that may be incorporated to overcome Ostwald ripening. For example, addition of a second oil phase with very low solubility and/or incorporation of polymeric surfactants that strongly adsorb at the o/w interface (which are also insoluble in the aqueous medium). 8. Fear of introduction of new systems without full evaluation of the cost and benefits. However, despite these difficulties, several companies have introduced nanoemulsions in

the market and, within the next few years, the benefits will be evaluated. Nanoemulsions have been used in the pharmaceutical field as drug delivery systems (15). The acceptance of nanoemulsions as a new type of formulation depends on customer perception and acceptability. With the advent of new high-pressure homogenizers and the competition between various manufacturers, the cost of production of nanoemulsions will decrease and may approach that of classical macroemulsions(16). Phospholipids, upon hydration, spontaneously form bilayer membrane vesicles (liposomes) or may act as surfactants in forming micro- or nanoemulsions or solid–lipid nanoparticles. Phospholipids, triglycerides, and cholesterol are the main ingredients of liposomes and lipid nanoparticles. They are natural components of biological membranes and lipoproteins and are, therefore, presumed to be highly biocompatible (17). Drugs and cell-targeting ligands can be incorporated into these structures by encapsulation (for hydrophilic molecules), lipid-phase solubilization (for lipophilic molecules), conjugation to a lipid anchor (as a lipid-derivatized prodrug), or electrostatic complexation (for poly-anionic molecules such as nucleic acids), depending on their specific physicochemical properties. Liposomes and lipid nanoparticles smaller than 300 nm are potentially suitable for systemic administration (18).

5.2 Structure of Nanoemulsion

"Nanoemulsions" are metastable dispersions of submicron droplets that have a significant surface tension, which form only when extreme shear is applied to fragment droplets strongly, and are kinetically inhibited against recombining by repulsive interfacial stabilization due to the surfactant. Nanoemulsions represent the extremely small limit of emulsions (19) of submicron droplets known as "miniemulsions"(20-22). Although nanoemulsions have extreme Laplace pressures, of order 10–100 atm, the droplets can remain stable against Ostwald ripening (23,24) if the liquid inside has very low solubility in the continuous phase outside the droplets. The strong Brownian motion of the tiny droplets in nanoemulsions makes them ideal for products in which gravitational creaming must be prevented to ensure a long shelf life. Although many investigations of the structure of hard particles have been made, (25) the study of the bulk structure of disordered deformable nanodroplets as a function of their volume fraction, φ, has not been extensively investigated and is fundamentally interesting. For nondeformable particles, the nature of the jamming point is known to depend on the shape of the particles, their interactions, and the driving forces present as the particles are concentrated to higher φ. (26,27) Nanoemulsions represent a promising system that facilitates systematic studies of the structure of deformable particles since, in principle, the dispersed phase can be continuously controlled from zero to nearly unity through an applied osmotic pressure.(28) At dilute φ, the droplets are spherical, whereas at high φ, the surfaces of the droplets are strongly repelled by the surfaces of neighboring droplets. This can cause the droplets to deform and become nonspherical, yielding a biliquid nanofoam. An appropriate surfactant can strongly inhibit the droplets from recombining through interfacial coalescence, making nanoemulsions long lived metastable liquid dispersions that have structures determined by the history of applied shear and osmotic pressure. By contrast, the structure of the dispersed phase in microemulsions is typically subject to morphogical changes (29) (e.g., from spherical droplet to a sponge structure or lamellar phase) when the volume fraction of the dispersed phase is changed appreciably. Such phase changes preclude investigations of microemulsions comprised spherical droplets over a wide range of droplet volume fractions. Studies of disordered dispersions of solid particles (25, 30–36) and of the liquid structures in foams (27, 37-43) have provided a better understanding of the fundamental structure of disordered soft materials. These studies can be linked to the structure of emulsions because emulsion droplets are spherical for φ below the jamming limit, and can have interfacial structures like foams for φ above the jamming limit. Due to the strong Brownian motion of the nanoscale droplets, nanoemulsions do not cream or drain in the manner typically observed for microscale and larger foams and emulsions. However, neither transmission nor scanning electron microscopy are well suited for examining liquid phases *in situ*, and cryofracture electron microscopy could potentially alter the morphology of the continuous phase structure in aqueous based emulsions. Since microscopy methods for studying nanoscale emulsions have limitations, scattering methods provide a better method for obtaining the average structure of nanoemulsions in reciprocal space over a wide range of environmental conditions. Small angle neutron scattering (SANS) is a good candidate for obtaining the bulk structure of nanoemulsions, since the neutron wavelengths are suitable for probing nanoscale structures. Moreover, the scattering contrast in SANS can be controlled through deuteration of the components. Although wide-angle static light scattering of index-matched uniform microscale emulsions has been used to determine structure factors at a few concentrated volume fractions, (38, 39) a clear systematic experimental investigation of the structure factor as a function of φ has not been made. Through repeated ultracentrifugal fractionation of a polydisperse nanoemulsion, we can obtain concentrated disordered nanoemulsions having a uniform size. It has been found that neutron scattering is an excellent probe of the structure of disordered nanoemulsions over a wide range of φ, even without enhancing the scattering contrast through selective deuteration. The scattering intensity $I(q,\varphi)$ of disordered nanoemulsions is measured over a wide range of droplet volume fractions using SANS (44).

5.3 Preparation of Nanoemulsions

The nanoemulsion is prepared by two main methods high energy emulsification and low energy emulsification methods. The low energy emulsification further consists of three methods the spontaneous nano-emulsification, the emulsion inversion point (EIP) method and Phase inversion temperature (PIT) method. These methods are discussed below.

5.3.1 High-Energy Emulsification Methods

In this section, we consider emulsification methods involving high (mechanical) energy used in the formation of nanoemulsion, that is, the use of devices to force the creation of huge interfacial areas. Nanoemulsion generation is very commonly performed with such high-energy emulsification methods, particularly exploited in nanoemulsion polymerization (45, 46). The formation of such nanometric scaled droplets is governed by directly controllable formulation parameters such as the quantity of energy, amount of surfactant and nature of the components, unlike the low-energy methods (presented in the following sections), governed by the intrinsic physicochemical properties and behavior of the systems. It follows therefore that high-energy methods present natural predispositions to preserve the formation processes of nanoemulsions droplets, against even the slightest potential modifications of the formulation like the addition of monomer, initiator, surfactant, etc.

5.3.1.1 Devices and processes

The mechanical processes generating nanometric emulsions include, as a first step of the drop creation, the deformation and disruption of macrometric initial droplets, followed by the surfactant adsorption at their interface to insure the steric stabilization. The challenge of these mechanical nano-emulsification methods is to combine these two steps, in order to allow and optimize nano-emulsion generation. Three main groups of devices used are: The rotor/stator devices, high-efficiency devices, including ultrasound generators and high-pressure homogenizers. Rotor/stator type apparatuses do not provide a good dispersion in terms of droplet size and monodispersity (47) in comparison with the nanoemulsions generated by the two others kinds of devices (and also with the low energy methods). Indeed, the energy provided is mostly dissipated, generating heat and being wasted in viscous friction (48,49). Therefore, the additional free energy ΔG_f necessary to create the huge interfacial area of nano-emulsions is not obtained. Nanoemulsions generated by sonifiers are generally attributed to a mechanism of cavitation (50,51), but are not as yet understood well enough. The ultrasound waves in liquid macroscopic dispersion, result in a succession of mechanical depressions and compressions, generating cavitation bubbles, which tend irremediably to implode. Subsequently, this shock provides sufficient energy locally to increase ΔA corresponding to nanometric-scaled droplets. Efficiency of nanoemulsification by sonication (considered as the final size of the nanoemulsion droplets as well as the time needed to attain this asymptotic size), depends both on the composition of the emulsion and the power device. Indeed, the addition of surfactants and/or monomers has been shown an important parameter to efficiently reduce droplet sizes (52). Sonication is thus the most popular way to produce nanoemulsions and nanoparticles for research purposes. It does not, however, appear practical for use on an industrial scale, for which high-pressure (53) devices (and low-energy methods) are often preferred. High-pressure homogenizers, generally Microfluidizer, or Manton– Gaulin devices, are designed in order to force macro-emulsions to pass through narrow gaps, by imposing high pressures. The fluid accelerates dramatically, reaching, in the micro-channels of Microfluidizers (46) for instance, a velocity of around 300 $m{\cdot}s^{-1}$. As a result, shear, impact, and cavitation forces are applied on very small volumes and generate nano-scaled nano-emulsion droplets (closely related to the phenomena involved in the use of sonifiers).

5.3.1.2 The choice of surfactants, monomers, aqueous and oily phases

The nature and amount of the surfactant, monomer, or hydrophobe used in the formulation completely determine the size distribution, structure, and stability of the resulting nano-emulsions and nanoparticles. Thus, the different components are chosen in function of the formulation strategies undertaken. For instance, Landfester (54–58) presented from nano-emulsions (by sonication), (i) the formulation of inorganic particles by playing on the physicochemical properties of molten salt droplets, (ii) the formulation of polymeric nanospheres by in situ polymer synthesis within nano-emulsion droplets, (iii) the combination of both these types of technology to provide hybrid nanoparticles, and finally (iv) the use of oil as a hydrophobe to generate core-shell nanocapsules, by polymer-specific synthesis and segregation to the oil/water region (57), or by interfacial nanoprecipitation (58).

5.3.1.3 On the potentialities, advantages and disadvantages of high-energy method

In general, high-energy nano-emulsification methods present a good potential for polymeric nanoparticle generation, since the formulation parameters are directly controllable. Thus the addition of monomers, initiators, or encapsulating molecules appears not to influence the emulsification process, governed by the high shear processes. If anything, it may be and additional molecules to be encapsulated, monomers, initiators, or stabilizing agents that interfere with the emulsification process, unlike for the low-energy methods in which nano-emulsification is totally governed by the physicochemical behaviors of the surfactants. However, when the purpose of the experiment is the encapsulation of fragile molecules such as peptides, proteins, or nucleic acid, often encountered in pharmaceutical or medical research, high-energy methods may give rise to drug degradation, denaturation or activity loss during processing. Moreover, in the case of an industrial scale-up, it is of importance to consider the energetic yield, which is incomparable between high and low-energy methods (59, 60). This is especially true for sonication, since during the emulsification process, only the near-volume of the sonifier nip is affected by ultrasonic waves, and for high volumes, a weak additional mechanical stirring is needed to homogenize the sizes and generate nano-emulsions. In concrete terms, the emulsification time (i.e. energy) to provide homogeneous nano-emulsions increases in function of the volume to nano-emulsify, which is fundamentally not the case for all low-energy methods.

5.3.2 Low-Energy Emulsification Methods

Nanometric-scaled emulsion droplets may be obtained by diverting the intrinsic physicochemical properties of the surfactants, co-surfactants and excipients composing the formulation Two groups of methods are proposed in the literature and developed below: (i) The first one describes emulsification as a spontaneous phenomenon (61–70), which uses the rapid diffusion of water-soluble solvent, solubilized first in the organic phase, moving towards the aqueous one when the two phases are mixed. In theory, the spontaneous nano-emulsification process can provide as much oili n- water as water-in-oil nano-emulsions, but the majority of the reported studies concern o/w generation. (ii) Secondly, the so-called phase inversion temperature (PIT) method (71–82), which uses the specific properties of polyethoxylated surfactants to modify their partitioning coefficient as a function of the temperature, and leads to the creation of bicontinuous systems when the temperature is close to the PIT, broken-up to generate nano-emulsions. Practically, it leads to o/ w nano-emulsions.

5.3.2.1 Spontaneous nano-emulsification

The diffusion mechanism and diffusion path theory describes the main principles of spontaneous emulsification, underlining the governing phenomena and mechanisms and considering the suitability of this method for nanoparticle generation. It is interesting to note first that the spontaneous features of such phenomena are simply the results of the initial non-equilibrium states of the two bulk liquids when they are brought into contact without stirring. It is only under specific conditions that spontaneous emulsification occurs and in some cases, nanometric-scaled droplets are generated. The spontaneous emulsification process itself increases entropy and thus decreases the Gibbs free energy of the system (83). Evolution of the system is basically promoted by diffusion of a solute into the phase in which it has greater solubility. Thus, spontaneous emulsification behaviors can potentially be predicted by following the diffusion pathway within the phase diagram. The different cases are described below. The source of energy of spontaneous emulsification reportedly stemmed mainly from interfacial turbulences, closely related to the surface tension gradient induced by the diffusion of solutes between two phases. Likewise, the interfaces are subject to capillary waves from thermal origins, gradually amplified as the surface tension decreases (68). The drops are created as a result of sufficiently large interfacial corrugations, similar to the dynamic behavior of the frontier between microemulsion and the bulk phase in multi-phase equilibrium systems, i.e. a continuous coalescence and break-off of emulsion droplets (84). Such a phenomenon has been called dispersion, spontaneously increasing the entropy and decreasing the Gibbs free energy of the system. The other (complementary) spontaneous emulsification mechanism, known as condensation, is also assumed to be intimately linked to the fluctuation of the interfacial amphiphile concentration. Owing to the region of local supersaturation (overconcentration of surfactant at interface) induced by the diffusion process, spontaneous interfacial expansion takes place, resulting in the nucleation and growth of drops. These conditions appear analogous to the system behavior in the two-phase microemulsion regions, for instance, where drops are continuously nucleated, grow, by similar spontaneous phenomena and disappear by coalescing (maintaining the two-phase equilibrium). The variation of composition in each phase (aqueous and oily) is directly represented on the ternary phase diagram by straight lines, from the semi-infinite reservoirs, to the interfacial concentration. Such a schematic representation of the evolution of the concentration within each phase is called a 'diffusion path'. The spontaneous emulsification only depends on the diffusion path with regards to the equilibrium phase diagram. On the other hand, stirring the two phases brought into contact has no influence on the own mechanism of spontaneous emulsification, even though it increases the rate of emulsification by increasing the interfacial area A. Example- the study of the water/alcohol/ oil ternary system, presented in Fig. 1 inspired from Ref. (68), where a pure water {w} phase (point 1) is brought into contact with an alcohol plus oil {a+o} phase (point 4). The local equilibrium at the interface is shown via the dotted segment (2–3) at the frontier of the two-phase equilibrium region (in the phase diagram). Depending on the initial composition and on the nature of the alcohol, the location of the interfacial equilibrium appears to condition spontaneous emulsification, as illustrated by the difference between the Fig. 1 a and b. Actually, it allows the diffusion path to cross the two-phase microemulsion equilibrium region (i.e. spontaneous emulsification (SE) region). This may indicate that the maximum intensity of spontaneous emulsification is not necessarily near the interface, but at the maximum depth within the SE region. Hence, the deeper the diffusion path within the SE region, the higher the interfacial turbulences and dispersion, diffusion or condensation phenomena. It is therefore easy to imagine a bridge with the droplet size of the forming emulsion intimately linked to the intensity of the spontaneous phenomenon. Nano-emulsion droplets appear to be formed in this way, with the use of a high quantity of diffusing solvent in the oily phase. Bouchemal et al. (85), for instance, proposed a study on the optimization of the solvent displacement method formulating nano-emulsions for cosmetic and pharmaceutical applications, in which the overall solvent/oil ratio was around 0.01. Thus the important influence on the nano-emulsification process, of oil viscosity, surfactant HLB, and the nature of the solvent (also as a function of its toxic potential) and miscibility with water was found. Of course, in all these optimized systems for nano-emulsion (and nanoparticle) formulation by such spontaneous emulsification methods, the systems are more complex than the three-component model described above. In fact, the above model only accounts for the droplet formation, none the less unstable and highly subject to destabilization after formation (even the nanometric-scaled droplets). Therefore, after creation, the newly formed interfaces have to be stabilized by surfactant adsorption. Hence, the initial phase diagrams are modified accordingly, and a more complex diffusion path has to be considered between the different phases potentially formed in the interfacial region. It is to be noted that the presence of liquid crystalline (LC) phases are acknowledged as playing a decisive role in these spontaneous-forming formulations. Two examples presenting the spontaneous emulsification of such quaternary systems are proposed in Fig. 2, for both surfactants having a negative and positive Winsor R ratio. R is defined as the ratio between the inter-molecular interactions per unit interfacial area, surfactant–oil/surfactant–water (86). In the case shown in Fig. 2a, of the rather hydrophilic surfactant (Rb1), the formation of the LC phase in the semi-infinite aqueous one (point 1) appears in equilibrium with a pure aqueous sub-phase, segment (2–3), before establishing the local interface equilibrium with the oil-rich phase, segment (4–5). The diffusion path is simpler and presents the spontaneous emulsification of hydrophilic droplets in the oil segment (5–5'). Symmetric phenomena are also conceivable in the case where the rather lipophilic surfactant (RN1) is used, presented in Fig. 2b, where an isotropic phase is generally formed in the surfactant/ oil-rich region. Subsequently, spontaneous emulsification of oily droplets in water arises within the segment (1'–2). The study of nano-emulsion formation using this method implies a thorough establishment of the phase diagram to disclose the potential feasibility diagram and optimization. In this context, it follows that the potential influence of additional components (like monomers and polymers, whether or not they are neutral in the formulation), need to be investigated, both on the phase diagrams and on the diffusion pathway. This may, to some extent imply restrictions in terms of ease of handling, modifying and adapting the nanoparticle formulation to the given needs. However, the generation of nanocapsules and nanospheres by nano-precipitation or in situ polymerization, from nano-emulsions using the solvent displacement method, has provided a great number of examples developed below, e.g. the works of Fessi et al. (87–91).

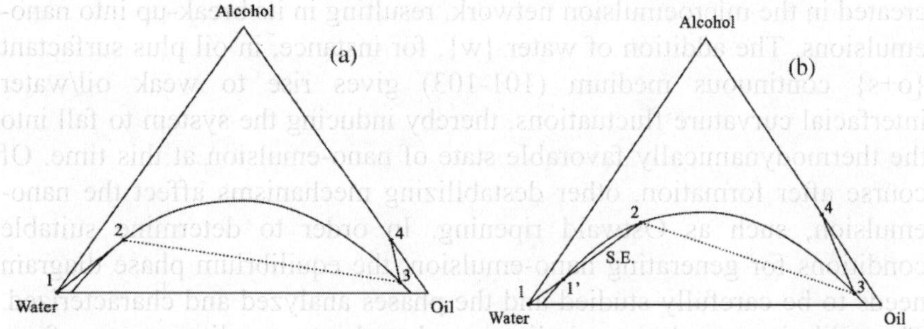

Figure 5.1 Diffusion path in a typical water/alcohol/oil system.Segment (1-2), diffusion path of the aqueous phase.Segment(3-4),diffusion path of the oily phase.Segment(2-3), interfacial local equilibrium. (a)Case where no spontaneous emulsification occurs. (b)Spontaneous emulsification occurs when the diffusion path crosses the two-phase equilibrium region, segment(1'-2).

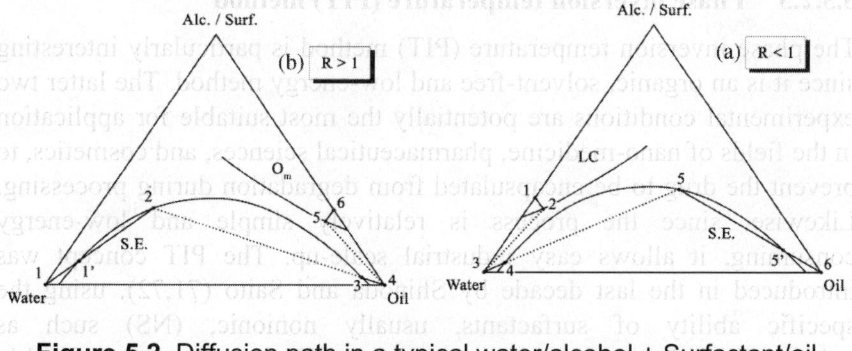

Figure 5.2 Diffusion path in a typical water/alcohol + Surfactant/oil system.

5.3.2.2 The emulsion inversion point (EIP) method

Another spontaneous emulsification method, known as the emulsion inversion point method, has been widely performed and reported in numerous works. At a constant temperature, it consists in diverting the intrinsic features of thermodynamically stable microemulsions D or Liquid crystals LC to be nano-structured by a progressive dilution with water or oil, in order to create thermodynamically unstable but kinetically stable, respectively direct or inverse nano-emulsions (92–100). By slightly changing the water or oil proportion within established microemulsion structures, the surfactant hydration is potentially changed, as well as their affinity for the aqueous phase, and thus instabilities are created in the microemulsion network, resulting in its break-up into nano-emulsions. The addition of water {w}, for instance, in oil plus surfactant {o+s} continuous medium (101-103) gives rise to weak oil/water interfacial curvature fluctuations, thereby inducing the system to fall into the thermodynamically favorable state of nano-emulsion at this time. Of course after formation, other destabilizing mechanisms affect the nano-emulsion, such as Ostwald ripening. In order to determine suitable conditions for generating nano-emulsion, the equilibrium phase diagram needs to be carefully studied and the phases analyzed and characterized. The dilution pathway indicates the best conditions to form nanoemulsions, i.e. the conditions for which the emulsion droplets formed are the smallest. The main results appear to show that nanometric-sized emulsion droplets are formed when the whole phase to be dispersed appears solubilized in the bicontinuous system in the phase diagram. These phases are reported to be either bicontinuous microemulsion, or lamellar liquid crystals L_α.

5.3.2.3 Phase inversion temperature (PIT) method

The phase inversion temperature (PIT) method is particularly interesting since it is an organic, solvent-free and low-energy method. The latter two experimental conditions are potentially the most suitable for application in the fields of nano-medicine, pharmaceutical sciences, and cosmetics, to prevent the drug to be encapsulated from degradation during processing. Likewise, since the process is relatively simple and low-energy consuming, it allows easy industrial scale-up. The PIT concept was introduced in the last decade by Shinoda and Saito (71,72), using the specific ability of surfactants, usually nonionic, (NS) such as polyethoxylated surfactants, to modify their affinities for water and oil in function of the temperature, and therefore to undergo a phase inversion. Indeed, the so-called transitional emulsion phase inversion occurs when, at fixed composition, the relative affinity of the surfactant for the different phases is changed, resulting in a gradual modification of the temperature. For example, an oil-in-water (o/w) emulsion is subjected to a phase inversion, giving rise to a water-in-oil (w/o) one, when the temperature rises. Within the transitional region between macro-emulsions, i.e. for the temperatures at which the nonionic surfactants exhibit a similar affinity for the two immiscible phases, the ternary system shows an ultralow interfacial tension and curvature, typically creating microemulsions, bicontinuous and nanoscaled systems (101-107). Therefore, the PIT method consists in suddenly breaking-up the chosen bicontinuous microemulsion maintained at the PIT, by a rapid cooling or by a sudden dilution in water or oil (108,109). Nanoemulsions are immediately generated. These bi-continuous systems have been thoroughly and widely characterized by establishing phase diagrams at equilibrium and formulation maps under dynamic conditions. The influence of the formulation (electrolyte concentration, temperature...) and composition parameters (surfactant amount or water/oil weight ratio, WOR=100× water / (water+oil)), has been largely reported on the potentialities to formulate nanometric-scaled emulsion droplets (110). During the emulsion inversion phenomena, the respective affinity of the NS for both immiscible phases is given by the difference between the chemical potentials of the surfactants in each phase. According to the physicochemical definition of De Donder, Eq. (5.1), the surfactant affinity difference (SAD) is defined with Eq. (5.2), at the physicochemical equilibrium, considering the activity coefficients close to the unity. It follows that the SAD is closely linked to the NS partitioning coefficient.

$$\mu_i = \mu_i^o + RT \ln (a_i C_i) \qquad .....(5.1)$$

μ_i is the chemical potential of the NS in phase i,

μ_i° are the standards, a the activity coefficients, and C the surfactant concentration.

$$SAD = \mu^o{}_{water} - \mu^o{}_{oil} = RT\ln (C_{oil}/C_{water}) \qquad .....(5.2)$$

In the case of ionic surfactants, the emulsion inversion corresponds to the SAD = 0, but this is not the case for nonionic surfactants and corresponds to a given reference noted SAD ^{ref}. Hence, this deviation with regards to the optimum formulation was defined with an a dimensional variable (111-115), known as the 'hydrophilic lipophilic deviation' (HLD), given for ionic and nonionic surfactants, and for a hydrocarbon n-alcane oily phase by the following Eqs. (3) and (4), respectively.

$$HLX = SAD/RT = \sigma + \ln S - kACN + t\Delta T + aA \qquad .....(5.3)$$

$$HLD = (SAD - SAD^{ref})/RT$$
$$= \alpha - EON + bs - kACN = t\Delta T + aA \qquad .....(5.4)$$

where EON is the number of ethylene oxide groups for NS, S is the weight percentage of electrolytes in the aqueous phase, ACN the amount of carbon numbers of the n-alcane composing the oily phase, ΔT the temperature difference from the reference temperature (25 °C), A the weight percentage of alcohol potentially added (not necessary for the PIT method), σ, α, k, t parameters in function of the used surfactant, *a* a constant given from the types of alcohol and surfactant, and finally *b* a constant function of the nature of the added electrolytes. Thus, the correlation between the HLD empirical expressions (3) and (4), and the SAD definition (2), gives the link between the temperature variation and the amphiphile partitioning coefficient (116), and thereby the surfactant behavior regarding the water/oil interface when using the PIT method. Hence, when NS is mainly used for generating nano-emulsions by the PIT method, formulation composition maps are typically built, as reported in Fig. 3a. Under constant stirring and for a fixed surfactant amount in the formulation, the emulsion gradually undergoes a phase inversion, as the HLD is changed by temperature variation. According to the HDL variation, at a constant WOR, the process is called transitional phase inversion. Moreover, for the lowest and highest WOR, emulsion inversion does not occur, due to the excessively rich water and oil regions. The emulsion morphology changes from the 'normal' to the 'abnormal' emulsion types, to form simple or multiple emulsions, respectively in accordance with Bancroft's rule and not. The illustration is provided in Fig. 3a with the transitions between (i) o/w and w/o/w, and (ii) w/o and o/w/o. In this case, even if the emulsion does not clearly exhibit a phase inversion, conditions are still suitable to perform the PIT method, where particular microemulsion structures can form at HLD=0, thus also leading to the generation of nano-emulsions, (117). The study of emulsion inversion involving only the variation of WOR at a constant HLD is known as catastrophic phase inversion, and has been extensively studied (118-120). It regards the transitions (horizontal pathways in Fig. 3a) between (i) w/o and w/o/w for HLD>0 and (ii) o/wand o/w/o for HLD<0, therefore this phenomenon is basically not included in the PIT method. Fig. 4b and c show the corresponding equilibrium phase diagrams, exhibiting the different thermodynamic equilibriums Winsor I to IV in function of the temperature. Kahlweit-fish diagram finally traces as well, such an evolution of the system morphology. For instance, in the case of WOR = 50, a rise in temperature crosses the fish body in Fig. 3b, and crosses the caudal fin in Fig. 3c. According to the comprehensive study proposed by Morales et al. (78), optimum conditions for nano-emulsion generation are closely linked to the ability of the microemulsion, precisely at the PIT, to solubilize all the phases to be dispersed. Indeed in most cases, this corresponds to the Winsor II and IV microemulsion formation, when the system is maintained at the PIT. In the basic cases of Fig. 3b and c, the nano-emulsions will be generated from the systems exhibiting W IV equilibrium microemulsions, or potentially W IV+LC, at the PIT, essentially for systems with higher surfactant amounts, Fig. 3c. Finally, the process implied in the PIT method of generating nano-emulsions, which suddenly breaks-up the microemulsions, can essentially be considered as irreversible since the nano-suspension formed is kinetically stable. It appears to go beyond the confines of the studied ternary system from the phase diagrams and formulation-composition maps, to create a kinetically stable nanoemulsion state. Of course, it should only be interpreted as a quasistable state, even if it is stable for months, and when achieved, the destabilization will provide the phase equilibrium considered above. Establishing the phase diagram is a requisite preliminary study in order to grasp and analyze the conditions suitable for nano-emulsion formulation. In this context, the link between EIP and PIT methods is clarified, highlighting the latter (PIT) as the one providing suitable experimental conditions to attain the nanometric structuring of the ternary system (in function of the formulation variables), similar to structures already established with mixing the components using the EIP method at room temperature (see above). Thus, the PIT method appears exclusively governed by the PEO surfactant phase behavior with regards to the formulation variables, and particularly the temperature. In this context, it is totally conceivable to add neutral components to the formulation, which influence neither the system phase behavior, nor the forming domain of nano-emulsions. For instance, some formulations based on the PIT method (117,121–122), include the addition of neutral amphiphiles for the formulation (such as phospholipids), in order to play on the final nanoparticle properties and structure: In these examples, the presence of phospholipids increases the lipid nanocapsule stability and acts as a framework on the final shell structure. To summarize, the phase inversion mechanisms of emulsions stabilized by PEO-surfactants, appear to establish close links between the NS partitioning coefficients and the temperature variation. In concrete terms, the water solubility and self association of NS are totally governed by the structuring state of water molecules, associated by hydrogen bonds into flickering clusters, wrapping the surfactant polar head. The low-energy and solvent-free PIT method generally appears relatively to be adaptable and, easy to handle. The PIT method generates nanoparticles at a low energetic cost, free from the toxicity of organic solvent, with a potentially low amount of surfactant (e.g. at 5wt. % in), making such a process essentially one of the most appealing methods(123).

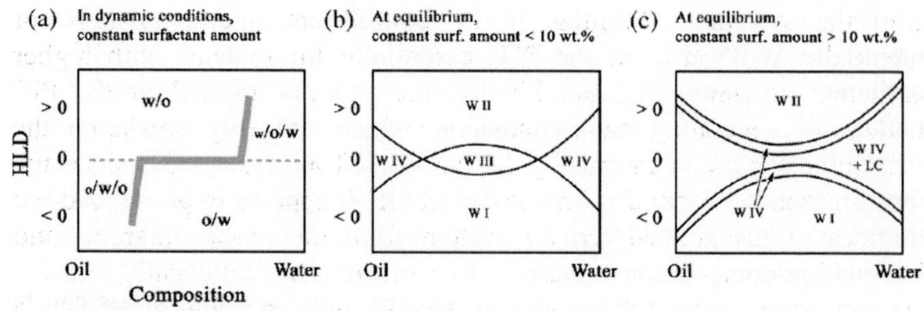

Fig 3. (a) Typical formulation composition map for water/nonionic surfactant/oil system, showing the emulsion inversion zones.Typical equilibrium phase diagram for the same system, HLD as a function of the composition, (b)for lower surfactants amounts,(c)for higher surfactants amounts.The frontier between both behaviour is roughly defined as 10wt %.

5.4 Drug Loading in Nanoemulsions

Drug loading depends on the type of nanoparticles (e.g., nanospheres, nanocapsules, solid lipid nanoparticles, dendrimers, polymeric micelles, nanoemulsions) and the preparation method. In general, two ways of drug loading can be distinguished: drug loading during nanoparticle formation and loading after the preparation of nanoparticles. If preformed nanoparticles are incubated in a drug solution for drug loading, drug can be extensively adsorbed to the large surface area of nanoparticles which in turn can result in initial burst release, which is more pronounced, the smaller the article. Furthermore, this method in general results in lower drug loading. Additionally, the time of incubation can also influence the drug loading, and the incubation time has to be sufficient to reach equilibrium for maximum loading. An important factor when loading drugs is the interaction between drug and carrier system because with increasing interaction the drug loading as well as entrapment efficiency increases; however, the release rate declines. For high entrapment efficiency, it is necessary that the drug interacts preferentially with the carrier system (polymer, lipid) rather than with the surrounding medium. Another factor influencing drug loading as well as entrapment efficiency is the nominal drug loading. Several studies showed that the drug loading can be enhanced by increased nominal drug loading. However, the entrapment efficiency does not necessarily increase with higher initial drug loading (Figure 5.1). If the maximum loading capacity of the nanocarrier is reached, further increase in nominal drug loading can even decrease the efficiency.

The drug loading of Solid lipid nanoparticles (SLN) and nanostructured lipid carriers (NLC) differs due to various preparation methods. In general, cold homogenization results in higher drug incorporation than hot homogenization .The type of surfactant, surfactant concentration as well as production temperature also affect drug loading during the hot homogenization process .The introduction of liquid lipid into the solid lipid matrix (forming nanostructured lipid carriers) is another approach to increase the drug entrapment in lipid nanoparticles. As the drug must be soluble in the lipid matrix, lipophilic drugs can be entrapped into SLN and NLC with higher efficiencies. As seen with lipid nanoparticles, polymeric micelle formation and drug incorporation also occur simultaneously. Lipophilic drugs especially are favorably incorporated into polymeric micellar carrier systems. Drug loading can be improved by increasing the polymer concentration employing polymers with higher molecular weights (MW) using grafted copolymers with functional end groups. However, the micelle production procedure can be optimized to obtain higher entrapment of drugs by careful selection of solvent, cosolvent and experimental setup.

Table 5.1 Examples of drug loading in nanoemulsions and nanocapsules.

Drug	Loading method	Time of loading	Entrapment efficiency (EE) and drug loading (DL)
Triclosan	• Solvent displacement method (NE) + chitosan (NC) Various oil amounts and chitosan at two different viscosity degrees	During	Emulsion: 86-99% (EE), 2-13% (DL) Capsules: 93-98% (EE), 2-14% (DL)
Ethionamide	• During: interfacial deposition (NC), nanoprecipitation (NS), spontaneous emulsification (NE) • After: incubation for different time periods	During vs. after	During: EE: 62.4% (NC), 53.0% (NS), 38.5% (NE) After (24 h): 56.2% (NC), 43.4% (NS)
Indomethacin	• Interfacial deposition (NC), nanoprecipitation (NS), spontaneous emulsification (NE)	During	EE. 94.5% (NC), 95.4% (NS), 90.0% (NE)

Table 5.1 *Contd...*

Drug	Loading method	Time of loading	Entrapment efficiency (EE) and drug loading (DL)
Metipranolol	• Interfacial polymerization (IP) vs. • Interfacial deposition (ID) with various polymer and oil types	During	EE: oil type: 34.6–49.9% (Migliol 840), 48.6–62.2% (Labrafil 1944 CS) Prepartion: 40.7–62.2% (IP), 34.6–60.2% (ID)

Table 5.1 shows examples of drug loading in nanoemulsions and nanocapsules, as well as the comparison between them and nanospheres. In general, higher drug loading is obtained for nanocapsules (prepared by interfacial deposition) compared to nanospheres (formed by nanoprecipitation) as the relative mass of the polymer used for capsules is reduced. The entrapment of drugs into nanoemulsions is reduced compared to nanospheres and nanocapsules. Moreover, the higher the solubility of the drug in the oil phase, the higher the drug entrapment in nanocapsules and nanoemulsions. In summary, no overall guideline can be given for drug loading as it depends on the drug that should be incorporated into the nanocarrier system; the intended administration route, the target, as well as the release profile (burst release at a specific target, sustained release, etc.). Based on the choice of the drug, the administration route, drug target and release profile, a nanocarrier system as well as loading technique can be selected using some general rules that can give optimum drug loading and entrapment efficiency (14).

5.5 Emulsion Destabilization

Emulsions in general, nanoemulsions included, can be destabilized by the following mechanisms: creaming (or sedimentation), flocculation, coalescence, or Ostwald ripening. Creaming is the separation of emulsion components based on the density of the droplets. The name is derived from the separation of the cream in unhomogenized milk. While creaming is usually considered to be undesirable, the process does not result in irreversible breaking of the droplets. Oil that is less dense than water will rise while oil that is more dense (such as with perfluorocarbon liquids) will settle to the bottom (sediment). It is routinely stated that the small size of nanoemulsions prevents creaming. While this is true in some instances, a more thorough explanation is necessary. Creaming is driven by gravitational forces. A theoretical treatment called the colloidal law of atmospheres has been developed to relate the gravitational potential energy of a droplet at height *h* above a surface with thermal energy:

$$mgh = k_B T \quad .....(5.5)$$

where m is the buoyant mass of a droplet, g is the acceleration of gravity, h is the height, k_B is Boltzmann's constant, and T is the absolute temperature. The mass of a droplet, m, is defined by

$$\frac{4}{3}\pi r^3 \Delta\rho \quad .....(5.6)$$

where r is the droplet radius and $\Delta\rho$ is the difference in density between the two phases. If $\Delta\rho$ is 0.1 g/cm^3, a particle with radius of 500 nm will have a gravitational height $\approx$ 0.01 mm, which means that creaming will occur. At the same density a particle with radius of 10 nm will have a gravitational height $\approx$100 cm, well above the height of most containers, and thus creaming will be prevented. However, with that same density difference, a particle with radius of 50 nm will have a gravitational height $\approx$ 0.8 cm so creaming will occur. Furthermore, this analysis only considers a situation where droplets repulse each other and there is no interaction with the solvent. If there is attraction between droplets, creaming can occur regardless of the size. Favorable electrostatic interactions between the droplets and the solvent, which are unaccounted for in this equation, also affect the rate of creaming. The end result is that the smallest nanoemulsions are stabilized against creaming, a significant advantage over macroemulsions, but only if densities of the dispersed phase and the continuous phase are fairly even. Flocculation refers to a process in which clusters of two or more droplets behave kinetically as a unit, but individual droplets still maintain their identity. It is reversible, but may lead to coalescence, which is irreversible. In systems stabilized by nonionic surfactants, the droplets are attracted by van-der Waals forces, but repulsed by steric interactions. The steric repulsion between emulsion droplets, W_s, can be represented by the following equation:

$$W_s \infty \; K_B T e^{-\pi D/l} \quad .....(5.7)$$

where k_B is the Boltzmann constant, T is the absolute temperature, D is the separation distance between droplet surfaces, and L is the film thickness of the adsorbed polymer. If the total interaction energy is smaller than the energy imparted from Brownian motion, ~ k_BT, the particles will remain unflocculated. An increase in the film thickness, L, will increase W_s and thus lead to more stable emulsions. The van der Waals attractive potential between two spherical droplets (with identical radius) is linearly dependent on the radius, with the following relationship:

$$W_{vdw} = \frac{-Ar}{12D} \quad .....(5.8)$$

where A is the Hamaker constant, r is the radius of the droplets, and D is the distance of separation between droplets. As the radius of the particles decreases, the attractive potential decreases. Therefore, in nanoemulsions with a small radius and large enough film thickness, flocculation can be prevented, another advantage over macroemulsions. Coalescence is the collision, and subsequent irreversible fusion, of two droplets. The ultimate end of coalescence is complete phase separation. Flocculation precedes coalescence, so the same methods that are appropriate for prevention of flocculation also prevent coalescence. A thick, surfactant film adsorbed at the interface is often sufficient to prevent coalescence, whether in nano- or macroemulsions. However, with the same polymeric thickness, the stabilization will be greater for a nanoemulsion because the polymer layer will be a greater percentage of the total diameter. Ostwald ripening is the growth in the size of emulsion droplets as the contents of one drop diffuse into another. The driving force for this growth is the difference in chemical potential between droplets, which is generally not substantial for droplets larger than 1 mm. Therefore, Ostwald ripening primarily affects nanoemulsions and is the most serious instability concern for nanoemulsions. This effect is related to the Laplace equation for spheres:

$$\Delta p = 2\gamma / r \quad .....(5.9)$$

where p is the pressure across an interface, γ is the interfacial tension, and r is the radius of the sphere. Kelvin adjusted this equation to describe the difference in vapor pressure between a small droplet of a liquid and the bulk liquid, the situation found in an emulsion:

$$RT\ln\frac{p}{p^o} = \frac{2\gamma\,[v]}{r} \quad .....(5.10)$$

where R is the gas constant, T is the absolute temperature, p is the vapor pressure of the bulk, p^0 is the vapor pressure of the droplet with radius r, γ is the interfacial tension, and (V) is the molar volume of the liquid. As the radius increases, the pressure difference is reduced and the dispersed droplets become more soluble in the continuous phase. If there is any diffusion of the contents of the dispersed phase, large droplets will grow larger at the expense of smaller droplets and the average size of the particle distribution will continually increase. According to Lifshitz–Slyozov–Wagner (LSW) theory, the rate of Ostwald ripening, ω, can be expressed by the following equation:

$$\omega = \frac{dr^3}{dt} = \frac{8DC_{\infty}\gamma M}{9\rho^2 RT} \qquad .....(5.11)$$

where r is the radius of the droplets, t is the time of storage, D is the diffusion coefficient of the molecules of the dispersed phase in the continuous phase, $C\infty$ is the bulk solubility of the dispersed phase in the continuous phase, γ is the interfacial tension between phases, M is the molar mass of the dispersed phase, ρ is the density of the dispersed phase, R is the gas constant, and T is the absolute temperature. However, the overall diffusion of the dispersed phase is affected by the diffusion across the interfacial layer in addition to the diffusion in the continuous medium, D. If the diffusion across the interface is slower than diffusion through the medium, then the overall rate of ripening will be slower than predicted. As can be seen, the cube of the particle radius varies linearly with time. For an o/w emulsion, the rate of ripening is directly related to the water solubility of the oil and ripening can even be seen in systems where the solubility is in the low nM range. It should be noted that in the ideal situation of a perfectly monodispersed distribution there would be no ripening because there would be no differences in solubility of droplets based on size. Thus, narrow distributions will be more resistant to Ostwald ripening than broader distributions. Higuchi and Misra suggested that the addition of a secondary, less water-soluble, component could slow ripening. The slower diffusion of the secondary component will lead to a heterogeneous distribution with smaller droplets enriched in the less soluble component and larger droplets enriched in the more soluble component. However, this internal segregation will be thermodynamically opposed as osmotic pressure will act to limit differences between droplets and equilibrium will eventually be reached. This principle has been successfully applied with hydrocarbon and fluorocarbon emulsions. The rate of ripening of a two-component disperse phase system is represented by the following equation:

$$\omega_{mix} = \left(\frac{\phi_1}{\omega_1} + \frac{\phi_2}{\omega_2}\right)^{-1} \qquad .....(5.12)$$

where φ represents the volume fraction and the subscripts 1 and 2 refer to the more and less water-soluble components, respectively. As φ_2 becomes larger, it becomes the dominant term until it solely controls the ripening rate. With a properly chosen additive, Ostwald ripening can be effectively eliminated. In some instances, the rate of ripening has increased as the amount of surfactant has increased, whether the excess surfactant is present in the form of vesicles. One of the justifications cited for such an effect is that the supramolecular aggregate (micelle or vesicle) provides a reservoir to solubilize excess oil, thus increasing the effective solubility of the oil in water. As can be seen in Eq. (11), as the solubility, C_1, increases, so too does the rate of ripening. However, a decrease in the ripening rate as the amount of surfactant increases has also been reported. In these cases, it has been proposed that the oil solubilized in the micelles is not dispersed in the continuous phase, and therefore is not subject to the same mass transfer between droplets. In this argument, C_1 is lowered as oil is withdrawn from the continuous phase into micelles, thus causing the ripening rate to decrease. An additional study showed that alkane emulsions stabilized by hexaethylene glycol dodecyl ether were unaffected by surfactant concentration. Though Ostwald ripening can be present in nanoemulsions, it has some advantages for pharmaceutical development. The ripening rate provides clear criteria for determining the acceptability of formulations. Commercialization of an emulsion mandates stability for at least 18 months. The rate of ripening allows the estimation of long-term stability, which in turn suggests guidelines for specifications and expiration dates. If the ripening rate is too rapid, then the nanoemulsion will be unacceptable for pharmaceutical use. The instability of emulsions can also have some drawbacks. For emulsions intended for parenteral injection the FDA requires sterilization. The most commonly employed form, terminal heat sterilization such as with an autoclave, can often affect the physical stability of the emulsion droplets. Additionally, emulsions are often stored at 58°C for greater stability, which places limits on the product use and storage (14).

5.6 Stability of Nanoemulsions

The main particularity of nano-emulsions, making them prime candidates for nanoparticle engineering, is their great stability of droplet suspension. A kinetic stability that lasts for months, stability against dilution or even against temperature changes, totally unlike the (thermodynamically stable) microemulsions. Emulsions are thermodynamically unstable systems, due to the free energy of emulsion formation (ΔG_f) greater than zero. The large positive interfacial energy term ($\lambda\Delta A$) outweighs the entropy of droplet formation ($T\Delta S_f$), also positive. The terms λ and ΔA respectively represent the surface tension and the surface area gained with emulsification. Emulsion instability is therefore induced by the positive sign of ΔG_f (Eq. (13)).

$$\Delta G_f = \gamma \Delta A - T\Delta S_f \qquad .....(5.13)$$

Accordingly, the physical destabilization of emulsions is related to the spontaneous trend towards a minimal interfacial area between the two immiscible phases. Therefore, a minimization of interfacial area is attained by two mechanisms: (i) Flocculation followed mostly by coalescence, and (ii) Ostwald ripening. In nano-emulsion systems, flocculation is naturally prevented by steric stabilization, essentially due to the sub-micrometric droplet size. In short, when interfacial droplet layers overlap, steric repulsion occurs, from two main origins. The first one is the unfavorable mixing of the stabilizing chain of the adsorbed layer, depending on the interfacial density, interfacial layer thickness δ, and Flory–Huggins parameter $\chi_{1,2}$ (which reflects the interactions between the interfacial layer and solvent). The second one is the reduction of the configurational entropy, due to the bending stress of the chains, which occurs when inter-droplet distance h becomes lower than δ. Generally, the sum of the energies of interaction U_T adopts a typical shape of systems wherein molecules repel and particles attract each other, showing a weak minimum, around h=2δ, and a very rapid increase below this value (see Fig. 4 for illustration).

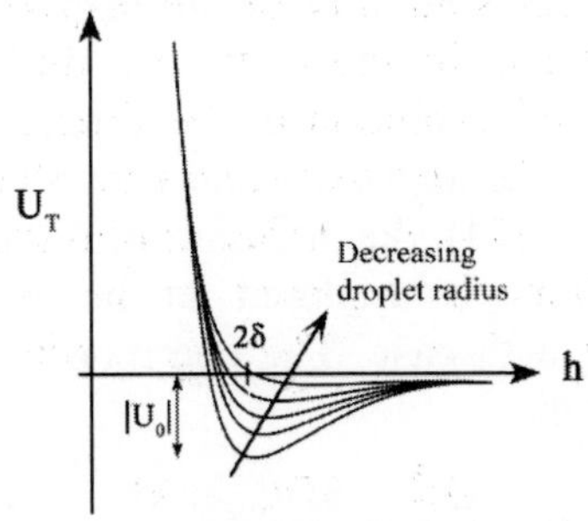

Fig 5.4 Diagram of the influence of emulsion droplet radius on steric stabilization.

The depth of the minimum $|U_0|$ will induce predispositions for coagulate in the colloidal suspension, that is to say, it is intimately linked to the stability of the suspension. $|U_0|$ is shown to be dependent on the particle radius r, the Hamaker constant A, and the adsorbed layer thickness δ, with the result that the higher the δ/r ratio, the lower the value of $|U_0|$. Now, in the case of nano-emulsion droplets, δ/r becomes very high in comparison with macro-emulsions, which in the end totally inhibits its ability to coagulate. On the other hand, it is worth noting that the small droplet sizes also induce stabilization against sedimentation or creaming, in so far as the droplets are solely under the influence of the Brownian motion. Taking all this into account, the destabilization of nano-emulsions is due only to a mass transfer phenomenon between the droplets through the bulk phase, popularly known as Ostwald ripening in emulsions. At the origin of this destabilization process, the differences, however slight, of the droplet radius induce differences in chemical potential of the material within the drops. The reduction of free energy in the emulsion will result in the decrease of the interfacial area, and therefore in the growth of the bigger emulsion droplets at the expense of the smaller ones. The dispersed phase migrates through the bulk from the smaller droplets to the bigger ones, owing to the higher solubility in the bulk of the smaller droplets. Ostwald ripening is initiated and will increase throughout the process. Under the assumption that only one component composes the dispersed phase, the solubility, C(r), of the dispersed material throughout the dispersion medium is expressed as a function of the droplet radius r, from the Kelvin equation , Eq. (14),

$$C(r) = C_{\infty} \exp\left(\frac{2\gamma M}{\rho RT\gamma}\right) \qquad \text{.....(5.14)}$$

where C_∞ is the bulk solubility of the dispersed phase, M its molar mass, and ρ its density. In most studies, the follow-up of Ostwald ripening as the temporal evolution of the droplet diameter still remains well fitted, even under the approximations involved in Eq. (14). Besides the consideration of Eq. (14), the diffusion of dispersed materials through the continuous medium is assumed to be diffusion-controlled, i.e. crossing the interface with ease. The commonly used expression of the ageing rate ω is given by,

$$\omega = \frac{dr_C^3}{dt} = \frac{8DC_{\infty}\gamma M}{9\rho^2 RT} \qquad \text{.....(5.15)}$$

where r_c is the critical radius of the system at any given time, at the frontier between the growth and decrease of the droplets. Consequently Ostwald ripening is reflected by a linear relationship between the cube radius and time. In processes involved in nanoparticle engineering, i.e. for multicomponent emulsion droplets, by adding monomer, polymer, or simply surfactant or co-surfactant, the above approximation is surpassed. The rate of ripening can be reduced by several orders of magnitude when the additive has a substantially lower solubility in the bulk phase than the main component of the droplet. This phenomenon has been widely studied, since it appears to be an efficient method to reduce the Ostwald ripening rate, even when using small amounts of additives. Indeed, up to now it has been considered that Ostwald ripening is a diffusion-controlled process, but this assumption does not take into account the fact that surfactants, polymeric emulsifiers or stabilizers can create a thick steric barrier at the droplet interface. As a consequence, the diffusion of the inner material of the droplets may be slowed down, reducing the ripening rate. The substantial difference in stability between nanoemulsions and nanocapsules for instance, appears essentially from such details.

5.7 Ostwald Ripening

One of the main problems with nano-emulsions is Ostwald ripening, which results from the difference in solubility between small and large droplets. The difference in chemical potential of dispersed phase droplets between different sized droplets was given by Lord Kelvin.

$$C(r) = C(\infty)\exp\left(\frac{2\gamma V_m}{\gamma RT}\right) \qquad \text{.....(5.16)}$$

where c(r) is the solubility surrounding a particle of radius r, $c(\infty)$ is the bulk phase solubility and V_m is the molar volume of the dispersed phase. The quantity ($2\gamma V_m/rRT$) is termed the characteristic length. It has an order of ~1 nm or less, indicating that the difference in solubility of a 1 μm droplet is of the order of 0.1% or less. Theoretically, Ostwald ripening should lead to the condensation of all droplets into a single drop (i.e. phase separation). This does not occur in practice since the rate of growth decreases with increasing droplet size.

For two droplets of radii r_1 and r_2 (where $r_1 < r_2$),

$$\left(\frac{RT}{V_m}\right)\ln\left[\frac{c(r_1)}{c(r_2)}\right] = 2\gamma\left(\frac{1}{r_1} - \frac{1}{r_2}\right) \qquad \text{.....(5.17)}$$

Equation (9.15) shows that the larger the difference between r_1 and r_2 the higher the rate of Ostwald ripening.

Ostwald ripening can be quantitatively assessed from plots of the cube of the radius versus time t

$$r^3 = \frac{8}{9}\left[\frac{c(\infty)\gamma V_{mD}}{\rho RT}\right]t \qquad \text{.....(5.18)}$$

where D is the diffusion coefficient of the disperse phase in the continuous phase and r is the density of the disperse phase.

Several methods may be applied to reduce Ostwald ripening:

1. Addition of a second disperse phase component that is insoluble in the continuous phase (e.g. squalene). Here, significant partitioning between different droplets occurs; the component that has low solubility in the continuous phase is expected to be concentrated in the smaller droplets. During Ostwald ripening in a two-component disperse phase system, equilibrium is established when the difference in chemical potential between different size droplets (which results from curvature effects) is balanced by the difference in chemical potential resulting from partitioning of the two components. If the secondary component has zero solubility in the continuous phase, the size distribution will not deviate from the initial one (the growth rate is equal to zero). For limited solubility of the secondary component, the distribution is the same as governed by Eq. (18), i.e. a mixture growth rate is obtained that is still lower than that of the more soluble component. The above method is of limited application since one requires a highly insoluble oil, as the second phase, which is miscible with the primary phase.
2. Modification of the interfacial film at the O/W interface: According to Eq. (17) reduction in γ results in a reduction of Ostwald ripening. However, this alone is not sufficient since one has to reduce γ by several orders of magnitude. By using surfactants that are strongly adsorbed at the O/W interface (i.e. polymeric surfactants) and which do not desorb during ripening, the rate could be significantly reduced. An increase in the surface dilational modulus and decrease in γ would be observed for the shrinking drops. The difference in γ between the droplets would balance the difference in capillary pressure (i.e. curvature effects). A-B-A block copolymers that are soluble in the oil phase and

insoluble in the continuous phase are useful in achieving the above effect. The polymeric surfactant should enhance the lowering of γ by the emulsifier. In other words, the emulsifier and the polymeric surfactant should show synergy in lowering γ (16).

5.8 Advantages of Nanoemulsion in Drug Delivery

5.8.1 Solubilization of Poorly Soluble Drugs

Solubilization of poorly soluble drugs is the most apparent application for nanoemulsion. Example: Lorazepam is injected intravenously for premedication and sedation before an operation. It is usually administered as a solution in organic solvents such as propylene glycol. The highest concentration that can be achieved in an aqueous diluent (5% dextrose in water) is 0.05 mg/mL. A phospholipid-stabilized soybean oil emulsion was able to stably emulsify lorazepam at 1 mg/mL, a 20-fold increase, which could significantly reduce the volume needed for injection. While emulsion solubilization has been applied to a host of lipophilic drugs, it traditionally cannot be employed if the drug has limited solubility in oils that have regulatory acceptance. One way to counter that deficiency is to position the drug directly in the interfacial lecithin layer. This has previously been achieved by dissolving the drug together with lecithin in an organic solvent, evaporating the solvent, and then using that mixture for de novo emulsification. Unfortunately, this method is impractical on an industrial scale. With this methodology, termed SolEmuls1, solid nanocrystals of the drug are homogenized with commercially available lipid emulsions, without the need of any organic solvents. In this way, they were able to stably emulsify carbamazepine, itraconazole , ketoconazole, and amphotericin B.

5.8.2 Reduced Pain/Irritation

At the direct site of intravenous injection, some drugs can cause local irritation. These drugs, as well as certain cosolvents in aqueous solutions, can also cause phlebitis, an inflammation of a vein that can lead to pain or redness. Nanoemulsions eliminate the need for cosolvents, as well as encapsulating drugs that might otherwise be irritants, and in both cases can reduce local irritation upon injection. Polyene antifungal agents, like amphotericin B, contain a large macrocycle with a series of conjugated double bonds and are typically insoluble in water. Recent efforts have focused on synthesizing new agents that retain the strong antifungal activity while improving the water solubility. One candidate, SPK-843, showed promising antifungal activity but gave mild phlebitis upon repeated intravenous injections. Therefore, it was proposed to study the intravenous injection of SPK-843 in an emulsified form. The drug was extemporaneously added to commercial Intralipid emulsions. After formation, the venous toxicity of the nanoemulsion was tested in the ear vein of rabbits and compared to a 5% glucose solution of SPK-843. A solution of 5% glucose alone (w/o drug) was tolerated for 17 infusions before vein occlusion due to phlebitis, but when the drug wasadded only three infusions could be tolerated. With the Intralipid alone, 13 infusions were tolerated, and when the drug was introduced the number remained at 13. Therefore, an emulsified form of the drug reduced the chance of phlebitis compared to an aqueous solution. In patients with ruptured aneurysms, vasospasm of the cerebral arteries can lead to delayed ischemic deficits, which are responsible for morbidity and mortality. To reduce morbidity and mortality caused by these reasons, nimodipine (NM) is the only available therapy. It is given orally, but its bioavailability is limited. Intravenous injection, as an ethanol solution, has been studied, but leads to irritation. A soybean oil nanoemulsion with lecithin and Tween-80 as the emulsifiers was proposed as a method to reduce irritation. The irritation was measured both by the rabbit ear vein test, as in the prior example, and by the rat paw lick test. Upon injection of the NM-ethanol solution 100% of the rats licked their paws and the average number of licks was 12, suggesting both a high frequency and a high intensity of pain. With the NM-nanoemulsion formulations, only 60% of the rats licked their paws, and the average number of licks decreased to five, indicating that the both occurrence and intensity of irritation decreased. Furthermore, the nanoemulsions did not alter the pharmacokinetic parameters, as compared to the ethanol solution. Therefore, the in vivo performance was maintained as the pain was reduced. In another study, an anti-cancer drug, norcantharidin, was also formulated in lecithin/Tween-80-stabilized nanoemulsions and the irritation was studied using the rat paw lick test. In this instance, the frequency of rats that licked their paws remained the same at 100% between a solution formulation and a nanoemulsion formulation, but the average number of licks decreased from 18.2 to 8.7, indicating that the intensity of pain was diminished.

5.8.3 Reduced Toxicity of Drug

Beyond the site of injection, drugs or their delivery vehicles can also cause irritation or toxicity once in the body. Paclitaxel is an important chemotherapeutic agent used in the treatment of breast, ovarian, colon, and non-small cell lung carcinomas. The commercially available product Taxol1 (Bristol-Myers Squibb) is formulated in a 1:1 v/v mixture of ethanol and polyoxyethylated castor oil (Cremophor EL). Cremophor EL has been associated with bronchospasms, hypotension, and other hypersensitive reactions. To reduce the toxicity associated with Cremophor EL, incorporation of paclitaxel into a wide variety of drug delivery vehicles, including liposomes, micelles, emulsions, and cyclodextrins, has been investigated. A representative nanoemulsion example will be described. Constantinides and coworkers created a nanoemulsion that did not employ any lecithin but used Vitamin E-TPGS (a-tocopherylpolyethyleneglycol- 1000 succinate) and Poloxamer 407 to emulsify Vitamin E (DL-a-tocopheryl) as the oil phase. The nanoemulsion droplets had a mean diameter of 67 nm, with 99% below 150 nm, meaning that the emulsions could be filter sterilized with a 0.22 mm filter, an objective of the study. The maximum tolerated dose (MTD) was determined in mice using a tail vein injection. For the commercially available Taxol1 formulation the MTD was approximately 20 mg/kg whereas for the nanoemulsion formulation it was approximately 70 mg/kg, over three times greater. The efficacy of the nanoemulsion formulation was assessed with B16 melanoma, a fast-growing solid murine tumor. Nanoemulsions showed increasing efficacy at increasing dosage amounts and were better than the commercial formulation in all cases.

5.8.4 Improved Pharmacokinetics

Pharmacokinetics is concerned with the fate of external substances introduced to the body, specifically the extent and rate of absorption, distribution, metabolism, and excretion of compounds. Improving these parameters for more favorable drug performance is a primary objective of drug delivery research in general and for nanoemulsions specifically. One specific parameter that will be mentioned multiple times is the area under the concentration–time curve, abbreviated AUC. Nalbuphine is a morphine-like drug and one of its advantages over morphine is that it lacks significant withdrawal symptoms. However, due to its short elimination half-life and poor oral bioavailability it needs to be injected every 3–6 h. Prodrugs of nalbuphine have been investigated for parenteral administration, and Fang and coworkers sought to use nanoemulsions for both nalbuphine and its prodrugs. Egg phospholipid was used as the main emulsifier, along with cosurfactants Brij 30, Brij 98, and stearylamine. Depending on the emulsifier composition the average size ranged from

167 to 314 nm. It was found that the stearylamine-containing emulsions had the highest prodrug encapsulation, while the incorporation of Brij 98 reduced prodrug entrapment, suggesting that the nature of the oil/water interface and the co-emulsifier may affect drug loading. In vivo pharmacokinetic profiles indicate that the plasma concentration of nalbuphine and its prodrugs was enhanced by incorporation into nanoemulsions. The nanoemulsion delivery of flunarizine, a drug used for migraine prophylaxis in which oral administration is shows marked low bioavailability and slow absorption on oral administration. A flunarizine solution (with 5% PEG 400 and 0.2% Tween-80 as stabilizers) and flunarizine-loaded nanoemulsions (both at 1 mg/mL) were compared. While all of the other pharmacokinetic parameters showed no significant difference, the AUC was 1.68 times greater for the nanoemulsion, demonstrating a prolonged circulation time in rats. Cerebral malaria is a medical emergency that requires treatment that can rapidly reach effective active drug concentrations in vivo. Halofantrine is a well-tolerated and effective antimalarial drug that acts more rapidly than quinine or mefloquine. However, it is given orally and its slow dissolution prevents the rapid therapeutic impact needed to treat cerebral malaria. An intravenous formulation could provide rapid delivery of the drug. A previously investigated parenteral formulation showed local irritation and toxicity. To improve upon those characteristics Barratt and coworkers investigated o/wwater nanoemulsions. They demonstrated that an emulsion stabilized by poly-D,L-lactide (PLA) and its copolymer methoxy-polyethylene glycol-co-poly-D,L-lactide (PLA-PEG) increased the AUC more than sixfold compared to the previous intravenous formulation. In a later work poloxamer 188 and lecithin were used in different surfactant combinations with the PLA and PLA-PEG. In all examples, Miglyoll 810 N, a medium chain triglyceride, was used as the oil phase and lecithin as one of the surfactants. Overall, particle diameters of the drug-loaded emulsion ranged from 200 to 350 nm. It was found that the addition of PEG-PLA copolymers increased the emulsion stability compared to those prepared with lecithin alone as well as providing a more consistent, sustained release, which may have application in the treatment of cerebral malaria cases. Other studies have also shown how PEG coating can affect nanoemulsion circulation and performance in vivo. Encapsulation in nanoemulsions stabilized by phosphatidylcholine and cholesterol reduced the clearance of the drug by 1.4 compared to the free drug. When the nanoemulsion also contained a DSPE-PEG(1,2-distearoylsn- glycero-3-phosphoethanolamine-N-polyethylene glycol 2000) conjugate in the stabilizing layer, the drug

clearance was reduced by a factor of 3. Because of the reduced uptake into tissues such as the liver and spleen, 7.5 times more of the PEG-coated nanoemulsions ended up at the site of inflammation compared to the uncoated nanoemulsion. Circulation times were also studied by Ueda and coworkers with nanoemulsions containing menatetrenone and stabilized by polyoxyethylatedhydrogenated castor oils (HCOs). They found that plasma half-lives and liver uptake of nanoemulsions stabilized by HCOs with 10 PEG units were similar to and larger than, respectively, emulsions stabilized by egg yolk phospholipids. However, when the length of the PEG chain was increased to 20 and 60 units there was a marked increase in the circulation time and decrease in the liver uptake, suggesting that there is minimum length of PEG necessary to see improved pharmacokinetic parameters. In another example, an o/wnanoemulsion of cyclosporin A was compared to commercially available formulations for both oral and intravenous delivery. When the nanoemulsion was delivered intravenously there was little difference with the commercial i.v. formulation (CIPOL Inj.1) and when delivered orally the AUC was actually less than the commercial oral formulation (Sandimmun Neoral1). However, with both routes of administration the pharmacodynamic efficiency was greater and pharmacodynamic availability was improved twofold versus the commercial formulations.

5.8.5 New Method of Delivery

There are many examples where a drug that is normally delivered orally can be incorporated into a nanoemulsion and injected intravenously, some of which are discussed in the pharmacokinetics section. Another instance where nanoemulsions can open up a new method of delivery is the case of volatile anesthetics for general anesthesia. Volatile anesthetics are low boiling liquids that are given as gases by inhalation. However, if injected directly into the bloodstream, the time for the anesthetic to equilibrate with the lungs is eliminated, which leads to a more rapid onset of anesthesia. Because direct i.v. delivery of the neat anesthetic causes pulmonary damage and death fat emulsions have been successfully utilized as a means of delivery for halothane, isoflurane, and sevoflurane . However, the modern volatile anesthetics (except for nitrous oxide) are all highly fluorinated, which reduces their solubility in classic hydrogenated oils and makes them more soluble in fluorinated oils. The solubility of sevoflurane in Intralipid (30%) is limited to a mere 3.5%. It has been found that the presence of Oxygent (a perfluorocarbon emulsion) greatly increases the blood:gas partition coefficient of isoflurane, sevoflurane, and desflurane compared to Intralipid. Building upon this information, it was found that a nanoemulsion stabilized by a fluorinated surfactant, with a fluorinated secondary additive to slow the Ostwald ripening, was capable of stably emulsifying up to 25% sevoflurane, a sevenfold increase over Intralipid. The efficacy and safety of this formulation for intravenous delivery was tested in rats with bolus dosing and found to safely induce anesthesia, from which recovery was smooth and rapid.

5.8.6 Drug Targeting

As with other drug delivery vehicles, targeted emulsions are of interest. Targeting can either be active, such as the inclusion of a secondary component in the stabilizing monolayer that has a recognition or functional element for a specific site, or passive, where the final destination is dependent on the size or surface characteristics of the particles.

5.8.6.1 Active Targeting

Rapidly dividing cells, such as cancer cells, require higher amounts of cholesterol to build cell membranes. Low-density lipoprotein (LDL) is the natural carrier of cholesteryl esters in the body and therefore certain tumors have elevated LDL-receptor activity. A phospholipid-stabilized nanoemulsion was developed to solubilize a cholesteryl ester of carborane, cholesteryl 1,12-dicarba-closo-dodecaborane- 1-carboxylate (BCH), which mimics the natural core of LDL and can be used for boron neutron capture therapy (BNCT). The mean emulsion particle size was 155 nm. If preincubated together, human LDL particles and the nanoemulsions only gave one band in agarose electrophoresis, demonstrating particle interaction and ability of BCH to transfer to the human LDL. Cell culture data showed sufficient uptake of BCH in rat 9L glioma cells for the levels necessary for BNCT. Cholesterol-rich nanoemulsions that mimic LDL have also been studied over the years. In an experiment Paclitaxel was solubilized in cholesteryl oleate nanoemulsions stabilized by egg phosphatidylcholine. The pharmokinetics of the nanoemulsion and its ability to concentrate the drug in tumors was studied in patients with gynecologic cancers (ovarian, cervix, endometrium). It was shown that paclitaxel in nanoemulsions was stable in the bloodstream and the pharmacokinetic profile was improved compared to the commercial formulation . Furthermore, on average 3.5 times more paclitaxel was concentrated in malignant tissues versus normal tissues. Together, these results pave the way for future clinical trials of paclitaxel nanoemulsions.

Nanoemulsions have also been used for cell targeting by surface modification with carbohydrates. Hashida and coworkers tested a series of glycosylated emulsions, including galactosylated, mannosylated, and fucosylated. In the first example, a galactose-cholesterol conjugate (Gal-C4-Chol) was inserted into the phosphatidylcholine monolayer of a soybean oil emulsion for hepatocyte-selective targeting in a weight ratio of 70:25:5 soybean oil:PC:Gal-C4-Chol. Nanoemulsions containing the Gal-C4-Chol had a higher rate of liver uptake than both emulsions without the Gal-C4-Chol and liposomes with it. Furthermore, a model lipophilic drug, probucol, showed efficient delivery to the liver as compared to Gal-liposomes. A subsequent study showed that the same effect could be achieved with both mannosylated and fucosylated cholesterol conjugates. Nanoemulsions which contained only 2.5% mannose showed no difference in liver accumulation versus the control emulsions, but 5.0 and 7.5% mannose showed an increasingly greater amount of uptake. Some amphiphiles, such as 1-O-alkylglycerols, can transiently open the blood brain barrier (BBB) to improve the brain delivery of some anticancer agents. Varieties of 1-O-alkylglycerols as cosurfactants were used along with lecithin to stabilize a soybean oil emulsion for the solubilization of carbamazepine (CBZ), used in the treatment of seizures. CBZ is available for oral and chewable tablets, but intravenous delivery could allow for more rapid treatment in acute seizures. The nanoemulsions were around 200 nm in diameter. In emulsions containing 1-O-decylglycerol, the brain/serum concentration ratio was 3.0 after 30 min, suggesting the potential for brain targeting.

5.8.6.2 Passive Targeting

Passive targeting is mainly accomplished by altering the size of injected particles and is a justification for many drug delivery vehicles in place of a free drug. With a particle diameter of 200–300 nm, traditional emulsions for parenteral nutrition rapidly enter the liver and are removed from circulation. Using the same ingredients found in nutritional emulsions (soybean oil and purified egg lecithin), Some nanoemulsions with a diameter between 25 and 50 nm and larger ones with a 200–300 nm diameter were developed. It was found that there was a lower liver uptake of the smaller particles, which enabled a longer plasma half-life. The incorporated drug, dexamethasone palmitate (DMP), is an anti-inflammatory agent. It is well known that inflammation sites have leaky capillary walls, which allow smaller particles to passively diffuse across. With a greater plasma half-life, the smaller nanoemulsions passed into the inflammation sites with greater efficiency, delivering more than three times greater the amount of DMP than with the larger particles. In the prior example, small size was desired so that the nanoemulsion particles could avoid liver uptake. However, in some situations the liver is the target destination of the drug, which makes a larger droplet size more desirable. All-trans retinoic acid (ATRA) is an anti-cancer agent that has been studied for the treatment of liver cancer metastasis. Oral administration has been studied, but bioavailability was highly variable, suggesting a possible improvement with intravenous delivery. Hashida and coworkers formulated a soybean oil nanoemulsion that was stabilized by either egg phosphatidylcholine (PC)/cholesterol or egg PC/DSPE-PEG and cholesterol to incorporate ATRA. The average particle diameter was133 nm. The nanoemulsions were stable in the presence of albumin in the blood and showed a statistically greater accumulation in the liver, compared to the free drug. In CT26 tumor cells, it was shown that emulsified ATRA reduced the number of metastatic nodules and liver weight, compared to a saline solution, as well as ATRA loaded in hydrogenated castor oil (HCO-60) micelles. Silymarin is a hepatoprotective agent that has a positive effect on metabolism and physiology of liver cells. Physical characterization studies showed enhanced release of the silymarin from the nanoemulsions, but no in vivo work has been performed yet.

5.8.7 Multiple Functionality

A recent development in nanoemulsions, as with other delivery systems, is the combination of therapeutic and imaging capabilities together in one system. This approach couples drug delivery with tissue imaging to allow simultaneous delivery of the drug and visualization of the physiological effects. A 20% pine nut oil emulsion stabilized by egg phosphatidylcholine for the simultaneous solubilization of paclitaxel and gadolinium ions (Gd3+) was developed, for enhanced tissue contrast for magnetic resonance imaging (MRI). The standard nanoemulsions had an average particle diameter of 90.4 nm. In addition to the phosphatidylcholine, a diethylenetriaminepentaacetic acid (DPTA)-phosphatidylethanolamine (PE) complex was added. The PE positions itself at the oil–water interface leaving the DPTA solvent exposed. DPTA is a known chelator for Gd3+, and the high affinity provides for tight binding in an aqueous environment. MRI T1 relaxation measurements showed similarity between the nanoemulsion and reported literature values for the commercial imaging agent Magnevist1, suggesting the appropriateness of the nanoemulsion as an imaging agent. The nanoemulsions also successfully delivered the paclitaxel to MCF-7 carcinoma cells in vitro, demonstrating the dual functionality, but the drug delivery performance was not superior to an aqueous paclitaxel solution.

5.8.8 New Therapies

Perhaps the most widely studied example of emulsions opening up a new therapeutic area, which would not be available otherwise, is the case of fluorocarbon emulsions for use as artificial blood substitutes. This topic has been extensively reviewed, and so will not be discussed in great detail here. Briefly, though, fully fluorinated molecules (perfluorocarbons, PFCs) have extremely low polarizability, which leads to low van der Waals interactions between molecules. The limited intermolecular forces in PFCs cause them to behave as nearly ideal, gas-like fluids, which allows them to dissolve significant amounts of gas such as O_2. If injected directly, these liquids could form an oil embolism that could ultimately lead to death, but as an emulsified form, these liquids show no adverse effects in vivo and are able to transport and deliver tremendous amounts of oxygen. Fluosol-DA is an example of a fluorocarbon emulsion that garnered FDA approval, though it was not a commercial success because of practical limitations (the emulsion had to be stored frozen because of stability concerns, thawed and mixed with two annex solutions prior to use, and discarded no more than 8 h after mixing).

5.9 Characterization of Nanoemulsion

5.9.1 Dynamic light scattering

Hydrodynamic diameter as much as poly-dispersity index, measuring, was carried out by dynamic light scattering using a Nano ZS apparatus, Malvern Instrument. The Helium-Neon laser, 4mW, operates at 633 nm, with the scatter angle fixed at 173°, and the temperature at 25 °C. The polydispersity index (noted PDI) only appears as a mathematical definition, accounting for the relative error between curve fit and experimental values. The PDI discloses the quality of the dispersion, from values lower than 0.1 for suitable measurements and good-quality of the colloidal suspensions, to values close to 1 for poor-quality samples, which in concrete terms either do not present droplets sizes in the colloidal range, or exhibit a very high polydispersity.

5.9.2 Electrical conductivity measurements

A conductimeter can be used in nonlinear temperature compensation mode. It helped in determining the location of the emulsion inversion zone following the conductivity variations according to the temperature. In fact, a conductivity value lower than 10μS cm^{-1} and essentially zero on the illustrated scales, means that the continuous phase is oil, whereas high steady state reached reflects that a water continuous phase is established.

5.9.3 Particle Size Disrtibution

Droplet size distribution of the nanoemulsion is determined by photon correlation spectroscopy (PCS), using a Zetasizer 1000 HS (Malvern Instruments). Light scattering was monitored at 25 °C at a scattering angle of 90°.

5.9.4 Rheology

The viscosity of the nanoemulsion can be determined using Brookfield DV III ultra V6.0 RV cone and plate rheometer, using spindle # CPE40 at 25±0.3°C.

5.9.5 Refractive index

Refractive index of nanoemulsion formulation was determined using an Abbes type refractrometer.

5.10 Nanoemulsion in Novel Drug Delivery

In principle, in the preparation of medicated nanoemulsions, the drug is initially solubilized or dispersed together with an emulsifier in suitable single oil or oil mixtures by means of slight heating. The water phase containing the osmotic agent with or without an additional emulsifier is also heated and mixed with the oil phase by means of high-speed mixers. Further homogenization takes place to obtain the needed small droplet size range of the nanoemulsion. A terminal sterilization by filtration or steam then follows. The SME thus formed contains most of the drug molecules within its oil phase. This is a generally accepted and standard method to prepare lipophilic drug-loaded SME for ocular use. This process is normally carried out under aseptic conditions and nitrogen atmosphere to prevent both contamination and potential oxidation of sensitive excipients. The average oil droplet size of the SME or nanoemulsions is above 50 nm, the emulsion exhibits a milky appearance, and the oil concentration is usually below 10%. Polar oils such as medium chain triglycerides (MCT) are commonly used as oil phase.

Delta-8-tetrahydrocannabinol, a lipophilic cannabinoid, was formulated in a negatively charged SME. The ocular hypotensive effect of this topical formulation was studied on rabbits with ocular hypertension and on normotensive rabbits. The mean droplet size of the emulsion was 130 +41 nm and the zeta potential was -57.1 mV. This formulation showed stability for up to 9 months in terms of pH and droplet size. An intense and long-lasting IOP depressant effect was observed following topical application in the lower conjunctival sac of ocular hypertensive albino rabbits, but less effect was observed in the ocular normotensive group. Similar results were obtained in a later study when HU-211, a nonpsychotropic synthetic cannabinoid, was formulated in a negatively charged SME and applied topically to the rabbit eye. The IOP reduction lasted 6 h with maximal magnitude of 24% of baseline in the treated eyes compared to 12.5% in the contralateral eye. On compared the IOP reduction effect of aqueous solution containing 2% pilocarpine hydrochloride and 1.7% pilocarpine base (equivalent to 2% of hydrochloride salt) incorporated in negatively charged SME. Maximum IOP reduction was 28.5% of baseline for the SME formulation compared to 18% for the aqueous solution. However the time to reach maximal effect was shorter for the aqueous solution compared to the SME formulation (2 and 5 h, respectively), but the IOP reduction effect was more sustained in the group receiving the SME formulation (29 compared to 11 h). The authors suggested that the sustained effect was related in part to the availability of pilocarpine in the oily phase. This advantage could be translated into less frequent applications for the SME formulation. After incorporation of pilocarpine in anionic SME, the interaction between the incorporated drug and the emulsion and the consequences of this interaction on both drug and emulsion stability was studied. In a later study from the same group, it was found that the bioavailability of the drug was pH-dependent with the best miotic activity at pH 5.0 and 8.5. However, high pH cannot be considered for clinical use because of pilocarpine degradation in the emulsion, which occurs at a similar rate as in aqueous solutions. Thus, a pH of 5.0 was suggested for better stability and bioavailability. In another report from the same research group, the partitioning of pilocarpine in the oil phase of anionic SME was increased by pilocarpine ion pairing. Surprisingly, the augmentation in pilocarpine content of the oil phase was not translated into a better ocular bioavailability. This finding was attributed to the components of the lipid emulsion (soybean oil and egg lecithin), which did not allow sufficient residence time on corneal surface. In a study of topical indomethacin penetration to the rabbit eye, a 300% increased indomethacin ocular bioavailability following instillation of the indomethacin submicron SME compared to the performance of a commercial solution was reported. The other studied colloidal systems (nanoparticles and nanocapsules) showed comparable performance. An endocytic mechanism of penetration into corneal epithelium cells was proposed based on confocal images obtained in this study. Adaprolol, a novel soft beta-blocking agent, was incorporated in anionic SME and safety and the sustained IOP reduction effect of this formulation following topical ocular application to healthy volunteers was studied. The IOP reduction effect of twice a day topical pilocarpine anionic SME with four times a day commercial pilocarpine solution in ocular hypertensive patients was studied. The two groups showed comparable results with 25% of baseline IOP reduction. This study has shown that the incorporation of pilocarpine in SME allows the achievement of the same clinical effect with less frequent administrations per day compared to pilocarpine solution. CsA, a highly lipophilic molecule, was formulated in an anionic castor oil-in-water SME. This formulation was found to incorporate up to 0.4% of CsA and was developed for the treatment of severe dry eye with inflammatory background. This indication requires only low concentrations of CsA in the ocular surface tissues for local immunomodulation without the need for deep intraocular penetration. The formulation was stable up to 9 months but caused mild discomfort when administered to rabbit eyes. Ocular pharmacokinetic data showed adequate penetration to ocular surface tissues and to the lacrimal gland whereas the penetration to the intraocular tissues was low and systemic absorption was minimal. This CsA formulation was challenged successfully in phase I and II clinical trials and formulations of 0.05% and 0.1% w/w of the drug were further evaluated in phase III trial. Patients treated with both concentrations have shown statistically significant improvement in subjective and objective dry eye indexes compared to the placebo arm. However, the results for 0.1% w/w concentration were not superior to those obtained with the 0.05% w/w concentration. This anionic SME with CsA concentration of 0.05% w/w was approved in December 2002 by the United States Food and Drug Administration (FDA). A positively charged SME favors electrostatic interaction with the negatively charged ocular surface and prolong the local residence time. Indeed, piroxicam was formulated in a positively charged SME and was shown to be the most effective formulation for the delivery of the lipophilic piroxicam to the rabbit cornea following alkali burn. The effect of emulsion charge on the ocular penetration of indomethacin was studied. Indomethacin was formulated in positively and negatively charged SMEs. The ocular penetration of both emulsions was evaluated and compared to a commercial ocular solution of indomethacin (Indocollyre hydro-PEG solution, Bausch & Lomb, Rochester, NY, USA). The positively charged SME achieved significantly higher drug levels than the negatively charged emulsion and the commercial solution in the aqueous humor and the sclera-retina. Ocular surface indomethacin levels were high and nondifferential to the tested formulation. The contact angle and the spreading coefficient of the different formulations on the isolated freshly excised rabbit cornea were studied to elucidate the effect of the formulation charge on its interaction with the corneal surface. Lower contact angle and higher spreading coefficient were exhibited by the positively charged SMEindicating better wettability properties on the cornea than either the saline or the negatively charged SME. Another study was conducted to investigate the effect of SME charge on the ocular penetration of the lipophilic and impermeable CsA. The drug was formulated in a positively charged SME and the optimization and characterization of this formulation was done. Following one single instillation to the rabbit eye, CsA incorporated in positively charged SME achieved higher drug availability on ocular surface, particularly on the conjunctiva, compared to CsA in negatively charged SME. The penetration to the intraocular tissues was limited whereas blood levels were extremely low. Following these encouraging results the CsA positively charged SME formulation was further developed and proceeded to phase I and II clinical trials. In conclusion, the approval of Restasis by the FDA is an important milestone in lipid emulsion research for ophthalmic application. This approval reflects the achievements of the last decade in terms of the availability of better ingredients, improved manufacturing processes, feasibility of sterilization, and better understanding of the optimization process. In all of the comparative studies done so far, positively charged SME achieved better ocular bioavailability regardless of the studied drug. Research efforts are underway to further explore the mechanism of interaction of positively charged SMEs with ocular tissues and to translate the results of this research into enhanced clinical performance.

Antimicrobial nanoemulsions, containing water and soybean oil with uniformly sized droplets in the 200 to 400 nm range, can destroy microbes effectively without toxicity or harmful residual effects. The nanoparticles fuse with the membrane of the microbe and the surfactant disrupts the membrane, killing the microbe. The classes of microbes eradicated are viruses (e.g., HIV, herpes), bacteria (e.g., *E. coli*,

Salmonella), spores (e.g., anthrax), and fungi (e.g., *Candida albicans*, *Byssochlamys fulva*).

Some systematic procedure were developed using a high-pressure microfluidic device to create bulk quantities of oil in water nanoemulsions with diameters as small as 30 nm . Although the raw nanoemulsions have a peaked monomodal size distribution, as determined by dynamic light scattering, ultracentrifugation is used to fractionate the droplets and make them more monodis- perse. Once fractionated, the nanoemulsions are an excellent model system for investigating the structure of concentrated dispersions of deformable droplets using small angle neutron scattering (SANS). The structure of a concentrated nanoemulsion are then compared with the structure of a concentrated suspension of uniform hard spheres..

The US Army, Natick Soldier Research, Development and Engineering Center has found that nanoemulsions generate significantly reduced particle size and increase zeta potential compared to a macro suspension emulsion control. Zeta potential measures the degree of repulsion and confers stability. When bioactive compounds are incorporated into nanoemulsions their bio-availability can be significantly enhanced. Many of the functional compounds of interest have problems with solubility, oral bioavailability or stability. They also aim to enhance the bio-availability and increase uptake of targeted performance optimizingcompounds when consumed in a military ration. Formulation testing of oil carriers and emulsifiers were optimized to develop nanoemulsions containing the bioactives quercetin, curcumin and tyrosine. Enhancement of absorption allows decrease of the concentrations of the bioactive when added to a ration component and still obtain the target dosage. Nanoemulsions produced with a Microfluidizer Processor containing curcumin, quercetin or tyrosine all showed a reduction in particle size (40, 3 and 12fold reduction respectively) as analyzed by a Malvern Zetasizer. The microfluidized nanoemulsionprocess has the capacity to increase the zeta potential for example, about 5 fold for curcumin and 3 foldfor quercetin. A further reduction in particle size was achieved for curcumin and quercetin (300 and 35fold decrease, respectively) using a Self-Assembly technique. Soybean oil was used to stabilize the microfluidized nanoemulsions for curcumin and quercetin, and rice bran oil for tyrosine. Coconut oil andrice bran oil were found to stabilize curcumin and quercetin nanoemulsions when using the self assembled procedure.

5.11 Various Parenteral Application of Nanoemulsion

While it is apparent that nanoemulsions (as with any drug delivery vehicle) will only be investigated if it is believed that an emulsion formulation can improve drug performance in some way, the nature of that enhancement can take many forms. Perhaps the most obvious improvement is for the solubilization of drugs with a low aqueous solubility. Other areas of study include the reduction of pain/irritation upon injection, reduced toxicity of the drug in vivo, improved pharmacokinetics of the drug, or the possibility of a new method of delivery (e.g., intravenous versus oral administration). Nanoemulsion formulations can also provide drug targeting or multiple functionalities with imaging coupled to therapy. Finally, nanoemulsions can create a new therapeutic area that would not be present without emulsions, such as with fluorocarbon emulsions intended for blood substitutes. The following paragraphs will discuss examples for each of the above categories. Of course, some examples of nanoemulsion formulation will encompass more than one of the above categories, so there will be some unavoidable overlap. Table 5.1 shows a representative list of drugs that have been investigated for intravenous delivery via nanoemulsions since 2000.

Table 5.1 List of drugs investigated uptill 2000 for delivery through intravenous route.

Drug	
All-*trans* retiboic acid	Menatetrenone
Amphotericin B	mTHPC
BCH	Nalbuphine
Carbamazepine	Nimodipine
Clomethiazole	Paclitaxel
Cyclosporin A	Mesylate
Dexamethasone palmitate	Probucol
Flunarizone	Prostaglandin E1
Halofantrine	Resveratrol
Indomethacin	Silymarin
Itraconazole	SPK-843
Lorazepam	Xanthone
	Zinc phthalocyanine

Conclusion

The development of new methodologies for the emulsification of o/w biphasic systems has allowed for the formation of nanoemulsions, which possess a stability and particle size that make them particularly suitable for the intravenous delivery of hydrophobic drugs. The combination of different surfactants, additives, and lipids has extended the utility of nanoemulsions to a wide array of drugs. Furthermore, the application of nanoemulsions in drug delivery has made possible the development of new therapies. The safety and efficacy of the use of nanosized emulsions in human patients has been demonstrated with a number of approved products, and current research is continually expanding the usefulness and applicability of nanoemulsion formulations for intravenous delivery. Specifically, the possibility of using nanoemulsions for both imaging and drug delivery has the potential of having a profound impact on the next generation of drug delivery systems.

References

1. T.G.Mason, S.M.Graves, J.N.Wilking, M.Y.Lin, Condensed Matter Physics 2006, Vol. 9, No 1(45), pp. 193–199
2. Bibette J., Leal-Calderon F., Poulin P., Rep. Prog. Phys. 1999, 62, 969.
3. Taylor G.I., Proc. R. Soc. A, 1934, 146, 501.
4. Rallison J.M., Ann. Rev. Fluid Mech., 1984, 16, 45.
5. Milliken W.J., Leal L.G., J. Non-Newtonian Fluid Mech., 1991, 40, 355.
6. Reiss H., J. Colloid Interface Sci., 1975, 53, 61.
7. Myers D., Surfaces, Interfaces, and Colloids. Wiley, New York, 1999.
8. Ugelstad J., Hansen F.K., Lange S., Die Makromolekulare Chemie, 1974, 175, 507.
9. Tang P.L., Sudol E.D., Silebi C.A., et al., J. Appl. Polymer Sci., 1991, 43, 1059.
10. Landfester K., Tiarks F., Hentze H., et al., Macromol. Chem. Phys., 2000, 201, 1.
11. Meleson K., Graves S., Mason T.G., Soft Materials, 2005, 2, 109.
12. Mason, T., Wilking, J., Meleson, K., Chang, C., & Graves, S. (2006). Nanoemulsions: formation, structure and physical properties. Journal of Physics: Condensed Matter, 18, R635–R666.

13. Solans, C., Izquierdo, P., Nolla, J., Azemar, N., & Garcia-Celma, M. J. (2005).Nano-emulsions. Current Opinion in Colloid & Interface Science, 10, 102–110.
14. Nanotechnology in Drug Delivery, Melgardt M. de Villiers,Pornanong Aramwit,Glen S. Kwon,pp-468-469
15. S. Benita, M. Y. Levy, J. Pharm. Sci.,1993, 82, 1069.
16. Tharwat F. Tadros, Applied Surfactants-Principles and Applications, pp 299-301
17. Allen TM.. Drugs 1998; 56:747.
18. Deepak Thassu,Michel Deleers,Yashwant Pathak, Nanoparticulate Drug Delivery Systems , pp-110.
19. J. Bibette, F. Leal-Calderon, and P. Poulin, Rep. Prog. Phys. 62, 969 (1999)
20. J. Ugelstad, F. K. Hansen, and S. Lange, Makromol. Chem. 175, 507 (1974)
21. P. L. Tang, E. D. Sudol, C. A. Silebi, and M. S. El-Aasser, J. Appl. Polym. Sci. 43, 1059 (1991).
22. K. Landfester, F. Tiarks, H. Hentze, and M. Antonietti, Macromol. Chem. Phys. 201, 1 (2000).
23. D. Myers, *Surfaces, Interfaces, and Colloids*, 2nd ed. sWiley, New York, (1999).
24. P. Taylor, Adv. Colloid Interface Sci. 106, 261 (2003).
25. J.-P. Hansen and I. R. McDonald, *Theory of Simple Liquids*, 2nd ed. sAcademic, London, (1990).
26. C. S. O'Hern, S. A. Langer, A. J. Liu, and S. R. Nagel, Phys. Rev. Lett. 86, 111 (2001).
27. C. S. O'Hern, S. A. Langer, A. J. Liu, and S. R. Nagel, Phys. Rev. Lett. 88, 075507 (2002).
28. T. G. Mason, M.-D. Lacasse, G. S. Grest, D. Levine, J. Bibette, and D. A. Weitz, Phys. Rev. E 56, 3150 (1997).
29. T. Tlusty and S. A. Safran, J. Phys.: Condens. Matter 12, A253 (2000).
30. W. B. Russel, D. A. Saville, and W. R. Schowalter, *Colloidal Dispersions* (Cambridge University Press, Cambridge, 1989).
31. P. N. Pusey and W. van Megen, Nature sLondond 320, 340 (1986).

32. W. van Megen and S. M. Underwood, Phys. Rev. E 49, 4206 (1994).
33. B. J. Maranzano and N. J. Wagner, J. Rheol. 45, 1205 (2001).
34. E. R. Weeks and D. A. Weitz, Phys. Rev. Lett. 89, 095704 (2002).
35. S. Torquato, T. M. Truskett, and P. G. Deenedetti, Phys. Rev. Lett. 84, 2064 (2000).
36. A.Donev, I. Cisse, D. Sachs, E. A. Variano, F. H. Stillinger, R. Connelly, S. Torquato, and P. M. Chaikin, Science 303, 990 (2003).
37. D. J. Durian, D. A. Weitz, and D. J. Pine, Science 252, 686 (1991).
38. T. G. Mason, A. H. Krall, H. Gang, J. Bibette, and D. A. Weitz, in *Encyclopedia of Emulsion Technology*, edited by P. Becher sMarcel Dekker, New York, (1996), Vol. 4, p. 299.
39. H. Gang, A. H. Krall, H. Z. Cummins, and D. A. Weitz, Phys. Rev. E 59, 715 (1999).
40. P. Hébraud, F. Lequeux, J. P. Munch, and D. J. Pine, Phys. Rev. Lett. 78, 4657 (1997).
41. A.M. Kraynik, D. A. Reinelt, and F. van Swol, Phys. Rev. E 67, 031403 (2003).
42. D. Weaire, S. Hutzler, S. Cox, N. Kern, M. D. Alonso, and W. Drenckham, J. Phys.: Condens. Matter 15, S65 (2003).
43. H. A. Stone, S. A. Koehler, S. Hilgenfeldt, and M. Durand, J. Phys.: Condens. Matter 15, S283 (2003).
44. S. Graves, K. Meleson, J. Wilking,M. Y. Lin and T. G. Mason, THE JOURNAL OF CHEMICAL PHYSICS 122, 134703 (2005)
45. M. Antonietti, K. Landfester, Prog. Polym. Sci. 27 (2002) 689–757..
46. J.M. Asua, Prog. Polym. Sci. 27 (2002) 1283–1346.
47. B. Abismail, J.P. Canselier, A.M. Wilhelm, H. Delmas, C. Gourdon, Ultrason. Sonochem. 6 (1999) 75–83.
48. P. Walstra, Chem. Engng. Sci. 48 (1993) 333.
49. S.E. Friberg, S. Jones, Othmer Encyclopedia of Chemical Technology, Kroschwith, J.I., 1994
50. C. Bondy, K. Söllner, Trans. Faraday Soc. 31 (1935) 835–842.
51. T.J. Mason, Ultrason Sonochem. 30 (1992) 147–196.

52. O. Behrend, K. Ax, H. Schubert, Ultrason Sonochem. 7 (2000) 77–85.
53. C.J. Samer, F.J. Schork, Ind. Eng. Chem. Res. 38 (1999) 1801–1807.
54. K. Landfester, Adv. Mater. 13 (2001) 765–768.
55. M. Willert, R. Rothe, K. Landfester, M. Antonietti, Chem. Mater. 13 (2001) 4681–4685.
56. K. Landfester, F. Tiarks, H.-P. Hentze, M. Antonietti, Macromol. Chem. Phys. 201 (2000) 1–5.
57. F. Tiarks, K. Landfester, M. Antonietti, Langmuir 17 (2001) 908–918.
58. U. Paiphansiri, P. Tangboriboonrat, K. Landfester, Macromol. Biosci. 6 (2006) 33–40.
59. P. Walstra, P.E.A. Smoulders, Modern Aspects of Emulsion Science, The Royal Society of Chemistry, Cambridge, 1998.
60. T.F. Tadros, P. Izquierdo, J. Esquena, C. Solans, Adv. Colloid Interface Sci. 108–109 (2004) 303–318
61. G. Quincke, Ueber emulsionbildung und den einfluss der galle bei der verdauung, Plüger Archiv für die Physiologie 19 (1879) 129–144 and references therein.
62. J.T. Davies, D.A. Haydon, Proc. 2nd Int. Congr. Surf. Act. London, vol. 1, 1957, p. 417.
63. P. Becher, Emulsions: Theory and Practice, Van Nostrand Reinhold, New York, 1966.
64. E.S.R. Gopal, Emulsion Science, Academic Press, New York, 1968.
65. M.J. Groves, Chem. Ind. 12 (1978) 417–423.
66. E. Rubin, C.J. Radke, Chem. Eng. Sci. 35 (1980) 1129–1138.
67. D.Z. Becher, Encyclopedia of Emulsion Technology, vol. 2, Marcel Dekker, New York, 1985.
68. C.A. Miller, Colloids Surf. 29 (1988) 89–102.
69. M.S. El-Aasser, C.D. Lack, J.W. Vanderhoff, F.M. Fowkes, Colloids Surf. 29 (1986) 103–118.
70. C.W. Pouton, Adv. Drug Deliv. Rev. 25 (1997) 47–58.
71. K. Shinoda, H. Saito, J. Colloid Interface Sci. 26 (1968) 70–74.
72. K. Shinoda, H. Saito, J. Colloid Interface Sci. 30 (1969) 258–263.

73. T. Forster, F. Schambil, H. Tesmann, Int. J. Cosmet Sci. 12 (1990) 217–227.
74. T. Forster, F. Schambil,W. von Rybinski, J. Disp. Sci. Technol. 13 (1992) 183–193.
75. T. Forster, W. von Rybinski, A. Wadle, Adv. Colloid Interface Sci. 58 (1995) 119–149.
76. A.J. Sing, A. Graciaa, J. Lachaise, P. Brochette, J.L. Salagers, Colloids Surf. A 152 (1999) 31–39.
77. P. Izquierdo, J. Esquena, T.F. Tadros, C. Dederen, M.J. Garcia, N. Azemar, C. Solans, Langmuir 18 (2002) 26–30.
78. D. Morales, J.M. Gutiérrez, M.J. García-Celma, Y.C. Solans, Langmuir 19 (2003) 7196–7200.
79. R. Pons, I. Carrera, J. Caelles, J. Rouch, P. Panizza, Adv. Colloid Interface Sci. 106 (2003) 129–146.
80. P. Izquierdo, J. Esquena, T.F. Tdros, J.C. Dederen, J. Feng, M.J. García-Delma, N. Azemar, C. Solans, Langmuir 20 (2004) 6594–6598.
81. J.L. Salager, A. Forgiarini, L. Marquez, A. Pena, M. Pizzino, P. Rodriguez, M. Rondon- Gonzalez, Adv. Colloid Interface Sci. 108–109 (2004) 259–272.
82. C. Solans, P. Izquierdo, J. Nolla, N. Azemar, M.J. Garcia-Celma, Colloid Interface Sci. 10 (2005) 102–110.
83. P.A. Rehbinder, V. Lichtman, Proc. 2nd Int. Congr. Surf. Act. London, vol. 58, 1957, p. 549.
84. M. Ostrovsky, R. Good, J. Colloid Interface Sci. 102 (1984) 206–226.
85. K. Bouchemal, S. Briançon, E. Perrier, H. Fessi, Int. J. Pharm. 280 (2004) 241–251.
86. P. Winsor, Solvent Properties of Amphiphilic Compounds, Butterworth, London, 1954.
87. H. Fessi, J.P. Devissaguet, F. Puisieux, Procédés de préparation de systems colloïdaux dispersibles d'une substance sous forme de nanocapsules, French Patent 8618444.
88. H. Fessi, F. Puisieux, J.P. Devissaguet, N. Ammoury, S. Benita, Int. J. Pharm. 55 (1989) 25–28.
89. H. Fessi, J.P. Devissaguet, F. Puisieux, Procédé de préparation de systems colloïdaux dispersibles d'une substance, sous forme de nanoparticules, EP 0275 796 B1 (1992).

90. K. Bouchemal, S. Briançon, E. Perrier, H. Fessi, I. Bonnet, N. Zydowicz, Int. J. Pharm. 269 (1) (2004) 89–100.

91. I.Montasser, H. Fessi, S. Briançon, J. Lieto, World Patent WO0168235.

92. L. Marszall, Nonionic Surfactants, Vol. 23 of Surfactant Sciences, Marcel Dekker, New York, 1987.

93. P. Taylor, R.H. Ottewill, Colloids Surf. A 88 (1994) 303–316.

94. P. Taylor, R.H. Ottewill, Prog. Colloid Polym. Sci. 97 (1994) 199–203.

95. A.Forgiarini, J. Esquena, C. González, C. Solans, Langmuir 17 (2001) 2076–2083.

96. H. Wu, C. Ramachandran, N.D. Weiner, B.J. Roessler, Int. J. Pharm. 220 (2001) 63–75.

97. M. Porras, C. Solans, C. González, A. Martínez, A. Guinart, J.M. Gutiérrez, Colloids Surf. A 249 (2004) 115–118.

98. N. Usón, M.J. García, C. Solans, Colloids Surf. A 250 (2004) 415–421.

99. Solè, A. Maestro, C.M. Pey, C. González, C. Solans, J.M. Gutiérrez, Colloids Surf. A 288 (2006) 138–143.

100. Solè, A. Maestro, C. González, C. Solans, J.M. Gutiérrez, Langmuir 22 (2006) 8326–8332.

101. M. Kahlweit, R. Strey, P. Firman, D. Haase, Langmuir 1 (1985) 281–288.

102. M. Kahlweit, Mikroemulsionen—eine qualitative beschreibung ihrer eigenschaften, Tenside Surf. Det. 30 (1993) 83–89.

103. K.V. Schubert, R. Strey, M. Kahlweit, J. Colloid Interface Sci.141 (1991) 21–29.

104. K. Shinoda, Prog. Colloid Polym. Sci. 68 (1983) 1–7.

105. K. Shinoda, S. Friberg, Adv. Colloid Interface Sci. 4 (1975) 281–300.

106. H. Kunieda, K. Shinoda, Bull. Chem. Soc. Jpn. 55 (1982) 1777–1781.

107. R. Aveyard, B.P. Binks, P.D.I. Fletcher, Langmuir 5 (1989) 1210–1217.

108. N. Anton, P. Gayet, J.P. Benoit, P. Saulnier, Int. J. Pharm. 344 (1–2) (2007) 44–52.

109. N. Anton, J.P. Benoit, P. Saulnier, J. Drug Del. Sci. Tech. 18 (2) (2008).

110. H. Kunieda, H. Fukui, H. Uchiyama, C. Solans, Langmuir 12 (1996) 2136–2140.

111. J.L. Salager, L. Marquez, A. Graciaa, J. Lachaise, Langmuir 16 (2000) 5534–5539.

112. J.L. Salager, R. Anton, J.M. Anderez, J.M. Aubry, Formulation des microemulsions pas la méthode du hld, Techniques de l'Ingénieur Génie des Procédés J2 (2001) 1–20.

113. J.L. Salager, Handbook of Detergents—Part A: Properties, vol. 82, 1999, pp. 253–302.

114. J.L. Salager, Pharmaceutical emulsions and suspensions, Ch. Formulation Concepts for the Emulsion Maker, Marcel Dekker, New York, 2000, pp. 19–72.

115. M. Bourrel, J.L. Salager, R.S. Schechter,W.H.Wade, J. Colloid Interface Sci. 75 (1980) 451–461.

116. J.L. Salager, Formulation Concepts for the Emulsion Maker, Marcel Dekker, New York, 2000, pp. 19–72, Ch. 2.

117. A. Béduneau, P. Saulnier, N. Anton, F. Hindré, C. Passirani, H. Rajerison, N. Noiret, J.P. Benoit, Pharm. Res. 23 (2006) 2190–2199.

118. M. Rondón-González, V. Sadtler, L. Choplin, J.L. Salager, Colloids Surf. A 288 (2006) 151–157.

119. M. Rondón-González, V. Sadtler, L. Choplin, J.L. Salager, Ind. Eng. Chem. Res. 45 (2006) 3074–3080.

120. E. Tyrode, J. Allouche, L. Choplin, J.L. Salager, Ind. Eng. Chem. Res. 44 (2005) 67–74.

121. E. Tyrode, I. Mira, N. Zambrano, L. Márquez, M. Rondón-González, J.L. Salager, Ind. Eng. Chem. Res. 42 (2003) 4311–4318.

122. Mira, N. Zambrano, E. Tyrode, L.Márquez, A. Pena, A. Pizzino, J.L. Salager, Ind. Eng. Chem. Res. 42 (2003) 57–61.

123. Nicolas Anton, Jean-Pierre Benoit, Patrick Saulnier, Journal of Controlled Release 128 (2008) 185–199.

6
Microspheres

A. V. Jithan and B. Chandrasekhar Reddy
Vaagdevi College of Pharmacy, Warangal, India

6.1 Introduction

Microspheres are spherical microparticles ranging in size from 1 to 6,000 μm. Microspheres have several advantages and are applied in several fields, apart from the pharmaceutical industry. They are made up of several material including polystyrene, polyethylene, glass, ceramics and other polymers. Polystyrene microspheres are typically used in biomedical applications due to their ability to facilitate procedures such as cell sorting and immuno precipitation. Proteins and ligands adsorb onto polystyrene readily and permanently, which makes polystyrene microspheres suitable for medical research and biological laboratory experiments. They are also used in diagnostic tests. The other types of commonly used microspheres include polyethylene microspheres. Polyethylene microspheres are commonly used as a permanent or temporary filler. Lower melting temperature enables polyethylene microspheres to create porous structures in ceramics and other materials. High sphericity of polyethylene microspheres, as well as availability of colored and fluorescent microspheres, makes them highly desirable for flow visualization and fluid flow analysis, microscopy techniques, health sciences, process troubleshooting and numerous research applications. Charged polyethylene microspheres are also used in electronic paper digital displays. Microspheres can be either solid or hollow. These two types vary a lot in density and, therefore, are used for different applications. Hollow microspheres are typically used as additives to lower the density of a material. Solid microspheres have numerous applications depending on what material they are constructed of and what size they are. The other types of microspheres used are made up of glass or ceramics. Glass microspheres are primarily used as a filler and volumizer for weight reduction, retro-reflector for highway safety, additive for cosmetics and adhesives, with limited applications in medical technology. Ceramic microspheres are used primarily as grinding media. Microspheres vary widely in quality, sphericity, uniformity, particle size and particle size distribution. The appropriate microsphere needs to be chosen for each unique application. Some pictures of the microspheres are shown in Figure 6.1. These microspheres have also been used in the pharmaceutical products. These are applied especially in drug delivery and offer several advantages. The pharmaceutical microspheres consist of natural or synthetic macromolecular materials in which drugs are encapsulated, entrapped, or dissolved or to which they are adsorbed. Several types of polymers have been used to prepare these microspheres. A wide variety of techniques are available to prepare the microspheres used in drug therapy. These are useful in delivering drugs into the body via various routes with added advantages. These microspheres offer unique advantage in controlled release drug delivery, targeted release drug delivery and also in enhancement of drug transport across the physiological membranes. Wide variety of pharmaceuticals including small molecular drugs, protein drugs, antisense oligonucleotide based drugs, genes have been incorporated into microspheres and further investigated. This chapter deals with the microspheres useful in the pharmaceutical and therapeutic applications.

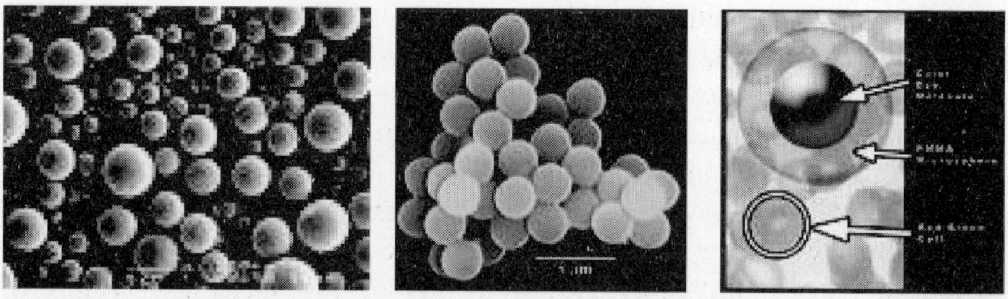

Figure 6.1 Some pictures of microspheres.

The terms microcapsules and microspheres are interrelated. The term microcapsules is defined as a spherical particle with the size varying between 1 μm to 2 mm (although extreme size of 6 mm is also described) containing a core substance, this being an active ingredient. On the other hand a microsphere may be of same size and contains the active ingredient dispersed in the microparticle. In other words, it is possible to encapsulate active agents in a matrix system or to produce microcapsules with a solid shell and a liquid core. The first process generally leads to the formation of microspheres or in other words the technique leads to the formation of matrices in the size range of microns and are spherical. The second process leads to the formation of a capsule like structure which is in the size range of microns and are generally spherical. Either of these processes is often termed as microencapsulation. However, the terms microcapsules and microspheres are often used synonymously. In addition, some related terms are also used as well. For example, microbeads and beads are used alternatively. Sphere and spherical particles are also employed for a large size and rigid morphology. In this chapter, because of convenience and popularity, all the particles of microsize whether they are micromatrices or microcapsules, are termed as microspheres and thus described. All these types of microspheres can be either administered orally for specific reasons or injected via parenteral routes including intravenous depending on the size of the particles, or administered into body crevices such as synovial fluids, administered into the site of the disease such as brain, skin cancers, or administered via pulmonary or nasal routes for a variety of needs, incorporated into dosage forms to alter the drug release, etc. In these modes of administration and the other applications, they offer several advantages.

Conventional oral drug administration does not usually provide rate-controlled release or target specificity. In many cases, conventional drug delivery provides sharp increase in drug concentration often achieving toxic level and following a relatively short period at the therapeutic level of the drug concentration eventually drops off until re-administration. In order to obtain maximum therapeutic efficacy, it becomes necessary to deliver an agent to the target tissue in the optimal amount for the required period of time, thereby causing little toxicity and minimal side effects. Desired drug release can be provided by rate-controlling membranes or by implanted biodegradable polymers containing dispersed medication. Microparticulate drug delivery systems are considered and accepted as a reliable one to deliver the drug to the target site with specificity, to maintain the desired concentration at the site of interest without untoward effects. The target sites for oral administration are very limited. Microencapsulation is a useful method which prolongs the duration of drug effect significantly and improves patient compliance. Eventually the total dose and few adverse reactions may be reduced since a steady plasma concentration is maintained. Different types of polymers are used in the preparation of microspheres intended for oral administration. Most of the times the polymers are either biodegradable or non-biodegradable. In recent years much research in drug delivery has been focused on biodegradable polymer microspheres. Administration of medication via such systems is advantageous because microspheres can be ingested or and in case parenteral administration injected, can be tailored for desired release profiles and in some cases it can provide organ-targeted release. The main aim of preparing microcapsules is to convert liquids to solids, altering colloidal and surface properties, providing environmental protection and controlling the release characteristics by using the coating materials. There are several methods of preparing microspheres. The success of any microencapsulation method depends on many factors such as the drug solubility, partition co-efficiency, polymer composition, molecular weight etc. Among the various microencapsulation methods, emulsion solvent evaporation technique is often widely used to prepare microcapsules of water insoluble drugs (within the water insoluble polymer). Microspheres are formed by the evaporation of an organic solvent from dispersed oil droplets containing both polymer and drug.

6.2 Pharmaceutical Applications of Microspheres

The range of techniques for the preparation of microspheres offers a variety of opportunities to control aspects of drug administration. The term "control" includes phenomena such as protection and masking, reduced dissolution rate, facilitation of handling, and spatial targeting of the active ingredient. This approach facilitates accurate delivery of small quantities of potent drugs; reduced drug concentrations at sites other than the target organ or tissue; and protection of labile compounds before and after administration and prior to appearance at the site of action. The characteristics of microspheres containing drug should be correlated with the required therapeutic action and are dictated by the materials and methods employed in the manufacture of the delivery systems. The behavior of drugs in vivo can be manipulated by coupling the drug to a carrier particle. The clearance kinetics, tissue distribution, metabolism, and cellular interactions of the drug are strongly influenced by the behavior of the carrier. Exploitation of these changes in pharmacodynamic behavior may lead to enhanced therapeutic effect. Because of these reasons microspheres have been applied for a variety of pharmaceutical situations. Some of the important applications of microspheres is described below.

6.2.1 Microparticulates for Gastrointestinal Delivery

Microparticles have been reported to be useful for sustainment or enhancement of gastrointestinal absorption. They are also used in the targeting to a specific location. The uniqueness of the microencapsulation is the smallness of the particles and their subsequent use and adaptation to a wide variety of dosage forms and product applications, which heretofore might not have been technically feasible. This is especially true with microspheres intended for gastrointestinal tract. Because of the smallness of the particles, drug moieties can be widely distributed throughout the gastrointestinal tract, thus potentially improving drug sorption. These microcapsules are generally distributed through out the gastrointestinal tract. The GIT applications of microencapsulation might well include sustained-release or prolonged action medications, taste-masked chewable tablets, powders and suspensions, single-layer tablets containing chemically incompatible ingradients, targeted oral delivery, gastro-retentive drug delivery systems, colon delivery.

For a variety of reasons aforementioned microspheres have been administered orally. Sustained release of zidovudine (AZT) was shown in vivo using ethylcellulose microspheres after oral administration in dogs (1). Encapsulated AZT showed significantly lower maximum concentration (Cmax) values and longer times to Cmax *(tmax)* values. And longer mean residence time (MRT) was also observed compared with AZT powder. In rats an oral delivery of γ-interferon was attempted in which it was encapsulated in polylactate microspheres (2). In this report, a quite different distribution of interferon level was observed in vivo at 15 and 240 min after oral administration, in contrast to the control group, which received equivalent doses of unencapsulated interferon. These findings would suggest the tentative conclusion that microencapsulation of proteins markedly affects oral uptake and possibly postabsorption pharmacokinetic parameters as well. This experiment stands on the hypothesis that particulate material itself could be absorbed from the gastrointestinal tract.

Despite many findings in the past 20 years of the uptake of micro- or nanoparticles through the intestine after peroral administration, there is much difference of opinion as to the site of absorption and its mechanism. In 1977, so-called persorption of large starch microspheres *(5-150 µm)* via the intestinal villae tips was reported in animals including humans (3). It was pointed out that this persorption was an infrequent occurrence and also a rather pathological phenomenon. Several investigators examined cytosis of particulates by the intestinal epithelial cells. In these examinations, phagocytic transport of 1 µm polystyrene microsphere in the intestines of rats and dogs was found. The uptake of polystyrene nanospheres up to 100 nm with the aid of endocytosis was also reported histologically. However, it was insisted that 2 µm microspheres of polystyrene could be taken up through the epithelial cells of the intestine.

In addition to these convenient routes of intestinal absorption, recent studies have revealed a predominant contribution of M cells on uptake of particulates at the Peyer's patch. Peyer's patch localizes in duodenum, jejunum, and ileum spottily and originally constructs a gut-associated lymphoid tissue in which there are M cells being specially compared with normal absorption cells. These M cells possess a high activity of endocytosis that is performed by specific receptors at the surface of its cell membrane. Macromolecules or particulates absorbed by M cells are released into extracellular space, being included by lymphocytes or macrophages, and thus appear in the intestinal lymph nodes. The effect of particulate size on their uptake through M cells has been argued by many investigators. In their research, various upper-limit sizes of particulates for uptake into Peyer's patch were reported: 15 μm, 10 μm, and 3 μm. It was also found that although particulates under 5 μm were transferred into the lymphatics, larger particles of 5-10 μm remained in the Peyer's patch. Recently, the usefulness of biologically erodable and adhesive particulates (0.1-10 mm) composed of a copolymer of fumaric acid with sebacic acid or lactide-co-glycotide was reported for potential oral drug delivery systems. These particulates were recognized microscopically to traverse both the intestinal absorptive cells as well as Peyer's patches and to reach the spleen and liver tissues after lengthy contact with gastrointestinal mucosa. An enhanced exertion of pharmacological effects of three model chemicals (dicumarol, insulin, and plasmid DNA) with widely different molecular weights loaded in these particulates was observed inside the intestinal tissue, in the liver and in the blood. As just mentioned, despite the very limited area of Peyer's patch, microparticles or nanoparticles probably enable the absorption of loaded drugs through M cells. Thus, for drugs designed for lymphatic absorption, uptake through the Peyer's patch is surely significant; however, further examinations are necessary for drugs that require a transfer into systemic circulation. Additionally, the route as well as mechanism of particulate uptake in the intestine is not completely clear, and thus such examinations also should be continued for clinical usage of micro and nanoparticles.

Microspheres for oral administration have also been developed so as to reduce the corrosiveness of the drugs to the GIT (4). A gastroresistant polymer film of ethylcellululose was used to prepare indomethacin microspheres that do not cause drug induced corrosion in the gastrointestinal tract. The prepared microspheres can act as a physical barrier and can keep the corrosive drug particles out of contact with the gastric mucosa, thus minimizing stomach irritation. Another purpose of choosing ethylcellulose is its water-insoluble character. The drug release can be modified and the prolonged action of this poor soluble antirheumatic agent can be achieved. Tao et al., investigated the mucoadhesive microspheres prepared using ethyl cellulose (5). Acyclovir-loaded mucoadhesive microspheres (ACV-ad-ms) using ethylcellulose as matrix and Carbopol 974P NF as mucoadhesive polymer were prepared for the purpose of improving the oral bioavailability of acyclovir. The AUC(0-t) and mean residence time (MRT) of ACV-ad-ms (6055.9 ng h/mL and 7.2 h) were significantly higher than that of ACV suspension (2335.6 ng h/mL and 3.7 h) ($P<0.05$), which indicated that the bioavailability of acyclovir was greatly improved due to the prolonged retention of ACV-ad-ms in gastrointestinal tract.

There are some exceptions to the oral route of drug administration. Most conventionally, drugs are delivered systemically via the oral route. Provided a drug has an adequate lipid solubility it can be well transported into the blood from the gastrointestinal tract by a process of passive diffusion. Further, a small number of drugs can be absorbed by an active transport mechanism. However, there are now an increasing number of drugs that cannot readily be given orally since they are highly polar in nature, are of a large size and/or have stability problems under the conditions prevalent in the gastrointestinal tract (e.g. acid pH in the stomach and/or endogenlous enzymes in the small intestines). Examples of agents which cannot readily be delivered via this route include the products of biotechnology (in the form of therapeutic proteins (such as granulocyte colony stimulating factor, erythropoietin, interferons and growth hormone)) as well as polypeptide drugs produced by synthesis (such as calcitonins, parathyroid hormone, desmopressin, LHRH analogues (buserelin, goserelin and nafarelin, cholecystokinin and atrial naturetic peptide). Insulin can be considered to be the best known drug which exhibits this problem. Other examples of compounds which demonstrate poor absorption from the intestines are polysaccharide materials such as heparin (and its low molecular weight derivatives), antisense agents, polar metabolites of opioid analgesics (morphine-6-glucuronide). Certain other drugs, when given orally, may be absorbed via the intestines, but are extensively metabolised in the wall of the intestines or the liver; this is termed "the first pass effect". In all these situations, other than oral route should be investigated.

6.2.2 Parenteral Microspheres

Nowadays, emphasis is being laid to development of controlled release dosage forms. Interest in this technology has increased steadily over the past few years. Although oral administration of drugs is a widely accepted route of drug delivery, bioavailability of drug often varies as a result of gastrointestinal absorption, degradation by first-pass effect, and hostile environment of gastrointestinal tract. Transdermal administration for percutaneous absorption of drug is limited by the impermeable nature of the stratum comeum. Ocular and nasal delivery is also unfavorable because of degradation by enzymes present in eye tissues and nasal mucosa. Hence, the parenteral route is the most viable approach in such cases. Of the various ways of achieving long-term parenteral drug delivery, biodegradable microspheres are one of the better means of controlling the release of drug over a long time. Because of the lipidic nature of liposomes, problems such as limited physical stability and difficulty of freeze-drying are encountered. Similarly, for emulsions, stability on long-term basis and in suspensions, rheological changes during filling, injecting, and storage poses limitation. Also, in all these systems, the release rate cannot be tailored to the needs of the patient. Parenteral controlled-release formulations based on biodegradable microspheres can overcome these problems and can control the release of drug over a predetermined time span, usually in the order of days to weeks to months. Various FDA-approved controlled-release parenteral formulations based on these biodegradable microspheres are available on the market, including Lupron Depot® Nutropin Depot® and Zoladex®.

Controlled release parenteral formulations can be considered safer than conventional parenteral dosage forms since less drug is required and since the drug may be targeted to the in vivo site, avoiding high systemic levels. Due to the lower dosing frequency, avoiding high systemic levels is a possibility. Due to the lower dosing frequency and simpler dosage regimens, patient compliance can be improved with these dosage forms. As mentioned previously microspheres can be used to modify release over periods of weeks and months. A controlled release parenteral dosage form is usually selected when there are problems associated with oral delivery (eg gastric irritation, first pass effects or poor absorption) and a need for extended release and/or targeted delivery (eg, rapid clearance, toxic side effects). Examples of applications for parenteral delivery include: fertility treatment, hormone therapy, protein therapy, infection treatments (antibiotics and antifungals), cancer therapy, orthopedic surgery and post operative pain treatment, chronic pain treatment, vaccination/immunization, treatment of CNS disorders, and immunosuppression. Parenteral microspheres are prepared for both water-soluble and water-insoluble drugs. There are special problems in the development of sustained release pharmaceutical compositions of biologically active macromolecules due to the size and complexity of the macromolecules, particularly proteins, which are susceptible to chemical and structural alteration upon mixing with pharmaceutical excipients, upon processing and upon storage. These problems are understood by those skilled in the art of pharmaceutical formulation and can be categorized as problems of chemical stability. Inadequate chemical stability of pharmaceutical compositions resulting from irreversible alteration of the structure of the macromolecules and/or interactions with the excipients can result in compositions that are either inactive or do not provide the expected level of biological response. Special care should be taken when protein drugs are encapsulated into parenteral microspheres. Generally the drugs for parenteral delivery are encapsulated into polylactide-co-glycolide and polycaprolactone like polymers. This possess special problems for encapsulation of proteins. U.S. Patent No. 4,837,381 discloses a microsphere composition of fat or wax or mixture thereof and a biologically active protein, peptide or polypeptide suitable for parenteral administration (6). The patent discloses the utility of the compositions for slow release of a protein, peptide or polypeptide in a parenteral administration, and discloses methods for increasing and maintaining increased levels of growth hormone in the blood of treated animals for extended periods of time and thereby increasing weight gains in animals and increasing milk production of lactating animals by the administration of compositions of the invention.

For parenteral injection, dosage unit forms may be utilized to accomplish intravenous, intramuscular or subcutaneous administration, and for such parenteral administration, suitable sterile aqueous or non-aqueous solutions or suspensions, optionally containing appropriate solutes to effectuate isotonicity, will be employed. Alternatively microspheres can be used in all the routes of administration. They can also be administered via the intravenous route for sustained release or active targeting or passive targeting. As a means to passively target pulmonary route, Chao and Sinko investigated the disposition of fluorescent microspheres after intravenous administration (7). Four mg of fluorescent microspheres with diameter ranging from 2 to 90 micrometers were administered intravenously to rats via the tail veins. At predetermined time intervals, animals were euthanized and the heart, lung, liver, kidney, and spleen were removed. After 24 hours of alkali digestion, the samples were filtered and the fluorescent microspheres were dissolved. The samples were analyzed using a fluorescence detector and corrected by an internal standard with different excitation and emission wavelength. The results indicated that more than 95% of the microspheres with diameter larger than 10 μm were retained in the lung for at least 2 days following administration. Two- and 3-μm diameter microspheres were not retained by the pulmonary capillary bed and were cleared from the circulation to the liver (~40% for 2-μm microspheres and ~70% for 3-μm microspheres) and spleen (~50% for 2-μm microspheres and ~20% for 3-μm microspheres). More than 80% of the 6-μm diameter microspheres were trapped in the lung at 1 hour after i.v. administration, and cleared from pulmonary bed to the liver (~70%) and spleen (~10%) 48 hours after administration. Microspheres with diameter of 10 μm or larger were mechanically filtered at pulmonary bed and were not cleared for the 48 hours of the study. The study concluded that 1. The efficacy of microsphere passive pulmonary targeting is size dependent. 2. The clearance of microspheres 6 μm in diameter or smaller from the circulation is probably due to the phagocytosis by cells of the reticulo-endothelial system (RES) of liver and spleen. The results suggest that microspheres can be injected intravenously with added advantages.

6.2.3 Microspheres for Nasal Administration

The highly permeable nasal epithelium allow the rapid drug absorption and to higher molecular mass of approximately 1000 Da. This is due to the high total blood flow, porous endothelial membrane, large surface area and escaping the first-pass metabolism. This method of delivery can eliminate the need for intravenous catheters, and effective drug levels can often be achieved rapidly with lower doses than traditional oral forms. Broad investigations into novel forms of drug delivery have supported nasal administration of drugs as a potential method of delivering medications directly to the bloodstream. The nasal route has been found efficient for systemic administration of many compounds.

Microsphere technology is one of the specialized systems becoming popular for designing nasal products, as it may provide prolonged contact with the nasal mucosa and thus enhances absorption and bioavailability. In the presence of microspheres, the nasal mucosa is dehydrated due to moisture uptake by the microspheres. This result in reversible shrinkage of the cells, providing a temporary physical separation of the tight (intercellular) junctions that increases the absorption of the drugs. Particularly important for the nasal drug absorption is the respiratory region, which contains three nasal turbinates and the deposition of the particles in this region will depend on their size. Classically, larger particles including droplets (>10 μm), are deposited in the nasal cavity after inhalation; the larger the particles, the more anterior the deposition. For smaller particles the site of deposition depends on the velocity at which the particles are inhaled and the turbulence in the air flow, however the particles of size smaller than 1 μm are not normally deposited in the nasal cavity but travel down to the trachea to reach the lung. The rationale behind the use of a microsphere system is that, the application of bioadhesive microspheres (in the powder form) with good bioadhesive properties would permit such microspheres to swell in contact with nasal mucosa to form a gel and control the rate of clearance from the nasal cavity, thereby giving poorly absorbed drugs a longer time to be available at the absorptive surface. Microspheres have been reported to be present up to 3-5 h in the nasal cavity depending upon the bioadhesive material used for formulation. The ideal microsphere particle size requirement for nasal delivery should range from 10 to 50 μm as smaller particles than this will enter the lungs.

This route of administration becomes attractive for protein drugs encapsulated in microspheres. A major problem in drug delivery is less effective absorption of high molecular weight material such as proteins and peptides across biological membranes. Normally such molecules are not taken up by the body if administered to the gastrointestinal tract, to the buccal mucosa, to the rectal mucosa, the vaginal mucosa or given as an intranasal system. These drugs can be encapsulated into the polymeric material to form microspheres. The microspheres can be administered via the nasal route using a nasal insufflator device or pressurized aerosol cannister. Example of these are already employed for commercial powder systems intended for nasal application. Microspheres for nasal administration can: 1. Increase the permeability of nasal mucosa by interaction of the formulation components with the nasal membrane in a safe, effective and reversible manner. 2. Increase in drug solubility and protects it against enzymatic degradation and 3. Increases the residence time of the drug in the nasal cavity. With the administration of nasal microspheres, systemic intranasal delivery of peptides and proteins usually produces peak concentrations at around 5-15 minutes because of small thickness of nasal epithelium and rapid mucociliary clearance. Adhesion of polymeric microspheres of dextran, starch and albumin to the nasal mucosa has been shown to raise the clearance time to 3 hours.

Gungor recently developed Ondansetron microspheres and evaluated the potential for nasal administration (8). The aim of this study was to develop chitosan microspheres for nasal delivery of ondansetron hydrochloride. Microspheres were prepared with spray-drying method using glutaraldehyde as the crosslinking agent. Using the methodology utilized in the study, microspheres produced had high encapsulation efficiency and the suitable particle size for nasal administration. Nasal absorption of ondansetron from crosslinked chitosan microspheres was evaluated in rats, and pharmacokinetic parameters of the drug calculated from nasal microsphere administration were compared with those of both nasal and parenteral administration of aqueous solution. In vivo data supported that drug-loaded microspheres were also able to attain a sustained plasma profile and significantly larger area under the curve values with respect to nasal aqueous solution of drug. Microspheres have considerably increased the bioavailability of insulin, desmopressin and gentamicin.

6.2.4 Microspheres for Pulmonary Administration

Drugs can be delivered to the lung for local effect, for example in treatment of asthma. Drugs which are known to act locally include bronchodilators, sodium cromoglycate and steroids. These substances are usually delivered to the central airways. The lungs can also be used to deliver drugs into the general blood circulation for systemic effect. Well known examples include the anaesthetic gases and nicotine (from inhaled tobacco smoke). In case of alternative to the oral route, pulmonary route can be used. The skilled person is presented with the problem of the provision of an alternative or a more effective delivery means for use with "challenging" therapeutic agents for oral administration. It is known to those skilled in the art that drugs may be well absorbed via the lungs; even polar molecules such as peptides and proteins are quite well absorbed (40% or better appearing in the blood) when delivered to the peripheral airways. The large surface area, good vascularization, immense capacity for solute exchange and ultra-thinness of the alveolar epithelium are unique features of the lung facilitating systemic drug delivery via pulmonary administration. However, this route possess some problems. First of all the physiology of the lung is not designed to absorb drug compounds systemically despite the very fragile and vulnerable nature of the lung epithelium. As an interface between the bodys circulation and the outside air its function is to exchange only gas molecules and it has built up a strong defence against any other components that may enter the lung by accident. The defence can be by either physical by mucosal clearance or physiological by a strong presence of macrophages and enzymes. The lung can also develop immunological reactions against unwanted entities such as viruses. A drug molecule needs to pass the physical clearance system represented by the upper airways and reach preferably the deep lung. Once deposited in the lung periphery, it should avoid the biological defence mechanism and be absorbed by the alveolar epithelium. This is how a drug may get absorbed into the systemic circulation after administration into the lungs (pulmonary route). The efficacy and safety of many new and existing inhaled therapies may be enhanced through advanced controlled-release systems by using polymer particles. Poly (D,L-lactic-co-glycolic acid) (PLGA) is well known by its safety in biomedical preparations which has been approved for human use by the FDA. The particles are inhaled to reach the deep lung (also called the lung periphery). The optimum aerodynamic particle size distribution for most inhalation aerosols has generally been recognized to be in the range of 1-5 microns. In the delivery of polar drugs to human patients, nebuliser systems are used, in which a mist of drug solution is inhaled into the lungs over an extended period (about 10-15 minutes). This is also an effective means of delivering a dose of drug to the lower (peripheral) airways. However, this method of dosing suffers from the disadvantage that it is unpopular with patients because of the time required to administer drug. Moreover, the process of nebulisation is also known to cause degradation of certain drugs. It is well known to those skilled in the art that solid particles intended for delivery to the lungs should be of an aerodynamic diameter of less than 10 microns and, preferably, of less than 5 microns. However, the particles should not be too small, or the particles will fail to deposit in the lung and be exhaled. A size range of 0.5 to 5 microns is thus preferred. However, this size is also challenged. It has been mentioned that for optimal lung deposition the aerodynamic diameter should be about 2 μm. Since it is defined to be dependent on the geometric diameter and the density, the increase in geometric diameter can be compensated by reducing the density of the particle. Thus, large porous particles can also be used. The so-called large porous particles are generated by spray drying of polymeric and non-polymeric nanoparticles into very thin-walled macroscale structures. For these particles, it is also claimed that, due to their large structure compared with small dense particles, attack by macrophage is less critical when deposited in the alveolar region of the lung. Complex drugs such as peptides and proteins, low molecular weight heparin, antisense agents, DNA, may be micronised in order to produce this size range, but this process is known to cause damage to labile molecules. Moreover, physical losses can occur during the active processing operation.

The microsphere powders produced for pulmonary administration should be used in a suitable dry powder device familiar to those skilled in the art. These include, but are not limited to, the Spinhaler™ (Fisons plc), Lyphodose™ (Valois S.A.), Monopoudre™ (Valois S.A.), Valois DPI™ (Valois S.A.), Turbospin™ (Phildeatech), multichamber powder inhaler

(Pfeiffer), Turbohaler™ (Astra-Draco AB), Rotahaler™ (Glaxo), Diskhaler™ (Glaxo), Pulvinal™ (Chiesi Farmaceutici SpA) and Ultrahaler™ (Fisons). Additionally, the microspheres may be lightly compacted to produce a solid compact from which a dose is taken via a mechanical method (e.g. Ultrahaler™, Fisons). If necessary the microspheres can also be mixed with a small amount of excipients such as lactose to improve flow properties. Apart from these several companies have developed their own devices. In order to reach the lung periphery as efficiently as possible, several state-of the art devices have been developed. The most prominent ones in case of insulin are developed by Nektar Therapeutics (Exubera Inhaler) and Aradigm (AERx), where a defined particle cloud is generated which can be efficiently administered through slow inhalation by the patient. Another method is utilized by Alkermes, where large particles are formulated that have a very low density. To deliver a drug formulation to the lung periphery the patient should breathe slowly (to avoid impaction) and as deeply as possible. Preferably, the aerosol cloud should stay for a few seconds in the lung before the patient starts to exhale so that the aerosol particles in the tiny alveolar capillaries can sink by gravitation and deposit on the alveolar surface.

Recently pulmonary microspheres of heparin were investigated. A study by Rawat et al., tested the feasibility of large porous particles as long-acting carriers for pulmonary delivery of low molecular weight heparin (LMWH) (9). Microspheres were prepared with a biodegradable polymer, poly(lactic-co-glycolic acid) (PLGA), by a double-emulsion-solvent-evaporation technique. The drug entrapment efficiencies of the microspheres were increased by modifying them with three different additives polyethyleneimine (PEI), Span 60 and stearylamine. The resulting microspheres were evaluated for morphology, size, zeta potential, density, in vitro drug-release properties, cytotoxicity, and for pulmonary absorption in vivo. Scanning electron microscopic examination suggests that the porosity of the particles increased with the increase in aqueous volume fraction. The amount of aqueous volume fraction and the type of core-modifying agent added to the aqueous interior had varying degrees of effect on the size, density and aerodynamic diameter of the particles. When PEI was incorporated in the internal aqueous phase, the entrapment efficiency was increased from 16.22+/-1.32% to 54.82+/-2.79%. The amount of drug released in the initial burst phase and the release-rate constant for the core-modified microspheres were greater than those for the plain microspheres. After pulmonary administration, the half-life of the drug from the PEI- and stearylamine-modified microspheres was increased by 5- to 6-fold compared to the drug entrapped in plain microspheres. The viability of Calu-3 cells (representation of respiratory cells) was not adversely affected when incubated with the microspheres. Overall, the data presented suggest that the newly developed porous microspheres of LMWH have the potential to be used in a form deliverable by dry-powder inhaler as an alternative to multiple parenteral administrations of LMWH.

6.2.5 Bioadhesive Microspheres

The term bioadhesion describes materials that bind to biological substrates, such as mucosal membranes. Adhesion of bioadhesive drug delivery devices to the mucosal tissue offers the possibility of creating an intimate and prolonged contact at the site of administration. This prolonged residence time can result in enhanced absorption, and in combination with controlled release of a drug, can also improve patient compliance by reducing the frequency of administration. For microspheres, it would be advantageous to have a means for providing intimate contact of the drug delivery system with the absorbing membranes. This can be achieved by coupling bioadhesion characteristics to microspheres and developing bioadhesive microspheres. In turn it leads to efficient absorption and enhanced bioavailability of the drug due to a high surface-to-volume ratio; a much more intimate contact with the mucous layer; and specific targeting of drugs to the absorption site achieved by anchoring plant lectins, bacterial adhesions, and antibiodies on the surface of the microspheres.

Bioadhesive microspheres can be tailored to adhere to any mucosal tissues, including those found in the eye, nasal cavity, urinary tract, colon, and gastrointestinal tract, thus offering the possibilities of localized as well as systemic controlled release of the drug. Application of bioadhesive microspheres to the mucosal tissues of the ocular cavity and gastric and colonic epithelium is used for administration of drugs and a reduction in frequency of drug administration to the ocular cavity can highly improve patient compliance. The latter advantage can be obtained for the drugs administered intranasally due to the reduction in mucociliary clearance of drugs adhering to the nasal mucosa. Microspheres prepared with bioadhesive and bioerodible polymers undergo selective reuptake by the M cells of Peyer patches in gastrointestinal (GI) mucosa. This uptake mechanism has been used for the delivery of protein and peptide drugs, antigens for vaccination, and plasmid DNA for gene therapy. Moreover, by keeping the drugs in close proximity to their absorption window in the GI mucosa, the bioadhesive microspheres improve the absorption and oral bioavailability of drugs like furosamide and riboflavin.

6.2.6 Microspheres for Special Purposes

Microspheres have been applied on several occasions for local delivery. Azouz investigated the use of paclitaxel loaded microspheres for prevention of local tumor growth (10). Poly-(D,L-lactic-co-glycolic acid) (PLGA) microspheres loaded with the antineoplastic agent paclitaxel were prepared and tested for antitumor efficacy in an in vitro cell proliferation assay for tumor inhibition and induction of apoptosis. The in vivo prevention of Lewis lung carcinoma cell establishment and growth in subcutaneous tissues of mice was assessed. Paclitaxel-loaded PLGA microspheres were found to effectively prevent growth of tumor cells in culture through the induction of apoptosis. Similarly, paclitaxel-loaded PLGA microspheres significantly inhibited tumor growth in vivo at both the 50×10^6 and 100×10^6 microsphere dose (0.497 ± 0.183 and 0.187 ± 0.083 g total tumor weight, respectively) compared with 2.91 ± 0.411 g for Lewis lung carcinoma cells with unloaded microspheres and 3.37 ± 0.433 g for untreated tumor ($P < .001$). Toxicity was not clinically apparent in any animal treated with paclitaxel-loaded PLGA microspheres.

Intrathecal injection of microspheres can be performed for various needs. Recently Baclofen microspheres were investigated (11). Since the first clinical studies published about 20 years ago, intrathecal baclofen remains the reference treatment for severe spasticity of spinal or cerebral origin. To treat this very disabling syndrome, very small quantities of baclofen have to be continually injected into the intrathecal space as a lifetime treatment. To date, the only solution found to solve this challenging drug delivery issue is to use electronic pumps surgically implanted in the patient's body and connected to indwelling catheters. The costs of surgical implantation and follow-up of these devices are significant, thus limiting the number of treated patients. Moreover, the risk of infection or catheter dysfunction is high, often resulting in the discontinuation of therapy. The development of long-lasting sustained release intrathecal baclofen dosage forms could solve many of the reported issues related to implanted pumps. These systems should provide a continuous release of baclofen for at least 3 months without any burst effect. Moreover, these formulations have to be well tolerated, biodegradable and easy to administer. Among all the formulations designed for spinal drug delivery, poly(lactide-coglycolide) (PLGA) microspheres could fulfill all these requirements. .

Microspheres were also applied previously to treat bone cancers. Bisphosphonates (BPs) are a class of drugs characterized by a P-C-P bond (12). Consequently, they are analogues of pyrophosphates that are more resistant to chemical and enzymatic hydrolysis. A number of bisphosphonates have been approved for clinical use in Paget's disease, hypercalcemia of malignancy, and osteoporosis; these conditions require continued BP administration. The BP action mechanism is not completely understood, but the most likely hypothesis is that the drug is incorporated in the bone matrix and absorbed in osteoclasts, blocking the resorption process. After oral administration in chronic drug therapy, poor absorption (1% of the oral dose) and adverse gastrointestinal effects in humans and high intra- and intersubject variability of absorption in both animal and human studies have been observed. Parenteral administration of BPs has several limitations: Intravenous administration must be performed by slow infusion of the drug, which is diluted in high amounts of solvent (200-500 mL) to avoid kidney failure, and intramuscular and subcutaneous administration, which lead to rapid and good BP absorption, can cause local tissue damage and irritation at the site of injection. For these reasons, BPs are good candidates for study, with the goals of improving their bioavailability and decreasing their side effects. Attempts have been made to use absorption enhancers and prodrugs, mainly for oral administration, but without any real improvement in drug absorption or lessening of side effects. Local implantation of a BP-loaded biodegradable microparticulate delivery system appears to warrant an assessment of the treatment of localized bone disease (ie, hypercalcemia in tumors). Chitosan microspheres loaded with BPs have had good results after local injection in the tibialis muscle, as described in the literature. The main advantage of these formulations is that the drug targets a specific site; moreover, the slow release of drug from microparticles would prevent local irritation and tissue necrosis associated with an intramuscular BP injection.

6.3 Polymers Used in the Preparation of Microspheres

The different polymers that were previously used to prepare microspheres include chitosan and its modifications, alginates, poly(adipic anhydride), gellan gums, poly lactide and co-glycolides, polycaprolactone, alginate-poly-L-lysine, aliphatic polyesters (DXO and D, L-LA), triglycerides, polypeptides, albumin, eudragits, etc. Some of these polymers with relevant examples and advantages are discussed below:

6.3.1 Cellulose based Microparticles

Methylcellulose, ethyl-cellulose, hydroxypropylmethylcellulose and cellulose acetate, the non-biodegradable and biocompatible polymers, are extensively studied in the encapsulation of material for controlled release of pharmaceuticals. Several researchers have investigated the utilization of ethyl-cellulose and cellulose-acetate as a polymer to prepare microparticulate drug delivery systems by solvent evaporation techniques, emulsification techniques etc Scanning electron microscopy of drugs-loaded ethylcellulose microspheres reveals that the microspheres posses a rough and rugged surface. It was also observed that after dissolution was performed, bigger and more pores are developed. The surface porosity is crucial for drug release in microspheres prepared with ethylcellulose. Since the polymer is not biodegradable, the release of the drugs from microspheres takes place by dissolution and diffusion through the pores. Ethylcellulose allows water to permeate through its surface without itself dissolving in it. Several researchers have investigated the utilization of ethylcellulose as a polymer to microencapsulate a drug by coacervation phase separation technique, emulsion solvent evaporation technique and spherical crystallization technique.

Recently, Yang et al., reported the preparation of isosorbide dinitrate microspheres using ethylcellulose as the polymer (13). Microcapsules for sustained release of poorly soluble isosorbide dinitrate (ISDN) were prepared based on ethylcellulose (EC) and/or blended with appropriate amounts of relatively hydrophilic hydroxypropyl cellulose (HPC) as matrix materials using the oil-in-oil emulsion evaporation method. The microspheres studied had three-mode sizes (100-150, 250-300 and 400-450 microm) and four polymer compositions (1, 0.833, 0.67 and 0.5 weight fraction EC). The microspheres were observed to contain essentially no drug crystalline domain and were of a porous morphology. The cumulative amounts of ISDN releasing from the microspheres as functions of mode fractions size and polymer compositions were measured in vitro. It was observed that the microspheres' size influenced the release behaviour of drug more obviously than the polymer composition. The smaller size and the higher hydrophilic HPC content show the faster release rate of drug and the smaller amount of drug residue. The kinetics of drug release depends on the size and polymer composition. The microspheres with 100-150 μm, of all polymer compositions, present one-stage diffusion kinetic with a lag period for drug release. On the other hand, the microspheres with the other two sizes exhibit two-stage diffusion kinetic with a lag period. According to the kinetic model, the microspheres obtained are surmised to have a core-shell like drug concentration distribution and/or a core-shell morphology.

6.3.2 Microspheres Based on Chitosan

Chitosan is a linear polysaccharide composed of randomly distributed β-(1-4)-linked D-glucosamine (deacetylated unit) and N-acetyl-D-glucosamine (acetylated unit). It has a number of commercial and possible biomedical uses. Several researchers have studied simple coecervation of chitosan in the production of chitosan beads. The chitosan beads have been used in various fields viz., enzymatic immobilization, chromatograph support, adsorbent metal ions, or lipoprotein and cell culture. Nishimura et al., investigated the possibilities of using chitosan microbeads as a cancer chemotherapeutic carrier for adriamycin (14). Recently chitosan microbeads were developed for oral sustained delivery of nefedipine, ampicillin and various steroids by adding to chitosan and then going through simple coacervation process (15). The release profiles of the drugs from all these chitosan delivery systems were monitored and it was shown in general that the higher release rates at pH 1-2 than at pH 7.2-7.4. The effect of the amount of drug loaded, the molecular weight of chitosan and the crosslinking agent on the drug delivery profiles have been discussed. Sazer and Akbuga recently investigated the delivery of a plasmid DNA using chitosan microspheres (16). Granulocyte-macrophage colony-stimulating factor (GM-CSF) is a cytokine used in the treatment of serious conditions resulting from chemotherapy and bone marrow transplantation such as neutropenia and aplastic anemia. Despite these effects, GM-CSF has a very short biological half-life, and it requires frequent injection during the treatment. Therefore, the cytokine production is possible in the body with plasmid-encoded GM-CSF (pGM-CSF) coding for cytokine administered to the body. However, the selection of the proper delivery system for the plasmid is important. In this study, two different delivery systems, encapsulated plasmid such as fucoidan-chitosan (fucosphere) and chitosan microspheres, were prepared and the particle physicochemical properties evaluated. Fucospheres and chitosan microspheres size ranges are 151-401 and 376-681 nm. The zeta potential values of the microspheres were changed between 8.3-17.1 mV (fucosphere) and +21.9-28.9 mV (chitosan microspheres). The encapsulation capacity of fucospheres changed between 84.2% and 94.7% depending on the chitosan molecular weight used in the formulation. In vitro plasmid DNA release from both delivery systems exhibited slower profiles of approximately 90-140 days. Integrity of released samples was checked by agarose gel electrophoresis, and any additional band was not seen. All formulations were analyzed kinetically. The calculated regression coefficients showed a higher r^2 value with zero-order kinetics. In conclusion, the characterizations of the microspheres can be modulated by changing the formulation variables, and it can be concluded that fucospheres might be a potential carrier system for the controlled delivery of GM-CSF encoding plasmid DNA.

Modifications of the chemical structure of chitosan to be used for the desired purpose has also been attempted. Goosen and coworkers attempted to develop very ideal microparticles based on chitosan (17). They aimed at extending the length of the cationic spacer arm on the chitosan main chain. In chemical modification, chitosan was first reacted with bromoacylbromide followed by reaction with an amine. The major problem in this procedure was the competing hydrolysis reaction of the bromoacylbromide. These synthetic polymer derivatives were used to form membrane coatings around the calcium alginate beads in which a blue dextran was entrapped. These microcapsules were prepared by extrusion of a solution of blue dextran in sodium alginate into a solution containing calcium chloride and the membrane polymer. The membrane integrity and the permeability were assessed by measuring the membrane elution of the blue dextran from the microcapsules. The encapsulation process of chitosan and calcium alginate as applied to encapsulation of hemoglobin was reported by Huguet et al (18). Two different processes were developed to prepare the capsules. Both procedures lead to beads containing high concentration of hemoglobin (more than 90% of the initial concentration (150 g/L) were retained inside the beads) provided chitosan concentration is sufficient. The molecular weight of chitosan (mol wt 245000 or 390000) and the pH (2,4, or 5.4) had only a slight effect on entrapment of hemoglobin, the best retention being obtained with beads prepared at pH 5.4. The release of hemoglobin during the bead storage in water was found to be dependent on the molecular weight of the chitosan.

In their studies on pharmaceutical applications of chitin and chitosan, Yao and coworkers reported chitosan/gelatin network polymer microspheres for controlled release of Cimetidine (19). The drug release studies were performed in hydrochloric acid solution (pH 1.0) and potassium dihydrogen phosphate (pH 7.8) buffer at ionic strength 0.1 m/L. A pH dependent pulsed-release of the drug was observed. Moreover, the release rate can be controlled via the composition of the mixture and the degree of deacetylation of chitosan. pH sensitive beads of chitosan were reported for controlled delivery of several drugs. Diclofenac sodium, Thyamine hydrochloride, chlorpheneramine maleate and Isioniazid were used as model drugs. In these studies, widely used products in medical and pharmaceutical areas Viz., glycine and polyehthylene glycol were employed as spacer groups to enhance the flexibility of the polymer networks and influence the swelling behaviour through macromolecular interactions. The procedure is based on adding drugs to chitosan solution and beads were prepared by simple coecervation. The swelling behaviour, solubility, hydrolytic degradation and drug loading capacity of the beads were investigated. Effect of the crosslinker was also studied by varying the amounts of the cross linker. The amounts and percent release in chitosan-PEG system is a bit higher, when compared to chitosan-glycine system, due to water diffusivity and pore forming properties of PEG. The effects of the amount of drug loaded and the crosslinking agent on the delivery profiles were reported as well.

Chitosan has also been used in the preparation of microspheres based on polyelectrolyte complexes. Polyelectrolyte complexes (PECs) are formed by the reaction of polyelectrolyte with an oppositely charged polyelectrolyte in an aqueous solution. Polysaccharides, which have bulky pyranose rings and highly stereoregular confirmation in their rigid linear backbone chains, have been frequently studied. PECs have numerous applications such as membranes, antistatic coatings, environmental sensors, and chemical detectors, medical prosthetic materials etc. Among these, their wide use as membranes for dialysis, ultrafiltration, and other solute separation processes are of special interest and also made it possible for the use in microcapsulcs membrane. Microcapsules can be used for mammalian cell cultures and the controlled release of drugs, vaccines, antibiotics and hormones. To prevent the loss of encapsulated material, the microcapsules should be coated with another polymer that forms a membrane at the bead surface. Few results have been reported about the formation of PECs of alginate with chitosan under acidic conditions. Although alginate/chitosan microcapsules have been studied a lot, the studies have been limited in a narrow pH region due to the solubility of chitosan. Because of this modifications of chitosan have been attempted. N-acylated chitosan were prepared. The N-acyl chitosan scarcely affected the formation behaviour of PECs with sodium alginates. For applications of PECs produced, the microencapsulation of a drug was performed and the release property of drug was tested (20). The microcapsules were prepared in one step by the extrusion of a solution of a guaifenesin and sodium alginate into a solution containing calcium chloride and chitosan through inter-polymeric ionic interactions. The N-acyl groups introduced to the chitosan enhanced the release rate remarkably.

6.3.3 Albumin microspheres

Albumin is a major plasma protein constituent, accounting for 55% of the total protein in human plasma. Since they were first described by Kramer, albumin microspheres have been extensively investigated in controlled release systems as vehicles for the delivery of therapeutic agents to local sites (21). The exploitable features of albumin include its reported biodegradation into natural products, its lack of toxicity, and its nonantigenicity. Albumin microspheres are metabolized in the body, and the size of particles, degree of stabilization, and site of metabolism are the main factors influencing the extent of metabolism. Drug release from the microspheres can be widely modulated by the extent and nature of cross-linking, size, the position of the drug, and its incorporation level in the microspheres. Colloidal forms of albumin have been considered as potential carriers of drugs for their site-specific localization or their local application to anatomically discrete sites. Recently, Gayakwad investigated albumin microspheres for the delivery of an antisense oligonucleotide (22). The aim of this study was to formulate and characterize microspheres containing antisense oligonucleotide to NF-kappaB using bovine serum albumin as the polymer matrix. Microspheres were prepared by spray-drying technique with 5, 10 and 15% drug loading. Glutaraldehyde was used as a cross-linking agent. The particle sizes ranged from 3-5 microns. Microspheres were smooth and spherical in shape, as determined by scanning electron microscopy (SEM). The yield of microspheres ranged from 70-75% and the encapsulation efficiencies were found to be in the range of 59-60%, as determined by a novel HPLC method. Zeta potential of the microspheres ranged between -39 to -53 mV, thus indicating good suspension stability in water. In-vitro release studies performed using phosphate buffer saline demonstrated extended drug release up to 72 h. Kinetic model fitting showed high correlation with the Higuchi model, suggesting that the drug release was primarily diffusion controlled. Okorowku recently investigated albumin microspheres and its modifications for DNA delivery (23). The study investigated the stability and transfection efficiency of plasmid DNA (pDNA) and sea urchin sperm histone H1 (Sp H1) complexes embedded in albumin microsphere formulations. Sp H1 increased the stability and transfection efficiency of pDNA, while providing a favourable sustained pDNA release profile. Encapsulating Sp H1-complexed pDNA into albumin microspheres further protected the pDNA from physical stress and heparin treatment. When compared with free pDNA encapsulated in albumin microspheres, the Sp H1-pDNA microsphere formulations exhibited decreased hydrophilicity, slower pDNA release profiles, protection against heparin-induced degradation of embedded pDNA and increased stability against physical stress. These results indicate that complex formation of pDNA with Sp H1 facilitates intracellular DNA transfer and that albumin microspheres-Sp H1-pDNA gene delivery formulations are suitable for controlled-release delivery of pDNA while offering protection of the pDNA from degradation and maintaining pDNA biological activity.

6.3.4 Eudragit Microspheres

Poly (meth) acrylates are popularly used in various pharmaceutical preparations. These polymers are popularly called as Eudragits. Eudragit acrylic polymers allow the active in the solid dosage form to perform during their passage in the human body. The flexibility to combine different polymers enables to achieve the desired drug release profile by releasing the drug at the right site, at the right time and over a desired period of time. Other important functions are protection from external influences (moisture) or taste/odor masking to increase patient compliance. The range of Eudragits available provide full flexibility for targeted drug release profiles by offering best performance for enteric, protective or sustained-release properties. Eudragit products are approved by the most important health authorities' world wide and are manufactured under GMP guidelines.

Recently Jain et al., prepared Eudragit S100 entrapped insulin microspheres and investigated their potential for oral delivery (24). The purpose of their research was to investigate whether Eudragit S100 microspheres have the potential to serve as an oral carrier for peptide drugs like insulin. Microspheres were prepared using water-in oil-in water emulsion solvent evaporation technique with polysorbate 20 as a dispersing agent in the internal aqueous phase and polyvinyl alcohol (PVA)/polyvinyl pyrrolidone as a stabilizer in the external aqueous phase. The use of smaller internal aqueous-phase volume (50 μL) and external aqueous-phase volume (25 mL) containing PVA in the manufacturing process resulted in maximum encapsulation efficiency (81.8% ± 0.9%). PVA-stabilized microspheres having maximum drug encapsulation released 2.5% insulin at pH 1.0 in 2 hours. In phosphate buffer (pH 7.4), microspheres showed an initial burst release of 22% in 1 hour with an additional 28% release in the next 5 hours. The smaller the volumes of internal and external aqueous phase, the lower the initial burst release. The release of drug from microspheres followed Higuchi kinetics. Scanning electron microscopy of PVA-stabilized microspheres demonstrated spherical particles with smooth surface, and laser diffractometry revealed a mean particle size of 32.51 ± 20 μm. Oral administration of PVA stabilized microspheres in normal albino rabbits (equivalent to 6.6 IU insulin/kg of animal weight) demonstrated a 24% reduction in blood glucose level, with maximum plasma glucose reduction of 76 ± 3.0% in 2 hours and effect continuing up to 6 hours. The area under the percentage glucose reduction-time curve was 93.75%. Thus, the results indicate that Eudragit S100 microspheres on oral administration can protect insulin from proteolytic degradation in the gastrointestinal tract and produce hypoglycemic effect.

In another study by Basu and Adhiyaman (25), microspheres made up of Eudragit polymers were prepared and investigated. The aim of the work was to prepare nitrendipne-loaded Eudragit RL 100 microspheres to achieve sustained release nitrendipine. Nitrendipne-loaded Eudragit RL 100 microspheres were prepared by an emulsion-solvent evaporation method using ethanol/liquid paraffin system. The resultant microspheres were evaluated for average particle size, drug loading, in vitro drug release and release kinetics. FTIR spectrometry, scanning electron microscopy, differential scanning calorimetry and x-ray powder diffractometry were used to investigate the physical state of the drug in the microspheres. The mean particle size of the microspheres was influenced by varying drug:polymer ratio and emulsifier concentration while drug loading was dependent on drug:polymer ratio. The results of FTIR spectrometry, differential scanning calorimetry and x-ray diffractometry indicated the stable character of nitrendipine in drug-loaded microspheres and also revealed absence of drug-polymer interaction. The drug release profiles of the microspheres at pH 1.2 showed poor drug release characteristics while at pH 6.8, drug release was extended over a period of 8 h; release was influenced by polymer concentration and particle size. Drug release followed the Higuchi model. The study indicated that nitrendipine-loaded Eudragit RL 100 microspheres prepared under optimized conditions demonstrate good sustained release characteristics and were stable under the conditions studied.

Mixture of Eudragit polymers based microspheres to obtain desired release were also fabricated and investigated (26). Microspheres were prepared using polymethacrylate polymers (Eudragit® RS 100 and RL 100) by solvent evaporation method and characterized for their micromeritic properties and drug loading, as well as by Fourier transform infrared spectroscopy (FTIR) and scanning electron microscopy. In vitro release studies were performed in phosphate buffer (pH 7.4). The resulting microspheres obtained by solvent evaporation method were white and free flowing in nature. The mean particle size of microspheres ranged from 420 -660 μm and the encapsulation efficiencies ranged from 40.27 - 86.67 %. The encapsulation efficiency was also found to be dependant on nature of polymer used in the formulation. From the in vitro drug dissolution studies it was found that the sustaining effect of microspheres depended on the polymer concentration, amount of dispersant used and the type of polymer used in the formulation. The mechanism of drug release from the microspheres was found to be non-Fickian type. When Eudragit® RL was used in combination with Eudragit RS, the drug released at a faster rate compared to Eudragit® RS alone. This is due to the fact that the amount of quaternary ammonium groups of Eudragit® RS is lower than that of Eudragit RL, which renders Eudragit® RS less permeable. Thus, a desired drug release can be obtained using a combination of the polymers.

6.3.5 Polylactide-co-glycolide Microspheres

PLGA or poly(lactic-*co*-glycolic acid) is a copolymer which is used in a host of Food and Drug Administration (FDA) approved therapeutic devices, owing to its biodegradability and biocompatibility. PLGA is synthesized by means of random ring-opening co-polymerization of two different monomers, the cyclic dimers (1,4-dioxane-2,5-diones) of glycolic acid and lactic acid. Common catalysts used in the preparation of this polymer include tin(II) 2-ethylhexanoate, tin(II) alkoxides or aluminum isopropoxide. During polymerization, successive monomeric units (of glycolic or lactic acid) are linked together in PLGA by ester linkages, thus yielding a linear, aliphatic polyester as a product. Depending on the ratio of lactide to glycolide used for the polymerization, different forms of PLGA can be obtained: these are usually identified in regard to the monomers' ratio used (e.g. PLGA 75:25 identifies a copolymer whose composition is 75% lactic acid and 25% glycolic acid). All PLGAs are amorphous rather than crystalline and show a glass transition temperature in the range of 40-60 °C. Unlike the homopolymers of lactic acid (polylactide) and glycolic acid (polyglycolide) which show poor solubilities, PLGA can be dissolved by a wide range of common solvents, including chlorinated solvents, tetrahydrofuran, acetone and ethylacetate. PLGA degrades by hydrolysis of its ester linkages in the presence of water. It has been shown that the time required for degradation of PLGA is related to the monomers' ratio used in production: the higher the content of glycolide units, the lower the time required for degradation. An exception to this rule is the copolymer with 50:50 monomers' ratio which exhibits the faster degradation (about two months). In addition, polymers that are end-capped with esters (as opposed to the free carboxylic acid) demonstrate longer degradation half-lives.

PLGA has been successful as a biodegradable polymer because it undergoes hydrolysis in the body to produce the original monomers, lactic acid and glycolic acid. These two monomers under normal physiological conditions, are by-products of various metabolic pathways in the body. Since the body effectively deals with the two monomers, there is very minimal systemic toxicity associated with using PLGA for drug delivery or biomaterial applications. It, however, can cause trouble for lactose intolerant people. Also, the possibility to tailor the polymer degradation time by altering the ratio of the monomers used during synthesis has made PLGA a common choice in the production of a variety of biomedical devices such as: grafts, sutures, implants, prosthetic devices, micro and nanoparticles. As an example, a commercially available drug delivery device using PLGA is Lupron Depot® for the treatment of advanced prostate cancer. Several groups investigated the preparation of microspheres using these polymers. Some of the formulations demonstrated success in the research and some products are already in the market. Depending on the solubility of the drug a process will be selected to prepare microspheres using these polymers. Further, there is a range of products available from different companies.

The different rates of degradation will help decide a suitable polymer for the controlled parenteral release of the drug for specific duration. Degradation begins with random hydrolysis in an aqueous environment through cleavage of its backbone ester linkages. Most of the literature indicates that these poly (ester) polymers do not involve any enzymatic activity and is purely through a hydrolytic mechanism. In vivo, however, enzymes are considered to enhance the initial degradation. The PLA, PGA and their copolymers biodegrades into lactic and glycolic acids. Lactic acid enters the tricarboxylic acid (TCA) cycle and is metabolized and excreted from the body as energy, carbon dioxide and water. Glycolic acid is either eliminated unchanged in the kidney or enter the TCA cycle and is excreted as energy, carbon dioxide and water. So they are biodegradable and bioresorbable polymers. In general, crystalline L-PLA is more resistant to hydrolytic degradation than the amorphous DL form.

Time required for L-PLA implants to be absorbed is relatively long and depends on polymer quality, processing conditions, implant site, and physical dimensions of the implant. The biodegradation rate of copolymers is dependent on the molar ratio of lactide/glycolide, molecular weight of the polymers, the degree of crystallinity and the Tg of the polymers. Altering the chemical composition by increasing the glycolide mole ratio in the copolymer increases the rate of biodegradation. For example, a copolymer of 50% DL-lactide and 50% glycolide degrades faster than other copolymers. Factors affecting the hydrolytic degradation behavior of PLA, PGA and their copolymers are as follows: chemical composition, molecular weight and molecular weight distribution, physical-chemical factor (pH, ionic strength, temp), additives (solvent, monomers, catalyst), morphology (crystalline, amorphous), glass transition temperature. (rubbery, glassy), device dimensions (shape, size, porosity, surface to volume ratio), mechanism of hydrolysis. (autocatalytic, noncatalytic, enzymatic), processing history (injection, molding, extrusion, sterilization), site of implantation. Some pictures showing microspheres based on PLGAs is presented in Figure 6.2.

Figure 6.2 Polylactide-co-glycolide microspheres.

Several scientists have investigated the application of polylactide-co-glycolide microspheres. Recently Yang et al., investigated interferon PLGA microspheres (27). By a double emulsion solvent evaporation method, interferon-alpha (IFN-alpha) microspheres were prepared with poly(lactide-co-glycolide) (PLGA) and their characteristics, such as morphology, drug loading, encapsulation efficiency, in vitro release and degradation were evaluated. The IFN-alpha microspheres were prepared by different viscosities from 0.17-1.13 dL g(-1) and concentrations between 5-25% of PLGA, which not only affected the drug loading and encapsulation efficiency of IFN-alpha microspheres, but also strongly influenced the in vitro release. With smooth and porous surface, the drug loading and encapsulation efficiency of the microspheres prepared by 15% 0.89 dL g(-1) PLGA were 7.736% and 77.38%, respectively. The

DSC curve of microspheres indicated IFN-alpha was loaded inside the microspheres. The degradation of microspheres was homogeneous and the mass loss was over 80% in 6 weeks. The release profile of microspheres showed a sustained fashion and the IFN-alpha released from microspheres maintained its bioactivity for 7 days.

6.3.6 Polycaprolactone Microspheres

Polycaprolactone (PCL) is a biodegradable polyester with a low melting point of around 60°C and a glass transition temperature of about −60°C. PCL is prepared by ring opening polymerization of ε-caprolactone using a catalyst such as stannous octanoate. PCL is one of the biocompatible and biodegradable aliphatic polyester polymers that degrades slowly and does not generate an acid environment unlike the polylactide (PLA) or polyglycolide (PLG) polymers. Although the permeability of macromolecules in PCL is low, such low permeability may be sufficient for drug delivery. Other advantages of PCL include hydrophobicity, in vitro stability and low cost. Therefore, many investigations have focused on the application of PCL microspheres to drugs in recent years. From formulation point of view also, polycaprolactone has several advantages. Vivek et al., investigated the efficacy of different biodegradable polymers in terms of encapsulation efficiency and drug release (28). Etoposide-loaded biodegradable microspheres of poly lactic-co-glycolide (PLGA) 50:50, PLGA 75:25, and polycaprolactone (PCL) were prepared by simple o/w emulsification solvent evaparation method and characterized by size analysis and microscopy. The influence of drug to polymer ratio on the entrapment of etoposide was studied. Of all the three types of microspheres, polycaprolactone microspheres (PCL MS) showed the highest entrapment efficiency (94.64%), followed by PLGA 75:25 microspheres (PLGA 75:25 MS) (88.64%) and PLGA 50:50 microspheres (PLGA 50:50 MS) (79.19%). The drug to polymer ratio of 1:20 gave the highest entrapment efficiency for all the three types of microspheres. The in vitro release of etoposide from the three microsphere formulations were studied in phosphate buffer pH 7.4 (pH 7.4 PB) containing 0.1% Tween 80. The microspheres showed an initial burst release, which was highest from the PLGA 50:50 MS and least from the PCL MS. PCL MS microspheres showed the lower and slow drug release than the remaining formulations. The release of etoposide from all the three microsphere formulations followed Higuchi's diffusion pattern. The microspheres in the dissolution medium for 28 days appeared irregular in shape and slightly fragmented.

6.3.7 Triglyceride Lipospheres

Lipospheres are an aqueous microdispersion of water insoluble, spherical microparticles each consisting of a solid core of hydrophobic triglycerides and drug particles that are embedded with phospholipids on the surface. The in vivo studies with lipospheres have shown that a single bolus injection can deliver antibiotics and anti-inflammatory agents for 3 to 5 days and also control delivery of vaccines. Recent reports were of bupivacaine-liposphere formulation, which produced 1-3 days reversible sensory and motor SLAB when applied directly to the rat sciatic nerve. The particle size of the lipospheres was in between 5 and 15μm, with over 90% surface phospholipid. Lipospheres released bupivacaine over two days under ideal sink conditions. Toonsuvan et al., prepared lipospheres for bupivacaine and investigated their efficacy (29). Bupivacaine lipospheres were prepared as a parenteral sustained-release system for post-operative pain management. Bupivacaine free base was incorporated into micron-sized triglyceride solid particles coated with phospholipids, which were formed via a hot emulsification and cold resolidification process. The bupivacaine liposphere dispersions were characterized with respect to drug loading, particle-size distribution, and morphology. Gelation of the fluid liposphere dispersions was observed at different time intervals upon storage. The type of phospholipids used in the formulation was found to have a major impact on the gelation of the dispersion. The use of synthetic phospholipids instead of the natural phospholipids in the formulation yielded bupivacaine liposphere dispersions exhibiting prolonged gelation time. The addition of a hydrophilic cellulosic polymer can further improve the physical stability of the dispersion.

6.4 Fabrication of Microspheres

There are several techniques of microencapsulation. Microcapsules are prepared using certain techniques while microspheres are prepared using different techniques, some techniques being common to the preparation of both microcapsules and microspheres. Emulsion solvent evaporation and its modifications, phase-separation methods such as coecervation-phase separation, spray drying/spray congealing, polymerization techniques are commonly used for the preparation of microspheres. The other techniques which are specific for the preparation of microcapsules include air suspension method, multiorifice centrifugal technique, and pan coating among others. Some of the popular techniques are discussed below:

6.4.1 Emulsion Solvent Evaporation and its Modifications

These techniques of preparation are also called as solvent evaporation techniques. One modified version of solvent evaporation technique is solvent diffusion technique. In these techniques, most of the times the preparation of emulsion becomes mandatory and a reason they are often called as emulsion solvent evaporation techniques. This methodology has been described by various companies. The processes are carried out in a liquid manufacturing vehicle. The microsphere coating is dissolved in a volatile solvent, which is immiscible with the liquid manufacturing vehicle phase. A core material to be encapsulated is dissolved or dispersed in the coating polymer solution. With agitation, the core coating material mixture is dispersed in the liquid manufacturing vehicle phase to obtain the appropriate size microsphere. The mixture is then heated (if necessary) to evaporate the solvent for the polymer. In case of water soluble polymers a modified version of the solvent evaporation called solvent diffusion is used. In these situations the solvent for the hydrophilic polymer and the drug is removed by diffusion. In the case in which the core material is dispersed in the polymer solution, polymer shrinks around the core. These are generally called as microcapsules. In the case in which the core material is dissolved in the coating polymer solution, a matrix-type microsphere is formed. Once all the solvent for the polymer is evaporated, the liquid vehicle temperature is reduced to ambient temperature (if required) with continued agitation. At this stage, the microcapsules can be used in suspension form, coated onto substrates or isolated as powders. Generally, formation of stable emulsion of the two immiscible phase is needed. Process variables which include, but not be limited to, methods of forming dispersions, evaporation rate of the solvent for the coating polymer, temperature cycles, and agitation rates. Important factors that must be considered when preparing microspheres by solvent evaporation techniques include choice of vehicle phase and solvent for the polymer coating, as these choices greatly influence microsphere properties as well as solvent recovery techniques. The solvent evaporation technique to produce microspheres is applicable to a wide variety of liquid and solid materials as in the form of the core or the dispersion. The material encapsulated may be either water-soluble or water insoluble. A variety of polymers have been used in the microencapsules techniques. These polymers include water soluble resins (gelatin, gum Arabic, starch, polyvinylpyrrrolidone, carboxymethylcellulose, hydroxyethylcellulose, methylcellulose, arabinogalactin, polyvinylalcohol, polyacrylic acid), water insoluble resins (ethylcellulose, polyethylene, polymethacrylate, polyamide/nylon, polyethylenevinylacetate, cellulose nitrate, silicones, polylactide-co-glycolides, polycaprolactone), waxes and lipids (paraffin, carnauba, spermaceti, beeswax, stearic acid, stearyl alcohol, glyceryl stearates), enteric resins (shellac, cellulose acetate phthalate), pluronics. There are several factors that influence the formation and characteristics of microspheres when solvent evaporation technique is employed. The factors include:

(i) Type of the emulsion used,
(ii) Type and concentration of the emulgent used.
(iii) Solubility of the drug and the polymer in the dispersed and the continuous phase.
(iv) Characteristics of the drug and the polymer and the drug/polymer ratio
(v) Stirring rate or Agitator characteristics
(vi) Inner water/oil ratio
(vii) Other factors.

(i) ***Type of the emulsion used:*** The emulsion type is selected depending on the characteristics of the drug and the polymer. Some times it is a single emulsion such as o/w or w/o or some times it is a multiple emulsion such as w/o/w. Generally, a o/w emulsion is used to encapsulate hydrophobic drugs. The other possibility is the o/o emulsion evaporation technique. Kim et al., recently used o/w and o/o techniques techniques to prepare felodipine microspheres using polycaprolactone as the polymer (30). The fabrication process is shown in Figure 6.3.

For the o/w-method, felodipine and polymers were dissolved in solvent or solvent mixture, then emulsified into aqueous solution containing emulsion stabilizer to form o/w emulsion. The volume ratio of oil: water phases were typically 10:500. In case of the o/o-method, similar to that of the o/w-method, felodipine and polymers were dissolved in solvent or solvent mixture and then emulsified into corn oil solution containing Span 80 to form an o/o emulsion. The volume ratio of oil1 : oil2 phases were typically 10:400. The mixture was mechanically stirred with a stirring blade until the complete evaporation of organic solvent was accomplished (3–6 h). After that, the microspheres were isolated by filtration, washed and vacuum-dried. In a previous study, the o/w-method was employed in order to compare its microsphere characteristics to those of microspheres prepared by the o/o-method using methacrylates and eudragits as the polymers (31). The spherical shape of the microspheres was examined by SEM. The microspheres prepared by the o/w-method were more spherical and uniform than those prepared by the o/o-method. The mean size prepared by the o/w-method was smaller than that prepared by the o/o-method. Encapsulation efficiency of the microspheres prepared by the o/w-method is a little higher than that prepared by the o/o-methods. This phenomenon is contributed to the hydrophobicity of the felodipine, since the hydrophobic corn oil was used as a continuous phase, the drug loss could be increased.

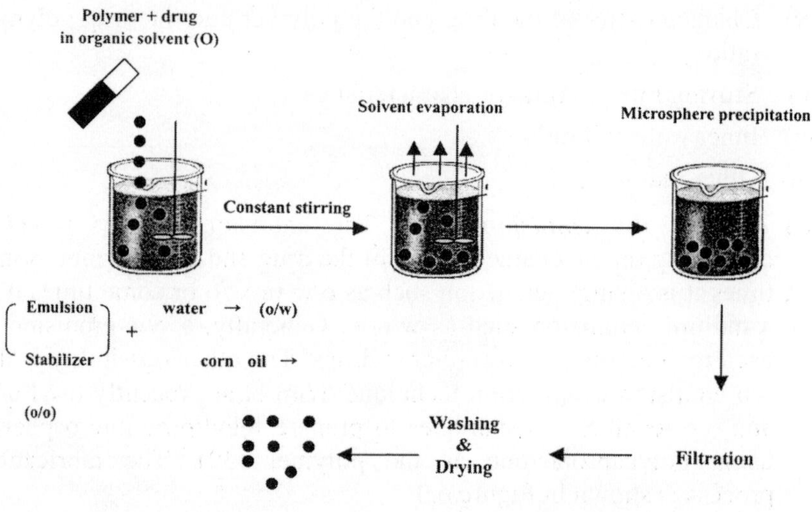

Figure 6.3 Schematic representation of preparation of microspheres using solvent evaporation techniques.

A w/o emulsion can be used to encapsulate hydrophilic drugs. For instance, the following method to prepare Losartan potassium microsphere was adopted (32) and the polymers selected were sodium alginate and acrycoat. The drug was dissolved in each polymeric aqueous solution. The solutions were emulsified into the oil phase by stirring the system in a 500 ml beaker. Constant stirring at 2000 rpm was carried out using mechanical stirrer the beaker and its content were heated by a hot plate at 80°C. Stirring and heating were maintained for 2.5 hour until the aqueous phase was completely removed by evaporation. The light oil was decanted and the microspheres were collected. Phromsofa and

Baimark prepared chitosan microspheres using a w/o emulsion solvent diffusion technique (33). Chitosan solutions (1.0% w/v) were prepared by dissolving chitosan flakes in 4% (v/v) acetic acid solution. Chitosan microparticles were prepared by the water-in-oil emulsion solvent diffusion method in 600 mL beakers with magnetic stirring. The 10 mL of chitosan solution was added drop-wise to 400 mL of ethyl acetate with stirring speed of 600 rpm for 45 min. The beaker was tightly sealed with aluminum foil to prevent evaporation of ethyl acetate during emulsification-diffusion process. The chitosan microparticles were recovered by centrifugation and finally dried in a vacuum oven at 30°C for 24 h. Insulin-loaded hydrophilic microspheres were previously prepared using a similar approach. The methodology allowed high encapsulation efficiency and the preservation of peptide stability during particle processing. The preparation method used the diffusion of water by an excess of solvent starting from a water-in-solvent emulsion. Briefly, the water dispersed phase containing albumin or lactose, or albumin-lactose in different weight ratios, and insulin was emulsified in water-saturated triacetin with and without emulsifiers, producing a water-in-triacetin emulsion. An excess of triacetin was added to the emulsion so that water could be extracted into the continuous phase, allowing the insulin-loaded microsphere precipitation.

Multiple emulsions have been used to prepare microspheres encapsulating water soluble drugs. The microencapsulation process in which the removal of the hydrophobic polymer solvent is achieved by evaporation has been widely reported in recent years for the preparation of microspheres and microcapsules. This is especially true with microspheres based on biodegradable polymers and copolymers of hydroxy acids. The encapsulation of highly water soluble compounds including proteins and peptides presents formidable challenges to the researcher. The successful encapsulation of such entities require high drug loading in the microspheres, prevention of protein degradation by the encapsulation method, and predictable release of the drug compound from the microspheres. To achieve these goals, multiple emulsion techniques and other innovative modifications have been made to the conventional solvent evaporation process. Lima et al., 2009 recently prepared microspheres containing bovine hemoglobin using this technique (34). Microspheres containing Hb were prepared by a W/O/W double emulsion technique. First, Hb aqueous solution was emulsified in organic solvent (methylene chloride) containing polymer by a high-speed homogenizer. Thereafter, the primary emulsion was poured into an aqueous solution containing 1% (w/v) PVA followed by a two-step re-emulsification for 25 s and 90 s, respectively. The double emulsion was subsequently added to 200 mL of a buffer solution (Tris.HCl 0.1 N pH 7.4). The microcapsules were recovered by partial evaporation, followed by centrifugation.

Other types of emulsions are also used to prepare microspheres of water soluble compounds. When the core material is hydrophilic, w/o/o emulsion type is preferable to w/o/w emulsion type. Therefore, the selection of the emulsion types may depend on the properties of core materials. The different factors that influence encapsulation using multiple emulsions is illustrated by the encapsulation of water soluble proteins. Many experimental parameters influence the properties of protein-loaded microspheres, manufactured by the double emulsion/solvent evaporation technique. These are listed below:

Steps of the process and the factors influencing

A. First emulsion preparation

- (a) Concentration of the polymer solution
- (b) Composition and molecular weight of the polymer
- (c) Organic solvent
- (d) Volume of the organic phase
- (e) Protein concentration in the aqueous inner phase
- (f) Volume of the aqueous inner phase
- (g) Emulsion equipment (e.g. for homogenization or sonication)
- (h) Rate at which the aqueous phase is added to the organic phase
- (i) Mixing time
- (j) Temperature and pressure

B. Second emulsion preparation

- (a) Volume of the outer aqueous phase
- (b) Nature and concentration of the emulsifying agent
- (c) Emulsion equipment
- (d) Rate at which the first emulsion is added to the aqueous phase

(e) Stirring rate
(f) Mixing time
(g) Temperature and pressure
(h) Solvent extraction volume of the extraction phase
(i) Presence of additives or stabilizers
(j) Stirring rate
(k) Extraction time
(l) Temperature and pressure

C. Collecting and washing collection system (e.g. filtration or centrifugation)

(a) Volume and composition of the washing solution
(b) Temperature and pressure
(c) Drying Method employed (e.g. lyophilization, fluid bed)
(d) Time
(e) Residual humidity
(f) Addition of excipients

Modifications of the solvent evaporation techniques also have been investigated. Morita et al., recently prepared microspheres using a novel methodology (35). A new method for preparing protein-loaded biodegradable microspheres by a process involving solid-in-oil-in-water (S/O/W) emulsion was established using poly(ethylene glycol) (PEG). In the first step, a protein solution was lyophilized with PEG, which resulted in the formation of spherical protein microparticles, less than 5 microns in diameter, dispersed in a continuous PEG phase. This process was well explained by the aqueous phase separation phenomenon induced by freezing-condensation. Since this lyophilizate could be directly dispersed in an organic phase containing biodegradable polymer by dissolving PEG with methylene chloride, a conventional in-water drying method could be adopted in the second step. Through this S/O/W emulsion process, horseradish peroxidase was effectively entrapped into monolithic-type microspheres of poly(DL-lactic-co-glycolic acid) (PLGA), without significant loss of activity. Bovine superoxide dismutase (bSOD), as another model protein, could be encapsulated into reservoir-type microspheres by the 'polymer-alloys method' using both poly(DL-lactic acid) (PLA) and PLGA. The initial release of bSOD from this reservoir-type microsphere was efficiently reduced. Further, the bSOD release kinetics could be suitably modified by adjusting the loading amounts of PEG or polymer composition. In this study, the multi-functional nature of PEG was successfully utilized in the preparation and designing of protein-loaded microspheres.

(ii) Type and concentration of the emulgent used.

The surfactant provides an interaction between the drug substance and solvent. This ensures the superiority of dispersion medium conditions, influencing the size of emulsion droplets. In preparing microspheres using o/w emulsion solvent evaporation technique several agents can be used for emulsification purposes. It is preferable to add an emulsifying agent to the water layer, and examples thereof include those which are able to form a stable O/W type emulsion, such as an anionic surfactant (sodium oleate, sodium stearate, sodium lauryl sulfate or the like), a nonionic surfactant (a polyoxyethylene sorbitan fatty acid ester, a polyoxyethylene castor oil derivative or the like), polyvinyl pyrrolidone, polyvinyl alcohol, carboxymethylcellulose, lecithin, gelatin and the like, which may be used alone or as a mixture of two or more. These agents may be used in a concentration of from about 0.01% to about 20%, more preferably from about 0.05% to about 10%.

In microencapsulation by solvent evaporation method, surfactants play an important part in the final characteristics of the microcapsules. Tween 80 (polysorbate 80) and Span 80 (sorbitan monooleate) are two of the most commonly surfactants used interchangeably by different authors. Two types of surfactants used have an influence on the particle size distribution of the microspheres. The hydrophobic surfactant Span 80 (Sorbitan monooleate, HLB 4.3) is found to produce smaller particle size microspheres compared to hydrophilic surfactant Tween 80 (Polyoxyethylene 20 sorbitan monooleate, HLB 14.9). Span 80 is oil soluble and produces a stable emulsion when the dispersion medium is oil. This may explain why smaller particle sizes are obtained with span 80. The concentration of surfactant/dispersing agents also affects the particle size. For both types of surfactants used, the higher concentration of surfactant resulted in production of smaller particle size. This is due to better stabilization of internal droplets with increase of surfactant concentration preventing coalescence. Also when more amount of surfactants are added, there is an accelerated dispersion of microcapsules in the microencapsulation system.

Different polysorbates are generally used in microsphere preparation. Polysorbate 80 tended towards microsphere formations with smaller average diameters as compared with the presence of other types, observed in preformulation studies performed with constant drug to polymer ratio and stirring rate. Besides the other variables, polysorbate 40 and 60 have influence on the formations of microspheres. Magnesium stearate was also used as a dispersing agent. The use of magnesium stearate as a dispersion agent decreased the interfacial tension between the lipophilic and hydrophilic phases of the emulsion and further simplified the formation of microcapsules. As the solvent evaporated, the viscosity of the individual droplets increased, and highly viscous droplets were observed to coalesce at a faster rate than they could be separated. Magnesium stearate formed a thin film around the droplets and thereby reduced the extent of coalescence, before hardening of the capsules, on collision of the droplets. The resultant microcapsules were free-flowing, and the use of magnesium stearate was deemed effective. The mean size of microspheres prepared by the o/w emulsion solvent evaporation method using different emulsifier, i.e. gelatin, Tween 80, PVA and Pol 237 is variable. The morphological analysis show that the surface of the microspheres prepared with all emulsion stabilizers appear to be smooth and spherical. PVA produces the smallest mean size with a broad particle size distribution. While Pol 237 yields the most narrowly size-distributed microspheres. However, the results may vary from drug to drug.

Several studies indicated that drug release from the microspheres is affected with the surfactant type and concentration. The faster drug release is generally noticed as the dispersing agent concentration rose may be due to decrease in particle size, which provided a larger surface for drug release. The effect of particle size on the drug release was also studied using microspheres of different size fractions (355 and 500 μm, respectively). The release profile was in line with the theory of the effect of the particle size on dissolution rate. As microsphere size decreased, the drug release increased as result of higher surface area.

(iii) Solubility of the drug and the polymer in the dispersed phase and continuous phase

In general, the encapsulation efficiency of the drug depends on the solubility of the drug in the solvent and continuous phase. An increase in the concentration of polymer in a fixed volume of organic solvent results in an increase in encapsulation efficiency. For instance, the encapsulation efficiencies of the microspheres prepared with Eudragit® RS 100 are higher than those of microspheres prepared with the Eudragit® RS/RL combination. This can be attributed to the high content of the ammonium group in Eudragit® RL, which might could facilitate the diffusion of some of the entrapped drug to the surrounding medium during formation of the microspheres. Higher entrapment of drug can be obtained when a lower concentration of magnesium stearate was used. This may be due to the formation of large-size microspheres at lower magnesium stearate levels, which provided less surface area for drug escape to the external processing medium. In case of propranolol HCl (PHCl) and verapamil HCl (VHCl) encapsulation in microspheres, liquid paraffin was selected as a bulk or outer phase, since PHCl, VHCl, and Eudragit RS/RL are only very slightly soluble in liquid paraffin (36). Acetone has a dielectric constant of 20.7 and was therefore chosen as the dispersed or inner phase, since solvents with dielectric constants between 10 and 40 show poor miscibility with liquid paraffin.

(iv) Characteristics of the drug and the polymer and the drug/polymer ratio

Different types of ethylcelluloses were used in the fabrication of microspheres. Ethylcellulose type and its added amount in formulations had a major effect on the organic phase viscosity, influencing the particle size distributions of gathered up microspheres shown by sieve analysis (4). The viscosity of the organic phase due to the concentration of polymer inside was effective on solvent diffusion and emulsification, while the shearing rate during stirring was kept constant. The geometric mean diameters of microspheres were found to be dependent on polymer concentration dispersed in the organic phase. But a variation in the initial drug loading up to 33.33 % (1:1 drug to polymer ratio) did not produce any significant change in mean particle size as can be seen from the laser diffraction patterns of formulations prepared with 1:2 and 1:3 drug to polymer ratios corresponding to 25.0 % and 20.0 % drug loadings. However, the change in particle size distribution was not proportional to polymer content.

The drug loading was affected by neither polymer content nor stirring rate or surfactant variables, but was consistently and slightly lower than the theoretical loading, with high encapsulation efficiencies close to 100 % in all cases. It was reported that ethylcellulose led to high encapsulation efficiencies, which might be related to a fast precipitation of ethylcellulose, resulting in reduced drug diffusion into the aqueous phase. Also the recovered amount of total microspheres demonstrated the adequacy of process variables during solvent evaporation. Various drug release patterns were obtained, which are biphasic i.e. an initial rapid drug release phase ('burst effect') was followed by a second, slower drug release phase. It could be seen from SEM pictures, that at thicker consistency of microspheres, less porous surface is obtained. These micropores may facilitate drug diffusion from microspheres during dissolution. Thus, the drug located closer to the surface, is accessible by the release medium, regarding to the coat thickness surrounding the drug particles. With increasing polymer amount and viscosity, the initial burst effect was significantly decreased in addition to the decrease in second drug release phase. The subsequent decreasing related to drug release rates can be attributed to the resulting decreased drug concentration gradients. In vitro drug release strongly depended on the type of polymer. This might be explained by the higher viscosity of the organic phase in the case of ethylcellulose. Different permeabilities of the drug within the polymers and/or drug distribution within the microspheres could be the reason of this fact. It has been reported that the release of the drug depends on viscosity grade of the ethylcellulose. With an increasing viscosity grade, the rate of release decreases. It has also been described that, in addition, the release rate depends on the overall viscosity of the system. This information has been corroborated by various studies. Microsphere formations with 1:3 drug to polymer ethylcellulose N100 ratio, which had the highest overall viscosity demonstrated slowest drug release. Additionally, when drug release profiles were carefully analyzed, it was observed that the rates were relatively fast and that more than 70 % of indomethacin was released in first minutes from microspheres prepared by lower amount and viscosity grade polymer, suggesting that viscosity is the major player in the fabrication and optimization of the microsperes. Increasing the polymer ratio and its viscosity grade, drug release rates were modified and incompleted and in this case, the cumulative amount of drug released within 9 h was approximately 35 %. It should be emphasized that the remaining indomethacin was actually located in the microspheres and did not undergo any degradation process during the kinetic experiment, which was reported to last almost 4 weeks and to depend on the retention capacity of the coating polymer. Incomplete dissolution rates of salbutamol sulphate from microcapsules have been also reported, because of impeding effect of ethylcellulose thickness surrounding the drug particles (37). In the indomethacin study, high encapsulation efficiency can be attributed to the probability for the drug to be entrapped within the microparticles based on the high solubility of indomethacin in organic phase when increasing the amount of ethylcellulose stepwise. The pores formed during preparation could be responsible for the initial fast release of drug. During this initial phase, the simultaneous swelling of polymer and the increase in distance that the drug must travel from has been completed and the drug release is thought to be dependent on the permeability of the swollen polymer. All microspheres presented a narrow particle size distribution and good flow characters according to USP 28-NF 23 criteria, besides microspheres were more spherical in shape in their manufacture with ethylcellulose N100 and higher ratio of both polymers.

Similar results were found when microspheres were prepared using Eudragit polymers to encapsulate the drugs VHCl and PHCl (36) (cited before). When 1:1 (w/w) drug/polymer concentrations were used for both the Eudragit RS and RL polymers, the quality of microcapsules formed was poor. These were irregularly shaped, not flowing, and presented with lots of indentation. Microcapsules were only formed when the polymer concentration was increased to ratios of between 1:2 and 1:6 (w/w) with respect to the drug concentration. Discrete, spherical, and uniform microcapsules were obtained with a 1:4 (w/w) drug/polymer ratio for both the RS and RL polymers. It is also evident that the microcapsules exhibited slightly porous surfaces, probably due to the high concentration of drug in the microcapsules. SEM was performed to determine whether microcapsules had been formed.

The use of SEM is important for establishing the encapsulating ability of different polymers, since the degree of porosity may be observed, and therefore, encapsulation ability of the polymers can be established in a qualitative manner. The formulations in which the drug/polymer ratio was 1:4 (w/w) produced uniform spherical particles that were harder to the touch than the ones manufactured at lower drug/polymer concentrations. Microspheres that were formulated with low concentrations of Eudragit RL and RS were irregular, non-spherical, soft, and had poor flowability; therefore, they were not considered suitable for analysis. The dissolution of PHCl and VHCl from the microspheres prepared with a 1:4 (w/w) drug/polymer ratio was investigated. Examination of the release profiles reveals that drug release was generally faster for VHCl and PHCl microspheres produced with the RL polymer despite the apparent similarity in particle size of the microcapsules. RS and RL are copolymers of partial esters of acrylic and methacrylic acids containing low amounts of quaternary ammonium groups, approximately 5% and 10% for RS and RL, respectively. The RS polymer is water-insoluble, and drug delivery systems prepared from it show pH-independent sustained drug release, attributed to the quaternary ammonium groups. The quaternary ammonium groups in the RS and RL chemical structures play an important role in controlling drug release because they relate to water uptake followed by the swelling of the polymers. This is most likely because the number of quaternary ammonium groups of RS is lower than that of RL, which renders RS less permeable. There was no significant difference between the dissolution profiles of PHCl and VHCl for the same polymer, since the solubility of each of the drugs is similar. The solubility of PHCl is greater than 150 mg/mL, and the solubility of VHCl is 123 mg/mL. The resulting microspheres formulated by solvent evaporation method was found to be spherical and free flowing in nature. The mean particle size of microspheres ranged from 420 - 660 μm. It was noticed that mean particle size increased with increase in polymer concentration and decrease in magnesium stearate concentration. The encapsulation efficiencies ranged from 40.27 - 86.67 %. The encapsulation efficiency was also found to be dependant on nature of polymer used in the formulation. From the in vitro drug dissolution studies it was found that the sustaining effect of microspheres depended on the polymer concentration, amount of dispersant used and the type of polymer used in the formulation. Drug release from microcapsules should theoretically be slower as the amount of polymer is increased because of an increase in the path length through which the drug has to diffuse. A relatively high encapsulation efficiency was observed for all microsphere formulations. The encapsulation efficiency was greater than 80% for all drug candidates investigated, and therefore, it is evident that both the RS and RL polymers are potentially useful materials for the encapsulation of relatively hydrophilic compounds such as PHCl and VHCl. The good flow properties (HR = 1.2) suggest that the microspheres can be easily handled during processing.

Felodipine loading in polycaprolactone microspheres prepared using o/w emulsion solvent evaporation was investigated. Felodipine loading concentrations at 1, 3 and 6% (w/v) were used. Encapsulation efficiency was highest at 3%. It was postulated that, at the high felodipine concentration (6%), the quantity of polymer present was insufficient to cover the felodipine completely (30). The effects of various process and formulation parameters on the AZT entrapment efficiency of microspheres was recently investigated (38). The highest (54%) entrapment efficiency was achieved by increasing polymer-drug ratio from 1:0.25 to 1:0.50. With further increase in polymer-drug ratio from 1:0.50 to 1:1, a significant decrease was observed on encapsulation efficiency of AZT. The higher drug loading typically results in lower encapsulation efficiency due to higher concentration gradients resulting the drug to diffuse out of the polymer/solvent droplets to the external processing medium. And also the viscosity of the polymer solution at higher drug loading was very high and is responsible for the formation of larger polymer/solvent droplets. It caused a decrease rate of entrapment of drug due to slower hardening of the larger particles, allowing time for drug diffusion out of the particles, which tends to decrease encapsulation efficiency. Among the different polymer-drug ratios investigated, 1:0.50 polymer-drug ratio had the optimum capacity for drug encapsulation. Keeping the drug-polymer ratio constant, there was a significant decrease in encapsulation efficiency of AZT with increasing the concentration of surfactant for secondary emulsification. This may be due to the fact that the increase in surfactant concentration proportionally increases miscibility of acetonitrile with light liquid paraffin (processing medium), which may increase the extraction of AZT into the processing medium. The volume of processing medium significantly influences the entrapment efficiency of AZT microspheres.

(v) Dispersed phase solvent

Dispersed phase solvent influences various parameters in the encapsulation of drugs. In case of AZT, due to its hydrophilicity it is likely to preferentially partition out into the aqueous medium, leading to low entrapment efficiency, when encapsulated using aqueous phase as the processing medium (38). Depending on the processing conditions as much as 80% of the AZT can partition out into the outer processing medium. However, sufficiently high entrapment efficiency of water soluble AZT can be obtained using a w/o/o double emulsion solvent diffusion method using a non-aqueous processing medium. The primary requirement of this method to obtain microspheres is that the selected solvent system for polymer should be immiscible with non-aqueous processing medium. Acetonitrile is a unique organic solvent, which is polar, water miscible and oil immiscible. All other polar solvents like methanol, ethanol, ethyl acetate, acetone and dimethyl sulfoxide are oil miscible and will not form emulsions of the polymer solution in oil. With oil as the processing medium, use of acetonitrile alone as a solvent did not ensure formation of primary emulsion of the aqueous phase in the polymer solution. Immediately on mixing, the water miscibility of the acetonitrile brought about the precipitation of the polymer (ethylcellulose). Hence, a non-polar solvent, namely dichloromethane was included with acetonitrile to decrease the polarity of the polymer solution. Additionally, it was also desirable that the second solvent be oil miscible, so that solvent removal is facilitated through extraction by processing medium leaving behind a viscous polymer solution. The optimal proportion of acetonitrile and dichloromethane was found to be 1:1, which enabled emulsion formation and yielded good microspheres. No surfactant was used for stabilizing primary emulsion, since ethylcellulose has the additional property of stabilizing w/o emulsion. Span 80, a representative of the nonionic dispersing agent, was used to stabilize the secondary emulsification process. It has the HLB value of 4.3 and is expected to have a high disparity for the present emulsion system by reducing the surface tension at the interface. This is one of the interesting methods of encapsulating water soluble drugs.

Interesting results have also been published with regard to encapsulating drugs in biodegradable polymers such as PLGAs and polycaprolactone. When felodipine was loaded into polycaprolactone, a profound influence of the dispersed phase solvent on microsphere characteristics prepared by the o/w and o/o emulsion solvent evaporation method was observed (30). The different solvents that were investigated were: methylene chloride (DCM), mixture of methylene chloride with either acetonitrile (DCM/ACN; 1 : 1 (v/v)), ethyl formate (DCM/EF; 1 : 1 (v/v)) or ethyl acetate (DCM/EA; 1 : 1(v/v)). In the case of the o/w-method, the mean size of microspheres prepared by using only DCM was larger than that prepared with a mixture of DCM and either ACN or EF. Evaporation through the aqueous phase depends on the solubility of the solvents in water. Practically, water solubility is in the range DCM< EF<ACN. Therefore, the rate of precipitation is in the same range. Since the higher the water solubility, the higher the diffusion rate before microsphere hardening. This is the major reason of smaller mean size of microspheres. Sah (2000) also showed that, although boiling point of the DCM is a little higher than that of EF, the high evaporating rate is contributed to the difference in water solubility (39). The differences in the evaporation rate of solvents may have influence on the encapsulation efficiency of the microspheres. The encapsulation efficiency of microspheres prepared only with DCM as dispersed phase was the highest of those prepared with solvent mixtures. These are also related to the water solubility of the solvents. Generally, drug partitioning into the aqueous phase can occur during the initial stages of microsphere formation prior to polymer precipitation. During evaporation of the solvent from the aqueous phase, the core materials can be migrated with solvent into aqueous phase. So, the faster the evaporation rate, the lower the encapsulation efficiency of the microspheres. However, in morphological analysis, no particular differences could be observed. In contrast to the o/w-method, the o/o-method showed

slightly lower encapsulation efficiencies and the bigger mean size. When the mixture of DCM with EA was used, the encapsulation efficiency was the highest and very narrowly distributed microspheres were obtained. This is in agreement with the findings of Li et al. (2000), who suggested that, when the volume ratio of DCM and EA was 1:1, a stable emulsion was achieved and microspheres with the highest encapsulation efficiency were attained (40). Morphological analysis showed that the microspheres prepared by mixture of DCM and EA had a few holes in surfaces, whereas other microspheres had many holes. This can also be the reason for the differences in the encapsulation efficiency.

(vi) Stirring rate

The stirring rate became an important factor to disperse the inner phase into the outer phase, since the viscosity of the outer phase was built up by the polymer amount, while the drug, surfactant and solvent amounts were kept constant (4). The speed of stirring affected mostly low viscosity ethylcellulose N10 emulsion phases, causing fractions with decreased particle size as could be seen from the sieve anaysis results of microspheres. During processing, it was observed that stirrer speeds of less than 500 rpm were not sufficient to produce microcapsules, and a huge coalesced mass was obtained. This is due in part to inadequate agitation of the media to disperse the inner phase in discreet droplets within the bulk phase. At stirring speeds above 1000 rpm, the turbulence caused frothing and adhesion of the microparticles to the container walls and propeller blade surfaces, resulting in high shear and a smaller size of the dispersed droplets. Spherical microspheres were obtained at a stirring rate of 500 rpm; therefore, this speed was used during manufacture of all microcapsules.

(vii) Effect of polymer type and molecular weight.

Effect of polymer type and the molecular weight were particularly noted with polycaprolactones (PCL) (30). Three different molecular weight PCLs were used as wall materials, i.e. 10 000 (PCLa), 65 000 (PCLb) and 80 000 (PCLc) Dalton. An increase in the molecular weight of PCL led to an increase in encapsulation efficiency. These results show that increasing the molecular weight of the PCL led to an increase in viscosity of organic phase, which reduced felodipine diffusion in the external aqueous phase before microsphere hardening. Similar results were obtained for PCL microspheres with an oily core. These can be described by other ways. Crystallinity of the polymer is very important for their barrier properties. Since the crystalline phase of the polymer is essentially impermeable to water, encapsulation is likely to occur in the amorphous region of the polymer, the higher the crystalline region, the lower the encapsulation efficiency and vice versa. The percentage of crystallinity was estimated using differential scanning calorimetry (DSC). Although three PCLs exhibited similar crystallinity, small molecular weight PCL (PCLa) exhibited a little higher crystallinity than those of higher ones (PCLb and PCLc). Therefore, the highest encapsulation efficiency was observed when PCLa was used, which indicates the crystallinity of the wall material is one of the most important factors for encapsulation efficiency. In contrast, encapsulation efficiency of PCLc was a little higher than that of PCLb, although the _H value of the PCLc was a little higher than that of PCLb, exhibiting the crystallinity of the wall material is not a dominant factor. Youan et al. (1999) reported similar results (41). In order to increase encapsulation efficiency, the balance between the molecular weight and crystallinity should be properly adjusted. In addition, the small size PCL (PCLa) microspheres have a sharp peak, indicating more homogeneous crystalline structure, in contrast, large size PCL (PCLb and PCLc) microspheres have a little broad peak, indicating more heterogeneous crystalline structure. Morphological analysis also revealed that the higher the molecular weight, the smoother and the less porous the surface of the microspheres. This confirms that the drug loss can be reduced with increase in the molecular weight of PCL.

(vii) Other factors

Several other factors are reported to be influencing the quality of microspheres formed. These include the temperature at which the microspheres is prepared, the rate of evaporation of the solvent, presence of sodium chloride in the external phase, etc.

6.4.2 Spray Drying

Spray drying is the continuous transformation of feed from a fluid state into dried particulate form by spraying the feed into a hot drying medium. The feed may be solution, slurry, emulsion, gel or paste, provided it is pumpable and capable of being atomized. It involves bringing together a highly dispersed liquid and a sufficient volume of hot air to produce evaporation and drying of liquid droplets. The air supplies the heat for evaporation and conveys the dried product to the collector; the air is then exhausted with the moisture. Three types of atomizers are commercially used namely rotary atomizer, pressure nozzle and two-fluid nozzle. The feed droplets while losing its moisture to hot air remain at temperatures much below the hot air temperature for a very short time. Hence spray drying is essentially known as "Low Temperature Drying". The dried product can be in the form of powders, granules, or agglomerates depending upon the physical and chemical properties of the feed, the dryer design and final powder properties desired. Schematic representation of the spray drying process is shown in Figure 6.4. Picture of a Laboratory scale spray drier is shown in Figure 6.5 while its industrial counterpart is shown in Figure 6.6. The detail of spray drying technique to prepare microspheres has been described previously (42,43). Recently, the process received great attention in the field of micro particles for the preparation of dried liposomes, amorphous drugs, mucoadhesive microspheres, drying of preformed microcapsules, Gastroresistant microspheres, and controlled-release systems. Comprehensive studies have been performed on the preparation of microspheres by spray drying techniques for different purposes, like modification of biopharmaceutical properties, formulation of dry emulsions, spray dried phospholipids, nanoparticle-loaded microspheres, for drug delivery, spray-dried powders formulated with hydrophilic polymers, biodegradable microspheres, and spray-dried silica gel microspheres. Eudragit RL microspheres containing vitamin C were prepared by Spray drying method. Spray-drying was useful for the preparation of Paracetamol encapsulating Eudragit RS/RL or Ethylcellulose microspheres. The spray drying technique has been widely applied to prepare micro-particles of drug with polymer. When a drug crystal suspension of a polymer solution is spray-dried, microcapsulated particles are prepared, whereas spray drying of solution of polymer containing dissolved drug leads to formation of drug-containing microspheres in which the drug can be dispersed in a molecular state or as micro crystals. In both cases, the particles tend to have a spherical shape and are free flowing.

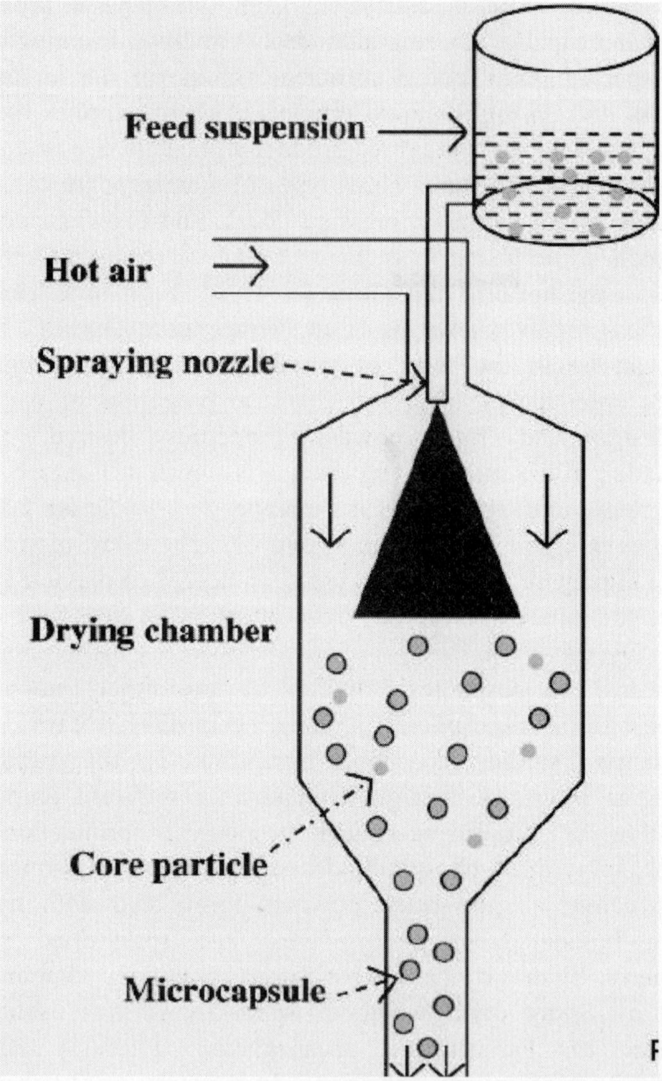

Figure 6.4 Schematic representation of spray drying technique.

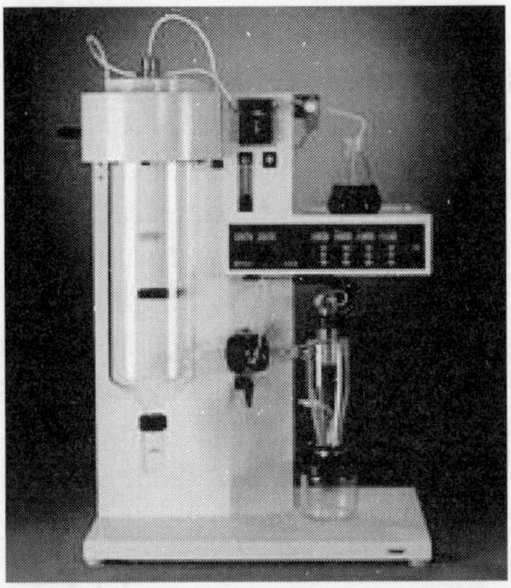

Figure 6.5 Minispray drier B-290

Figure 6.6 Aseptic production of microspheres using spray drying technique

Gaekwad et al., 2009 investigated the preparation of spray dried microspheres containing an antisense oligonucleotide (22). The aim of this study was to formulate and characterize microspheres containing antisense oligonucleotide to NF-kappaB using bovine serum albumin as the polymer matrix. Microspheres were prepared by spray-drying technique with 5, 10 and 15% drug loading. Glutaraldehyde was used as a cross-linking agent. The particle sizes ranged from 3-5 microns. Microspheres were smooth and spherical in shape, as determined by scanning electron microscopy (SEM). The yield of microspheres ranged from 70-75% and the encapsulation efficiencies were found to be in the range of 59-60%, as determined by a novel HPLC method. Zeta potential of the microspheres ranged between -39 to -53 mV, thus indicating good suspension stability in water. In-vitro release studies performed using phosphate buffer saline demonstrated extended drug release up to 72 h. Kinetic model fitting showed high correlation with the Higuchi model, suggesting that the drug release was primarily diffusion controlled. Biodegradable polymers belonging to polylactate-co-glycolide were also prepared and evaluated for microsphere formulation using spray drying technique. Wang and Wang investigated a spray drying technique for the development of etanidizole microspheres (44). A spraying technique was used to encapsulate etanidazole (a hypoxic radiosensitizer) into different poly(lactide/glycolide) polymers. The properties of the obtained microspheres, especially the particle size and distribution, morphology and release rate were investigated. Unexpectedly, poly(L-lactide) (PLLA) shows a fast release rate, comparable to PLGA 50: 50, due to the dissociation of the microspheres although the release rate of the spray-dried microspheres of other polymers decreases with increasing lactide ratio. It is also interesting to note that, contrary to the viscosity sequence of the polymer solutions, the particle size of the microspheres decreases in the order PLGA 50: 50, PLGA 65: 35, PLGA 85: 15 and PDLA. The morphology of microspheres can be affected by polymer properties (e.g. lactide/glycolide ratio, molecular weight, crystallinity and Tg) and fabrication conditions (e.g. solvent and polymer concentration to be sprayed). Although most of the microspheres fabricated by EA have a doughnut-like shape with smooth surface, it is possible to obtain spherical particles by choosing proper polymer type and polymer concentration. A further examination of the mechanisms of the atomization process and the solvent evaporation process reveals their respective effect on droplet formation and particle formation, both of which are essential for the spray-drying technique. It is found that polymer phase transition (affected by the polymer solubility) and its subsequent solvent evaporation processes can finally determine the morphology and the particle size of the spray-dried particles made from different polymers. In essence, the lactide/glycolide ratio of the polymers plays a more important role in affecting the properties of the spray-dried microspheres.

Although the process of spray drying is simple, it requires sophisticated equipment. There are three fundamental steps involved in spray drying. 1. Atomization of a liquid feed into fine droplets. 2. Mixing of these spray droplets with a heated gas stream, allowing the liquid to evaporate and leave dried solids. 3. Dried powder is separated from the gas stream and collected.

Spray drying involves the atomization of a liquid feedstock into a spray of droplets and contacting the droplets with hot air in a drying chamber. The sprays are produces by either rotary (wheel) or nozzle atomizers. Evaporation of moisture from the droplets and formation of dry particles proceed under controlled temperature and airflow conditions. Powder is discharged continuously from the drying chamber. Operating conditions and dryer design are selected according to the drying characteristics of the product and powder specification.

The atomizing device, which forms the spray, is the 'heart' of the spray drying process. Equipment that breaks bulk liquid into small droplets, forming a spray is called an atomizer. Prime functions of atomization are: a. A high surface to mass ratio resulting in high evaporation rates, b. Production of particles of the desired shape, size and density. The aim of atomizing the concentrate is to provide a very large surface, from which the evaporation can take place. The smaller droplets, the bigger surface, the easier evaporation, and a better thermal efficiency of the dryer are obtained. The ideal from a drying point of view would be a spray of drops of same size, which would mean that the drying time for all particles would be the same for obtaining equal moisture content. Over the years several researchers have studied the mechanism by which atomization takes place and several theories have evolved. The most widely accepted are based on the liquid jet theory described in 1878 by Lord Rayleigh. A liquid stream accelerated by the force of gravity is pulled apart or disintegrated into teardrop-shaped droplets. The surface tension of the liquid causes the droplet, suspended in air, to form itself into a sphere. In order to produce top-quality products in the most economical manner, it is crucial to select the right atomizer. Three basic types of atomizers are used commercially: a. Rotary atomizer (atomization by centrifugal energy); b. Pressure nozzle (atomization by pressure energy); c. Two-fluid nozzle (atomization by kinetic energy). Ultrasonic energy and vibrations have also been studied, but as yet have found few commercial applications. The selection of a specific atomizer is made based on the properties of the feed, the desired powder properties, the dryer type and its capacity and the atomizer capacity.

Once the liquid is atomized it must be brought into intimate contact with the heated gas for evaporation to take place equally from the surface of all droplets within the drying chamber. The heated gas is introduced into the chamber by an air disperser, which ensures that the gas flows equally to all parts of the chamber. The air disperser uses perforated plates or vaned channels through which the gas is directed, creating a pressure drop and, thereby, equalizing the flow in all directions. It is critical that the gas entering the air disperser is well mixed and has no temperature gradient across the duct leading into it. As a result, it is important that any type of heater used inherently produces a well-mixed gas stream, or that a mixing section is placed between the heater and the air disperser. The air disperser is normally built into the roof of the drying chamber and the atomization device is placed in or adjacent to the air disperser. This arrangement allows instant and complete mixing of the heated drying gas with atomized cloud of droplets. To fully understand the characteristics of spray-dried powders, one needs to examine the mechanism for drying within a single droplet. Typically, there are many very small particles suspended in a sphere of liquid. When the droplet is first exposed to hot gas, rapid evaporation takes place. Material dissolved in the liquid will tend to form a thin shell at the surface of the sphere. Although the evaporation has kept the particle itself quite cool, as the liquid concentration decreases, the particle will begin to heat. Evaporation then takes only as quickly as the liquid can diffuse to the surface of the sphere. This phase of the drying process is called first-order drying or is said to be diffusion-rate-limited.

Fortunately, this phase occurs in the cooler part of the dryer where the drying gas is at or near the outlet temperature of the dryer. As a result the solids in each particle are never heated above the outlet temperature of the dryer, even though the dryer inlet may be considerably higher. The final dried powder will be at a temperature approximately 20°C lower than the air outlet temperature. The thermal energy of the hot air is used for evaporation and the cooled air pneumatically conveys the dried particles in the system. The contact time of the hot air and the spray droplets is only a few seconds, during which drying is achieved and the air temperature drops instantaneously. The dried particle never reaches the drying air temperature. This enables efficient drying of heat sensitive materials without thermal decomposition. Turbulence within the dryer, which is necessary for good drying, does cause some particles to be exposed to elevated temperature. This sometimes causes a loss in activity or modification of additives such as binders. Therefore, test work is recommended on each formulation, and the best combination of inlet and outlet temperatures needs to be established relatively to activity and performance of the powder in further processing.

The largest and most obvious part of a spray-drying system is the drying chamber. This vessel can be taller and slander or have large diameter with a short cylinder height. Selecting these dimensions is based on two process criteria that must be met. First, the vessel must be of adequate volume to provide enough contact time between the atomized cloud and the heated glass. The second criterion is that all droplets must be sufficiently dried before they contact a surface. This is where the vessel shape comes into play. Centrifugal atomizer requires larger diameter and less cylinder height. Nozzles are just the opposite. Most spray dryer manufacturers can estimate, a given powder's mean particle size, what dimensions are needed to prevent wet deposits on the drying chamber walls. Drying chambers are usually constructed of stainless steel sheet metal, with stiffeners for structural support and vessel integrity. Sheet steel finish and weld polish can be specified to meet any requirement. Insulation is usually applied to the outside of the vessel, and stainless steel wrapping is seam-welded over the entire vessel. This provides a thermally efficient and safe system that is easy to clean has no crevice areas that might become contaminated. In almost every case, spray-drying chambers have cone bottoms to facilitate the collection of the dried powder. When the coarse powder is to be collected, they are usually discharged directly from the bottom of the cone through a suitable airlock, such as a rotary valve. The gas stream, now cool and containing all the evaporate moisture, is drawn from the center of the cone above the cone bottom and discharge through a side outlet. In effect, the chamber bottom is acting as a cyclone separator. Because of the relatively low efficiently of collection, some fines are always carried with the gas stream. This must be separated in high-efficiency cyclones, followed by a wet scrubber or in a fabric filter (bag collector). Fines are collected in the dry state (bag collector) are often added to the larger powder stream or recycled.

Process control variables include feed material properties such as viscosity, uniformity, and concentration of core and coating material, feed rate, method of atomization, and the drying rate, which is normally controlled by the inlet and outlet temperatures and the air stream solvent concentration. The process produces microspheres approaching a spherical structure in the size range of 5 to 600 μm. Characteristically, spray drying yields products of low bulk density, owing to the porous nature of the coated particles. Low active contents are normally required to provide the necessary protection desired. For instance, the adequate retention of volatile, liquid core materials is difficult to achieve without maintaining low active content levels, perhaps below 20%.

6.4.3 Air Suspension

Microencapsulation by air suspension techniques is generally ascribed to the inventions of Professor Dale E. Wurster during his tenure at the University of Wisconsin. The process is described in detail elsewhere (43). Basically, the Wurster process consists of the dispersing of solid, particulate core materials in a supporting air stream and the spray-coating of the air suspended particles. Within the coating chamber, particles are suspended on a upward moving air stream as each pass through the coating zone, the core material receives an increment of coating material. The cyclic process is repeated, perhaps several hundred times during processing, depending on the purpose of microencapsulation, the coating thickness desired, or whether the core material particles are thoroughly encapsulated. The supporting air stream also serves to dry the product while it is being encapsulated. Drying rates are directly related to the volume temperature of the supporting air stream.

Processing variables that receive consideration for efficient, effective encapsulation by air suspension techniques include the following:

1. Density, surface area, melting point, solubility, friability, volatility, crystallinity and flowability of the core material.
2. Coating material concentration (or melting point if not a solution).
3. Coating material application rate.
4. Volume of air required to support and fluidize the core material.
5. Amount of coating material required.
6. Inlet and outlet operating temperatures.

The air suspension process offers a wide variety of coating material candidates for microencapsulation. The process has the capacity of applying coatings in the form of solvent solutions, aqueous solutions, emulsions, dispersions, or hot melts in equipment ranging in capacities from one pound to 990 pounds. The coating material selection appears to be limited only in that the coating must form a cohesive bond with the core material. The process generally is considered to be applicable only to the encapsulation of solid core materials. Indirectly, however, liquids can be encapsulated by the process at relatively low active levels by coating solid sorbents that have been pretreated with liquid sorbents. In regard to particle size, the air suspension technique is applicable to both microencapsulation and macroencapsulation coating process. The practical particle size range for microencapsulation, however, is considered to be in excess of 74 microns. Under ideal conditions, particles as small as 37 microns can be effectively encapsulated as single entities. Core materials comprised of micron or submicron particles can be effectively encapsulated by air suspension techniques, but agglomeration of the particles to some larger size is normally achieved.

6.4.4 Coacervation-Phase Separation

In water, organic chemicals do not necessarily remain uniformly dispersed but may separate out into layers or droplets. If the droplets which form contain a colloid, rich in organic compounds and are surrounded by a tight skin of water molecules, then they are known as coacervates. These structures were first investigated by the Dutch chemist H.G. Bundenberg de Jong in 1932. The process is described in detail previously (43). A wide variety of solutions can give rise to them; for example, coacervates form spontaneously when a protein, such as gelatin, reacts with gum arabic. They are interesting not only in that they provide a locally segregated environment but also in that their boundaries allow the selective absorption of simple organic molecules from the surrounding medium. In other words coacervation refers to the phase separation of a liquid precipitate, or phase, when solutions of two hydrophilic colloids are mixed under suitable conditions. The general outline of the processes consists of three steps carried under continuous agitation:

Step 1: Formation of three immiscible chemical phases

The immiscible chemical phases are (i) a liquid manufacturing vehicle phase (ii) a core material phase and (iii) a coating material phase. To form the three phases, the core material is dispersed in a solution of the coating polymer, the solvent for the polymer being the liquid manufacturing vehicle phase. The coating material phase, an immiscible polymer in a liquid state, is formed by utilizing one of the methods of phase separation coacervation, that is,

- By changing the temperature of the polymer solution
- By adding a salt
- By adding a non-solvent
- By adding incompatible polymer to the polymer solution
- By inducing a polymer-polymer interaction.

Step 2: Depositing the liquid polymer coating upon the core material

This is accomplished by controlled, physical mixing of the coating material (while liquid) and the core material in the manufacturing vehicle. Deposition of the liquid polymer coating around the core material occurs if the polymer is adsorbed at the interface formed between the core material and the liquid vehicle phase, and this adsorption phenomenon is a prerequisite to effective coating. The continued deposition of the coating material is promoted by a reduction in the total free interfacial energy of the system, brought about by the decrease of the coating material surface area during coalescence of the liquid polymer droplets.

Step 3: Rigidizing the coating

This is usually done by thermal, cross linking or desolvation techniques, to form a self sustaining microcapsule.

Several authors used several polymers in this process of preparing microspheres.

Nihant et al., (2003) investigated this technique to prepare polylactide-co-glycolide microcapsules (45). Protein microencapsulation by coacervation of poly(lactide-co-glycolide) solutions in CH_2Cl_2 induced by the addition of silicone oils of various viscosities was employed. This coating technique proceeds along three steps: phase separation of the coating polyester, adsorption of the coacervate droplets around the protein phase, and hardening of microparticles. Size distribution, surface morphology and internal porosity of the final microspheres clearly depend on the main characteristics of the coacervate, particularly the viscosity, in a direct connection with the CH_2Cl_2 content. Indeed, the whole porosity (which may be as high as 80%), average pore size and broadness of pore size distribution decrease as the coacervate is more viscous. Hardening of the coacervate droplets is thus so fast that the organic solvent is entrapped within the polymer matrix and predetermines the internal porosity. Finally, size distribution of microspheres is bimodal in a clear relation with the coacervate viscosity. A less viscous coacervate favours smaller microspheres (within the 7-90 μm range), contaminated with a minor population of microparticles below 4 μm.

Gokale and Jonnalagadda (2008) recently prepared microspheres encapsulating a protein using a coecervation-phase separation technique (46). Microspheres were prepared using coacervation/phase separation technique using dicholoromethane (DCM) as solvent and hexane as nonsolvent. Infliximab (~ 10 mg) was suspended in PLGA 50/50 solution

(5% w/v) in DCM with continuous stirring at ~ 300 rpm. Hexane was added to precipitate the polymer, followed by 30 minutes hardening and air-drying for 24 hours. Drug to polymer ratio was maintained at 1:5 (A), 1:10 (B) or 1:20 (C). Particle size distribution was measured by laser light scattering at 660 nm. A controlled release formulation was obtained with PLGA 50/50 with drug to polymer ratio as low as 1:5, while maintaining physical and chemical stability of Infliximab. Honary et al., recently prepared microspheres of chitosan encapsulating prednisolone using a coecervation-phase separation technique (47). Alginate-chitosan microparticles were prepared by complex coacervation using sodium alginate as a gel core. All alginate solutions (2%, w/v) were prepared by dissolving sodium alginate in de-ionized water. Micronized prednisolone powder (2–4%w/v) was then suspended thoroughly in the alginate solutions by vigorous stirring for 10 min. Two procedures were used. In the one step method, the alginate-prednisolone mixture was directly sprayed into the calcium chloride solution (0.5–1.0%, w/v) containing chitosan (0.5–1.5% w/v). In the two-step method, the alginate-prednisolone mixture was sprayed into calcium chloride solution (0.5–1.0%w/v), followed by a membrane forming step where the prednisolone containing calcium alginate particles were suspended in chitosan solution (0.5–1.5%w/v). The microparticles were allowed to harden for at 2 h before washing them twice with distilled water and dried at room temperature and the alginate-prednisolone mixture was sprayed into the gelling solutions using two types of spray equipment: Casals 17500 Ripoll (Spain) and Inter Eko 1.5 lit, Pmax, 3 bar and rate, 0.61/min (Czech Republic) were used. The latter was used in the one-step method. In the methodology, a ternary phase diagram for sodium alginate, chitosan and calcium chloride illustrating phase separation/coacervation by the addition of an incompatible polymer was generated. The optimum concentration range required for each compound to form microparticles was identified in a triangular diagram. This region was chosen by comparison of microparticles in terms of particle size, shape, reproducibility, symmetry and absence of free prednisolone crystals. The diagram was then used to select the various ratios of the materials in the formulation. The effects of stirring rate and time, spray apparatus type, distance of spray gun from chitosan solution surface, and the preparation method (one- or two-step) were evaluated. This work has demonstrated that when prednisolone alginate/chitosan microparticles were prepared with chitosans of varying MW using a coacervation method, smaller and more uniform microparticles were produced by high MW chitosan.

6.5 Characterization of Microspheres

Routinely, a group of parameters are determined to characterize a microsphere sample, including: morphology, particle size, encapsulation efficiency, loading, release profile, residual solvent, among others. In case of microspheres intended for oral administration several other parameters such as carr's index, micromeritic properties, etc. are routinely investigated. This is to mainly ensure the tabletability of the microspheres. In case of protein drugs and other biotechnology based drugs, characterization of the active at the end of the fabrication of the microspheres may become mandatory. This is due to their lability, thereby experimenting structural changes that affect its physicochemical and biological properties, and ultimately its function as a therapeutic agent. These methodologies will be briefly described:

Morphology

Morphological studies of microparticles reveal their relevant properties, such as shape, surface regularity, membrane continuity (for microcapsules), pore size and the uniformity of their distribution, particle size homogeneity, defects and/or aggregation, and give data on size, although it is not the most appropriate method to determine this parameter. These properties affect microparticle applications. For example, aggregation is undesirable, affecting the homogeneity of the product and blocking the syringe needles when injecting the particles. At the same time, the amount and size of pores could influence the release of the encapsulated protein or even modify the release mechanism. The morphology of particles is studied by microscopy techniques, commonly by transmission electron microscopy, and also by atomic force microscopy to characterize the surface of nanospheres; confocal microscopy, to obtain evidence on the acidity of microsphere cores derived from PLGA degradation or to study the distribution of the protein inside the particle; and fluorescence microscopy is used, to unravel the inner structure of the particle. The morphological properties of microparticles can be determined by the microencapsulation technique, and particularly, by experimental conditions; for example, the type of solvent and evaporation rate. By adding salts to one or both aqueous phases in a double emulsion, particles with differential inner structures can be obtained.

Particle size

Particle size is a relevant parameter for microspheres; it should not be longer than 180 μm when it is to be administered by the parenteral route.

Besides, their size must be reproducible batch-to-batch, influencing the release profiles and encapsulation efficiency, among other properties. Several experimental conditions affect particle size, irrespective of the method used, such as polymer type and molecular weight, polymer/drug ratio, polymer concentration in the organic phase, polyvinyl alcohol concentration in the outer aqueous phase and stirring rate. Therefore, microsphere size and its distribution must be carefully determined. There are several methods available, including: centrifugation, analytical ultracentrifugation, sedimentation, electrical conductivity, optical and electron microscopy, light scattering and laser diffraction, among others. However, differences in their respective measuring principles and their requirements to build models from the experimental data generate inconsistent results between methods for a given sample. Only electron microscopy covers the whole range of sizes obtained. Burgess and coworkers offered an extensive analysis on this topic, recommending the appropriate selection of the method according to the production process, particle size required for clinical use and particle segregation during formulation manufacturing and storage.

Loading and encapsulation efficiency

The efficiency of encapsulation is the fraction, expressed in percentage, of the drug encapsulated in respect to the total amount of drug used in the process. This is an essential parameter, indicating the quality of the process, which is better or more efficient when a larger fraction of the drug is encapsulated. In the same way, the load is the amount of drug encapsulated per microsphere mass, expressed in percentage. This parameter comprises a wide range of values, according to the drug dose required for administration, and that needs to be precisely determined, because this defines the amount of microspheres to administer in a single formulation dose. To determine these parameters, the drug must be extracted by several methods to an aqueous phase, for proper quantification. The aqueous two-phase extraction method, which uses two immiscible liquid phases, the organic solvent precipitation method followed by filtration, and the accelerated hydrolysis of the polymer by incubating microspheres in NaOH. Of these three procedures, the latter is the most widely applied, because it generates aqueous phases that are neutralized and further analyzed using analytical methods such as HPLC. All these parameters are profoundly affected by the encapsulation method and the experimental conditions where the particles are generated, including the volume of different phases and the concentration of the drug on it, polymer type and concentration in the organic phase, emulsification time, and additives in the different emulsion phases, which are among the relevant factors.

Residual humidity

The presence of water inside the particles can cause undesirable events, such as changes in the polymeric matrix by the hydrolysis of ester bonds in the polymer, or changes in the drug favored by the damp media. For this reason, it is important to determine residual humidity in microspheres. The method of choice is Karl-Fisher's method, commonly used to determine humidity in lyophilized products. Residual humidity derives mainly from the method used to generate them and also the drying procedure. Lyophilization is the most common laboratory-scale procedure to dry protein-loaded microspheres, because of being amenable to preserve the structure and properties of the encapsulated drug. Additionally, the vacuum-drying process in specific devices has been used for the industrial production of microsphere batches. Microspheres with low humidity have been obtained by both procedures.

Residual solvent

Solvents remaining in the pharmaceutical products are defined as volatile, organic chemical products used or produced while manufacturing excipients and drugs. Due to their toxicity, these substances have to be avoided in any medicine, but unfortunately, most of the time they cannot be completely removed during the technological manufacturing processes. Therefore, they must be quantified as part of the quality control of formulations, their content having to be below the limit established by regulatory agencies. Gas Chromatography is the most common technique used to determine these substances in microspheres, although alternative procedures are being introduced for such purposes. The encapsulation method could notably influence the amount of solvent remaining in microspheres, and also the type of solvent. Another important factor is the drying process, including the conditions where it is carried out.

Release profile and Bioavailability Studies

The release profile is a highly relevant parameter when designing microspheres for the therapeutic administration of drugs. It is studied *in vitro* and could be related or not to the *in vivo* properties of the drug release process. Nevertheless, the *in vitro* study offers an idea of the potential of the system, indicating the affordability for the controlled. On the other hand, the knowledge of bioavailability of the drug from the microsphere formulations used in all the routes of administration is essential. In this regard, the recent trend in the application of in vitro release is to also have a in vitro in vivo correlation. In these situations, the in vitro release in conjunction with in vivo pharmacokinetics is helpful. The comprehension of in vitro in vivo correlation (IVIVC) of the delivery is essential to better tailor the delivery system for the future needs A predictive IVIVC can empower in vitro dissolution as a surrogate for *in vivo* bioavailability/biotheraequilance. IVIVCs can decrease regulatory burden by decreasing the number of biostudies required in support of a drug product. Additionally, IVIVC is also helpful in the product development including the development of depot microspheres. The development of an IVIVC is a dynamic process starting from the very early stages of development program through the final step. Different types of IVIVCs are used in the regulatory terminology. These include assumed IVIVC, retrospective IVIVC, and prospective IVIVC. An assumed IVIVC is essentially one that provides the initial guidance and direction for the early formulation development activity. Thus, during stage 1 and with a particular product concept in mind, appropriate *in vitro* targets are established to meet the desired *in vivo* profile specification. This assumed model can be the subject of revision as prototype formulations are developed and characterized *in vivo*, with the results often leading to a further cycle of prototype formulation and in vivo characterization. Out of this cycle and *in vivo* characterization and, of course, extensive *in vitro* testing is often developed what can be referred to as retrospective IVIVC. With a defined formulation that meets the *in vivo* specification, Stage 2 commences. At this stage based on a greater understanding and appreciation of defined formulation and its characteristics, a prospective IVIVC is established through a well defined prospective IVIVC study. Once the IVIVC is established and defined it can be then used to guide the final cycle of formulation and process optimization leading into Stage 3 activities of scale-up, pivotal batch manufacture, and process validation culminating in registration, approval and subsequent post-approval scale-up and other changes. Thus, in vitro release testing has to be carefully performed.

As in vitro dissolution is one of the main tests performed to characterize the microsphere formulations, several official method have been developed to understand in vitro drug release. The USP recommends seven different types of apparatus for in vitro release testing.

USP Apparatus 1 (basket) and 2 (paddle) were designed for the evaluation of immediate-release (IR) and modified-release (MR) oral formulations, whereas USP Apparatus 5 (paddle over disc), 6 (cylinder), and 7 (reciprocating holder) were designed for the assessment of transdermal products. USP Apparatus 3 (reciprocating cylinder) and 4 (flow-through cell) were designed for the evaluation of extended-release (ER) oral formulations. It has been recently come to an understanding that the last two methods useful for ER oral formulations are also suitable for CR parenterals, however, with some modifications. Some researchers have noted evaporation problems with apparatus 3. Alternative apparatus, such as small sample vials and vessels, with and without agitation, are currently used for CR parenterals. However, problems are associated with these alternative apparatus, including lack of sink conditions and sample aggregation. USP apparatus 4 was found to be most suitable of the currently available USP apparatus for controlled and sustained release parenterals such as microspheres. This apparatus allows flexibility in volume, sample cell, flow rate and can be modified for specific product applications (such as avoidance of aggregation problems and of potential violation of sink conditions).

It is also aimed to determine the mechanisms of drug release using the release testing methodology. The release of the drug from microspheres is either first-order drug release or zero-order drug release. This is first investigated followed by the equations for mechanisms of drug release. The curvilinear nature of the cumulative percentage drug released versus time plots suggest that drug release from the microcapsules is not following zero-order kinetics. This is confirmed by low correlation coefficients obtained in all cases when these data are fitted to a zero-order model. In the usual protocol, various kinetic equations, like zero order (percentage release Vs time), first order (log percentage of cumulative drug remaining Vs time), Higuchi's model (percentage drug release Vs square root of time) and Korsemeyer peppas model(log cumulative amount Vs log time) will be plotted and evaluated. Correlation coefficient (r^2) values will be calculated for the linear curves obtained by regression analysis of the above plots. Further, to confirm the mechanism of drug release, the first 60% of drug release was fitted in Korsemeyer-Peppas model with the following equation: M_t/M_∞. = Kt, where M_t/M_∞ is the fraction of the drug release at time t, the n value is used to characterize different release mechanisms and is calculated for the slope of the plot of the log of fraction of drug released vs log of time (t). The 'n' value of all the formulations was between 0.5 and 1 indicating that the mechanism of drug release was non-Fickian type diffusion. Dissolution data is also fit to the Kopcha matrix model or Makoid-Banakar model. There are several other models used to characterize the drug release. The popular ones are only mentioned here. The diffusion release can be further established using Kopcha model. As the ratio of exponents A/B derived from the Kopcha model is greater than 1, the conclusions of Korsemeyer-Peppas is validated. The Kopcha model can easily be used to help quantify the contribution of diffusion and polymer relaxation. The value of A if is far greater than the value of B. It suggests that drug release occurs mainly as a result of Fickian diffusion. When the parameter c of the Makoid–Banakar model is equal to zero, this model becomes the Korsmeyer–Peppas power law ($e\text{-}0t = 1$).

Drug release is very interesting with regard to protein drugs. The proteins encapsulated in PLGA microspheres are commonly released following a pattern of three main steps. First, the burst release phase, usually occurring during the first day and mainly determined by the protein in the surface, channels and pores of the microspheres, which were filled by the incubation media for a few hours at the beginning of the trial. Secondly, the slow release phase, releasing few or no protein at all. The third and last phase comprises a faster release of protein due to the erosion of particles. Occasionally, the release can occur in two steps and the profile shows an asymptotic pattern. Several processes contribute to the release of the encapsulated proteins, such as the diffusion through pores and channels, and exposure of protein molecules to the incubation media, due to the superficial erosion of particles, also derived from the degradation of the polymeric matrix. The channels and pores are formed during the assembly of the particles or result from polymeric degradation. Therefore, factors influencing the release profile include the properties of the polymeric matrix and the protein used, the structure of the microparticle, the encapsulation technique and the experimental conditions, as well as the co-encapsulation of additives for several purposes. Factors other than those of microparticle properties that determine the release profile are related to the assay conditions, like composition and volume of the incubation media, temperature, the profiling device, the procedure and stirring rate, and the method used to change the incubation media (partially or completely). When conducting release profile studies in protein loaded microspheres, highly variable results are usually obtained due to biomolecule degradation by the acidic media, which arises from polymeric matrix degradation products and their exposure to the aqueous media. This inconvenience can be solved by frequently changing the incubation media or measuring the amount of proteins retained by microspheres, instead of measuring the concentration of proteins in the incubation solution.

Sterility

Sterility of microspheres is important on several occasions. Most of the times, for other than oral route sterility of the formulations is essential. Thus sterility is investigated at the end of the fabrication of the microspheres. The formulations prepared with these systems cannot be sterilized by steam or irradiated, because the properties of the polymeric matrix and the encapsulated molecule can be undesirably modified. Besides, the size of the particles hampers the use of sterilizing filtration for the final product. These limitations make it necessary to manufacture these pharmaceutical products under aseptic conditions, also requiring internal and external sterility verifications.

Drug polymorphism and Drug-Polymer Interaction

The physical states of the active ingredients are characterized using thermal methods such as DSC and TGA. The changes in the peaks in the formulation compared to the original drug indicate changes in the polymorphism. For instance, in one study, the broadened and shifted peaks suggested the presence of the crystalline form of the active agents and these were not observed for the drug-loaded chitosan microspheres, as an indication of the molecular dispersion of the drug in the matrix. Drug polymer interaction is generally studied using HPLC, DSC or FTIR.

Characterization of the encapsulated protein

In case of microspheres incorporated with protein drugs, characterization of the encapsulated protein becomes important. The encapsulated protein should be characterized according to the nature of each particular molecule, and should reflect its functioning properties. Usually, chromatography, polyacrylamide gel electrophoresis and immunoenzymatic and biological activity assays are carried out for these purposes. PLGA insolubility in water demands the design of a procedure to extract the encapsulated protein and to obtain aqueous samples that are appropriate to run the previously mentioned techniques, also preserving protein properties. It has been previously mentioned that there are several methods based on extracting the protein by using systems composed by two immiscible liquid phases, extraction by precipitation with organic solvents where the polymer is soluble and electrophoretic extraction. Specifically, the extraction in the two-phase systems has been used to evaluate the properties of encapsulated peptides, with good results, while other authors have found that protein recovery can be negatively affected because the protein tends to be distributed between the interphase and the aqueous phase. This can also cause changes in the protein extracted and alter the results. Nevertheless, this extraction variant could be used, while demonstrating its applicability, to the system under study. A similar case comprises the precipitation methods, where proteins can experiment interactions with polymeric materials under the extraction conditions, leading to nonquantitative yields. In general, all these methods have specific advantages and disadvantages, and are selected according to the protein of interest. Another group of techniques, potentially useful to study the properties of the encapsulated protein without extracting it is available; however, only the Fourier transform infrared spectroscopy method has yielded successful results.

Micromeretic properties

The micromeritic properties of microspheres such as particle size distribution, bulk and tapped density, surface topography, tangent of angle of repose, compressibility index, Hausner ratio and flow rates are evaluated at the end of fabrication. These are mainly determined for the microspheres intended for oral delivery. The flow properties of the microspheres are expressed in terms of Carr's Index. The Carr's index for all formulations should be less than 16, which indicate excellent flow properties and suggests that the microspheres can be easily handled during processing. Compressibility index and Hausner ratio are indirect measures of bulk density, size and shape, surface area, moisture content and cohesiveness of microspheres. All microspheres should show good flow characteristics according to USP 28-NF 23 criteria, with Hausner ratio less than 1.18 and percentage of compressibility less than 15. Characterizing the flow property, angle of repose values of all microspheres should not exceed 30-35° and thus formed microspheres are accepted as free-flowing. Flow rates should be faster to confirm its powder properties.

6.6 Conclusions

The range of techniques available to fabricate microspheres offers a wide variety of opportunities for drug delivery. The development of newer polymers is enhancing the sphere of microsphere formulation. The enhanced need for a variety of sustained release formulations for controlled and targeted release along with the awareness of the applications of this dosage form among drug delivery scientists is taking microsphere formulations in forefront of drug delivery research. The result can be noticed with the number of such formulations administered via the pulmonary route, nasal route and other invasive routes. Also, in future research into very important practical applications of microspheres along with their commercialization will be actively initiated.

References

1. Abu-Izza K, Tambrallo L and Lu Dr. In vivo evaluation of zidovudine (AZT)-loaded ethylcellulose microspheres after oral administration in Beagle dogs, J Pharm Sci, 85(5):554-559 (1997).
2. Eyles JE, Alpar HO, Conway BR and Keswick M. Oral delivery and fate of poly (lactic acid) microsphere-encapsulated interferon in rats. J Pharm Pharmacol, 49(7):669-674 (1997).
3. Volkheimer G. Persorption of microparticles. Pathologie, 14(5):247-252 (1993).
4. Yuce M and Canefe K. Indomethacin-loaded microspheres: preparation, characterization and in-vitro evaluation regarding ethylcellulose matrix material. Turk J Pharm Sci, 5(3):129-142 (2008).
5. Tao Y, Lu Y, Sun Y, Gu B, and Pan J. Development of mucoadhesive microspheres of acyclovir with enhanced bioavailability. Inter J Pharmaceutics, 378(1-2):30-36 (2009).
6. Steber W, Fishbein R and Cady S. Compositions for parenteral administration and their use. US patent 4837381. American Cynamid company (1989).
7. Chao P and Sinko PJ. Passive pulmonary targeting and tissue distribution of microspheres. AAPS Abstract, AAPS Journal, 002665 (2002).
8. Gunger S, Okyar A, Toker S, Baktir G and Ozsoy Y. Ondansetron-loaded biodegradable microspheres as a nasal sustained delivery system: In vitro/in vivo studies. Pharmaceutical Development and Technology, doi. 1080/10837450903148257 (2009).
9. Rawat A, Majumder QH and Ahsan F. Inhalable large porous microspheres of low molecular weight heparin: in vitro and in vivo evaluation. J Controlled Release, 128(3):224-232 (2008).
10. Azouz SM, Walpole BS, Amirifeli S, Taylor KN, Grinstaff MW, Colson YL. Prevention of local tumor growth with paclitaxel-loaded microspheres. J Thorac Cardiovasc Surg, 135:1014-1021 (2008).

11. Lagarce F, Renaud P, Faisant N, Nicolas G, Cailleux R, Richard J, Menei P, Benoit JP. Baclofen-loaded microspheres: preparation and efficacy testing in a new rabbit model. Eur J Pharm Biopharm, 59(3):449-459 (2005).

12. Perugini P, Genta I, Conti B, Modena T and Pavenetto F. Long term release of Clodronate from biodegradable microspheres. AAPS PharmSciTech, 2(3):article 10 (2001).

13. Yang G, Kuo J and Woo E. Preparation and control-release kinetics of isosorbide dinitrate microspheres. J Microencapsulation, 23(6):622-631 (2006).

14. Nishimura K, Nishimura S, Seo H, Nishi N, Tokura S and Azuma I. Microphage activation with multi-porous beads prepared from partially deacetylated chitin. J Biomed Materials Res, 20(9):1359-1372.

15. Ravi Kumar MNV. Nano and microparticles as controlled drug delivery devices. J Pharm Pharmaceut Sci, 3(2):234-258 (2000).

16. Sezer AD and Akbuga J. Comparison on in vitro characterization of fucospheres and chitosan microspheres encapsulated plasmid DNA (pGM-CSF): formulation design and release characteristics. AAPS PharmSciTech, 10(4):1193-1199 (2009).

17. Dunn EJ, Zhang X, Sun D and Goosen MFA. Synthesis of N-(aminoalkyl) chitosan for microcapsules. J Appl Polym Sci, 50:353-365 (1993).

18. Huguet ML, Groboillot A, Neufeld RJ, Poncelet D and Dellacherie E. Hemoglobin encapsulation in chitosan/calcium alginate beads. J Applied Polymer Sci, 51(8):1427-1432.

19. Yao KD, Xu MX, Yin YJ, Zhao JY, Chen XL. pH sensitive chitosan/gelatin hybrid polymer network microspheres for delivery of cimetidine. Polymer International, 39(4):333-337 (1999).

20. Lee KY, Park WH and Ha WS. Polyelectrolyte complexes of sodium alginate with chitosan or its derivatives. J Appl Polym Sci, 63:425-432 (1997).

21. Kramer PA. Albumin microspheres as vehicles for achieving specificity in drug delivery. J Pharm Sci, 63(10):1646-1647 (1974).

22. Gayakwad SG, Bejugam NK, Akhavein N, Uddin NA, Oettinger CE, Dsouza MJ. Formulation and in vitro characterization of spray-dried antisense oligonucleotide to NF-kB encapsulated albumin microspheres. J Microencapsulation, 26(8):692-700.

23. Okoroukwu ON, Green GR, D'Souza MJ. Development of albumin microspheres containing SpH1-DNA complexes: A novel gene delivery system. J Microencapsulation, 27(2):142-149 (2010).

24. Jain D, Panda AK and Majumder DK. Eudragit S100 entrapped insulin microspheres for oral delivery. AAPS PharmSciTech, 6(1):E100-E107 (2005).

25. Basu SK and Adhiyam R. Preparation and characterization of Nitrendipine-loaded EudragitRL 100 microspheres prepared using solvent evaporation method. Tropical J Pharm Res, 7(3):1033-1041 (2008).

26. Behera BC, Sahoo SK, Dhal S, Barik BB and Gupta BK. Characterization of Glipizide-loaded polymethacrylate microspheres prepared by an emulsion solvent evaporation method. Tropical J Pharm Res, 7(1):879-885 (2008).

27. Yang F, Song F, Pan Y, Wang Z, Yang Y, Zhao Y, Liang S and Zhang Y. Preparation and characteristics of interferon-alpha poly(lactic-co-glycolic acid) microspheres. J Microencapsulation, 27(2):133-141 (2010).

28. Vivek K, Reddy HL, Murthy RSR. Comparative study of some biodegradable polymers on the entrapment efficiency and release behavior of etoposide from microspheres. Pharmaceutical Development and Technology, 12:79-88 (2007).

29. Toongsuwan S, Li L, Erickson BK and Chang H. Formulation and characterization of bupivacaine lipospheres. Int J Pharmaceutics, 280(1-2): 57-65 (2004).

30. Kim BK, Hwang SJ, Park JB and Park HJ. Characteristics of felodipine-located polycaprolactone microspheres. J Microencapsulation, 22(2):193-203 (2005).

31. Kim BK, Hwang SJ, Park JB and Park HJ. Preparation and characterization of drug-loaded polymethacrylate microspheres by an emulsion solvent evaporation method. J Microencapsulation, 19(6):811-822 (2002).

32. Prasant Kumar R and Shankar Nayak B. Formulation design, preparation of losartan potassium microspheres by solvent evaporation method and its in vitro characterization. Arch Pharm Sci & Res, 1(1):166-170 (2009).

33. Phromsopha T and Baimark Y. Chitosan microparticles prepared by the water-in-oil solvent diffusion method for drug delivery, 9:61-66 (2010).

34. Lima FF, Andrade CT, Wang SH and Drumond WS. Microencapsulation of bovine hemoglobin: entrapment efficiency using w/o/w double emulstion technique, Abstract, 11th International Conference on Advanced Materials, Rio de Janairo (2009).

35. Morita T, Sakamura Y, Horikiri Y, Suzuki T and Yoshino H. Protein encapsulation into biodegradable microspheres by a novel S/O/W emulsion method using poly(ethylene glycol) as a protein micronization adjuvant. J Control Release, 69(3):435-444 (2000).

36. Khamanga SM, Parfitt N, Nyamuzhiwa T, Haidula H and Walker RB. The evaluation of eudragit microcapsules manufactured by solvent evaporation using USP Apparatus 1. Dissolution Technologies, 16(2): 15-23 (2009).

37. Amperiadou A and Georgarakis M. Controlled release salbutamol sulphate microcapsules prepared by emulsion solvent-evaporation technique and study on the release affected parameters. Inter J Pharmaceutics, 115(1):1-8 (1995).

38. Das MK and Rao KR. Evaluation of zidovudine encapsulated ethylcellulose microspheres prepared by water-in-oil-in-oil (w/o/o) double emulsion solvent diffusion technique. Acta Poloniac Pharmaceutica – Drug Research, 63(2):141-148 (2006).

39. Sah H. Ethyl formate – alternative dispersed solvent useful in preparing PLGA microspheres. Inter J Pharmaceutics, 195(1-2): 103-113 (2000).

40. Li X, Zhang Y, Ran Y, Jia W, Yuan M, Deng X and Huang Z. Influence of process parameters on the protein stability encapsulated in poly(dl-lactide)-poly (ethylene glycol) microspheres. J Controlled Release, 68(1):41-52 (2000).

41. Youan BBC, Benoit MA, Baras B and Gillard J. Protein-loaded polycaprolactone microparticles. I. Optimization of the preparation by (water-in-oil)-in water emulsion solvent evaporation. J Microencapsulation, 16:587-599 (1999)

42. Gohel MC, Parikh RK and Nagori SA. Spray drying: a review. http://www.pharmainfo.net/reviews/spray-drying-review (Dated: 28/02/2010).

43. Bakan JA. Microencapsulation in The Theory and Practice of Industrial Pharmacy, 3rd Edition (Eds: Leon Lachman, Harbert A Liberman and Joseph L Kanig), pp. 412-429 (1986).

44. Wang F and Wang C. Sustained release etanidazole from spray dried microspheres prepared by non-halogenated solvents. J Controlled Release, 81(3):263-280 (2002).

45. Nihant N, Stassen S, Grandfils C, Jerome R, Teyssie and Goffinet G. Microencapsulation by coacervation of poly(lactide-co-glycolide). III. Characterization of the final microspheres. 34(3):289-299 (2003).

46. Gokale K and Jonnalagadda S. Evaluation of microsphere processing techniques on the physical and chemical stability of influximab. AAPS Abstract (2007).

47. Honary S, Maleki M and Karami M. The effect of chitosan molecular weight on the properties of alginate/chitosan microparticles containing prednisolone. Tropical J Pharm Res, 8(1):53-61 (2009).

7
P-Glycoprotein and CYP3A Limiting Oral Drug Absorption

Y. Shravan Kumar, A. Bhargavi Latha and Y. Madhusudan Rao[a*]
Centre for Biopharmaceutics and Pharmacokinetics, University College of Pharmaceutical Sciences, Kakatiya University, Warangal – 506 009, A.P. India.

7.1 Introduction

The goal of a delivery system is to achieve and sustain therapeutic blood levels of drug, except for those targeted to specific sites. Various routes of administration are exploited for efficient delivery of drugs, in which peroral and parenteral routes are predominant. Unlike parenteral administration, peroral delivery poses many hurdles starting from drug dissolution in gastrointestinal fluid to first pass metabolism due to various physicochemical and biopharmaceutical problems (1). L. Z. Benet and V. Wacher, et al, 1996 were first to hypothesize that for many drugs poor oral bioavailability could be due to biochemical processes in the intestine rather, and often in addition to, the physico-chemical problems. Based on a series of cellular, animal and human studies, it was hypothesized that intestinal metabolic enzymes and efflux transporters, working coordinately as a protective mechanism, may be responsible for the poor bioavailability of a number drugs. The importance of CYP3A and P-Glycoprotein in limiting oral drug delivery was suggested by: a) their joint presence in small intestine enterocytes, b) the significant overlap in their substrate specificities and c) the poor oral bioavailability of drugs that are substrates for both CYP3A and P-glycoprotein. These enzymatic and drug transporter protein are induced or inhibited by man with the same compounds (2, 3).

The concept of poor absorption of drugs due to intestinal metabolism was not considered as clinically significant. But, Watkins and co-workers were the first to report that a major Cytochrome P-450 enzyme system, CYP 3A, is relatively abundant in the intestinal mucosa (4, 5) and the substrates for this enzyme may have poor oral bioavailability due to extensive first pass metabolism in the intestine (6). The high levels of this specific Cytochrome P-450 in the intestine becomes of even greater importance when it is recognized that more than 50 percent of human drugs may be substrates for this enzyme. In addition to intestinal metabolism, P-gp, a transmembrane protein, further supports the poor oral bioavailability of certain drugs. It is capable of the active transportation of the drug from intestinal, renal and hepatic cells. It has received increasing attention as a significant factor in the elimination of a number of drugs. Thus, the first pass effect in the gut may be affected by the inhibition or induction of CYP3A and/or P-gp caused by drug-drug interactions. It is believed that further understanding the physiology and biochemistry of the interactive nature of the intestinal CYP3A and P-glycoprotein will be important in defining, controlling and improving oral bioavailability of CYP3A/P-glycoprotein substrates.

7.2 Factors Influencing Oral Bioavailability of Drugs

The bioavailability (F) of a drug is defined as the fraction of the dose that reaches intact in the systemic circulation. The various factors that influence the systemic bioavailability following oral administration can be described by the following equation:

$$F = fa.\ (1 - E_G).\ (1 - E_H)$$

where, fa is the fraction of the dose absorbed over the mucosal membrane of the enterocyte, $(1\text{-}E_G)$ is the fraction of the drug that escapes metabolism in the gut wall, and $(1\text{-}E_H)$ is the fraction of the drug that escapes metabolism and biliary excretion in the liver (7, 8).

From historical perspective, only fa and $(1 - E_H)$ were considered to have significant impact on F. However, intestinal metabolism and active extrusion of absorbed drug have been recognized as major determinants of oral drug bioavailability (2). Cytochrome P 450 (CYP) 3A and the multidrug efflux pump, P- Glycoprotein, are present at high levels in the villus tip of enterocytes in the gastrointestinal tract, the primary site of absorption for orally administered drugs. The importance of CYP 3A and P-gp in limiting oral drug delivery is suggested to us by their joint presence in small intestinal enterocytes. These proteins are induced or inhibited by many drug compounds administered orally.

7.3 Efflux Mechanisms as Barriers to Intestinal Absorption

The small intestine represents the principal site of absorption for any ingested compound, whether dietary, therapeutic, or toxic. Oral administration is the most popular route for drug administration since dosing is convenient and non-invasive and many drugs are well absorbed by the gastrointestinal tract. As well as degrading and absorbing nutrients and solutes from the intestinal lumen, intestinal enterocytes form a selective barrier to drugs and xenobiotics. This barrier function depends largely upon specific membrane transport systems and intracellular metabolising enzymes. The extent to which a compound is absorbed by the intestinal epithelium is therefore a critical factor in determining its overall bioavailability.

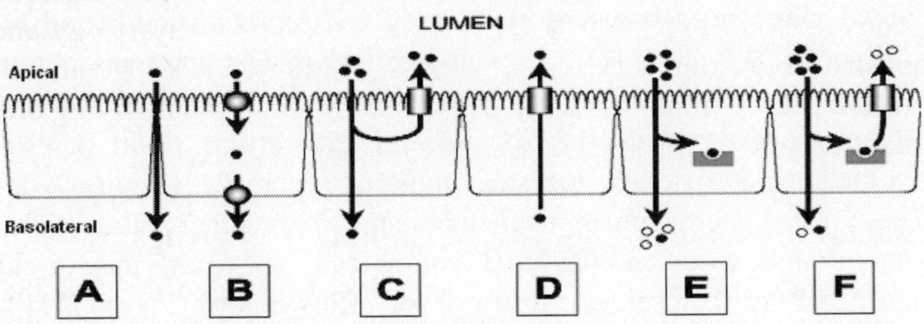

Courtesy: Lauretta M.S. Chan, Barry H. Hirst et al., *European Journal of Pharmaceutical Sciences* 21 (2004) 25–51

Figure 7.1 The intestinal epithelium forms a selective barrier against the entry of compounds into blood. (A) Absorption of compounds via the paracellular route is restricted by intercellular tight junctions. (B) Carrier-mediated mechanisms at the apical and/or basolateral membranes facilitate the transcellular absorption of certain compounds. (C) Efflux transporters at the apical membrane may actively drive compounds back into the intestinal lumen thus restricting their absorption into blood. (D) Apical efflux transporters may also facilitate the intestinal clearance of compounds that are already present in blood. (E) Intracellular metabolising enzymes may modify compounds before they enter the blood. (F) Apical efflux transporters and intracellular metabolising enzymes may co-ordinately metabolise and excrete compounds, forming an effective barrier against intestinal absorption.

There are two principal routes by which compounds may cross the intestinal epithelium: paracellular or transcellular. A number of small hydrophilic, ionised drugs are absorbed via the paracellular pathway (Figure 7.1A) (9). However, absorption via this route is generally low since intercellular tight junctions restrict free transepithelial movement between epithelial cells. Transcellular absorption from lumen to blood requires uptake across the apical membrane, followed by transport across the cytosol, then exit across the basolateral membrane and into blood. The transcellular absorption of hydrophilic drugs may be facilitated via specific carrier-mediated pathways by means of utilizing the same route of absorption followed by nutrients and micronutrients (Figure 7.1B). Many orally administered drugs are lipophilic and undergo passive transcellular absorption (10). Drugs that cross the apical membrane may be substrates for apical efflux transporters, which extrude compounds back into the lumen (11, 12, 13, 14, 15) (Figure 7.1C). These apical efflux transporters are principally ABC proteins such as P-gp and MRP2, and are ideally situated to act as the first line of defense by limiting the absorption of potentially toxic 'foreign' compounds. Compounds that are already present in the blood may undergo active blood-to-lumen secretion facilitated by these transporters (Figure 7.1D). As well as efflux pumps, the transcellular route of absorption exposes drugs to intracellular metabolic systems; small intestinal enterocytes provide the first site for cytochrome P450 (CYP)-mediated metabolism of orally ingested drugs and xenobiotics (Figure 7.1E) (16). The CYP system (phase I metabolism), plus other intracellular metabolic systems, such as phase II conjugating enzymes, may yield metabolites that are themselves substrates for efflux pumps (Figure 7.1F), thus providing additional possibilities for interactions (17).

The fraction of drugs that escapes this first, intestinal line of defense to reach the blood then passes to the liver via the portal system, where it is subject to further metabolism and biliary excretion, often by a similar system of enzymes and P-gp transporter to that present in the intestine. The portal system may be bypassed by the fraction of drug that travels in the lymphatics. Drugs are specifically removed from the portal blood into hepatocytes across the sinusoidal (basolateral) membrane, where they are exposed to an environment rich in metabolising enzymes. Compounds, metabolised or unaltered, may then be excreted across the canalicular (apical) membrane into bile and ultimately into the intestine, which completes the first cycle of enterohepatic circulation. Drugs that reach the systemic circulation following first pass extraction by the liver, or through the lymphatics, will meet the kidneys, which are also well equipped for the active excretion of waste products. Thus, apical efflux systems and intracellular metabolism in the intestine and liver are critical determinants of overall oral bioavailability.

Location and Vocation

P-gp shows extremely broad substrate specificity, with a tendency towards lipophilic, cationic compounds. The list of substrates/inhibitors is continually growing and includes anticancer agents, antibiotics, antivirals, calcium channel blockers, and immunosuppressive agents. Naturally occurring substrates for P-gp include biologically active compounds found in normal diet, such as plant chemicals (10, 18), consistent with P-gp acting as part of a detoxification and excretion pathway. Immunocytochemical studies using specific monoclonal antibodies against P-gp confirm high expression in solid tumours of epithelial origin, including those of the colon (19), kidney (20) and breast (21) indicating the pivotal role of P-gp in resistance to anticancer therapy. Of equal significance, however, is its constitutive expression in a wide range of normal tissues; P-gp is present at high levels in kidney and adrenal gland, at intermediate levels in liver, small intestine, colon and lung, and at low levels in prostate, skin, spleen, heart, skeletal muscle, stomach and ovary (22, 23). P-gp is also expressed in brain therefore ideally positioned to limit the absorption of compounds by driving efflux back into the lumen (19, 24). The first evidence indicating that P-gp acts as a secretory detoxifying system to limit drug absorption came from studies in the human intestinal epithelial cell lines Caco-2, HT29, and T84 (13, 14, 25). Polarised, apical P-gp expression in these cells was accompanied by secretory (basolateral-to-apical; blood-to-lumen) transport of the cytotoxic anticancer drug, vinblastine, which was reduced in the presence of MRK16, an inhibitory monoclonal antibody directed against P-gp, and the P-gp inhibitors/substrates verapamil and nifedipine (13, 14, 25). As expected, P-gp secretes a host of other drugs in Caco-2 cell monolayers, including celiprolol (26), digoxin(27), erythromycin (28), ranitidine, cimetidine (29), saquinavir (30), and fluoroquinolone antibiotics (31, 32). Importantly, there is a large body of *in vivo* evidence to support the notion that P-gp influences oral drug bioavailability. P-gp-knockout mice were developed by Schinkel et al., to investigate P-gp function *in vivo*. Mice have two *mdr1* genes, *mdr1a* and *mdr1b*, which together appear to perform the same function as human *MDR1 (33)*. Knockout animals were generated for both genes, as well as *mdr1a/1b* (-/-) double knockouts (33, 34). The *mdr1a* (-/-), *mdr1b* (-/-), and *mdr1a/1b* (-/-) mice all display normal viability, fertility, and lifespan, with no obvious physiological abnormalities (33, 34). However, under stress, such as after exposure to toxins, the knockout animals show striking differences to their wild-type counterparts; *mdr1a* (-/-) mice, for example, are 50- to 100-fold more sensitive to toxicity caused by the pesticide ivermectin, a neurotoxic P-gp substrate (33). Knockout animal models have proved invaluable both in determining the contribution of P-gp to the disposition of a given drug and in explaining certain clinical drug–drug interactions. For example, one unexplained observation in humans was that co-administration of digoxin with either quinidine (35) or verapamil (36), all three of which are P-gp substrates, reduces total digoxin clearance. A study by Fromm et al., used knockout mice to show that where as quinidine increased plasma digoxin concentrations in wild-type animals, there was no interaction in *mdr1a* (-/-) mice, strongly suggesting that competition for P-gp mediated transport is responsible for the clinical drug–drug interactions observed with these compounds (37). Additionally, plasma concentrations of paclitaxel (38) and of the HIV-1 protease inhibitors indinavir, nelfinavir and saquinavir (30), were shown to be significantly higher in *mdr1a* (/) mice compared to wild-type mice following oral administration (38). The knockout mouse model also provides important insight as to the contribution of P-gp to drug disposition in individual tissues. As described above, work with cell lines has demonstrated P-gp mediated secretion by intestine. In bile duct-ligated mice, intestinal excretion of digoxin was reduced from 16% of the administered dose in *mdr1a/1b* (+/+) controls to 2% of the dose in *mdr1a/1b* (-/-) animals (34). Although the data from *mdr1* knockout mice suggest that intestinal P-gp may play a direct clinical role in the disposition of orally administered drugs, this is difficult to demonstrate in humans. Consequently there is a paucity of information available at present. Nevertheless, a recent study directly investigated intestinal secretion of digoxin using human volunteers (39). Segments of intestine were isolated by inflating balloons at desired points inside the lumen. After intravenous administration of 1mg digoxin, samples of the intestinal perfusate were collected within 3 h of digoxin administration, 0.45% of the dose was found in a 20 cm segment of jejunum (39). By scaling up their findings from the isolated intestinal segments to whole intestine, Drescher et al., estimated conservatively that 11% of the intravenous dose would be eliminated by the gut in 3 h. Perfusing the intestine with quinidine, an inhibitor of P-gp, reduced intestinal digoxin secretion by half (39). Another study using human volunteers suggested that intestinal P-gp has an even greater potential to affect drug disposition following its up-regulation. Greiner et al., (40) showed that rifampin, when co-administered orally with digoxin, not only increased intestinal P-gp levels, it also concomitantly decreased the area under the curve (AUC) for digoxin (relative to digoxin administered alone). When both drugs were administered intravenously, however, rifampin produced a substantially less-pronounced effect, suggesting that direct intestinal exposure to rifampin was responsible for the effects observed following oral administration (40). Thus it appears that intestinal P-gp has the capacity to have a major influence on the drug disposition of orally administered compounds.

Structure of P-glycoprotein

P-gp is a type of ATPase and an energy dependent transmembrane drug efflux pump which belongs to members of ABC transporters. P-gp is a 1280 amino acid long (molecular weight 170KDa) glycoprotein, is expressed as a single chain containing two homologous portions of equal length, each containing six transmembrane domains and two ATP binding regions separated by a flexible linker polypeptide region between the walker A and B motifs (3,4) (Figure 7.2).

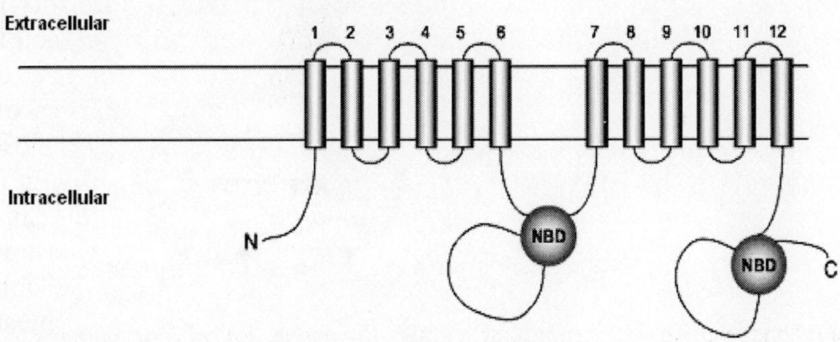

Courtesy: Lauretta M.S. Chan, Barry H. Hirst et al., *European Journal of Pharmaceutical Sciences* 21 (2004) 25–51

Figure 7.2 Transmembrane arrangement of P-gp efflux proteins

The Nucleotide binding sites

A functional unit of an ABC transporter requires two NBDs or ABCs which are composed of several conserved sequence motifs, the A-loop (an aromatic residue 25 amino acids upstream of the Walker A), the Walker A, the Walker B, the signature motif (LSGGQ motif, linker peptide or C motif) and the D, H and Q-loops. Crystal structures of the ABC domains of several ABC transporters indicate that a functional ATP site is formed by the interaction of residues from both halves of the protein (41, 42, 43, 44, 45). Moreover, the structure of NBDs of several ABC and closely related proteins show that the two NBDs form a 'nucleotide-sandwich dimer' with ATP bound along the dimer interface, anked by the Walker A and B motifs of one subunit and the signature motif and D-loop of the other (42, 46). In addition, the adenine ring of ATP interacts with an aromatic residue (A-loop) upstream of Walker A and the three phosphate and magnesium moieties interact with the Walker A (P-loop) and B motifs, and possibly the H-loop (46, 47, 48). The crystal structures of the NBDs of several ABC transport proteins suggest specific roles for several highly conserved amino acid residues within the ABC (46, 48, 49) for comprehensive maps of these residues.

The drug substrate binding sites

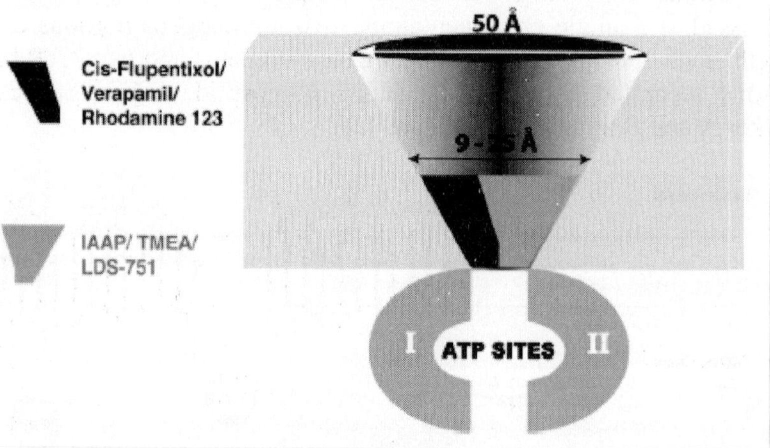

Courtesy: Suresh V. Ambudkar et al., *European journal of pharmaceutical sciences* 2 7 (2 0 0 6) 392–400

Figure 7.3 Schematic showing the drug-substrate binding pocket of P-gp. The schematic is based on biochemical data presented in (50, 51, 52, 53, 54). The drug binding pocket appears to be funnel shaped narrow at the cytoplasmic side. A central, putative drug-substrate binding region has a diameter of 9–25 A° and is possibly the high affinity site or "ON" site. The funnel is approximately 50 A° at its widest and represents either a low affinity site or the "off" site for the release of the substrate. Two drugs are depicted as simultaneously bound in the drug-binding pocket. Three substrate pairs are shown: (i) Verapamil and tris-2,maleimidoethyl amine (TMEA); (ii) Rhodamine 123 and 2-[4-(4-[dimethylamono]phenyl)-1,3,-butadienyl]3-ethylbenzothiazolium percholate (LDS-751) and (iii) cis(*z*).upentixol and IAAP. Drugs shown in orange appear to have a greater affinity for Pgp than the ones depicted in red.

Drugs belonging to diverse chemical classes and having different targets interact with P-gp (55). The most direct approach to elucidating the regions of P-gp that interact with drugs is the use of photoaffinity analogs of drug substrates (for reviews see (56, 57)). The methods used to localize the regions of P-gp interacting with these photoaffinity analogs have significantly increased in sophistication and accuracy over the last fifteen years. While the earliest studies used antibodies raised against specific regions of P-gp (58, 59), subsequent studies used HPLC in conjugation with Edman-MS (60, 61, 62), and more recently MALDI-MS (63). The results of these studies and analyses of site-directed mutants of P-gp (see (55) for review) that affect substrate binding or specificity support the view that drug-substrate binding occurs principally in the trans-membrane domains. A significant limitation of these approaches however, is that it only permits localization of the TM helices that interact with drugs and provides no information about the tertiary structure of the drug binding pocket. Information about the proximity of the TM helices involved in drug binding and the amino acid residues that line the drug binding pocket has largely come from cysteine scanning mutagenesis (64, 65, 66). The proximity of amino acid residues in 3D-space were identified by screening hundreds of mutants each with a unique pair of cysteines reintroduced into a cys-less mutant of P-gp. These experiments provided evidence that a thiolreactive substrate of P-gp, dibromobimane, reacted with cysteines reintroduced in TMs 4, 5, 6, 10, 11 and 12 suggesting that these regions contribute to drug binding (67, 68). Moreover, using methanethiosulfonate cross linkers with spacers of different lengths, it was possible to model the drug binding domain of P-gp (69, 70), which has been proposed to be funnel shaped and narrow at the cytoplasmic side (Figure 7.3). The central, putative drug-substrate binding region has a diameter of 9–25 A° (also designated as the high affinity site) and approximately 50 A° at its widest (low affinity site). These dimensions are consistent with a study that used dimers of stipiamide (a synthetic antibiotic that is a modulator of P-gp) that differ in the length of a polyethylene glycol ether (53). Improved binding was found with spacers of 11–35 A°. Shorter spacers either do not allow the stipiamide dimer sufficient flexibility or do not permit the interactions between both the stipiamides and the requisite amino acid residues in the drug binding site. Very large spacers (>35 A°), on the other hand, increase the bulk of the molecule to the extent that it no longer fits in the drug binding site. Though the mammalian P-gp was the first MDR protein described, it soon became apparent that such pumps are widespread across all taxonomic groups (71). Biochemical data and the structures of ABC transporters suggest the presence of a large hydrophobic cavity. The QacR structures reveal the remarkable plasticity of this cavity and provide insights into the conformational changes that accompany drug binding. In addition, the simultaneous binding of ethidium and proflavin has been demonstrated by Schumacher (72). These results are broadly consistent with the structure of the *E. coli* AcrB pump solved with four different substrates (73). All these structures show the use of a single large cavity comprising a "drug binding pocket" where individual "drug-binding sites" are generated by each ligand using a different subset of residues for drug-binding. These structural insights are consistent with biochemical studies with P-gp that suggest that there are at least two and possibly several overlapping drug binding sites in P-gp (50, 51, 52, 54) and Figure 7.3 depicts an emerging view of the drug-binding pocket of P-gp. The high affinity binding site or the narrow region of the funnel appears large enough to accommodate two different drugs at the same time (see Figure 7.3 and (51, 52)). It must however be emphasized that QacR is the multidrug-binding regulatory protein and not the transport protein per se. Also, while these studies provide an understanding of how the drug binding pocket may be occupied, we do not understand how the transport of these drugs occurs under conditions where more than one drug or modulator is present.

Mechanism of drug efflux

Various models were proposed to explain the mechanism of xenobiotic extrusion by P-gp, however, the exact site of substrate interaction with the protein is not well resolved.

The three prevalent models, pore model, flippase model and hydrophobic vacuum cleaner (HVC) model, explains the efflux mechanism to certain extent (see Figure 7.4). Among these HVC model has gained wide acceptance in which P-gp recognizes substrates embedded in the inner leaflet of plasma membrane and transported through a protein channel (71). Recently, Rosenberg et al., reported that three-dimensional conformation of P-gp changes upon binding of nucleotide to the intracellular nucleotide-binding domain (74). In the absence of nucleotide, the two transmembrane domains form a single barrel with a central pore that is open to the extracellular surface and spans much of the membrane depth, while upon binding nucleotides, the transmembrane domains reorganize into three compact domains that open the central pore along its length in a manner that could allow access of hydrophobic drugs directly from the lipid bilayer to the central pore of the transporter. ATP binding and hydrolysis was found to be essential for functioning of P-gp, where one molecule of drug is effluxed at the expense of two molecules of ATP (75). Sauna et al., elucidated the catalytic cycle of P-gp, which expands the opportunity for the development of P-gp inhibitors, comprises of two cycles where drug and nucleotide binding sites coordinately function to efflux out the substrates by an ATP driven energy-dependent process. The drug and ATP initially binds to the protein at their own binding sites, where nucleotide hydrolyses to ADP yields energy for the extrusion of drug. The release of ADP from nucleotide binding site ends the first catalytic cycle followed by a conformational change that reduces affinity for both substrate and nucleotide. Further, the second catalytic cycle starts by hydrolysis of another molecule of ATP and released energy is utilized to reorient the protein to its native conformation. Subsequent release of ADP completes another catalytic cycle, bringing P-gp molecule back to the original state, where it again binds to both substrate and nucleotide to initiate the next cycle (57).

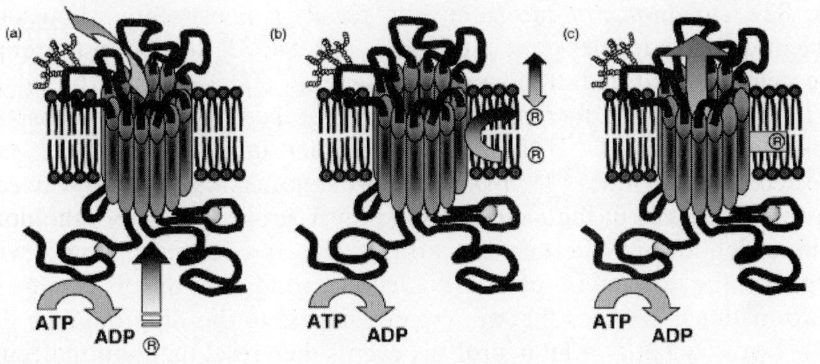

Courtesy: Manthena V.S. Varma et al., *Pharmacological Research* 48 (2003) 347–359.

Figure 7.4 Models proposed to explain the mechanism of drug efflux by P-gp. (a) Pore model, (b) flippase model and (c) hydrophobic vacuum cleaner model. In pore model, drugs associate with P-gp in the cytosolic compartment and are transported out of the cell through a protein channel. In flippase model, drugs embed in the inner leaflet of the plasma membrane, bind to P-gp within the plane of membrane and are translocated to the outer leaflet of the bilayer from which they passively diffuse into extracellular fluid. The hydrophobic vacuum cleaner model combines the features of 'pore' and 'flippase' models.

The transport cycle of P-glycoprotein

Understanding how the protein–substrate interactions at the NBDs and the drug-substrate bindings sites are coupled to bring about drug transport is the ultimate goal of studies with ABC transport proteins. They have exploited photoaffinity analogs of drug-substrates and nucleotides to simultaneously monitor events at the drug-substrate site and the ATP sites (57, 76, 77). Higgins and coworkers have monitored changes in the 2D structure of the molecule (78, 79). Loo and Clarke have used a cys-less mutant of P-gp in which pairs of cysteines are reintroduced to report changes in the cross-linking pattern in the presence of drugs and nucleotides (80). The essential features of the two alternative models for the transport cycle of P-gp are illustrated in Fig 7.5. The drug and ATP first bind to P-gp (Fig 7.5A, Steps I and II; Fig 7.5B, Step I). Higgins and Linton have postulated that drug-substrate binding lowers the activation energy for the formation of a closed NBD dimer and increases the affinity for ATP (78). While such dimerization was first shown in the Rad50 structure (42) and subsequently demonstrated in MJ0796 (46) and MalK (81, 82), stoichiometry measurements for P-gp consistently show one nucleotide occluded per protein molecule (83, 84, 85, 86). Several groups also demonstrated that drug-substrates have no effect on the affinity for ATP (87, 88, 89). Furthermore, the binding of the photoaffinity transport substrate of P-gp, [^{125}I] Iodoaryl azidoprazosin (IAAP) (90) is not affected by nucleotides (76, 77). The most significant difference between the two models is in the nature of the power stroke that drives the drug from a high affinity site to a low affinity site. In one model (Fig 7. 5A, Step II), the formation of the nucleotide sandwich dimer results in conformational changes that are communicated to the drug binding site (78). Two sequential ATP hydrolysis events then reset the P-gp molecule (Steps IV–VI; Fig 7.5B). The alternate model (57, 76, 77) also requires two ATP hydrolysis events, but one powers the efflux of the drug (Fig 7.5B, Step II) and the other resets the protein to its ground state (Fig 7.5B, Step VI). Both these models essentially subscribe to the "alternating catalytic sites" scheme propounded by Senior and coworkers (91), viz. that only one of the two NBDs hydrolyses ATP at any given time and the two NBDs alternate during the catalytic cycle. Based on studies with MJ0796, it has been hypothesized that in ABC transporters the energy of ATP hydrolysis is necessary to drive apart the cassette formed by the ATP-mediated dimerization of the two NBDs and thus reset the molecule (46). In the case of P-gp, however, the ATP switch model (78, 79) invokes the sequential hydrolysis by the two ATP sites to incorporate the considerable biochemical data that supports the alternating catalysis scheme (for reviews see (55, 57, 91, 92, 93, 94)). The model depicted in Fig. 7.5B, on the other hand, is based on experimental data that shows that when P-gp is trapped in the Vi-induced post-hydrolysis transition state, P-gp·ADP·Vi both IAAP, the photoaffinity substrate of P-gp and [^{32}P]8azido ATP, an analog of ATP show drastically reduced affinity for P-gp.

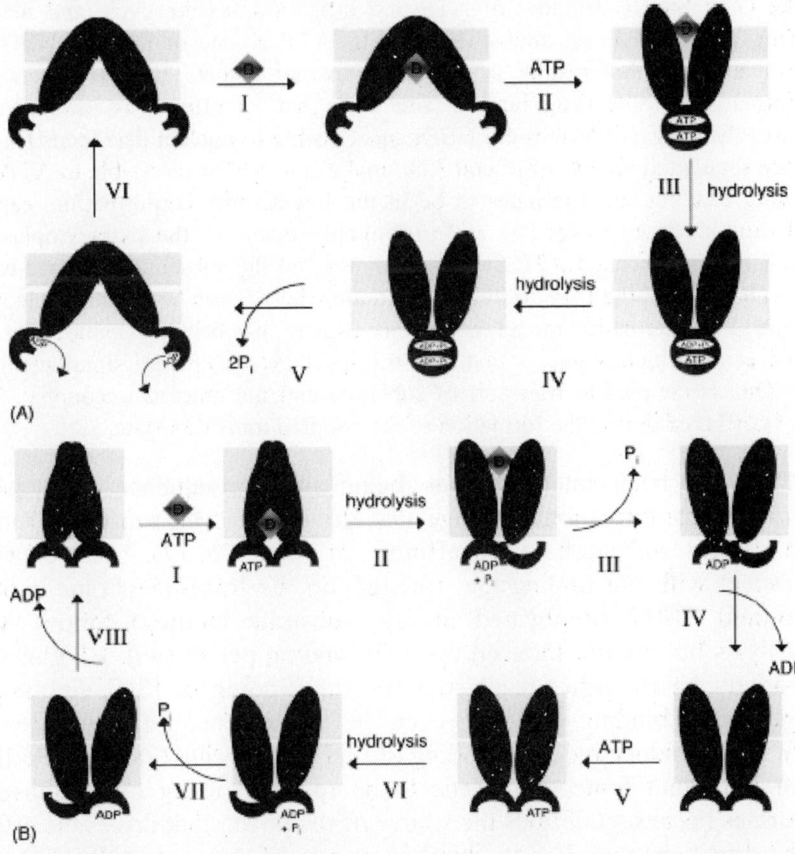

Courtesy: Suresh V. Ambudkar et al., *European journal of pharmaceutical sciences* 2 7 (2 0 0 6) 392–400

Figure 7.5 Models describing the transport cycle of P-gp. The blue rectangles represent the plasma membrane (dark blue, outer leaflet; light blue, inner leaflet); the red circles and semi-circles represent nucleotide binding domains; the red ovals depict drug-binding sites composed of TM domains and the drug molecule is represented as a 'D enclosed in an orange diamond. (A) Model proposing the dimerization of the two ATP sites as the power stroke (78): Step I: drug binds to a high affinity binding site in the TM domain. Step II: binding of drug (i) reduces the activation energy and (ii) increases the affinity for ATP. This results in the dimerization of the two NBDs and ATP is tightly bound at the interface. The conformational changes driven by this dimerization move the drug to a low affinity extracellular location from where it is released. Steps III&IV: Two sequential ATP hydrolysis events provide the energy to break apart the nucleotide dimer-ATP sandwich and the energy is used to reset the P-gp to its ground state (Steps V–VI). (B) Model proposing ATP hydrolysis as the power stroke (77): Step I: Binding of ATP and drug initiates the cycle and neither binding event influences the other. Step II: ATP at one of the two NBDs is hydrolyzed and this results in a conformational change that; (i) alters the conformation of the drug binding site such that its affinity is lowered and releases the drug. (ii) Makes the NBDs inaccessible to nucleotides. Steps III and IV: the sequential release of Pi and ADP makes the NBDs accessible to ATP but the drug binding site continues to be in the low-affinity conformation, cannot bind drug-substrate molecules and presumably opens to the extracytoplasmic face. Step VI: a second ATP hydrolysis event and the subsequent release of Pi and ADP restores the P-gp to the ground state where it can bind both nucleotide and drug. An alternative model, not discussed here, has been propounded by Al-Shawi et al. which suggests that two transport cycles operate simultaneously (96). One is coupled to transport of substrate and the other is uncoupled. The drug is effluxed during the formation of the coupled transition state.

This has been interpreted as being the consequence of a post-hydrolysis conformational change that drives the drug-substrate from a high affinity "on" site to a low affinity "off" site (76, 77). These data are consistent with the finding that rotations in the trans-membrane helices TM6 and TM12 (implicated in drug-substrate binding) follow ATP hydrolysis but are not induced by ATP binding per se (86). Higgins and coworkers on the other hand studying the binding of [3H] vinblastine suggest that binding of ATP even in the absence of hydrolysis is sufficient to induce reduced binding of [3H] vinblastine (78, 79, 95). This is an important issue that needs to be resolved using other transport substrates because it defines the source of the power that drives the efflux of the drug substrate. Is it the binding energy of the nucleotide sandwich that generates the closed dimer or is ATP hydrolysis per se required? In addition, the subsequent steps in this model (IV–VIII, Fig.7.5B) are based on the binding that while ADP release is sufficient for an additional ATP hydrolysis event, it is not sufficient to bind the drug substrate (77). The details of these models have been discussed and the experimental details that support them can be found in previous reviews (57, 78). It is clear that additional kinetic and thermodynamic studies are required to differentiate conformational changes resulting from ATP binding and hydrolysis.

7.4 CYP 3A and Oral Bioavailability

The cytochrome P450 (CYP3A) mixed function oxidases are a family of enzymes, which account for the majority of oxidative biotransformations of xenobiotics and endogenous biochemicals. Over 30 different human CYP enzymes have been identified. CYP 3A4 appears to be one of the most important human enzymes as approximately 60% of oxidized drugs are biotransformed, at least in part, by it (97). The isoforms of CYP3A in human include CYP3A3, CYP3A4, CYP3A5 and CYP3A7. Each of these enzymes share at least 85% amino acid sequence homology (98). CYP3A3 is so similar to CYP3A4 that the distinction between the two may be artificial; no attempts have been made to distinguish between them (99). CYP3A5 is the predominant isoform in the lung (100) and stomach (98) and is present in the small bowel and renal tissue its contribution to drug metabolism is uncertain. CYP3A7 is in fetal liver but does not appear to be present in adults.

CYP3A4 is the predominant isoform of CYP3A in adult humans. It can catalyse a remarkable number of metabolic process including aliphatic oxidation, aromatic hydroxylation, N-dealkylation, S-demethylaton, Oxidative deamination, sulfoxide formation, N - oxidation and N-hydroxylation, to mention a few. This usually produces inactivation and elimination of most pharmaceuticals. However, it can also activate carcinogenic substances such as aflatoxins and polycyclic aromatic hydrocarbons. Although CYP3A4 drug metabolizing activity varies widely among individuals, it has a unimodal population distribution and does not appear to be subject to genetic polymorphism as is seen with other CYP isoforms (CYP2D6, 2C9, and 2C19). The wide inter individual variability is likely, in part, to be caused by ethnic or cultural differences, perhaps related to an interaction between race and diet. Other factors known to play a role in activity are age (101) and the presence of small bowel or liver disease (102). There may be modest gender differences, perhaps related to the hormonal milieu in which the enzyme functions, although this is controversial (103).

CYP3A4 content is highest in liver. Hepatic CYP3A4 content has been shown to vary at least 20- fold among individuals and activity, as measured by the erythromycin breath test, to range10-fold. In small bowel CYP3A4 is found in the apical enterocytes and its content varies 11-fold among individuals. Hepatic and enteric CYP3A4 content appear to be regulated independently of each other. The other location of CYP3A4 in the small bowel and liver makes it well suited to play a significant role in first-pass (or presystemic) drug metabolism.

Noteworthy is high concentration of terfinadine, astemizole and cisapride when these drugs are taken concomitantly with azole antifungals or macrolide antibiotics or fluoxetine or fluvoxamine. High concentration of these drugs leads to life threatening cardiac arrhythmias like *torsade de points* and ventricular fibrillation. Grapefruit juice (GFJ) is often taken at breakfast in western countries when drugs are taken .GFJ contains bioflavonoids mainly naringin and furanocoumarin which cause mechanism based inhibition of presystemic elimination of a number of drugs and increase their bioavilability (104). An important interaction involving induction of CYP3A is the reduction in efficacy of oral conrtraceptives by rifampicin and rifabutin , because of a induction of CYP3A mediated metabolism of estradiol and norethisrerone, the components of oral contraceptives (105).

The apical recycling theory

CYP3A4 and P-gp have almost complete overlap in substrate specificity and they are commonly localized near the apical membrane of the enterocyte (106, 107, 108). They are furthermore both regulated by the orphan nuclear pregnane X receptor (PXR) (109, 110, 111). It has been suggested, in the apical recycling hypothesis, that CYP3A4 and the P-gp cooperate in the transport and metabolism of drugs in the intestinal membrane.

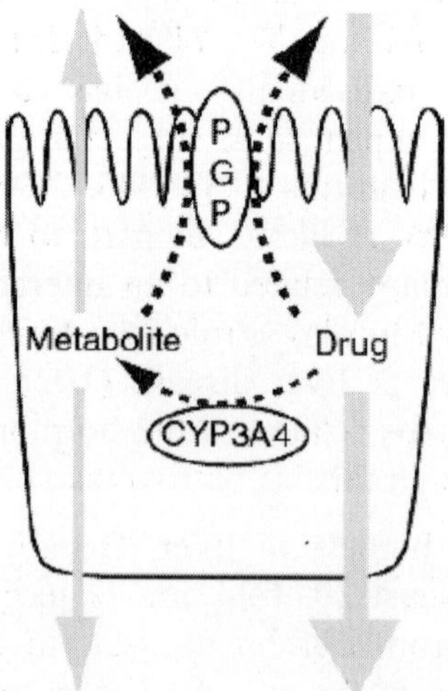

Figure 7.6 Schematic illustration of how CYP3A4 and the P-glycoprotein (P-gp) might cooperate in the enterocyte. P-gp could slow down the drug presentation rate to the enzyme by apical recycling. In addition P-gp might aid the apically directed transport of the metabolite formed.

According to this theory, P-gp slows down the drug presentation rate and increases the drug exposure to the CYP3A4 enzyme by recycling the drug in the enterocyte (2,107, 112, 113). It has also been suggested that this interplay includes P-gp mediated active transport of the metabolites formed to the intestinal lumen (Figure 7.6) (114, 115).

7.5 Models to Screen CYP3A and/or P-gp Modulators as Oral Bioavailability Enhancers

There have been several *in vitro* and *in vivo* models are available. These methods include, everted sac transport studies in rats, transepithelial transport across Caco-2 cell monolayer using various P-gp and CYP3A substrates as probes at various induced and inhibitory conditions. Most known probes include rhodamine 123, digoxin, erythromycin, terfenadine, indinavir, verapamil etc.

In vivo methods in rats include intestinal absorption and exsorption studies using various CYP3A and/or P-gp modulators as probes. In these studies an *in situ* single pass perfusion studies are performed and the rate of absorption, rate of exsorption and rate of elimination are determined. The studies like tissue distribution and pharmacokinetic determinations in knockout mice in which mdr1a or mdr1b or both genes are disrupted can also be used.

In vivo pharmacokinetic studies in experimental animals and human volunteers can be successfully applied for oral bioavailability estimations. In these both intravenous and oral pharmacokinetic parameters like area under the plasma concentration Vs time curve (AUC), peak plasma concentration (Cmax), time to reach peak plasma concentration (Tmax), percent oral bioavailability (%F) etc., will be calculated using various CYP3A and P-gp substrates and/or inhibitors as probes like cyclosporine, midazolam in human volunteers.

7.5.1 Experimental Models

The involvement of P-gp in the absorption and consequent distribution of orally administered xenobiotics has been extensively studied in *in vitro, in situ* and *in vivo* models. Some routinely used systems include cultured cell lines, isolated intestinal segments, everted sacs, and brush border membranes. Organ (brain, liver, and kidney) perfusion and gene knockout mice have also been used. Each of these models has certain advantages and disadvantages. A brief description of the models that have been used to evaluate the role of P-gp in the disposition of drug molecules follows.

7.5.1.1 *In Vitro* Models

Caco-2 Cells

Of the many cell types utilized to model drug behavior in the human intestine, the immortalized human colorectal carcinoma derived cell line, Caco-2, is the most widely accepted *in vitro* model to date. This cell line has demonstrated several advantages over others that have made it the cell line of choice in both academia and the pharmaceutical industry. The most attractive features of the Caco-2 cell line include, the spontaneous differentiation into mature enterocytes, the expression of several biochemical and anatomical features common to normal intestinal enterocytes and these monolayers become polarized and display a well defined brush border membrane located in the apical domain. The brush border contains several transporters, metabolic enzymes, and efflux pumps such as P-gp, whose expression is both stable and functional (116).

Western blot analysis demonstrated that P-gp was expressed as early as day 7 of culturing (117). Biochemical barrier posed by P-gp is not fully functional until day 17. This is most likely due to the amount of expression of P-gp per cell and the subsequent increase in this number to day 17. Expression of specific proteins can be easily induced in Caco-2 cells using simple culturing techniques. For example, the induction and over expression of cytochrome P450 3A4 (CYP3A4) was achieved by culturing the cells with 1α-25-dihydroxy vitamin D3 beginning at confluence, and this expression was shown to be dose and duration dependent (118). Expression of P-gp can also be easily induced in Caco-2 cell line by culturing with vinblastine, verapamil, or celiprolol. Conversely, metkephamid has been used to decrease the level of P-gp expression. No morphological differences were noticed for vinblastine-cultured cells with respect to appearance, formation of tight monolayers, or transepithelial resistance (116).

Madine–Darby Canine Kidney Cells

The most significant advantage the MDCK cell line has over the Caco-2 cell line is much shorter culture time. Studies by Simons et al., have shown that these cells are polarized and contain well-defined apical brush border membrane with a membrane composition similar to that of the intestine (119, 120). The spontaneous differentiation MDCK into polarized cell monolayers with defined apical and basolateral domains make studying the actions of transporters in polarized fashion facile. In addition, this cell line has also been transfected with other drug-effluxing transporters (expressed in either apical or basolateral domain) to study their effects on altering the flux of a compound as it crosses a polarized monolayer (121).

Brain Microvessel Endothelial Cells

One of the most extensively used *in vitro* models to study drug behavior at the BBB are cultured brain microvessel endothelial cells (BMECs), a primary culture that forms confluent monolayers 9-12 days after initial seeding (122). These cultured cells have been shown to retain many morphological and biochemical properties of their *in vivo* counterparts, including distinguishable luminal and abluminal membrane domains that are functionally and biochemically distinct. One of the major advantages of BMECs is that these cells can be grown on collagen coated or fibronectin-treated polycarbonate membranes, and thus this system can be used to the transport across the monolayer by various mechanisms (i.e., passive diffusion, transcytosis, endocytosis, inwardly directed carrier proteins, polarized efflux, and uptake in both luminal and abluminal directions). One limitation of this system is that the tight junctional complexes of BMECs are not as developed as those seen *in vivo*, and thus the contribution of paracellular permeability to the overall permeability of a compound is much greater in this *in vitro* system than that would be seen for a compound crossing the BBB *in vivo* (116).

Membrane Vesicles

Membrane vesicles are typically formed from intact cells and require some skill for their preparation. Given this limitation, the use of membrane vesicles as a rapid screen for P-gp efflux activity has not been extensive, and has proven a better tool for studying

Isolated Intestinal Segments

In these studies, the intestine is removed and either mounted in diffusion apparatus (Ussing chamber) or everted to make an everted sac. Factors affecting the transport of drugs (i.e., metabolism and efflux) can be studied by determining the fate of the test compound as it crosses the intestinal epithelium.

The transport characteristics of verapamil were determined for each region of the rat intestine as well as the colon with this model system. The duodenum and jejunum showed the most P-gp activity followed by lower activity in the colon and, surprisingly, none in the ileum (123). Various experiments were conducted with these segments using different drugs. The results of these experiments provided evidence that P-gp is active in limiting tissue exposure drugs and also that the intestinal metabolism of certain compounds can be significant the microscopic aspects of P-gp-mediated efflux.

Everted Sac Study

Male Wister rats were fasted overnight with free access to water before the experiments. The whole small intestine was flushed with 50 ml of ice-cold saline with the animal under anesthesia with pentobarbital (30mg/kg/ip). The rat was exsanguinated, and the small intestine isolated was divided into 3 segments of equal length. Each segment was everted, and a 10 cm long everted sac was prepared. 10 mg/ml of Probe drug was dissolved in pH 7.4 isotonic Dulfecco's PBS (D-PBS) containing 25 mM glucose and 4% of DMSO. The probe drug solution (1ml) was introduced into the everted sac (serosal side), and both ends of the sac were ligated tightly. The sac containing probe drug solution was immersed into 40 ml of D-PBS containing 25 mM glucose and the same concentration of DMSO as that in the serosal side. The medium was pre-warmed at 37 ^{o}C and pre-oxygenated with 5% CO_2/ 95% O_2 for 15 minutes. Under bubbling with a CO_2/O_2 mixture gas, the transport of the probe drug from serosal to mucosal surfaces across the intestine was measured by sampling the mucosal medium periodically for 90 minutes.

In an inhibition study, the inhibitor ketoconazole at 200 μM in DMSO was added to the mucosal medium. Using these media the transport of probe drug in the absence (control) or presence of inhibitor (200 μM ketoconazole) and under induced conditions after pre-treatment with rifampicin (60 mg/kg/po for 7 days) and sodium butyrate (0.5 mg/kg/ip for 5 days) probe drug was measured (124, 125).

Noneverted Intestinal Sac Study

Male wistar rats weighing 200 ± 25 gm are selected for experiments. Phytochemicals to be screened were administered to rats at a defined dose for seven days. Ketoconazole (standard inhibitor) suspension was prepared by suspending 1 gm in 0.25 % w/v of sodium carbaoxy methyl cellulose and was administered at a dose of 50 mg Kg^{-1} to rats for seven days. Untreated rats were used as control.

The rats were fasted overnight with free access to water before the experiments. Control rats and pretreated rats on seventh day were sacrificed using anesthetic ether, the intestine was surgically removed and flushed with 50 mL of ice cold saline. The small intestine was cut into 3 segments, duodenum, jejunum, ileum and colon of equal length (10 cm).

The probe drug was dissolved in pH 7.4 isotonic Dulbecco's PBS (D-PBS) containing 25 mM glucose. The probe drug solution (1 mL) was filled in the normal sac (mucosal side), and both ends of the sac were ligated tightly. The sac containing probe drug solution was immersed in 40 mL of D-PBS, containing 25 mM glucose in the mucosal side. The medium was pre-warmed at 37° C and pre-oxygenated with 5 % CO_2/ 95 % O_2 for 15 minutes, under bubbling with a CO_2/O_2 mixture gas, the transport of the probe drug from mucosal to serosal surfaces across the intestine was measured by sampling the serosal medium periodically for 120 minutes. The samples of 1 mL were collected at predetermined time intervals from the serosal medium and replenished with fresh buffer. The drug transported was measured using high performance liquid chromatography (HPLC) method (126, 127).

Experimental Methods and Design

The use of appropriate experimental design can provide definitive evidence that P-gp mediated efflux is altering the transport of a compound, and can provide further mechanistic information regarding the transport of a compound. Many of the following techniques can be applied to any of the *in vitro* model systems described above.

Transport Assay

The most direct method of identifying the effect of P-gp on drug absorption is to measure the transport of drug molecules in both apical-to-basolateral and basolateral-to-apical directions. A significantly larger effective permeability in the basolateral-to-apical direction provides evidence that some form of secretory pumps such as P-gp is enhancing the transport of test compound in the secretory direction above what is expected from simple passive diffusion. As a consequence of this secretory mechanism, the apical-to-basolateral transport is reduced, whereas the basolateral-to-apical transport is enhanced. For a typical P-gp substrate, a plot of flux in the secretory direction versus concentration has both a passive diffusion component and a saturable (Michaelis - Menten) component. Demonstration of saturable efflux in the secretory direction provides direct evidence that an efflux pump such as P-gp, which has the finite capacity, is active in the test compound. By inhibiting P-gp efflux, the additional secretory component is removed and transport is expected to resemble a passive diffusion process i.e., transport in each direction should converge to a common value (116).

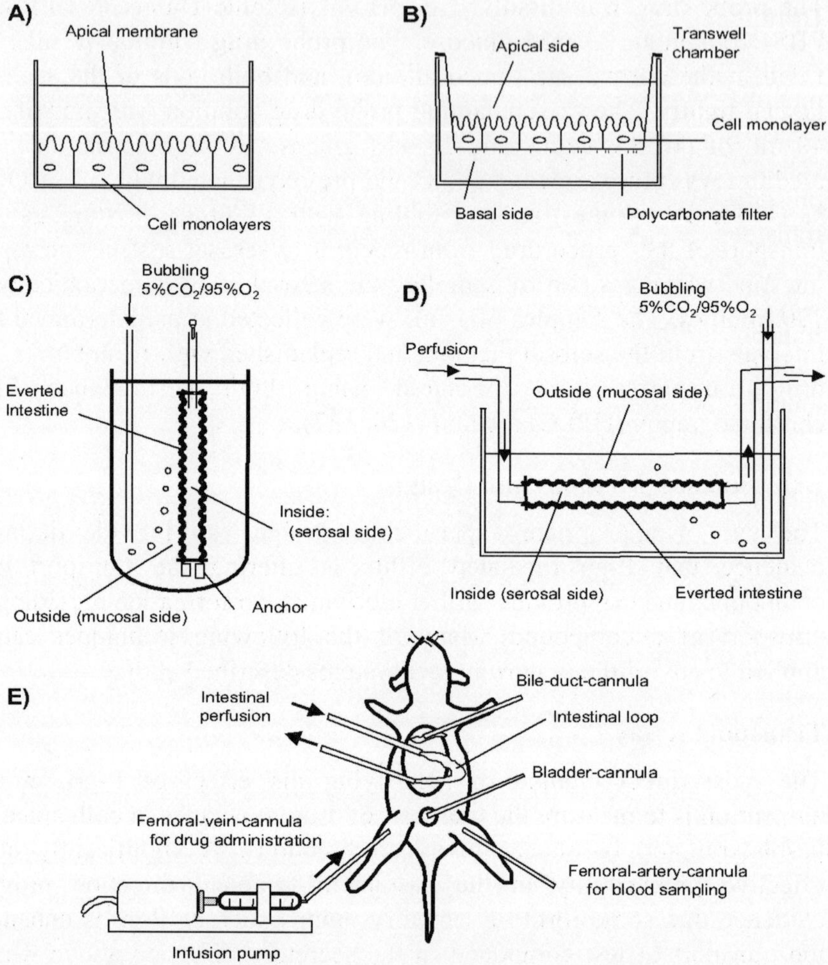

Mikihisa Takano et al., *Pharmacology & Therapeutics* 109 (2006) 137 – 161

Schematic representation of some experimental methods for the functional analysis of intestinal efflux transporters located on apical membrane: culture cells (A, B), everted intestine (C, D), and whole animal (E).

Competition Assay

Fluorescent dyes such as Calcein-AM and rhodamine derivatives have been demonstrated to be P-gp substrates. These compounds can be used in competition assays in which the test compound is added with these dyes. Any reduction in the dye efflux would be indicative of the inhibitory properties of the test compounds toward P-gp.

Uptake Assay (Uptake or Accumulation Assay)

Accumulation and efflux studies can be performed on cell suspension cell monolayer, or membrane vesicle preparations. For accumulation studies, uptake of a probe, typically either fluorescent or radio labeled, into the cell or membrane vesicle is examined under control conditions and in the presence of inhibitors of the drug efflux transporter system. For transfected cells or drug efflux transporter, accumulation studies can be compared to the wild type or parental cell line that does not have as high levels of drug efflux transporter expression. As the drug efflux transporters are outwardly directed systems, in the cellular accumulation studies, inhibition of the transporters would result in an increase in the amount of probe in the cell. For efflux studies, cells are preloaded with the probe of interest and resulting transport of the probe out into extracellular environment is measured under various conditions known to influence transporter activity (128).

ATPase Activity Assay

Binding of ATP to its nucleotide binding site on the protein is essential for substrate transport as is the hydrolysis of ATP by P-gp ATPase for restoring the transporter to its active conformational state. Thus monitoring of ATPase activity in cell membrane preparations or purified membrane proteins represents a method of identifying those compounds that interact drug efflux transporters. The essential components of ATPase assay include a cell membrane preparation enriched with the drug efflux transport protein of interest, ATP, an analytical method for detecting inorganic phosphate generated from ATP hydrolysis, and a mechanism for discriminating between general ATPase activity and ATPase activity related to the transporter of interest. The ATPase activity associated with the activation of the transporter is distinguished from other ATPase activity through the inclusion of various inhibitors such as oubain, EGTA, and azide to inhibit Ca- ATPases, Na^+, K-ATPases, and mitochondrial ATPases, respectively (128).

7.5.1.2 *IN SITU* MODELS

Some efforts have been made to determine the effect of P-gp on the disposition of its substrates by use of *in situ* perfusion methods, including intestinal perfusion, liver perfusion, kidney perfusion, and brain perfusion. These experiments allow the researcher to study the transport of compounds in a physiologically relevant environment in which the integrity of the organ is preserved with regard to cell polarity and representation of all cell types seen in the organ.

Intestinal Perfusion

In situ intestinal perfusion studies are typically done with live animals in which a perfusion loop has been inserted into the intestine. Depending on the experimental protocol, the system can offer a relatively unbiased view of intestinal transport with respect to the normal expression of transporters in healthy animals. One limitation of this protocol is that the disappearance rather the appearance of a compound is determined (appearance can be determined by collection of blood in the vessels perfusing the section of intestine studied, a process requiring significant surgical skill). Estimates of the polarity of transport imparted by P-gp are difficult to assess and typically can be determined only by using an inhibitor or antibody to P-gp, each of which may have unknown effects on the passive transport of the test compound. Often the animal is anesthetized, and the anesthetic agent can further affect the results (altered membrane fluidity, possible inhibitory effects on P-gp mediated efflux activity). There are some other obvious limitations. Using the intact intestine adds more levels of complexity, which can further confound studies meant to elucidate the role of transporters, which act at the cellular level. It is possible that results will differ by intestinal region and also due to the presence of the Payer's patches, which have different physiological roles from enterocytes. These studies suffer from an interspecies variability (rats are typically the test subjects). Despite certain disadvantages, these studies, if conducted with appropriate controls involving known P-gp substrates, can provide valuable insights on how to correlate the effect of P-gp observed in cellular transport studies to that expressed in the absorption of drugs *in vivo*.

These *in situ* techniques can be powerful tools to gauge the actual extent of P-gp efflux that can be expected *in vivo*. However, there are confounding factors that must be addressed when interpreting data obtained from these studies. As with all biological models, the appropriate controls must be used to ensure the observed effects are in fact due to P-gp mediated efflux activity.

7.5.1.3 *In Vivo* Models

While *in vitro* assays can be used to characterize the various drug transporter proteins and provide information concerning the interaction of compounds with these transporters, the ultimate determination of the impact of the drug efflux transporters on drug absorption, distribution and elimination requires *in vivo* examination. In this regard, the transgenic and mutant animal models have provided important tools for assessing drug efflux transporter activity.

Transgenic Models

The development of transgenic animals has been an important tool in evaluating genes and their products, proteins. Unlike humans, which have only one gene that encodes the P-gp involved in multiple drug resistance (MDR1), mice have two, mdr1a and mdr1b. Consequently, one and /or both can be knocked out to evaluate P-gp function in drug absorption, distribution, and excretion. Both mdr1a and mdr1a/b knockout mice have been used to evaluate the contribution of P-gp in tissues to drug disposition, efficacy, toxicity, as well as cellular developmental processes (128).

Mutant Models

Currently there are two mutant animal models used in evaluating drug efflux activity. Unlike transgenic mice that are produced through genetic engineering, these mutant models are naturally deficient in the expression of a drug efflux transport protein. These mutants offer an additional tool in evaluating the role of specific drug efflux transporters in drug pharmacokinetics, efficacy, and toxicity (128).

Table 7.1 Advantages and limitations *in-vitro* methods (128).

Assay type	Advantage	Limitation
Uptake / efflux	Can use whole cells (adherent or in suspension), or membrane vesicles. Readily adaptable for flow cytometry. Easy and fast method for high throughput screening of drug efflux interactions (could be either substrate or inhibitor)	Cannot determine localization of transporters. Tends to underestimate substrate activity of low permeability compounds
Transport / permeability	Can be used to help identify cellular localization of drug efflux transporters in plasma membrane. Is definitive method for determining the effect of drug efflux transporter of a drug (i.e. determines substrate activity)	Requires confluent epithelial or endothelial cells with tight cellular cells junctions. Tends to underestimate substrate activity of high permeable compounds

Contd...

Assay type	Advantage	Limitation
ATPase	Commercially available P-gp membrane preparation. Does not require whole cells. Readily adaptable to high throughput screening of drug efflux interactions	High intra and inter assay variability. Tends to underestimate substrate activity of low permeable compounds. Localization of transporters is difficult.

7.6 Correlation of *In Vivo* and *In Vitro* Screening Models

In vitro models have provided invaluable information about properties of compounds that affect their *in vivo* transport and absorption. Regardless of how closely these *in vitro* systems model *in vivo* conditions, they do not completely represent what may be seen *in vivo*. It is important to compare the results obtained from some key *in vitro* and *in vivo* experiments so that the magnitude of certain processes seen *in vitro* can be gauged properly and so that any dissimilarity between the *in vitro* and *in vivo* systems can be identified. Although it is certain that these relationships will not hold for all drug compounds, a comparison of the data for a limited set of compounds is useful.

In studies by Yamazaki et al., a good correlation (r^2 =0.93) was observed between brain accumulation of drugs in CF-1 mutant mice and the transcellular transport ratios L-mdr1a epithelial cells.

P-glycoprotein inhibitors: an overview

Screening studies for P-gp–drug interactions identified a number of clinically important drugs as P-gp substrates, which are as diverse as anthracyclines (doxorubicin, daunorubicin), alkaloids (reserpine, vincristine, vinblastine), specific peptides (valinomycin, cyclosporine), steroid hormones (aldosterone, hydrocortisone) and local anaesthetics (dibucaine). Even dye molecules (Rhodamine 123) and pharmaceutical excipients exhibited P-gp substrate activity (Table 7.2). Few of them were identified to inhibit P-gp, setting off an opportunity in MDR reversal. Improved clinical efficacy of various drugs observed by P-gp inhibition, especially drug subjected to MDR, lead to the design and development of modulators, which specifically block P-gp efflux and having improved toxicity profiles. P-gp inhibitors are gaining recognition to improve bioavailability by inhibiting P-gp in intestine, brain, liver and kidneys, which has been hypothesised and emphasized by many researchers in recent years (129).

Based on the specificity and affinity, P-gp inhibitors are classified to three generations. First-generation inhibitors are pharmacological actives, which are in clinical use for other indications but have been shown to inhibit P-gp. These include calcium channel blockers such as verapamil; immunosuppressants like cyclosporin A; anti-hypertensives, reserpine, quinidine and yohimbine; and antiestrogens like tamoxifen and toremifena. The usage of these compounds is limited by their toxicity due to the high serum concentrations achieved with the dose that is required to inhibit P-gp. A great deal of research by industrialist and academicians in the direction of improving toxicity profile resulted in second and third-generation inhibitors that specifically modulate P-gp. Second-generation modulators are agents that lack the pharmacological activity of the first-generation compounds and usually possess a higher P-gp affinity. However, inhibition of two or more ABC transporters leads to complicated drug–drug interactions by this class of compounds, which include non-immunosuppresive analogues of cyclosporin A, PSC 833; D-isomer of verapamil, dexverapamil; and others such as biricodar (VX-710), GF120918 and MS-209. On the other hand, several other novel third-generation P-gp blockers are under development, however, primarily with the purpose to improve the treatment of multidrug resistant tumours and to inhibit P-gp with high specificity and toxicity. Modulators such as LY335979, OC144093 and XR9576 are identified to be highly potent and selective inhibitors of P-gp with a potency of about 10-fold more than the first and second-generation inhibitors.

In general, P-gp can be inhibited (i) by blocking drug binding site either competitively, non-competitive or allosterically; (ii) by interfering ATP hydrolysis; and (iii) by altering integrity of cell membrane lipids (130). Although most of the drugs inhibit P-gp function by blocking drug binding sites, presence of multiple binding sites complicate understanding as well as hinder developing a true, conclusive SAR for substrates or inhibitors. However, the mode of handling of substrates and inhibitors are same by P-gp if the protein transport and/or inhibition are mediated only through binding sites. Then the issue to be addressed is how the substrates and inhibitors are discriminated at the molecular level. In this regard, Eytan et al., proposed a plausible explanation that the modulator or inhibitor 'flipped' by P-gp can 'flop' back into the inner leaflet of the membrane, for further transport, which is very rapid creating a large difference between the rate of efflux of the substrate and inhibitor (131). Thus, the P-gp modulator is cycled repeatedly, preventing efflux of substrates, which depends on the hydrophobicity of the compound. This concept had been proved from the drug delivery point of view that absorption of high affinity drugs to the protein need not necessarily be limited by P-gp, e.g. verapamil, if it is highly permeable whereas less permeable drugs though weak substrates may undergo a substantial extrusion mediated by P-gp, e.g. tanilol (132). Compounds inhibiting ATP hydrolysis could serve as better inhibitors, since they are unlikely to be transported by P-gp, and these kinds of agents will require at low dose which is well desirable to use locally at gut lumen. Quercitin, a naturally occurring flavanoid, has been proposed to block P-gp function by an unknown mechanism but in general by interfering ATPase activity (133). Since none of the substrates till now had been found to interact with the nucleotide binding sites to interfere the P-gp ATPase catalytic cycle, further research in exploring the detailed mechanism of inhibition of ATP hydrolysis would provide newer and better inhibitors with potent and specific activity.

Commonly used pharmaceutical surfactants are emerging as a different class of P-gp inhibitors, which act by altering integrity of membrane lipids. The change in secondary and tertiary structure is found to be the reason for loss of P-gp function due to disturbance in hydrophobic environment by surfactants. In a series of studies by Hugger and co-workers (134, 135) it was observed that the change in fluidity of cell membrane facilitates influx of P-gp substrates by surfactants like polyethylene glycol, cremophor EL and Tween 80, demonstrated in Caco-2 cell line.

Early studies on recemic verapamil to reverse P-gp mediated resistance to vincristine and vinblastine (136) and its established record of safety provided the rationale for its clinical usefulness as P-gp inhibitor. In addition to this, orally administered verapamil has been shown to increase peak plasma level, prolong elimination of half-life and increase volume of distribution of doxorubicin after oral administration (137). The total body clearance of paclitaxel and digoxin has been found to decrease substantially after co-treatment of verapamil in human subjects (138). However, verapamil being a potent cardiovascular drug showed serious toxicities at the plasma levels needed for effective MDR1 reversal. As a result, dexverapamil emerged as a second-generation inhibitor which does not have any effect on cardiovascular system. Combination therapy of dexverapamil and paclitaxel in metastatic breast cancer patients showed increased mean peak paclitaxel concentration and delayed clearance (139). The effects of cyclosporine A on the pharmacokinetics of etoposide have been demonstrated to be dose-dependent in a phase I clinical trial (140), however, its use for long-term oral dosing may be hindered by the immunosuppressive effect. This lead to the design of valspodar (PSC 833), an analogue of cyclosporine D with no immunosuppressive activity. *In vitro* studies demonstrated that PSC 833 may be as much as 20 times more potent inhibitor of P-gp as cyclosporine A (141). In a comparative study of pharmacokinetics of paclitaxel in mdr1a(−/−) mice, enhanced oral absorption of paclitaxel was observed when wild-type mice were co-treated with PSC 833. Similarly, PSC 833 found to enhance oral bioavailability of anticancer drug etoposide in rats (142). GF120918, an acridonecarboxamide derivative, was shown to be a potent blocker of P-gp in tumour cells *in vitro* and *in vivo* (143). Further, it has been taken up to demonstrate its role in improving intestinal absorption of drugs like paclitaxel (144). The plasma concentration of paclitaxel in wild-type mice receiving GF120918 was found to be similar to that observed in mdr1a-knockout mice for at least 12 h after i.v. administration suggesting that GF120918 blocks P-gp during this entire period.

Third-generation modulators, LY335979, XR9576 and OC144093 are highly potent and selective inhibitors of P-gp with less of drug interactions and toxicity. LY335979 is among the most potent modulators of P-gp. Treatment of mice, bearing P388/ADR murine leukaemia with LY335979 in combination with doxorubicin or etoposide, administered intravenously/intraperitonially, showed a significant increase in life span without significant effect on the pharmacokinetics of these anticancer agents (145). In contrast, LY335979 reduced paclitaxel clearance by approximately 19% when administered by infusion (146). This may be due to its poor inhibition of four major cytochrome P450 isozymes important in metabolizing doxorubicin and etoposide. It was found that LY335979 is around 60-fold more specific to P-gp over cytochrome P450 3A4, the major metabolizing enzyme. The selectivity and potency of this modulator avoids complicated drug interactions, which make pharmacokinetic optimization difficult. In spite of wide acceptance of MDR modulators for the cancer chemotherapy, the following issues should be addressed for a meaningful translation of MDR modulators to bioavailability enhancers. The inherent pharmacological action and higher concentration of drug required to inhibit the protein function may not be compromised for bioavailability enhancement. In addition, a local effect at GIT would be desirable rather a systemic effect and whole body burden.

Table 7.2 Agents that interact with P-glycoprotein

Pharmacological category	Examples
Antiarrhythmics	Amioderone, lidocaine, quinidine
Antibiotics and antifungals	Cefoperazone, ceftriazone, erythromycin, itraconazole, ketaconozole, aureobasidin A
Antimalarials and antiparasites	Chloroquin, emetine, hydroxychloroquin, quinacrine, quinine
Calcium channel blockers	Bepridil, diltiazem, felodipine, nifedipine, nisoldipine, nitrendipine, tiapamil, verapamil
Calmodulin antagonist	Chlorpromazine, trifluperazine
Cancer chemotherapeutics	Actinomycin D, colchicines, daunorubicin, doxorubicin, etoposide, mitomycin C, mitramycin, podophyllotoxin, puromycin, taxol, topotecan, triamterene, vinblastine, vincristine
Fluorescent dyes protease inhibitors	BCECF-AM, fluro-2, fura-2, rhodamine 123, HIV Indinavir, nelfinavir, ritonavir, saquinavir
Hormones	Aldosterone, clomiphene, cortisol, deoxycorticosterone, dexamethosone, prednisone, progesterone analogs, tamoxifen, hydrocortisone, testosterone
Immunosuppressants	Cyclosporin A, cyclosporin H, tacrolimus, sirolimus
Indole alkaloids	Reserpine, yohimbine
Local anaesthetics	Bupivacaine
Surfactants/solvents	Cremophor-EL, triton X-100, tween 80
Toxic peptides	N-Acetyl-leucyl-leucinal, gramicidine D, valinomycin
Tricyclic antidepressants	Desipramine, trazadone
Miscellaneous	Components of grape and citrus fruit juice, ethidium bromide, GF120918, ivermectin, MS-209, liposomes, LY335979, quercetin, SDZ PSC 833 (valspodar), terfindine, tumour necrosis factor, Vitamin A

7.7 MDR and CYP3A4 Mediated Drug–Herbal Interactions

Many drug substances along with a variety of naturally occurring dietary or herbal components are capable of interacting with the CYP enzyme system and P-gp efflux pump in several ways:

1. A herbal component can be a substrate of one or several isoforms of CYP enzymes and/or efflux systems (P-gp, MRP and BCRP). Therefore, one substrate can compete for another substrate for either metabolism by the same CYP isozyme and/or efflux system resulting in higher plasma concentrations due to competitive inhibition.
2. A herbal constituent can also be an inducer of one or several CYP isoforms and/or efflux systems, thereby lowering plasma concentrations due to either higher metabolism and/or higher efflux. Such interactions may produce subtherapeutic plasma drug concentrations.
3. A compound can also be an inhibitor of CYP450 enzymes resulting in reduced activity of one or several isoforms of CYPs. If a compound is an inhibitor of efflux system, it will reduce drug efflux resulting in improved absorption. However, induction is a slow process, dependent on the rate of protein synthesis. Expression of specific mRNA may be possible within a few hours, but functional expression and maturation of such proteins may require longer duration. In contrast, inhibition is more rapid and can produce results within a very short period of time, particularly if the inhibition is competitive in nature.

Role of CYP450 in drug–herbal interaction

The most versatile enzyme system involved in the metabolism of xenobiotics is cytochrome P450. The CYP3A family of enzymes constitutes the most predominant phase-I drug metabolizing enzymes and accounts for approximately 30% of hepatic CYP and more than 70% of intestinal CYP activity. Moreover, CYP3A is estimated to metabolize between 50% and 70% of currently administered drugs (4). A congener of CYP family is CYP3A4, the most abundant form (5). This CYP3A4 enzyme is present primarily in the hepatocytes and enterocytes (147, 148). It is now fairly established that naturally occurring dietary supplements can modulate hepatic and entrocytic CYP activity. Perhaps the best documented clinically relevant drug interaction is observed with grapefruit juice. Simultaneous consumption of grapefruit juice with a number of therapeutic agents that are subject to first pass intestinal/hepatic metabolism, resulted in higher plasma levels with subsequent adverse effects (149, 150). Grapefruit juice acts through inhibition of intestinal CYP3A4, which regulate pre-systemic metabolism (151). Although hepatic biotransformation can make a major contribution to systemic drug elimination, a combination of hepatic and intestinal drug metabolism may cause significant pre-systemic or first-pass drug loss. Preliminary studies that have directly investigated SJW interactions with CYPs indicate that it may modulate CYP, particularly the 3A4 isoform. Several *in vitro* studies revealed that crude extract of SJW inhibits CYP3A4 and such inhibition is of a competitive nature (152, 153). These studies also identified hyperforin, hypericin, quercetin and 13,118-biapigenin as components primarily responsible such inhibitory interactions. Since HIV protease inhibitors, macrolide antibiotics and azole antifungals along with many herbal agents are substrates of the same CYP3A4, these compounds can affect oral bioavailability of therapeutic agents indicated in the treatment of immunosuppression, cancer, AIDS and other opportunistic infections. Depending on the mechanisms of herbal interactions with therapeutic agents, these substances could lower blood levels of the anti-HIV drugs, thus possibly putting people at risk for the development of resistance. Even though many clinical studies reported that St. John's wort had an inductive action on CYP3A4, one study suggested that SJW had no statistically significant effect on CYP3A4 induction (154). This latter report is consistent with another recent study describing CYP3A4 metabolic interaction (155). This study reported that 4-day treatment of SJW extract or its constituents, hypericin and hyperforin, in mice did not result in any CYP3A4 induction. In contrast, an inhibitory effect of the major constituents of St. John's wort was also reported in the CYP transfected cells (152). In line with this observation, quercetin one of the major constituents in St.John's wort was reported to have an inhibitory effect on CYP3A4 (156). The discrepancy so far remains unresolved.

Since the herbal products can also competitively inhibit CYP, simultaneous intake of these agents may raise blood levels of therapeutic agents, thus exposing patients to a greater risk of serious side effects. A recent report revealed pure herbal constituents (quercetin, hypericin and kaempferol) inhibit CYP3A4-mediated cortisol metabolism (153). However, silibinin did not inhibit any cortisol metabolism.

Approaches to assessing drug–herb interactions

Overlapping substrate specificities of these proteins result in complex and sometimes perplexing pharmacokinetic profiles of multidrug regimens.

Saquinavir undergoes extensive first pass metabolism by the major metabolizing isozyme CYP3A4.

Ketoconazole (a selective CYP3A4 inhibitor) inhibited the formation of all saquinavir metabolites. Also, saquinavir lowers the metabolism of terfenadine and causes formation of 6-β- hydroxylation products of testosterone, indicating its specificity towards CYP3A4 (157). Metabolism of ritonavir on the other hand is caused by both CYP3A4 and CYP2D6. It significantly inhibits the metabolism of CYP3A4 and CYP2D6 substrates like nifedipine and dextromethorphan, respectively, when administered in combination (158). The major isozyme responsible for indinavir metabolism is CYP3A4. However, metabolism of nelfinavir is caused by several isozymes including CYP3A4 followed by CYP2C19, CYP2D6 and possibly CYP2C9 and CYP2E1 (159, 160, 161).

Role of efflux proteins on drug-herbal interaction

Multidrug resistance (MDR) proteins play an important role in protecting cells against cytotoxic drugs (162). It has recently become apparent that MDR gene products also need to be considered in drug absorption. The multidrug resistance phenotype in tumors is associated with over expression of ATP binding cassette (ABC) efflux pumps termed MDR proteins. Pglycoprotein (P-gp, MDR1, ABCB1) (163, 164, 165) is considered a versatile xenobiotic pump. Substrate specificity and tissue distribution of MDR proteins vary widely. MRP1 is almost ubiquitously expressed, while the expression of P-gp is more restricted to tissues involved in absorption and secretion. Although P-gp was initially discovered in cancer cells, it was later observed that a number of normal tissues such as intestine, liver, kidney, pancreas and adrenal gland constitutively express P-gp. High levels of P-gp is expressed in blood–brain barrier and the choroid plexus (55, 75,166). All multidrug transporters are localized predominantly in the plasma membrane. In polarized cells, P-gp is localized in the apical (luminal) membrane surface (e.g., in the epithelial cells of the intestine and the proximal tubules of kidney, and in the biliary canalicular membrane of hepatocytes).

P-gp mediated drug–herb interactions

In the last 5 years, St. John's wort has been on the top of the list of herbal usage. Indinavir and saquinavir concentrations were reduced to 57% by SJW and 51% by garlic, respectively (167). However, the fact that its interaction could alter outcome of anti-HIV therapy was realized very recently (167). Such reduction in indinavir and saquinavir exposure may lead to the development of drug resistance strains and may cause treatment failure in HIV patients. Such reduction was attributed to the induction of P-gp and CYP3A4 expression by SJW (168). Several flavanoids, which constitute one of the primary classes of active constituents in most herbs, appear to be capable of modulating P-gp. SJW was shown to induce P-gp (169, 170, 171). Quercetin and kaempferol were found to induce P-gp (54, 172, 173). Ten days treatment with pure herbal constituents (hypericin, kaempferol, quercetin and silibinin) can cause a significant increase in P-gp-mRNA expression (153). Quercetin and kaempferol have also been reported to induce P-gp (54, 172, 173). Also, four- to seven-fold elevation in the expression of P-gp in LS180 intestinal carcinoma cells was caused by hypericin or SJW treatment (170). *In vivo* studies also indicated that long-term (14 days) exposure of SJW leads to higher expression of MDR1 in rat intestine (169). SJW taken as 900 mg/day for 14 days resulted in 1.4-fold increase of P-gp expression in healthy volunteers (169). In another clinical study, a 4.2-fold increase of P-gp expression was observed in human peripheral blood lymphocytes after chronic treatment with SJW for 16 days (174).

Since HIV protease inhibitors, macrolide antibiotics, azole antifungals and herbals are substrates of same the metabolizing enzymes and transporters, herbal agents can adversely affect the course of HIV treatment and other opportunistic infections. *In vitro* studies from laboratory indicated that concomitant administration of erythromycin with SJW and/or ketoconazole can enhance erythromycin oral absorption. Depending on the mechanism by which herbal compounds interact with CYP450 and efflux proteins, these agents could lower plasma levels of the anti-HIV drugs, thus possibly reducing efficacy and enhancing the risk for the development of drug resistance. Reversibly, these compounds could raise the blood levels of the antiretrovirals, thus placing patients at greater risk of serious side effects. Herbs with reported effects on the CYP3A4 and P-gp include SJW, garlic, ginseng, milk thistle and skullcap.

Employing MDR1 transfected MDCK cells expressing high amount of P-gp, has been recently demonstrated that ritonavir uptake was enhanced by five to eight-fold in the presence of 100ìM pure herbal constituents (allicin, kaempferol, quercetin and hypericin) (153). In other *in vitro* studies, it was also observed several folds increase in erythromycin and ritonavir uptake in the presence of SJW extract. Inhibitory effect of MRP-mediated ritonavir efflux by SJW was also noted. These results also demonstrate that simultaneous administration of SJW with drugs like erythromycin, saquinavir and ritonavir, can enhance drug absorption due primarily to competitive inhibition of P-gp/ MRP-mediated efflux. These *in vitro* results demonstrated the inhibitory properties of herbs towards P-gp-mediated efflux on short-term exposure. Thus, all these *in vitro* and *in vivo* studies have demonstrated that SJW, upon chronic exposure induces intestinal P-gp resulting in reduced intestinal absorption possibly through enhanced drug efflux.

Coordinated function of efflux and metabolism

In addition to oxidative metabolism, conjugation reactions may play an important role in the detoxification of xenobiotics from small intestine. Several drug molecules are effluxed into intestinal lumen after being conjugated with a glucuronide or sulfate moiety. The transport system responsible for the cellular extrusion of thc conjugated metabolites and organic anions has been characterized recently (175). Herbs can pharmacokinetically act as inhibitors or inducers, when anti- HIV medication or any other conventional therapeutics taken simultaneously. An understanding of the increased/decreased bioavailability of one drug in the presence of herbals may greatly aid in the design of appropriate drug regimen.

Coadministration of herbal and therapeutic drugs may lead to increased absorption due to inhibition of P-gp-mediated efflux and CYP-mediated metabolism leading to potential toxic effects. In contrast, chronic administration of certain herbal products (SJW, garlic, etc.) will enhance the production of MDR proteins (P-gp) and CYP enzymes resulting in lower bioavailability and subtherapeutic plasma concentrations of drugs. Finally, this process may lead to the emergence of drug resistance. Activation of the mammalian xenobiotic-sensing nuclear receptors PXR and CAR: P-gp/MDR1 is regarded as one of the major factor in the development of cellular drug resistance. Upregulation of CYP enzymes in response to herbals may also play an important role in lowering drug concentration. A close chromosomal location of P-gp and CYP3A4 genes, their expression in mature enterocytes and similar substrate specificities suggest that the function of these two proteins may be complementary in nature and may form a coordinated intestinal barrier (176).

This view was further validated by the stimulatory effect of SJW on pregnane X receptor, which regulates many CYP isoforms in rats (177). A recent study also demonstrated that SJW induces CYP3A4 by negatively acting on interleukin-6, which is known to inhibit the pregnane X receptor that in turn may be involved in the expression of the CYP class of enzymes (178). Molecular mechanisms of induction of P-gp and drug metabolizing enzymes are still being investigated by researchers. Recently, the role of two members of nuclear receptor superfamily of transcription factors, the pregnane X receptor (PXR) and the constitutive androstane receptor (CAR) have been discovered.

References

1. Manthena, V. S. Varma, Yasvanth Asokraj, Chinmoy, s. Dey, Ramesh panchagula. P-glycoprotein inhibitors and their screening: a perspective from bioavailability enhancement. *Pharmacol.* Res. 48 347-359 (2003).
2. Benet, L. Z., Wu, C. Y., Hebert, M. F., Wacher, V. J. Intestinal drug metabolism and antitransport processes: a potential paradigm shift in oral drug delivery. *J Control Release* 39: 139-143 (1996).
3. Benet, L. J. Carolyn, L. Cummins. The drug efflux – metabolism alliance: biochemical aspects. *Adv. Drug. Del. Rev.* 50 S3-S11 (2001).
4. Watkins, P.B., Wrighton, S.A., Schuetz, E.G., Molowa, D.T., Guzelian, P.S. Identification of glucocorticoid-inducible cytochromes P-450 in the intestinal mucosa of rats and man. *J. Clin. Invest.* 80 (4), 1029–1036 (1987).
5. Kolars, J.C., Schmiedlin-Ren, P., Schuetz, J.D., Fang, C., Watkins, P.B. Identification of rifampin-inducible P450IIIA4 (CYP3A4) in human small bowel enterocytes. *J. Clin. Invest.* 90 (5), 1871–1878 (1992).
6. Kolars, J. C. Awni, W. M. Watkins, P.B. et al. First pass metabolism of cyclosporin by gut. *Lancet.* 338 1488-1490 (1991).
7. Helena Engman. Intestinal barriers to oral drug absorption-Cytochrome P 450 3A and ABC–transport proteins (Comprehensive summaries of Uppsala Dissertations from the Faculty of Pharmacy) (2003).
8. Yuanchao Zhang and. Benet, L.Z. The gut as a barrier to drug absorption; combined role of Cytochrome P4503A and P-Glycoprotein. *Clin. Pharmacokinet* .40 (3) 159-163 (2001).
9. Hayashi, M., Tomita, M., Awzu, S. Transcellular and paracellular contribution to transport processes in the colorectal route. *Adv. Drug Deliv. Rev.* 28, 191–204 (1997).

10. Hunter, J., Hirst, B.H. Intestinal secretion of drugs. The role of P-glycoprotein and related drug efflux systems in limiting oral drug absorption. *Adv. Drug Del. Rev.* 25, 129–157 (1997).

11. Evers, R., Kool, M., Van Deemter, L., Janssen, H., Calafat, J., Oomen, L.C.J.M., Paulusma, C.C., Oude Elferink, R.P.J., Baas, F., Schinkel, A.H., Borst, P. Drug export activity of the human canalicular multispecific organic anion transporter in polarised kidney MDCK cells expressing cMOAT (MRP2) cDNA. *J. Clin. Invest.* 101, 1310–1319 (1998).

12. Fromm, M.F., Kauffman, H.M., Fritz, P., Burk, O., Kroemer, H.K., Warzok, R.W., Eichelbaum, M., Siegmund, W., Schrenk, D. The effect of rifampin treatment on intestinal expression of human MRP transporters. *Am. J. Pathol.* 157, 1575–1580 (2000).

13. Hunter, J., Jepson, M., Tsuruo, T., Simmons, N.L., Hirst, B.H. Functional expression of P-glycoprotein in apical membranes of human intestinal Caco-2 cells. Kinetics of vinblastine secretion and interaction with modulators. *J. Biol. Chem.* 268, 14991–14997 (1993a).

14. Hunter, J., Hirst, B.H., Simmons, N.L. Drug absorption limited by P-glycoprotein-mediated secretory drug transport in human intestinal epithelial Caco-2 cell layers. *Pharm. Res.* 10, 743–749 (1993b).

15. Walgren, R.A., Karnaky, K.J., Lindenmayer, G.E., Walle, T. Efflux of dietary flavonoid quercetin 4'_β_glucoside across human intestinal Caco-2 cell monolayers by apical multidrug resistance-associated protein-2a. *J. Pharmacol. Exp. Ther.* 294, 830–836 (2000).

16. Watkins, P.B. Drug metabolism by cytochromes P450 in the liver and small bowel. *Gastrointest. Pharmacol.* 21, 511–526 (1992).

17. Keppler, D., Cui, Y., König, J., Leier, I., Nies, A. Export pumps for anionic conjugates encoded by MRP genes. *Adv. Enzyme Regul.* 39, 237–246 (1999).

18. Evans, A.M. Influence of dietary components on the gastrointestinal metabolism and transport of drugs. *Ther. Drug Monit.* 22, 131– 136 (2000).

19. Cordon-Cardo, C., O'Brien, J.P., Casals, D., Bertino, J.R., Melamed, M.R. Expression of the multidrug resistance gene product (P-glycoprotein) in human normal and tumor tissues. *J. Histochem. Cytochem.* 38, 1277–1287 (1990).

20. Fojo, A.T., Shen, D.W., Mickley, L.A., Pastan, I., Gottesman, M.M. Intrinsic drug resistance in human kidney cancer is associated with expression of a human multidrug-resistance gene. *J. Clin. Oncol* 5, 1922–1927 (1987a).
21. Merkel, D.E., Fuqua, S.A.W., Tandom, A.K., Hill, S.M., Buzdar, A.U., McGuire, W.L. Electrophoretic analysis of 248 clinical breast cancer specimens for P-glycoprotein overexpression of gene amplification *J. Clin. Oncol.* 7, 1129–1136 (1989).
22. Fojo, A.T., Ueda, K., Slamon, D.J., Poplack, D.G., Gottesman, M.M., Pastan, I. Expression of a multidrug-resistance gene in human tumors and tissues. *Proc. Natl. Acad. Sci.* U.S.A. 84, 265–269 (1987b).
23. Gatmaitan, Z.C., Arias, I.M. Structure and function of P-glycoprotein in normal liver and small intestine. *Adv. Pharmacol.* 24, 77–97 (1993).
24. Begley, D.J., Lechardeur, D., Chen, Z.D., Rollinson, C., Bardoul, M., Roux, F., Scherman, D., Abbott, N.J. Functional expression of P-glycoprotein in an immortalised cell line of rat brain endothelial cells RBE4. *J. Neurochem.* 67, 988–995 (1996).
25. Hunter, J., Hirst, B.H., Simmons, N.L. Epithelial secretion of vinblastine by human intestinal adenocarcinoma cell (HCT-8 and T84) layers expressing P-glycoprotein. *Br. J. Cancer* 64, 437–444 (1991).
26. Karlsson, J., Kuo, S.-M., Ziemniak, J., Artursson, P. Transport of celiprolol across human intestinal epithelial (Caco-2) cells: mediation of secretion by multiple transporters including P-glycoprotein. *Br. J. Pharmacol.* 110, 1009–1016 (1993).
27. Cavet, M.E., West, M., Simmons, N.L. Transport and epithelial secretion of the cardiac glycoside, digoxin, by human intestinal epithelial (Caco-2) cells. *Br. J. Pharmacol.* 118, 1389–1396 (1996).
28. Takano, M., Hasegawa, R., Fukuda, T., Yumoto, R., Nagai, J., Murakami, T. Interaction with P-glycoprotein and transport of erythromycin, midazolam, and ketoconazole in Caco-2 cells. *Eur. J. Pharmacol.* 358, 289–294 (1998).
29. Collett, A., Higgs, N.B., Sims, E., Rowland, M., Warhurst, A. Modulation of the permeability of H2 receptor antagonists cimetidine and ranitidine by P-glycoprotein in rat intestine and the human colonic cell line Caco-2. *J. Pharmacol. Exp. Ther.* 288, 171–178 (1999).

30. Kim, R.B., Fromm, M.F., Wandel, C., Leake, B., Wood, A.J.J., Roden, D.M., Wilkinson, G.R. The drug transporter P-glycoprotein limits oral absorption and brain entry of HIV-1 protease inhibitors. *J. Clin. Invest.* 101, 289–294 (1998).

31. Lowes, S., Simmons, N.L. Multiple pathways for fluoroquinolone secretion by human intestinal epithelial (Caco-2) cells. *Br. J. Pharmacol.* 135, 1263–1275 (2002).

32. Yamaguchi, H., Yano, I., Hashimoto, Y., Inui, K.I. Secretory mechanisms of grepafloxacin and levofloxacin in the human intestinal cell line Caco-2. *J. Pharmacol. Exp. Ther.* 295, 360–366 (2000).

33. Schinkel, A.H., Smit, J.J.M., Van Tellingen, O., Beijnen, J.H., Wagenaar, E., Van Deemter, L., Mol, C.A.A.M., Van Der Valk, M.A., Robanus-Maandag, E.C., Te Riele, H.P.J., Berns, A.J.M., Borst, P. Disruption of the mouse mdr1a P-glycoprotein gene leads to a deficiency in the blood-brain barrier and to increased sensitivity to drugs. *Cell* 77, 491–502 (1994).

34. Schinkel, A.H., Mayer, U., Wagenaar, E., Mol, C.A., Van Deemter, L., Smit, J.J., Van Der Valk, M.A., Voordouw, A.C., Spits, H., Van Tellingen, O., Zijlmans, J.M., Fibbe, W.E., Borst, P. Normal viability and altered pharmacokinetics in mice lacking mdr1-type (drug-transporting) P-glycoproteins. *Proc. Natl. Acad. Sci.* U.S.A. 94, 4028–4033 (1997).

35. Hager, W.D., Fenster, P., Mayersohn, M., Perrier, D., Graves, P., Marcus, F.I., Goldman, S. Digoxin-quinidine interaction: pharmacokinetic evaluation. N. *Engl. J. Med.* 300, 1238–1241 (1979).

36. Pedersen, P.L. Multidrug resistance—a fascinating, clinically relevant problem in bioenergetics. *J. Bioenerg. Biomem.* 27, 3–5 (1995).

37. Fromm, M., Kim, R., Stein, C., Wilkinson, G., Roden, D. Inhibition of P-glycoprotein-mediated drug transport: a unifying mechanism to explain the interaction between digoxin and quinidine. *Circulation* 99, 552–557 (1999).

38. Sparreboom, A., Van Asperen, J., Mayer, U., Schinkel, A.H., Smit, J.W., Meijer, D.K.F., Borst, P., Nooijen, W.J., Beijnen, J.H., Van Tellingen, O. Limited oral bioavailability and active epithelial excretion of paclitaxel (Taxol) caused by P-glycoprotein in the intestine. *Proc. Natl. Acad. Sci.* U.S.A. 94, 2031–2035 (1997).

39. Drescher, S., Glaeser, H., Mürdter, T., Hitzl, M., Eichelbaum, M., Fromm, M.F. P-glycoprotein-mediated intestinal and biliary digoxin transport in humans. *Clin. Pharmacol. Ther.* 73, 223–231 (2003).
40. Greiner, B., Eichelbaum, M., Fritz, P., Kreichgauer, H.P., Von Richter, O., Zundler, J., Kroemer, H. The role of intestinal P-glycoprotein in the interaction of digoxin and rifampin. *J. Clin. Invest.* 104, 147–153 (1999).
41. Chang, G., Roth, C.B. Structure of MsbA from *E. coli*: a homolog of the multidrug resistance ATP binding cassette (ABC) transporters. *Science* 293, 1793–1800 (2001).
42. Hopfner, K.P., Karcher, A., Shin, D.S., Craig, L., Arthur, L.M., Carney, J.P., Tainer, J.A. Structural biology of Rad 50 ATPase: ATP-driven conformational control in DNA double-strand break repair and the ABC-ATPase superfamily. *Cell* 101, 789–800 (2000).
43. Locher, K.P. Structure and mechanism of ABC transporters. *Curr. Opin. Struct. Biol.* 14, 426–431 (2004).
44. Locher, K.P., Lee, A.T., Rees, D.C. The *E. coli* BtuCD structure: a framework for ABC transporter architecture and mechanism. *Science* 296, 1091–1098 (2002).
45. Reyes, C.L., Chang, G. Structure of the ABC transporter MsbA in complex with ADP-vanadate and lipopolysaccharide. *Science* 308, 1028–1031 (2005).
46. Smith, P.C., Karpowich, N., Millen, L., Moody, J.E., Rosen, J., Thomas, P.J., Hunt, J.F. ATP binding to the motor domain from an ABC transporter drives formation of a nucleotide sandwich dimer. *Mol. Cell.* 10, 139 –149 (2002).
47. Hung, L.W., Wang, I.X., Nikaido, K., Liu, P.Q., Ames, G.F., Kim,S.H. Crystal structure of the ATP-binding subunit of an ABC transporter. *Nature* 396, 703–707 (1998).
48. Zaitseva, J., Jenewein, S., Jumpertz, T., Holland, I.B., Schmitt, L. H662 is the linchpin of ATP hydrolysis in the nucleotide-binding domain of the ABC transporter HlyB. *EMBO J.* 24, 1901–1910 (2005).
49. Davidson, A.L., Chen, J. ATP-binding cassette transporters in bacteria. *Annu. Rev. Biochem.* 73, 241–268 (2004).
50. Dey, S., Ramachandra, M., Pastan, I., Gottesman, M.M., Ambudkar, S.V. Evidence for two nonidentical drug-interaction sites in the human P-glycoprotein. *Proc. Natl. Acad. Sci.* U.S.A. 94, 10594–10599 (1997).

51. Loo, T.W., Bartlett, M.C., Clarke, D.M. Simultaneous binding of two different drugs in the binding pocket of the human multidrug resistance P-glycoprotein. *J. Biol. Chem.* 278, 39706–39710 (2003).

52. Lugo, M.R., Sharom, F.J. Interaction of LDS-751 and rhodamine 123 with P-glycoprotein: evidence for simultaneous binding of both drugs. *Biochemistry* 44, 14020–14029 (2005).

53. Sauna, Z.E., Andrus, M.B., Turner, T.M., Ambudkar, S.V. Biochemical basis of poly valency as a strategy for enhancing the efficacy of P-glycoprotein (ABCB1) modulators: Stipiamide homodimers separated with defined-length spacers reverse drug efflux with greater efficacy. *Biochemistry* 43, 2262–2271 (2004).

54. Shapiro, A.B., Ling, V. Positively cooperative sites for drug transport by P-glycoprotein with distinct drug specificities. *Eur. J. Biochem.* 250, 130–137 (1997).

55. Ambudkar SV, Dey S, Hrycyna CA, Ramachandra M, Pastan I, Gottesman MM. Biochemical, cellular and pharmacological aspects of multidrug transporter. *Annu Rev Pharmacol Toxicol.* 39:361–98 (1999).

56. Ambudkar, S.V., Kimchi-Sarfaty, C., Sauna, Z.E., Gottesman, M.M. P-glycoprotein: from genomics to mechanism. *Oncogene* 22, 7468–7485 (2003).

57. Sauna ZE, Smith MM, Muller M, Kerr KM, Ambudkar SV. The mechanism of action of multidrug-resistance-linked P-glycoprotein. *J Bioenerg Biomembr.*33:481–91 (2001).

58. Bruggemann, E.P., Germann, U.A., Gottesman, M.M., Pastan, I. Two different regions of P-glycoprotein are photoaf.nity-labeled by azidopine. *J. Biol. Chem.* 264, 15483–15488 (1989).

59. Bruggemann, E.P., Currier, S.J., Gottesman, M.M., Pastan, I. Characterization of the azidopine and vinblastine binding site of P-glycoprotein. *J. Biol. Chem.* 267, 21020–21026 (1992).

60. Demmer, A., Thole, H., Kubesch, P., Brandt, T., Raida, M., Fislage, R., Tummler, B. Localization of the iodomycin binding site in hamster P-glycoprotein. *J. Biol. Chem.* 272, 20913–20919 (1997).

61. Demmer, A., Andreae, S., Thole, H., Tummler, B. Iodomycin and iodipine, a structural analogue of azidopine, bind to a common domain in hamster P-glycoprotein. *Eur. J. Biochem.* 264, 800–805 (1999).

62. Isenberg, B., Thole, H., Tummler, B., Demmer, A. Identification and localization of three photobinding sites of iodoarylazidoprazosin in hamster P-glycoprotein. *Eur. J. Biochem.* 268, 2629–2634 (2001).
63. Ecker, G.F., Csaszar, E., Kopp, S., Plagens, B., Holzer, W., Ernst, W., Chiba, P. Identification of ligand-binding regions of P-glycoprotein by activated-pharmacophore photoaffinity labeling and matrix-assisted laser desorption/ionization-time-of-flight mass spectrometry. *Mol. Pharmacol.* 61, 637–648 (2002).
64. Loo, T.W., Clarke, D.M. Covalent modi.cation of human P-glycoprotein mutants containing a single cysteine in either nucleotide-binding fold abolishes drug-stimulated ATPase activity. *J. Biol. Chem.* 270, 22957–22961 (1995a).
65. Loo, T.W., Clarke, D.M. Membrane topology of a cysteine-less mutant of human P-glycoprotein. *J. Biol. Chem.* 270, 843–848 (1995b).
66. Loo, T.W., Clarke, D.M. Do drug substrates enter the common drug-binding pocket of P-glycoprotein through "gates"? *Biochem. Biophys. Res.* Commun. 329, 419–422 (2005).
67. Loo, T.W., Clarke, D.M. Identification of residues in the drug-binding site of human P-glycoprotein using a thiol-reactive substrate. *J. Biol. Chem.* 272, 31945–31948 (1997).
68. Loo, T.W., Clarke, D.M. Identification of residues within the drug-binding domain of the human multidrug resistance P-glycoprotein by cysteine-scanning mutagenesis and reaction with dibromobimane. *J. Biol. Chem.* 275, 39272–39278 (2000).
69. Loo, T.W., Clarke, D.M. Defining the drug-binding site in the human multidrug resistance P-glycoprotein using a methanethiosulfonate analog of verapamil, MTS-verapamil. *J. Biol. Chem.* 276, 14972–14979 (2001a).
70. Loo, T.W., Clarke, D.M. Determining the dimensions of the drug-binding domain of human P-glycoprotein using thiol cross-linking compounds as molecular rulers. *J. Biol. Chem.* 276, 36877–36880 (2001b).
71. Higgins, C.F. ABC transporters: from microorganisms to man. Annu. Rev. *Cell Biol.* 8, 67–113 (1992).
72. Schumacher, M.A., Miller, M.C., Grkovic, S., Brown, M.H., Skurray, R.A., Brennan, R.G. Structural basis for cooperative DNA binding by two dimers of the multi drugbinding protein QacR. *EMBO J.* 21, 1210–1218 (2002).

73. Yu, E.W., McDermott, G., Zgurskaya, H.I., Nikaido, H., Koshland, D.E. Structural basis of multiple drug-binding capacity of the AcrB multidrug efflux pump. *Science*. 300, 976–980 (2003).

74. Rosenberg MF, Kamis AB, Collaghan R, Higgins CF, Ford RC. Three-dimensional structures of the mammalian multidrug resistance P-glycoprotein demonstrate major conformational changes in the transmembrane domains upon nucleotide binding. *J Biol Chem* (2003).

75. Gottesman MM, Pastan I. Biochemistry of multidrug resistance mediated by the multidrug transporter. *Annu Rev Biochem*.62: 385–427 (1993).

76. Sauna, Z.E., Ambudkar, S.V. Evidence for a requirement for ATP hydrolysis at two distinct steps during a single turnover of the catalytic cycle of human P-glycoprotein. *Proc. Natl. Acad. Sci.* U.S.A. 97, 2515–2520 (2000).

77. Sauna, Z.E., Ambudkar, S.V. Characterization of the catalytic cycle of ATP hydrolysis by human P-glycoprotein: the two TP hydrolysis events in a single catalytic cycle are kinetically similar but affect different functional outcomes. *J. Biol. Chem.* 76,11653–11661 (2001).

78. Higgins, C.F., Linton, K.J. The ATP switch model for ABC transporters. *Nat. Struct. Mol. Biol.* 11, 918–926 (2004).

79. Rosenberg, M.F., Velarde, G., Ford, R.C., Martin, C., Berridge, G., Kerr, I.D., Callaghan, R., Schmidlin, A., Wooding, C., Linton, K.J., Higgins, C.F. Repacking of the transmembrane domains of P-glycoprotein during the transport ATPase cycle. *EMBO J.* 20, 5615–5625 (2001).

80. Loo, T.W., Clarke, D.M. Vanadate trapping of nucleotide at the ATP-binding sites of human multidrug resistance P-glycoprotein exposes different residues to the drug-binding site. *Proc. Natl. Acad. Sci.* U.S.A. 99, 3511–3516 (2002).

81. Davidson, A.L. Mechanism of coupling of transport to hydrolysis in bacterial ATP-binding cassette transporters. *J. Bacteriol.* 184, 1225–1233 (2002).

82. Fetsch, E.E., Davidson, A.L. Vanadate-catalyzed photocleavage of the signature motif of an ATP-binding cassette (ABC) transporter. *Proc. Natl. Acad. Sci.* U.S.A. 99, 9685–9690 (2002).

83. Tombline, G., Bartholomew, L., Gimi, K., Tyndall, G.A., Senior, A.E. Synergy between conserved ABC signature ser residues in P-glycoprotein catalysis. *J. Biol. Chem.* 279, 5363–5373 (2004a).

84. Tombline, G., Bartholomew, L.A., Urbatsch, I.L., Senior, A.E. Combined mutation of catalytic glutamate residues in the two nucleotide binding domains of P-glycoprotein generates a conformation that binds ATP and ATP tightly. *J. Biol. Chem.* 279, 31212–31220 (2004b).

85. Tombline, G., Muharemagic, A., White, L.B., Senior, A.E. Involvement of the "occluded nucleotide conformation" of P-glycoprotein in the catalytic pathway. *Biochemistry* 44, 12879–12886 (2005).

86. Urbatsch, I.L., Sankaran, B., Weber, J., Senior, A.E. P-glycoprotein is stably inhibited by vanadate-induced trapping of nucleotide at a single catalytic site. *J. Biol. Chem.* 270, 19383–19390 (1995).

87. Al-Shawi, M.K., Senior, A.E. Characterization of the adenosine triphosphatase activity of chinese hamster P-glycoprotein. *J. Biol. Chem.* 268, 4197–4206 (1993).

88. Ambudkar, S.V., Lelong, I.H., Zhang, J., Cardarelli, C.O., Gottesman, M.M., Pastan, I. Partial purification and reconstitution of the human multidrug-resistance pump: characterization of the drug-stimulatable ATP hydrolysis. *Proc. Natl. Acad. Sci.* U.S.A. 89, 8472–8476 (1992).

89. Sarkadi, B., Price, E.M., Boucher, R.C., Germann, U.A., Scarborough, G.A. Expression of the human multidrug resistance cDNA in insect cells generates a high activity drug-stimulated membrane ATPase. *J. Biol. Chem.* 267, 4854–4858 (1992).

90. Maki, N., Hafkemeyer, P., Dey, S. Allosteric modulation of human P-glycoprotein: inhibition of transport by preventing substrate translocation and dissociation. *J. Biol. Chem.* 278, 18132–18139 (2003).

91. Senior, A.E., Al-Shawi, M.K., Urbatsch, I.L. The catalytic cycle of P-glycoprotein. *FEBS Lett.* 377, 285–289 (1995).

92. Gottesman, M.M., Pastan, I., Ambudkar, S.V. P-glycoprotein and multidrug resistance. *Curr. Opin. Genet. Dev.* 6, 610–617 (1996).

93. Senior, A.E. Catalytic mechanism of P-glycoprotein. *Acta Physiol. Scand.* Suppl. 643, 213–218 (1998).

94. Sharom, F.J. The P-glycoprotein efflux pump: How does it transport drugs? *J. Membr. Biol.* 160, 161–175 (1997).

95. Martin, C., Berridge, G., Mistry, P., Higgins, C., Charlton, P., Callaghan, R. Drug binding sites on P-glycoprotein are altered by ATP binding prior to nucleotide hydrolysis. *Biochemistry* 39, 11901–11906 (2000).

96. Al-Shawi, M.K., Polar, M.K., Omote, H., Figler, R.A. Transition state analysis of the coupling of drug transport to ATP hydrolysis by P-glycoprotein. *J. Biol. Chem.* 278, 52629–52640 (2003).

97. Guengerich FP. Cytochrome P4503A4: regulation and role in drug metabolism [review]. *Annu Rev Pharmacol Toxicol.* 39: 1-17 (1999).

98. Kolars JC, Lown KS, Schmiedin RP. CYP3A gene expression in human gut epithelium. *Pharmacogenetics.* 4:247-259 (1994).

99. Wilkinson GR. Cytochrome P4503A (CYP3A) metabolism: prediction of *in vivo* activity in humans. *J Pharmacokinet Biopharm.* 24: 475-490 (1996).

100. Antila S, Hukkanen J, Hakkola J. Expression and localization of CYP3A4 and CYP3A5 in human lung. *Am J Respir Cell Mol Biol.* 16: 242-249 (1997).

101. Tanaka E. *In vivo* age-related changes in hepatic drug oxidizing capacity in humans. *J Clin Phar Ther.* 23: 247-255 (1998).

102. Paintaud G, Bechtel Y, Brientini MP. Effects of liver disease on drug metabolism. *Therapie*; 51:383-389 (1996).

103. Kashuba AD, Bertino JSJ, Rocci MLJ. Quantification of 3-month intraindividual variability and the influence of sex and menstrual cycle phase on CYP3A activity as measured by phenotyping with intravenous midazolam. *Clin Pharmacol Ther.* 64: 269-277 (1998).

104. Van A, Gupta B, Vander GCA, Van Boxtel CJ. The effect of grapefruit juice on the time-dependent decline of artemether plasma levels in healthy subjects. *Clin Pharmacol Ther.* 31: 158-159 (1999).

105. Crovo PB, Trapnell CB, Ette E, Zacur HA, Coresh J, Rocco LE. The effect of rifampicin and rifabutin on the pharmacokinetics and pharmacodynamics of a combination oral contraceptive. *Clin Pharmacol Ther.* 65: 428-438 (1999).

106. Wacher, V. J., Wu, C.Y., Benet, L. Z. Overlapping substrate specificities and tissue distribution of cytochrome p450 3a and p-glycoprotein: implications for drug delivery and activity in cancer chemotherapy. Mol Carcinog 13: 129-34 (1995).

107. Watkins, P. B. The barrier function of cyp3a4 and p-glycoprotein in the small bowel. Adv Drug Deliv Rev 27: 161-170 (1997).

108. Cummins, C. L., mangravite, L.M., Benet, L. Z. Characterizing the expression of cyp3a4 and efflux transporters (p-gp, mrp1, and mrp2) in cyp3a4-transfected caco-2 cells after induction with sodium butyrate and the phorbol ester 12-o-tetradecanoylphorbol-13-acetate. Pharm Res 18: 1102-9 (2001).

109. Kliewer, S. A., Moore, J. T., Wade, L., Staudinger, J. L., Watson, M. A., Jones, S. A., Mckee, D. D., Oliver, B. B., Willson, T.M., Zetterstrom, R. H., Perlmann, T., Lehmann, J.M. An orphan nuclear receptor activated by pregnanes defines a novel steroid signaling pathway. Cell 92: 73-82 (1998).

110. Lehmann, J.M., Mckee, D. D., Watson, M. A., Willson, T.M., Moore, J. T., Kliewer, S. A. The human orphan nuclear receptor pxr is activated by compounds that regulate cyp3a4 gene expression and cause drug interactions. J Clin Invest 102: 1016-23 (1998).

111. Synold, T. W., Dussault, I., Forman, B.M. The orphan nuclear receptor sxr coordinately regulates drug metabolism and efflux. Nat Med 7: 584- 90 (2001).

112. Ito, K., Kusuhara, H., Sugiyama, Y. Effects of intestinal cyp3a4 and p-glycoprotein on oral drug absorption--theoretical approach. Pharm Res 16: 225-31 (1999).

113. Johnson, B. M., Charman, W.N., Porter, C. J. The impact of pglycoprotein efflux on enterocyte residence time and enterocyte-based metabolism of verapamil. J Pharm Pharmacol 53: 1611-9 (2001a).

114. Gan, L. S., Moseley,M. A., Khosla, B., Augustijns, P. F., Bradshaw, T. P., Hendren, R.W., Thakker, D. R. Cyp3a-like cytochrome p450- mediated metabolism and polarized efflux of cyclosporin a in caco-2 cells. Drug Metab Dispos 24: 344-9 (1996).

115. Hochman, J.H., Chiba,M., Yamazaki,m., Tang, C., Lin, J. H. P-glycoprotein- mediated efflux of indinavir metabolites in caco-2 cells expressing cytochrome p450 3a4. J. Pharmacol Exp Ther 298: 323-30 (2001).

116. Mathew, D. T. et al. The role of P glycoprotein in drug disposition: significance to drug development, in Drug- Drug interactions, edited by A. David Rodrigues. Marcel Dekker, Inc. New York. 295-336 (2002).

117. Hosoya, K. I. Et al, Age dependant expression of P-glycoprotein 170 in Caco- 2 cell monolayers. *Pharm Res.* 13 (6) 885-890 (1996).

118. Schmieedlin –Ren, P. et al. Expression of enzymatically active CYP3A4 by Caco-2 cells grown on extracellular matrix –coated permeable supports in the presence of 1 alpha, 25-dihydraxyvitamin D3. *Mol. Pharmacol.* 51 (5) 741-754 (1997).

119. Simon, K. Van Meer, G. Lipid sorting in epithelial cells. *Biochemistry* 27 (17) 6197-6202 (1998).

120. Simon, K. Fuller S. D. Cell surface polarity in epithelia. *Annu Rev Cell Biol.* 1 243- 288 (1985).

121. Zhang, Y. Benet, L. Z. Characterization of P glycoprotein mediated transport of KO2, a novel vinylsulfone peptidomimetic cysteine protease inhibitor, across MDR1- MDCK and Caco-2 cell monolayer. *Pharm. Res.* 15 (10) 1520-1524 (1998).

122. Bradbury, M. W. The blood – brain barrier. *Exp Physiol.* 78 (4) 453-472 (1993).

123. Saith, H. Aungst, B.J. Possible involvement of multiple P-glycoprotein mediated efflux systems in the transport of verapamil and other organic cations across rat intestine. *Pharm Res.* 12 (9) 1304-1310 (1995).

124. Yumoto, R. Murakami, T. *J Pharmacol Exp Ther.* 289: 149-152 (1999).

125. Shravan Kumar. Y, Ramesh. S, Madhusudhan Rao. Y and Paradkar A.R. Effect of Rifampicin Pretreatment on the transport across rat intestine and oral pharmacokinetics of ornidazole in healthy human volunteers. *DMDI*, Vol: 22, No. 2-3 (2007).

126. Ruan, L.P, Chen, S, Yu, B.Y., Zhu, D.N., Cordell, G.A and Qiu, S.X. Prediction of human absorption of natural compounds by the non-everted rat intestinal sac model. *Eur J Med Chem.* 41: 605–610 (2006).

127. Shravan Kumar.Y, Gannu, R, Adukondalu D, Shiva kumar R, Sarangapani M and Madhusudan Rao Y. Effect of Silymarin and Pomegranate juice on intestinal transport of Buspirone across rat intestine. *Acta Pharmaceutica Scienica,*51:289-296 (2009).

128. Yan Zhang, Corbin Bachmeier, Donald W. Miller. In vitro and in vivo models for assessing drug efflux transporter activity. *Adv. Drug. Del. Rev.* 55: 31-51 (2003).

129. Van Asperen J, Mayer U, van Tellingen O, Beijnen JH. The functional role of P-glycoprotein in the blood-brain barrier. *J Pharm Sci; 86:881*–4 (1997).

130. Drori S, Eytan GD, Assaraf YG. Potentiation of anticancer drug cytotoxicity by multidrug-resistance chemosensitizers involves alterations in membrane fluidity leading to increased membrane permeability. *Eur J Biochem.* 228:1020–9 (1995).

131. Eytan GD, Regev R, Oren G, Assaraf YG. The role of passive transbilayer drug movement in multidrug resistance and its modulation. *J Biol Chem* 271:12897–902 (1996).

132. Doppenschmitt S, Spahn-Langguth H, Regardh CG, Langguth P. Role of P-glycoprotein mediated secretion in absorptive drug permeability: an approach using passive membrane permeability and affinity to P-glycoprotein. *J Pharm Sci.* 88:1067–72 (1999).

133. Shapiro AB, Ling V. Effect of quercetin on Hoechst 33342 transport by purified and reconstituted P-glycoprotein. *Biochem Pharmacol.*53:587–96 (1997).

134. Hugger ED, Audus KL, Brochardt RT. Effects of poly(ethylene glycol) on efflux transporter activity in Caco-2 cell monolayers. *J Pharm Sci.*91:1980–90 (2002).

135. Hugger ED, Novak BL, Burton PS, Audus KL, Brochardt RT(2002). A comparison of commonly used polyethoxylated pharmaceutical excipients on their ability to inhibit P-glycoprotein activity in vitro. *J Pharm Sci.*91:1991–2002 (2002).

136. Tsuruo T, Iida H, Tsukagoshi S, Sakurai Y. Overcoming of vincristine resistance in P388 leukemia in vivo and in vitro through enhanced cytotoxicity of vincristine and vinblastine by verapamil. *Cancer Res.*41:1967–72 (1981).

137. Kerr DJ, Graham J, Cummings J, Morrison JG, Thompson GG, Brodie MJ, et al. The effect of verapamil on the pharmacokinetics of adriamycin. *Cancer Chemother Pharmacol.*18:239–42 (1986).

138. Berg SL, Tolcher A, O'Shaughnessy JA, Denicoff AM, Noone M, Ognibene FP, et al. Effect of R-verapamil on the pharmacokinetics of paclitaxel in women with breast cancer. *J Clin Oncol.*13:2039–42 (1995).

139. Tolcher AW, Cowan KH, Solomon D, Ognibebe F, Goldspiel B, Chang R. Phase I crossover study of paclitaxel with R-verapamil in patients with metastatic breast cancer. *J Clin Oncol.*14:1173–84 (1996).

140. Lum BL, Kaubisch S, Yahanda AM, Adler KM, Jew L, Ehsan MN et al. Alteration of etoposide pharmacokinetics and pharmacodynamics by cyclosporine in a phase I trial to modulate multidrug resistance. *J Clin Oncol.*10:1635–42 (1992).

141. Glisson B, Gupta R, Hodges P, Ross W. Cross-resistance to intercalating agents in an epipodophyllotoxin-resistant Chinese hamster ovary cell line: evidence for a common intracellular target. *Cancer Res.* 46:1939–42 (1986).

142. Keller RP, Altermatt HJ, Donatsch P, Zihlmann H, Laissue JA, Hiastand PC. Pharmacologic interactions between the resistancemodifying cyclosporine SDZ PSC 833 and etoposide (VP 16–213) enhance in vivo cytostatic activity and toxicity. *Int J Cancer* 51: 433–8 (1992).

143. Hyafil F, Vergely C, Du Vignaud P, Grand-Perret T. In vitro and in vivo reversal of multidrug resistance by GF120918, an acridonecarboxamide derivative. *Cancer Res.*53:4595–602 (1993).

144. Bardelmeijer HA, Beijnen JH, Brouwer KR, Rosing H, Nooijen WJ, Schellens JH, et al., Increased oral bioavailability of paclitaxel by GF120918 in mice through selective modulation of P-glycoprotein. *Clin Cancer Res.* 6:4416–21 (2000).

145. Dantzig AH, Shepard RL, Cao J, Law KL, Ehlhardt WJ, Baughman TM, et al., Reversal of P-glycoprotein-mediated multidrug resistance by a potent cyclopropyldibenzosuberane modulator, LY335979. *Cancer Res.*56:4171–9 (1996).

146. Schellens JHM, Kruytzer MF, Vasey PA, Harris AL, et al., Phase I and pharmacokinetics study of the P-glycoprotein inhibitor LY335979 and paclitaxel in patients with solid tumors. Am Assoc *Cancer Res.* 42:535 (2001).

147. Guengerich, F.P., Martin, M.V., Beaune, P.H., Kremers, P., Wolff, T., Waxman, D.J. Characterization of rat and human liver microsomal cytochrome P-450 forms involved in nifedipine oxidation, a prototype for genetic polymorphism in oxidative drug metabolism. *J. Biol. Chem.* 261 (11), 5051–5060 (1986).

148. Parkinson, A. An overview of current cytochrome P450 technology for assessing the safety and efficacy of new materials. Toxicol. Pathol. 24 (1), 48–57 (1996).

149. Bailey, D.G., Malcolm, J., Arnold, O., Spence, J.D. Grapefruit juice– drug interactions. Br. J. Clin. Pharmacol. 46 (2), 101–110 (1998).

150. Fuhr, U. Drug interactions with grapefruit juice. Extent, probable mechanism and clinical relevance. *Drug Safety* 18 (4), 251–272 (1998).

151. Guo, L.Q., Taniguchi, M., Xiao, Y.Q., Baba, K., Ohta, T., Yamazoe, Y. Inhibitory effect of natural furanocoumarins on human microsomal cytochrome P450 3A activity. *Jpn. J. Pharmacol.* 82 (2), 122–129 (2000).

152. Obach, R.S. Inhibition of human cytochrome P450 enzymes by constituents of St. John's wort, an herbal preparation used in the treatment of depression. *J. Pharmacol. Exp. Ther.* 294 (1), 88–95 (2000).

153. Patel, J., Buddha, B., Dey, S., Pal, D., Mitra, A.K. In vitro interaction of the HIV protease inhibitor ritonavir with herbal constituents: changes in P-gp and CYP3A4 activity. *Am. J. Ther.* 11 (4), 262–277 (2004).

154. Markowitz, J.S., DeVane, C.L., Boulton, D.W., Carson, S.W., Nahas, Z., Risch, S.C. Effect of St. John's wort (Hypericum perforatum) on cytochrome P-450 2D6 and 3A4 activity in healthy volunteers. *Life Sci.* 66 (9), PL133–PL139 (2000).

155. Bray, B.J., Perry, N.B., Menkes, D.B., Rosengren, R.J. St. John's wort extract induces CYP3A and CYP2E1 in the Swiss Webster mouse. *Toxicol. Sci.* 66 (1), 27–33 (2002).

156. Zou, L., Harkey, M.R., Henderson, G.L. Effects of herbal components on cDNA-expressed cytochrome P450 enzyme catalytic activity. *Life Sci.* 71 (13), 1579–1589 (2002).

157. Vella, S., Floridia, M. Saquinavir. Clinical pharmacology and efficacy. *Clin. Pharmacokinet.* 34 (3), 189–201 (1998).

158. Hsu, A., Granneman, G.R., Bertz, R.J. Ritonavir. Clinical pharmacokinetics and interactions with other anti-HIV agents. *Clin. Pharmacokinet.* 35 (4), 275–291 (1998).

159. Li, X., Chan, W.K. Transport, metabolism and elimination mechanisms of anti-HIV agents. *Adv. Drug Deliv. Rev.* 39 (1–3), 81–103 (1999).

160. Malaty, L.I., Kuper, J.J. Drug interactions of HIV protease inhibitors. *Drug Safety* 20 (2), 147–169 (1999).

161. Williams, G.C., Sinko, P.J. Oral absorption of the HIV protease inhibitors: a current update. *Adv. Drug Deliv. Rev.* 39 (1–3), 211–238 (1999).

162. Borst, P., Evers, R., Kool, M., Wijnholds, J. A family of drug transporters: the multidrug resistance-associated proteins. *J. Natl. Cancer Inst*. 92 (16), 1295–1302 (2000).

163. Chen, C.J., Chin, J.E., Ueda, K., Clark, D.P., Pastan, I., Gottesman, M.M., Roninson, I.B. Internal duplication and homology with bacterial transport proteins in the mdr1 (P-glycoprotein) gene from multidrugresistant human cells. *Cell* 47 (3), 381–389 (1986).

164. Juliano, R.L., Ling, V. A surface glycoprotein modulating drug permeability in Chinese hamster ovary cell mutants. *Biochim. Biophys. Acta* 455 (1), 152–162 (1976).

165. Ueda, K., Cardarelli, C., Gottesman, M.M., Pastan, I. Expression of a full-length cDNA for the human "MDR1" gene confers resistance to colchicine, doxorubicin, and vinblastine. *Proc. Natl. Acad. Sci.* U. S. A. 84(9), 3004–3008 (1987).

166. Lin, J.H., Yamazaki, M. Role of P-glycoprotein in pharmacokinetics: clinical implications. *Clin. Pharmacokinet*. 42 (1), 59–98 (2003).

167. Piscitelli, S.C., Burstein, A.H., Chaitt, D., Alfaro, R.M., Falloon, J. Indinavir concentrations and St. John's wort. *Lancet* 355 (9203), 547–548 (2000).

168. Roby, C.A., Anderson, G.D., Kantor, E., Dryer, D.A., Burstein, A.H. St. John's wort: effect on CYP3A4 activity. *Clin .Pharmacol. Ther*. 67 (5), 451–457 (2000).

169. Durr, D., Stieger, B., Kullak-Ublick, G.A., Rentsch, K.M., Steinert, H.C., Meier, P.J., Fattinger, K. St. John's wort induces intestinal Pglycoprotein/ MDR1 and intestinal and hepatic CYP3A4. *Clin. Pharmacol. Ther*. 68 (6), 598–604 (2000b).

170. Perloff, M.D., von Moltke, L.L., Stormer, E., Shader, R.I., Greenblatt, D.J. Saint John's wort: an in vitro analysis of P-glycoprotein induction due to extended exposure. *Br. J. Pharmacol.* 134 (8), 1601–1608 (2001).

171. Ruschitzka, F., Meier, P.J., Turina, M., Luscher, T.F., Noll, G. Acute heart Schinkel AH, Kemp S, Dolle M, Rudenco G, Wagenaar E, 1993. N- Glycosylation and deletion mutants of the human MDR1 P-glycoprotein. *J Biol Chem*.268:7474–81 (2000).

172. Chieli, E., Romiti, N., Cervelli, F., Tongiani, R. Effects of flavonols on P-glycoprotein activity in cultured rat hepatocytes. *Life Sci*. 57 (19), 1741–1751 (1995).

173. Scambia, G., Ranelletti, F.O., Panici, P.B., De Vincenzo, R., Bonanno, G., Ferrandina, G., Piantelli, M., Bussa, S., Rumi, C., Cianfriglia, M., et al. Quercetin potentiates the effect of adriamycin in a multidrug-resistant MCF-7 human breast-cancer cell line: P-glycoprotein as a possible target. *Cancer Chemother. Pharmacol.* 34 (6), 459–464 (1994).

174. Hennessy, M., Kelleher, D., Spiers, J.P., Barry, M., Kavanagh, P., Back, D., Mulcahy, F., Feely, J. St Johns wort increases expression of Pglycoprotein: implications for drug interactions. *Br. J. Clin. Pharmacol.* 53 (1), 75–82 (2002).

175. Leslie, E.M., Deeley, R.G., Cole, S.P. Toxicological relevance of the multidrug resistance protein 1 MRP1 (ABCC1) and related transporters. *Toxicology* 167 (1), 3–23 (2001).

176. Mitra, A.K., Patel, J. Strategies to overcome simultaneous P-glycoprotein ediated efflux and CYP3A4 mediated metabolism of drugs. *Pharmacogenomics* 2 (4), 401–415 (2001).

177. Moore, L.B., Goodwin, B., Jones, S.A., Wisely, G.B., Serabjit-Singh, C.J., Willson, T.M., Collins, J.L., Kliewer, S.A. St. John's wort induces hepatic drug metabolism through activation of the pregnane X receptor. *Proc. Natl. Acad. Sci.* U. S. A. 97 (13), 7500–7502 (2000).

178. Fiebich, B.L., Hollig, A., Lieb, K. Inhibition of substance P-induced cytokine synthesis by St. John's wort extracts. *Pharmacopsychiatry* 34 (Suppl. 1), S26–S28 (2001).

8
Liquid Crystals

Mahesh R. Dabhi, Milan D. Limbani and Navin R. Sheth
Department of Pharmaceutical Sciences, Saurashtra University, Rajkot, India

8.1 Introduction

Liquid crystals are state of order between liquids and crystals. They can be fluid like liquid and they can have anisotropic properties like crystals. These are the substances that flow like liquid, but maintain the ordered structure characteristic of crystalline solids (Liquid + Crystal = Liquid Crystal) (1). Liquid crystal was first introduced by George-Luis LeClerc, on observing myelin figures consisting of concentric cylindrical phospholipid bilayers growing from lecithin in water, compared them to writhing eels. The credit for discovery of liquid crystals goes to Austrian botanist Friederich Reinitzer, who, in 1888, observed “two melting points” in cholesterol benzoate, which he extracted from plants (2).He shared his observations, that “At 145.5 °C it melts, forming a turbid but completely fluid liquid, that suddenly becomes completely clear at about 178.5 °C” with the German physicist Otto Lehmann. In 1922, French crystallographer Georges Friedel convincingly argued that liquid crystals represented new states of matter, intermediate mesomorphic between solid crystals and ordinary liquids. A new phase of matter, liquid crystal phase, was born.

The distinguishing characteristic of liquid crystalline state is tendency of molecules (mesogens) to point along a common axis, called the Director. The characteristic orientational order of liquid crystal state is between solid and liquid phases and this is origin of term mesogenic state, used synonymously with liquid crystal state. The average alignments of molecules for each phase are shown in Figure 8.1.

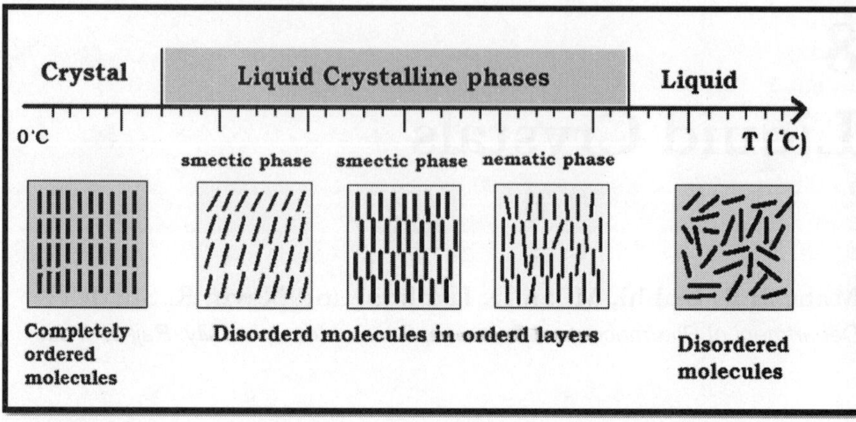

Figure 8.1 Average alignment of the molecules for each phase.

Crystalline materials characterized by long range periodic order in three dimensions. Liquid crystals are characterized by long range orientational order but not positional order. Isotropic liquid has no orientational order (3). Positional and orientational order of various state of substance was shown in Figure 8.2.

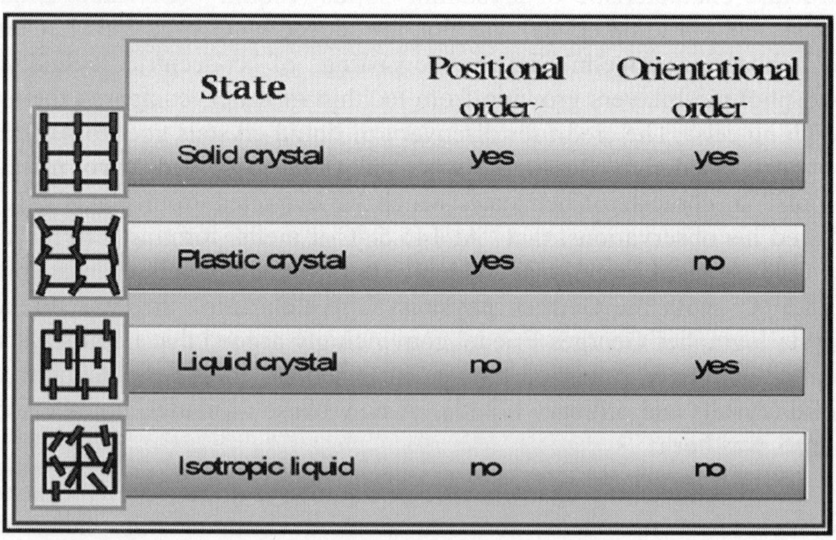

Figure 8.2 Positional and orientational order of various state of substance.

Liquid crystal situated between the solid and the liquid phase, so it contain characteristics of both. As with a solid and a liquid, the organization of molecules or molecular aggregates plays a key role in defining the liquid crystal phase. In most solids, for example, molecules arrange themselves into a rigid lattice structure, with ordered position and orientation. Thermal energy may cause these molecules to vibrate, but on the whole, they do not change their orientation dramatically. In liquids, on other hand, molecules have many degrees of freedom, allowing them to move in any direction and rotate freely. This jump from zero to six degree of freedom make it not at all surprising that liquid crystals, situated as they are between solid and liquid have an intermediate number of degrees of freedom. The molecules in a liquid crystal contain orientational order and sometimes a degree or two of positional order as well. This means that, while individual molecules will vary slightly in their orientation, on average the molecules will orient in the same general direction.

There are many types of liquid crystal states, depending upon amount of order in material. Liquid crystals can be classified into two main categories: Thermotropic liquid crystals and Lyotropic liquid crystals (4). These two types of liquid crystals are distinguished by mechanisms that drive their self-organization, but they are also similar in many ways. Thermotropic phases are initiated by changes in temperature, while lyotropic phases can also be initiated by changes in concentration. Thermotropic liquid crystalline state can be obtained by raising temperature of solid and/or lowering temperature of liquid. Lyotropic liquid crystal occurs as result of solvent-induced aggregation of constituent mesogens into micellar structures (5, 6). Lyotropic liquid crystals or mesogens are typically amphiphilic, meaning that they are composed of both lyophilic (solvent attracting) and lyophobic (solvent repelling) part. This causes them to form into micellar structure in presence of solvent since lyophobic end will stay together as lyophilic ends extended outwards towards the solution (7).

Liquid crystals have both properties of solid and liquid state. It is very important to understand properties of liquid crystals and how these relate to molecular structure. Various factors such as concentration, temperature, energy and entropy, pH and salt etc. which affect properties like phase behaviour, structure, size, shape and rheology of liquid crystal (8). Liquid crystals can be characterized by various methods such as Optical Polarizing Microscopy, Differential Scanning Calorimetry (DSC), Fourier Transfer Infrared Spectroscopy (FT-IR), X-ray diffraction (XRD) etc. (9). Recent issues in liquid crystal science are associated with applications of liquid crystals in area of biology and medicine in which liquid crystal concepts are important. Lyotropic liquid crystals are found in countless everyday situations. Soaps and detergents form Lyotropic liquid crystals when they combine with water. Examples of naturally occurring Lyotropic liquid crystals that are essential for life include lecithin, DNA, cholesterol, cellulose, esters, gangliosides and paraffin. Biological membranes also exhibit liquid crystalline behaviour (2).

Lyotropic liquid crystals can be used for solubility enhancement of poorly soluble drugs, controlled drug release/sustained drug release and to modify stability of drug molecules in GIT (10). For example, cyclosporine is practically insoluble drug and having toxic side effects. To solve these problems associated with cyclosporine, Vitamin E ∂-alpha-Tocopheryl Polyethylene Glycol-1000 Succinate (Vitamin E-TPGS) formulation can be utilized. It form liquid crystalline structure with gastric fluid to enhance drug solubility, and further employs dosage forms with impermeable or semi-permeable barriers to control drug release over time, thereby providing means for lowering dose within therapeutic window (11, 12). Cubic phases of liquid crystal are also applied for protection of labile moieties like enzymes and other drugs sensitive to hydrolysis and oxidation (13). Shah and Paradkar had reported glyceryl monooleate matrices for oral administration of serratiopeptidase, which undergo transformation to highly viscous cubic phase and provide sustained release of drug (10). Lyotropic liquid crystals that represent novel drug candidates for treatment of wide range of diseases. Liquid crystals were used as controlled release drug delivery systems with help of thermoresponsive cellulose nitrate membrane which controlled rate of drug transport through them and it was dependent on the amount of liquid crystal deposited on the membrane (14). Thermoresponsive membrane was developed by adsorbing the binary cholesteric liquid crystals, 36% cholesteryl oleyl carbonate and 64% cholesteryl nonanoate solved in an organic solvent into the cellulose nitrate membranes (15). Such thermoresponsive liquid crystal embedded membranes are much permeable to drug molecules at temperatures above 41.5 ◦C. This triggerable drug delivery system was released drug only above a certain temperature as a drug reservoir system. One of the rationales for such thermoresponsive system may be potential for their use in chemotherapeutics delivery under local hyperthermia.

Amphiphilic substances spontaneously tend to self-associate and with increasing concentration they can form highly ordered aggregates, such as lamellar, hexagonal and cubic phases. Using these lyotropic liquid crystalline phases as topical drug delivery systems is favourable because of their high solubilization capacity, thermodynamical stability or broad range of rheological properties. Sustained drug release of hydrophilic substances from cubic and inverse hexagonal structures was observed by several authors (16).

Lamellar lyotropic liquid crystalline systems are thermodynamically stable, optically isotropic systems, which are formed with low energy input. New possibilities for development of controlled drug delivery systems are inherent in these systems due to their stability and special skin-similarly structure. Lamellar lyotropic liquid samples contained: Brij 96 (poly-oxyethylene-10-oleyl ether) with water, liquid petrolatum and glycerol in particular concentration range.A prolonged drug release was observed in case of very water-soluble ephedrine hydrochloride and same phenomena were observed in case of tenoxicam, which is practically insoluble in water (17).

A new class of amphiphiles with a glycerate headgroup, recently shown to form reverse hexagonal phase in excess water. The application of these novel materials to development of a new injectable formulation of irinotecan was investigated. The formulation of irinotecan with small percentage of oleic acid in oleyl glycerate permitted a clinically relevant dose of irinotecan to be dissolved in glycerate surfactant and dispersed in aqueous medium to form an injectable particle-based dose form of irinotecan (18).

In cosmetics, Lyotropic liquid crystals are useful in eye shadow and lip gloss preparation. Lyotropic liquid crystals are also used in cosmetic gels and emulsions to stabilize the structure and to retain moisture. Liquid crystal also used in laptops, televisions and cell phones as Liquid Crystal Display (LCD) and Liquid Crystal Spatial Light Modulators (LCSLMs) technology useful in photolithography, optical tweezers, optical processing, beam shaping, active and adaptive optics etc.

8.2 Structure

The molecules which make up Lyotropic liquid crystals are surfactants consisting of two distinct parts:

(i) A polar head (often ionic) and

(ii) A nonpolar hydrocarbon tail

Lyotropic liquid crystal molecules belong to class of substances called amphiphilic compounds (e.g. sodium laurate). They follow the rule of "like dissolves like", head is attracted to water (or hydrophilic) and tail is repelled by water (or hydrophobic). When molecules dissolved in high enough concentrations, molecules arrange themselves so that polar heads are in contact with polar solvent and/or nonpolar tails are in contact with nonpolar solvent (19). As concentration of molecules increases in solution, they take on different arrangements or phases.

At low concentrations, solution looks like any other particles of solute distributed randomly throughout water. When concentration gets high enough, molecules begin to arrange themselves in hollow spheres, rods, and discs called micelles. Spherical micelle and its cross section are shown in Figure 8.3.

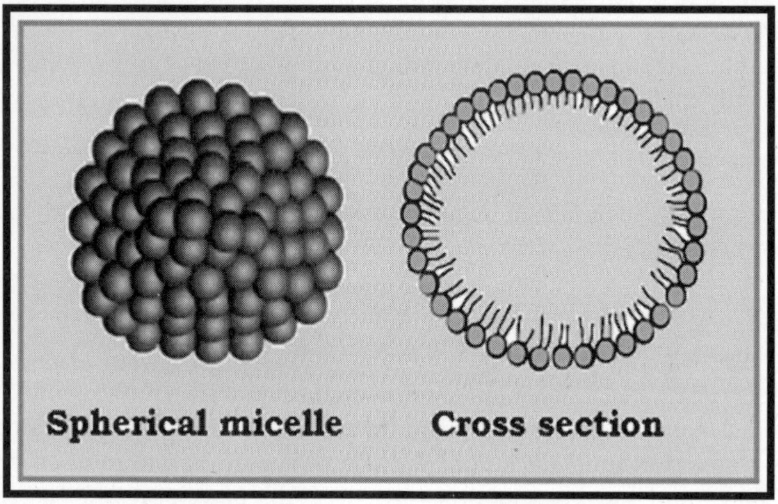

Figure 8.3 Spherical micelle and its cross section.

In some reactions, type of micelle affects the reaction rate, most likely because parts of molecule involved in reaction are more likely to be exposed in some formations than in others.The surface of micelle is layer of polar heads dissolved in water, while inner portion consists of hydrophobic tails screened from water by hydrophilic heads (shown in Figure 8.4). Micelles come in varied sizes, but the smallest ones have a diameter about twice as long as the length of a hydrocarbon chain with all trans- bonds.As concentration of amphiphiles increases, micelles become increasingly able to dissolve nonpolar substances. When this occurs, micelles become large and swollen. If they reach large enough size, solution becomes cloudy and is called an emulsion. At lower concentrations, swollen micelles are not large enough to interfere with light, but they are still extremely stable and exist in equilibrium. This phase is referred to as microemulsion (20).

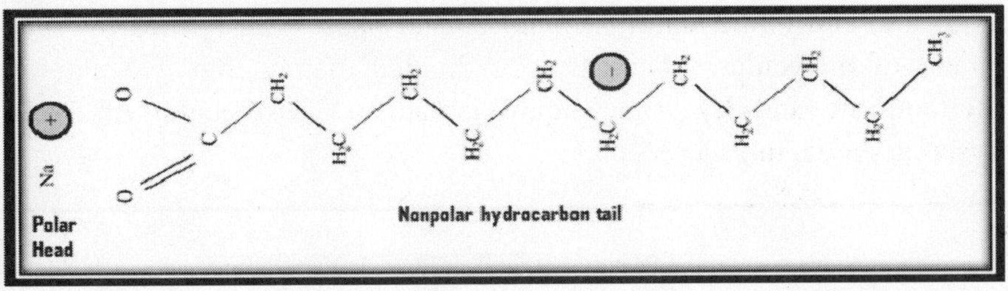

Figure 8.4 Micelle with polar head and nonpolar hydrocarbon tail.

As the concentration increases, micelles begin to arrange themselves into loose patterns. These patterns are actual liquid crystal aspects of the molecular behaviour. One of the first liquid crystal phases has micelles forming a structure similar to a face-centred or body-centred cubic crystal lattice. The illustration below shows a body-centred cubic crystal structure. Micelles take the place of individual atoms, ions, or molecules. It should also be noted that the pattern is not as stable or as rigid as that of a solid crystal like graphite or table salt, hence the term liquid crystal. Rod-shaped micelles often form into hexagonal arrays made out of six rods groupcd around a central one for a total of seven, as illustrated in Figure 8.5. In the enlargement of a single rod, notice that micelle surface is composed of hydrophilic heads. The hydrophobic tails are isolated inside the micelle. Hexagonal liquid crystals generally exist in solutions that are 40-70% amphiphiles. The liquid crystals may come apart if too much water or salt is added to the solution, but many varieties can absorb oil by expanding the diameter of rod-shaped micelle.

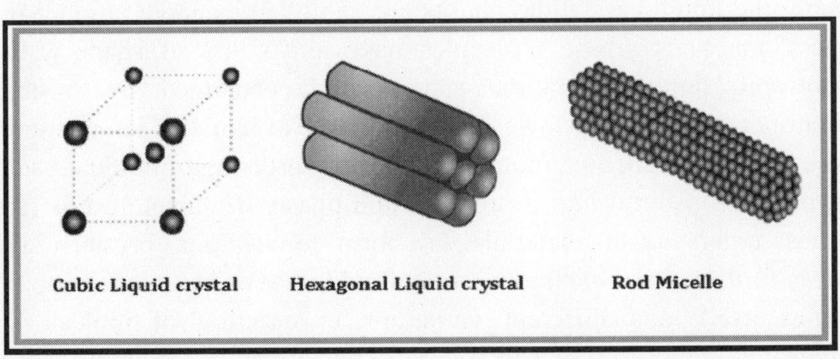

Figure 8.5 Cubic liquid crystal and Hexagonal liquid crystal.

8.3 Type of Liquid Crystal

The liquid crystal state is fourth phase of matter observed between crystalline and liquid states. Liquid crystal phases are formed by wide variety of molecules. They can be divided mainly into two classes, Thermotropic and Lyotropic liquid crystal. Classification of liquid crystal is shown in Figure 8.6.

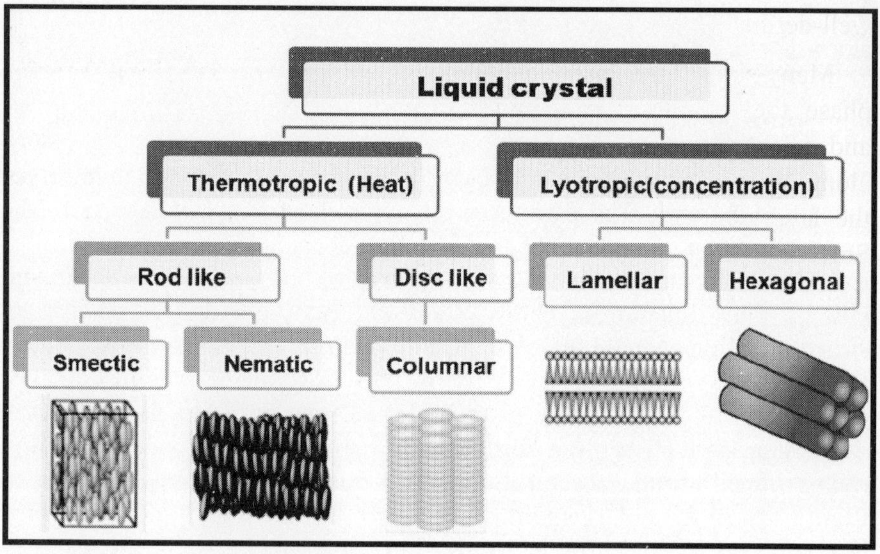

Figure 8.6 Classification of liquid crystal.

8.3.1 Thermotropic Liquid Crystals

Thermotropic liquid crystalline phases are exhibited by large number of organic compounds whose molecules have anisotropy of shape (21). Thermotropic liquid crystalline state can be obtained by raising temperature of solid and/or lowering temperature of liquid. If temperature increase is too high, thermal motion will destroy ordering of liquid crystal phase, pushing material into isotropic liquid phase. If temperature is too low, most liquid crystal materials will form conventional crystal (21). Scientists further subdivide thermotropic liquid crystals based on shape of molecules itself, and different variations in ordering of molecules. Thermotropic liquid crystals can generally be formed by calamitic (Rod like) molecules or discotic (Disc like) molecules (22).

8.3.1.1 Smectic

The word "smectic" is derived from Greek word for soap. The thick and slippery substance often found at bottom of soap dish is actually type of smectic liquid crystal. Smectic liquid crystals have layered arrangement of orientationally ordered rod-like molecules, i.e. they are characterized by translational order (23). The smectic state is more "solid-like" than nematic because it found at lower temperatures than the nematic, it has well-defined layers that can slide over one another (21).

Many compounds are observed to form more than one type of smectic phase, they all characterized by different types and degrees of positional and orientational order. In Smectic A phase, molecules are oriented along layer normal, while in Smectic C phase they are tilted away from the layer normal. These phases are liquid-like within the layers (24). Smectic A and Smectic C phase are shown in Figure 8.7.

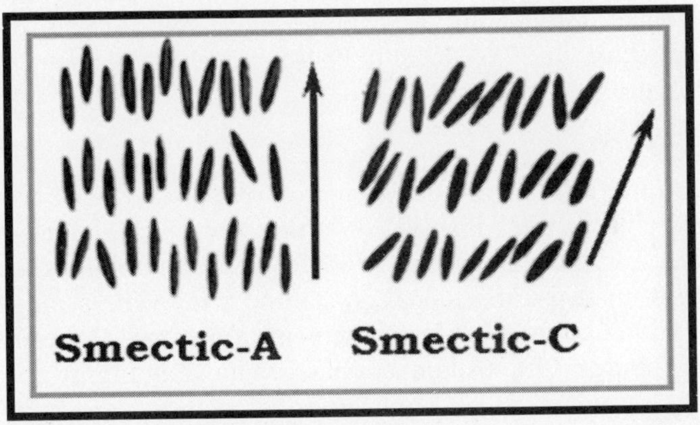

Figure 8.7 Smectic A and C.

8.3.1.2 Nematic

The simplest liquid crystal phase is called nematic phase (N) and is close to liquid phase. The molecules float around as in liquid phase, but are still ordered in their orientation (25). It is characterized by high degree of long range orientational order but no translational order. The nematic phase is seen as the marbled texture.When molecules are chiral and in nematic phase, they arrange themselves into strongly twisted structure that often reflects visible light in different bright colours which depend on temperature (26). They can therefore be used in temperature sensors (thermometers).

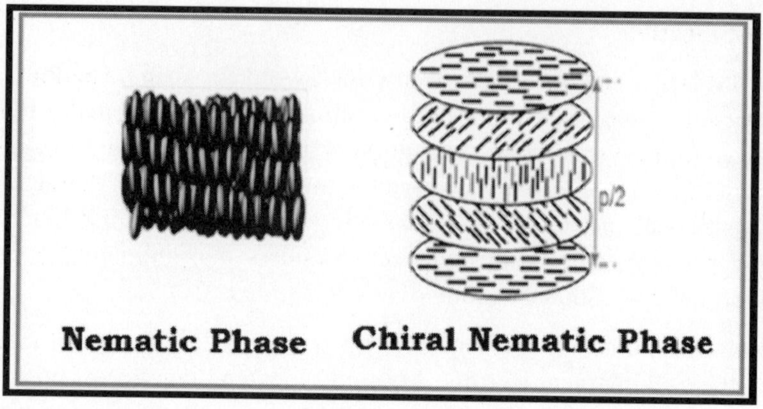

Figure 8.8 Nematic Phase and Chiral Nematic Phase.

Chiral molecules can also form nematic phases called chiral nematic (or cholesteric) phases. This phase shows nematic ordering but preferred direction rotates throughout the sample. The axis of this rotation is normal to director (1, 27). Nematic Phase and Chiral Nematic Phase are shown in Figure 8.8.

8.3.1.3 Columnar

Columnar liquid crystals are different from previous types because they are shaped like disks instead of long rods. This mesophase is characterized by stacked columns of molecules. The columns are packed together to form two-dimensional crystalline array. The arrangement of molecules within columns and arrangement of columns themselves leads to new mesophases (28). Columnar phase is shown in Figure 8.9.

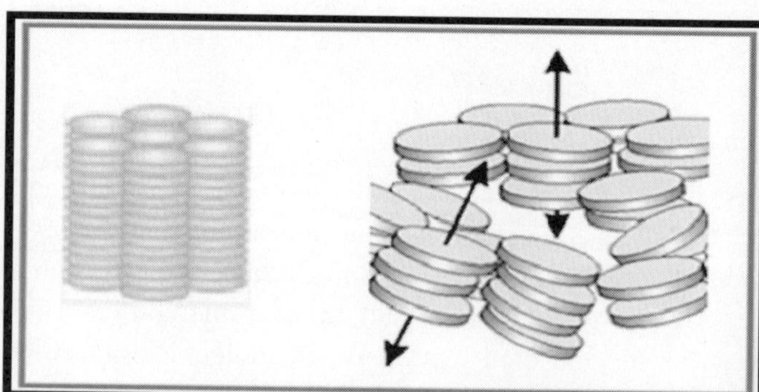

Figure 8.9 Columnar Phase.

8.3.2 Lyotropic Liquid Crystals

Lyotropic liquid crystals have several fundamental characteristics that make them dramatically different than thermotropic liquid crystals. First, these types of liquid crystals form in solutions rather than in pure substances. Secondly, individual molecules do not align by themselves to create anisotropy; instead, the molecules come together to form anisotropic aggregates, which themselves align along director (29). For these reasons, additional conditions besides temperature determine whether the liquid crystal phase forms. Temperature still affects phase in the same ways as thermotropic liquid crystals, but concentration of substance also has a strong effect on liquid crystalline behaviour. Lyotropic liquid crystal phases are formed by amphiphilic molecules. A schematic diagram of an amphiphilic molecule is shown in Figure 8.10. These often consist of polar head group attached to one or more non-polar chains and are often known as surfactants (5). In order to form a liquid crystal phase, the concentration must be high enough both to induce formation of aggregates if they do not form at all concentrations, and to force these aggregates to align together. As a result, some solutions may be macroscopically isotropic, but still contain anisotropic aggregates. When these are dissolved in an appropriate solvent they self-assemble so the polar (hydrophilic) heads protect non-polar (hydrophobic) tails. Sometime, scientists thought that only amphiphilic molecules formed this type of liquid crystal (29).Amphiphilic molecules, such as detergents and phospholipids, contain a hydrophilic head attached to a long hydrophobic tail, which cause them to aggregate in aqueous solutions. This entropy-driven aggregation process results in two types of aggregates known as micelles and vesicles (29). In micelles and vesicles, hydrophilic head groups shield hydrophilic tails from the solution (shown in Figure 8.11). Because of the way the amphiphilic molecules organize themselves, they cannot form at arbitrarily low concentrations. For example, a micelle cannot form two molecules, since the polar heads could not appropriately shield the inner hydrophobic tails. At some point, known as the critical micelle concentration, the number of molecules becomes large enough to form these aggregates.

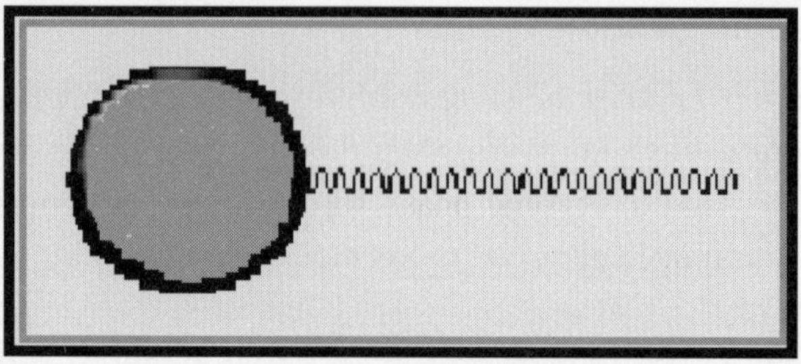

Figure 8.10 Schematic diagram of an amphiphilic molecule.

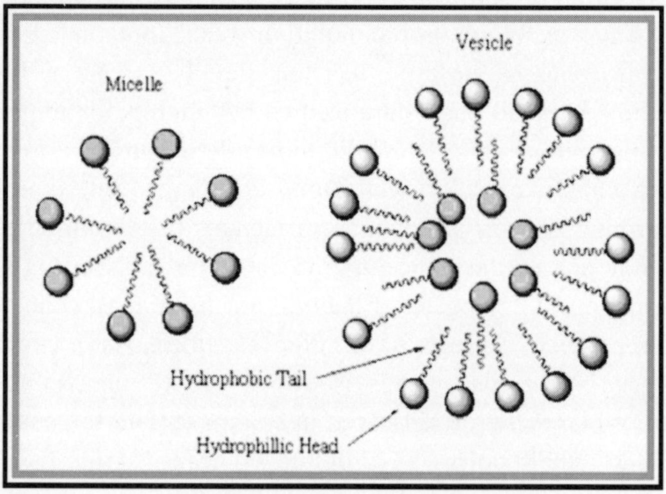

Figure 8.11 Aggregates in a lyotropic liquid crystal formed from amphiphilic molecules.

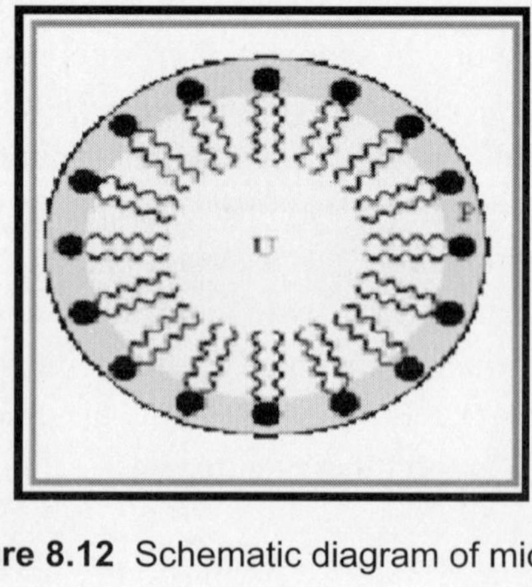

Figure 8.12 Schematic diagram of micelles.

At low surfactant concentrations these are roughly spherical, as shown in Figure 8.12. As surfactant concentration increases then other phases are formed (6). These include lamellar phase where amphiphiles form bilayer structure and hexagonal phase where amphiphiles form cylinders that pack in hexagonal array.

8.3.2.1 Lamellar

Lamellar phase is also known as 'Neat phase'. In lamellar phase, Amphiphiles molecules are arranged in bilayers and separated by water layers (shown in Figure 8.13). Generally, lamellar phase exist down to 50%. If surfactant is less than 50%, transition from lamellar to hexagonal phase. In lamellar phase parallel layers slide over each other freely during apply shear and hence lamellar phase is less viscous than hexagonal phase (6).

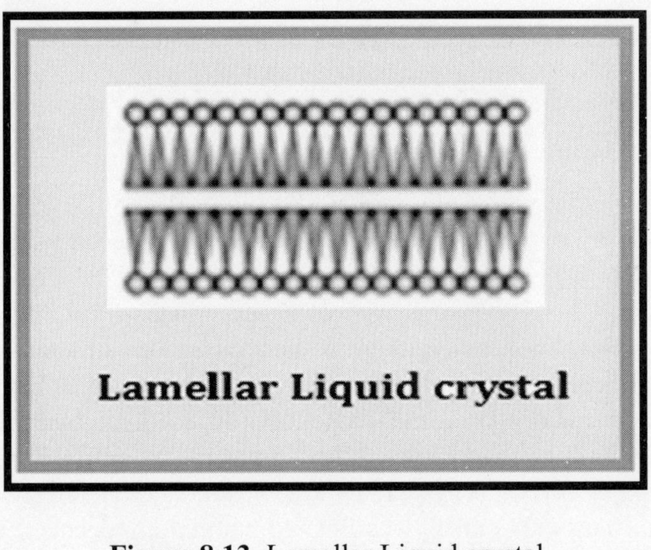

Figure 8.13 Lamellar Liquid crystal.

8.3.2.2 Hexagonal

In hexagonal phase, Amphiphiles molecules are aggregate in hexagonal array (Shown in Figure 8.14). In hexagonal phase, water content is high in comparison to lamellar phase so it is more viscous than lamellar phase (6).

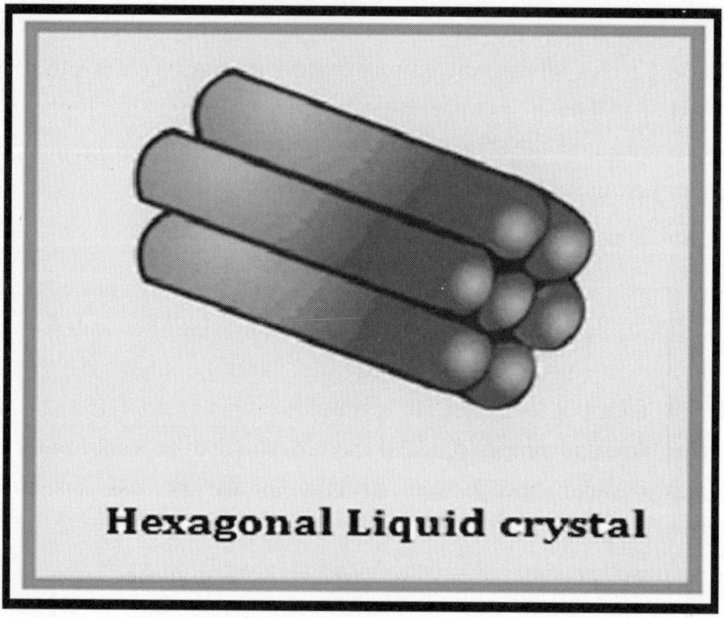

Figure 8.14 Hexagonal Liquid crystal.

8.3.3 Chromonic Liquid Crystals

Despite the initial focus on amphiphilic molecules as the only type of matter that forms lyotropic liquid crystal phases, a second subcategory, known as chromonic liquid crystals, has also emerged. Named after asthma drug, Chromolyn, this type of liquid crystal differs dramatically from amphiphilic molecules, both in their molecular structure and their aggregation behaviour (30). Chromonic liquid crystals often have a plank or disk shape rather than a rod shape, and do not have hydrophilic and hydrophobic groups clustered on either end of the molecule. Furthermore, while changes in entropy catalyse the aggregation of amphiphilic molecules, changes in enthalpy associated with intermolecular bonds also promote the aggregation of chromonic molecules (30, 31). Instead of clumping together to form a micelle, chromonic aggregates often have a tendency to stack in columns, generating anisotropic aggregates which can align to form a liquid crystal. Because any two molecules can bond together and begin formation of an aggregate, traditional chromonic molecules will have aggregates at very low concentrations, and contain a variety of aggregate sizes (30). Consequently, chromonic aggregates lack the critical micelle concentration characteristic of amphiphilic aggregates

(30, 32). Since chromonic liquid crystals form aggregates even at low concentrations, a solution may contain anisotropic aggregates, but they may not align and form a macroscopically anisotropic phase. Thus phase transition will only occur when aggregates grow large enough that interactions between aggregates force them to align (Shown in Figure 8.15).

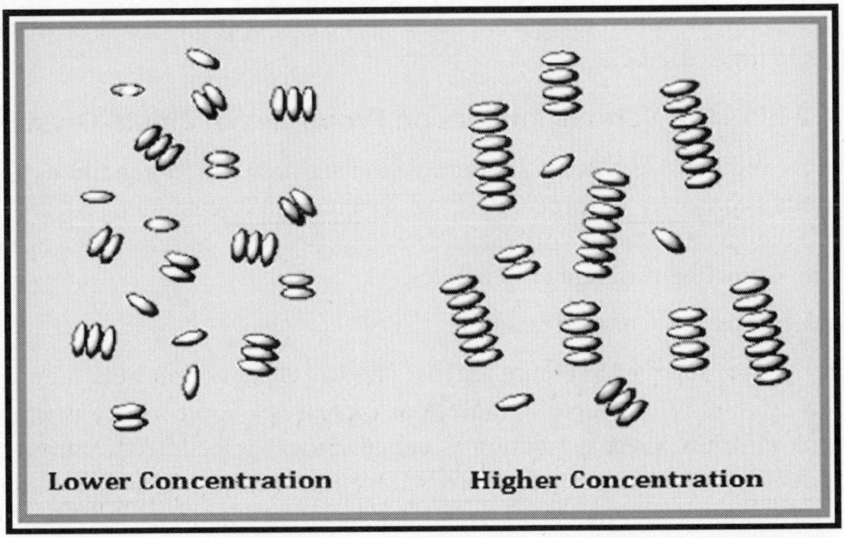

Figure 8.15 Aggregates in a simple chromonic liquid crystal.

At the lower concentration, aggregates form but do not form the liquid crystal phase. At the higher concentration, aggregates are large enough to align together. In both cases, there are a range of aggregate sizes.

8.4 Physicochemical Properties of Liquid Crystal

8.4.1 Properties of Liquid Crystal

Liquid crystals have all the properties of liquid as well as solid. Its physical properties are optical properties, dielectric properties, diamagnetic properties, viscous properties, elastic properties, electrical conductance, thermal conductance, density measurement etc. (33). Here we discuss brief information regarding some important physical properties of liquid crystal are as follow, Optical property (Birefringence) of liquid crystal can be defined as the resolution or splitting of a light wave into two unequally reflected or transmitted waves by an optically anisotropic medium such as calcite or quartz.

Dielectric property of liquid crystal can be defined as the material that does not conduct electricity readily.

Diamagnetic property of liquid crystal can be defined as the property exhibited by substances with a negative magnetic susceptibility, that is, by substances which magnetize in a direction opposite to that of an applied magnetic field.

8.4.2 Effect of Various Factors on Properties of Liquid Crystal

In this section, we discussed some of the most common conditions for various geometries to form. However, we changes few factors, properties of liquid crystal can be affected. This can be done by changing head, chain, or mixing different amphiphiles.

8.4.2.1 Effect of Concentration

The phase diagram (Figure 8.16) shows changes in structure as concentration of amphiphilic molecules increases. The concentration, at which micelles form in solution is called critical micelle concentration, showed as dotted line. Also shows dark line below which few liquid crystals form. This line represents boundary temperature, referred to as Krafft temperature. Below Krafft temperature, few liquid crystals may be suspended in solution, but for most part amphiphilic molecules stay widely distributed (34).

If the concentration by weight of amphiphilic molecules is higher than that of water, the molecules form sort of matrix with water droplets scattered inside, in contact with polar heads. If molecules are dissolved in nonpolar solvent, their behaviour is similar to that when dissolved in water, except that now nonpolar tails are in contact with solvent and polar heads are isolated in centre of micelles and bilayers.If the solution contains both water and a higher concentration of nonpolar solvent, similar inverse micelles form with water droplets quarantined inside the micelle and nonpolar solvent on the outside (35). Finally, if weaker amphiphilic molecules and simple salts are dissolved together in water, they form "lyotropic nematic phases." In these crystals, as in thermotropic nematic, the director orientation can be changed by applying a magnetic field.

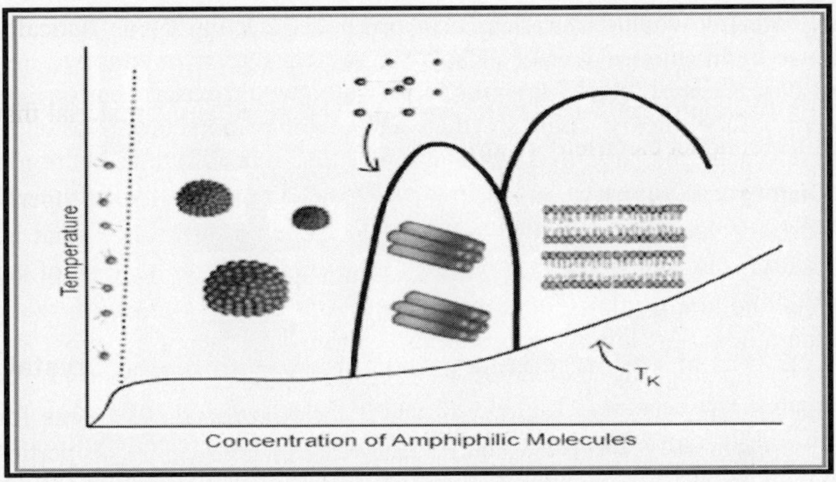

Figure 8.16 Phase diagram for effect of concentration.

If water, hydrocarbon and surfactant are mixed together, it is possible to get microemulsion known as ringing gel (36, 37). This phase forms when micelles shift from rod to sphere shapes in presence of hydrocarbon. If gel is placed in container and container is tapped, gel will vibrate with an audible resonance frequency.

8.4.2.2 Effect of Temperature

Although lyotropic liquid crystals are characterized by fact that concentration is determining factor in their phase transitions, temperature also plays an important role. The amphiphilic molecules have temperature boundaries which depend on specific kind of molecule (shown in Figure 8.16). When temperature is above this boundary, micelles and liquid crystals can form. Below it, solution remains clear, although some small crystals may be suspended in solvent (38).

One theory for role of temperature points out that forming of micelles requires hydrocarbon chains to bend and fold in all directions. This bending and folding behaviour is similar to what happens when hydrocarbons melt. Then one possibility is that temperature boundary is related to melting point for hydrocarbon tails (39).

8.4.2.3 Effect of Energy and Entropy

The rule "like dissolves like" can be understood by looking at thermodynamic aspects of trying to dissolve hydrocarbons in water. In most cases, dissolving hydrocarbons in water would lead to reduction in enthalpy. This is good for the reaction - it leads to an energy reduction since enthalpy would be negative. However this reaction does not occur, because hydrocarbons do not dissolve in water (reason is entropy). In most cases, dissolving hydrocarbons in water would reduce entropy of system. One theory suggests that when water molecules rearrange themselves around large hydrocarbon molecules to restore their hydrogen bonds, process increases order in that area. The more hydrocarbons dissolve, more order would increase.

These concepts relate to lyotropic liquid crystals by fact that, for amphiphilic molecules, dominant form of entropy changes with concentration. At low concentrations, entropy is increased most by allowing amphiphilic molecules to mix thoroughly with water. As concentration becomes higher, though, order created by allowing organized structures becomes less important than order created by forcing water molecules to rearrange themselves around dissolve hydrocarbons (20). At that point, structures begin to form in which hydrocarbon tails are kept away from water, preventing local order from increasing.

The strong role of entropy makes lyotropic liquid crystals different from most other substances. In many substances, order exists at low temperatures when the low enthalpy is enough to reduce free energy. Disorder arises at higher temperatures when high entropy is needed to reduce free energy. In lyotropic liquid crystals, though, structure exists at high concentrations when order created by dissolving hydrocarbons would be larger than disorder of having them randomly distributed through the water. At low concentrations, entropy plays its usual role of encouraging complete solvation and structures do not form.

8.4.2.4 Effect of Salt

Chromonic liquid crystals are formed by addition of aromatic molecules such as disodium chromoglycate to water. The addition of sodium and potassium salts shifts the isotropic–nematic phase boundary upward by more than 10 °C, so that samples that were isotropic at room temperature are transformed into nematic phases (31).

Salt effects are predominantly dictated by cation, not the anion, and appear to differ based on cation size. In contrast to small, hydrated cations like sodium, large, weakly hydrated cations such as tetraethylammonium and tetrabutylammonium shift the phase boundary downward, thus stabilizing isotropic phase at the expense of nematic one. The phase behaviour results are highly correlated with viscosity measurements, with upward shift in phase boundary correlating with an increase in solution viscosity and vice versa (31).

8.4.2.5 Effect of pH

With addition of salt to the solution or lowering the pH, hydrophilic interaction of head is reduced and optimal interface area is lowered. The additional ions in solution also reduce repulsive interactions between head groups, reducing radius of curvature. This makes molecules more likely to form bilayers or inverse micelles. Using these methods to reduce interface area often has additional effect of straightening hydrocarbon chains (40).

When hydrocarbon chains are unsaturated or branched, their length is reduced. This increases the volume to length ratio and again makes bilayers and inverse micelles more likely. When two different kinds of amphiphiles are mixed, characteristics of solution are similar to average characteristics of individual solutions types, provided the two types can mix freely in solution (41). Furthermore, carefully adding more of one kind of molecule can cause solution to form structures of different shapes or sizes than either molecule would form alone.

8.4.3 Method for Characterization of Liquid Crystal

8.4.3.1 Macroscopic View

Optical polarizing microscopy is standard tool in identification of liquid crystal phases and phase transitions but requires considerable experience, particularly in study of new and less familiar materials.

8.4.3.2 X-Ray Diffraction

X-Ray diffraction is very useful tool in identification of liquid crystal phases and phase transitions. It is useful in determining surface texture of liquid crystal during phase transition process (42).

8.4.3.3 Differential Scanning Calorimetry

Differential Scanning Calorimetry (DSC) is useful tool which complements optical methods in study of liquid crystal phase transitions. Its utilization in determining heat supplied or extracted during process such as phase transition (42).

A phase diagram is shown in Figure 8.17. This includes crystal to liquid crystal (smectic A) transition at 55° C followed by barely detectable smectic A to nematic transition at 67° C and finally nematic to isotropic transition near 80° C. The upper, cooling curve shows slight displacement of nematic to isotropic transition, partially due to super cooling and partially instrumental hysteresis attributable to temperature scan rate. The smectic A to crystal transition is depressed strongly due to super cooling of smectic A phase. Thus, phase diagram for cooling process would not be identical to that for heating.

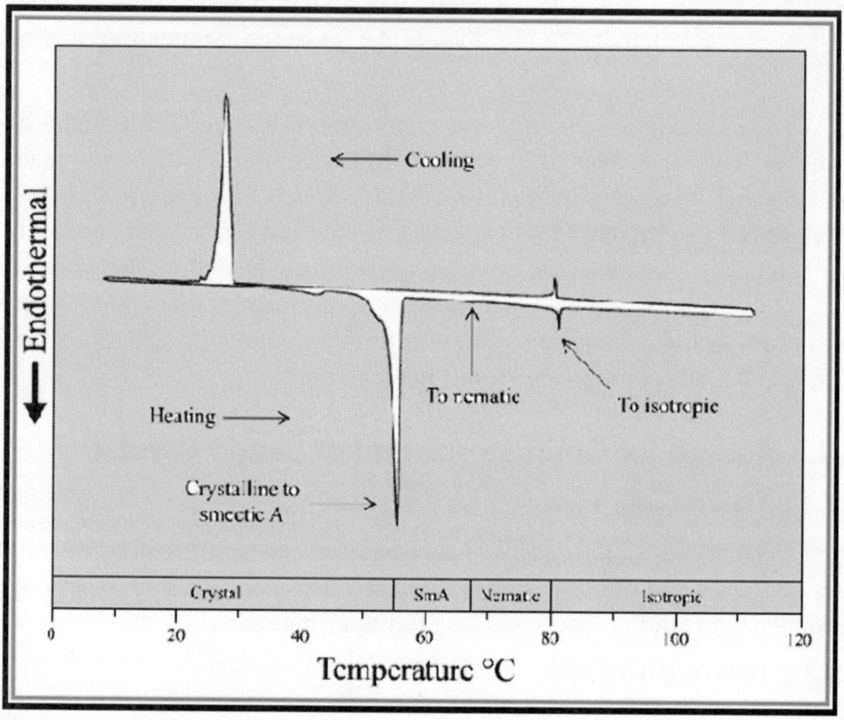

Figure 8.17 DSC plot with phase diagram.

8.5 Application

8.5.1 Pharmaceutical Application

There are many applications exist for amphiphilic molecules, because of their ability to dissolve in both polar and nonpolar medium. In medical field, lyotropic liquid crystal can coat a drug to keep it from being destroyed in GIT. The drug can then be taken orally, and after it reaches to proper location in body, liquid crystal breaks down and drug is released.

The liquid crystals and vesicles prepared herein have a number of applications. For example, they can be used

(i) As delivery vehicles for a number of active ingredients in pharmaceutical and cosmetic applications
(ii) As precursors to templated nano-structured materials
(iii) In areas that thermotropic liquid crystals have been used such as variable focus lens applications
(iv) As polymer dispersed liquid crystals (PDLC)
(v) In piezoelectric applications
(vi) In non-linear optical applications
(vii) In sensing technologies
(viii) In targeted release in controlled chemical environments such as plants, microchips, fermentors, and bioreactors
(ix) In the manipulation of optical axes of birefringence by mechanical means

8.5.1.1 Solubility Enhancement of Poorly Soluble Drugs

Many substances are more soluble in lyotropic liquid crystals. One example is hydrocortisone. It is often taken in topical applications, but its uses have been limited because highest concentration possible has been only 1%. When hydrocortisone was blended into liquid crystal of lecithin and water, concentration went up to 4%. In time, liquid crystals may become a primary solvent for topical medications.

Another example for improvement of solubility is pH-induced nanosegregation of Ritonavir to lyotropic liquid crystal of higher solubility than crystalline polymorphs. Birefringent spherical vesicles of ritonavir are formed by increasing pH of aqueous solutions from 1 to 3 or to 7 and by addition of water to ethanol solutions at room temperature. Increasing pH creates super saturation levels of 30–400. Upon this change in pH, solutions become translucent, implying that some kind of ritonavir assembly was formed. Small spherical vesicles of narrow size distribution are detectable only after a few hours by optical microscopy. Ritonavir self-organizes into various phases (lyotropic liquid crystal form-I, crystal form-II) as a result of super saturation created in aqueous solutions. The lyotropic liquid crystal vesicles do not fuse but slowly transform to polymorphs of Ritonavir (in days), liquid crystal Form I and finally liquid crystal Form II. Amorphous Ritonavir in aqueous suspension also undergoes a transformation to a mesophase of similar morphology. The dissolution and solubility of lyotropic liquid crystal is slightly lower than that of amorphous phase and about 20 times higher than that of Form II (amorphous > liquid crystal Form I > Form II). Ritonavir also self-assembles at air/water interface as indicated by decrease in surface tension of aqueous solutions. This behaviour is similar to that of amphiphilic molecules that induce lyotropic liquid crystal formation (43).

8.5.1.2 Controlled/Sustained Release of Drug

Applications of mesophases (liquid crystal) in pharmaceutical drug delivery systems depend upon their properties. Lipid-based liquid crystalline materials, formed by swelling of certain polar lipids when exposed to aqueous environments, have received much recent attention in drug delivery field for their ability to sustain the release of a wide range of molecules (44). The liquid crystalline matrices possess distinct lipidic and aqueous domains, and may exhibit a number of well-defined geometric arrangements depending on the chemical structure of the lipid, the aqueous content of the system, the presence of other additives, and solution conditions such as pH, temperature, pressure. Most often this arrangement consists of lamellar bilayer structures, but for a relatively small subset of lipids, the exhibited phase structures may include the viscous reverse hexagonal phase or bicontinuous cubic phase (45, 46).

Recently, there have been only a few materials known to exhibit this kind of phase behaviour; glyceryl monooleate (GMO), and other closely related unsaturated monoglycerides are the best known molecules of this class (47, 48).

The relatively low-viscosity lamellar phase has good syringeability and bioadhesivity. Norling et al have reported lamellar phase injectable dental gels containing glyceryl monooleate (GMO), which upon administration was transformed into highly viscous cubic phase. The cubic phase, in contrast, has lower bioadhesivity than lamellar phase (49, 50).

The cubic arrangement provides high viscosity and is widely used for controlled drug release systems (51). Cubic phases are also applied for protection of labile moieties like enzymes and other drugs sensitive to hydrolysis and oxidation (13). However, because of their highly viscous nature, cubic phases are restricted to certain applications. This restriction stimulated research into design of precursors that are transformed into liquid crystals upon hydration.

Another example for In situ sustain release formulation is glyceryl monooleate matrices for oral administration of serratiopeptidase, which undergo transformation to the highly viscous cubic phase in situ and provide sustained release from the microenvironment-controlled highly viscous cubic phase (10).

A new class of lipids that display finite swelling of hexagonal and cubic phases, based on glycerate and urea headgroups has recently been of these materials, phytanyl glycerate (PG) and oleyl glycerate (OG) were found to form reverse hexagonal phase in excess water at physiological temperature disclosed (52).

The finite swelling phenomenon is important for the potential use of lipid-based liquid crystals in drug delivery, as it provides an opportunity to form a lipidic matrix which, on exposure to excess bodily fluids following administration, forms a persistent matrix for both lipophilic and hydrophilic drugs from which drug can be released (44, 53, 54). The hydrophobic domains can also provide a medium for enhancing the solubility of poorly water-soluble drugs.

A second important consequence of the finite swelling phenomenon is that the bulk lyotropic liquid crystals can be dispersed into sub-micron particles in excess aqueous solution, and still retain their internal structure (55). The dispersion of Hexosome and cubicphases results in particles termed Cubosomes and Hexosomes, respectively. Cubosomes and Hexosomes have been almost exclusively prepared using fatty acid based amphiphiles such as GMO (56, 57) until the recent report of submicron dispersions of hexosomes prepared with glycerate surfactants (58), only non-fatty acid based liquid crystalline particle system was that described by Abraham, et al. for beta-XP (59).

Lyotropic liquid crystal preconcentrates for treatment of periodontal disease

This study was done for development of water-free lyotropic liquid crystalline preconcentrates, which consist of oils and surfactants with good physiological tolerance and spontaneously form lyotropic liquid crystalline phase in aqueous environment. In this way these preconcentrates having low viscosity can be injected into the periodontal pocket, where they are transformed into highly viscous liquid crystalline phase, so that the preparation is prevented from flowing out of pocket due to its great viscosity, while drug release is controlled by liquid crystalline texture. Lyotropic liquid crystal was formed by using mixture of surfactants (Cremophor EL, Cremophor RH40) and oil (Miglyol 810) up on addition of water (60).

Example Nicotine Formulation

Polar lipid formulations of nicotine in liquid crystals and colloidal dispersions thereof and precursors or offspring thereof which when in contact with body fluid and/or by the influence of body temperature, are transformed to a liquid crystal or a mixture of liquid crystals, which functions as a controlled release matrix for nicotine suitable in e.g. smoking cessation and/or replacement therapies. Disclosed compositions of said liquid crystals or dispersions thereof, their precursors or offspring containing nicotine and anti-irritants or a local analgesic, or any combination of these reduce local irritation of nicotine and masks its taste. Applicable routes of administration and devices includes buccal, using chewing gum in which liquid crystals, dispersions thereof, or precursor or offspring thereof are formulated, buccal adhesives, gels and patches, using mucoadhesive liquid crystal, dispersions thereof or precursor thereof, and mouth spray using dispersion of said liquid crystals, their precursors or offspring, nasal spray and gel using said liquid crystals or their dispersions, their precursors or offspring, topical, using said liquid crystal or their precursors or offspring in adhesive patches and gels.

Example of Lamellar Liquid Crystals Containing Glycerol

Lamellar lyotropic liquid crystalline (LLC) systems are thermodynamically stable, optically isotropic systems, which are formed with low energy input. These are new possibilities for development of controlled drug delivery systems is inherent in these systems due to their stability and special skin-similarly structure.This study was done for formulate multicomponent lamellar lyotropic liquid crystal systems with relatively low surfactant content, composed of materials official in the European Pharmacopoeia 4th. Lamellar lyotropic liquid crystal was formed by using Brij 96 (poly-oxyethylene-10-oleyl ether) with water, liquid petrolatum (LP) and glycerol. A prolonged drug release was observed in case of the very water-soluble ephedrine hydrochloride and the same phenomena were observed in the case of tenoxicam, which is practically insoluble in water (17).

Example of Vitamin E TPGS/Drug Composition

This invention relates to field of pharmaceutical drug compositions and delivery devices, and particularly to use of substances and devices that form liquid crystal structures to enhance the solubility and control the release of active drug compounds in patient. The present invention solves solubility and toxic side effect problems associated with prior art immediate release formulations of certain highly insoluble drugs, e.g., cyclosporine. In particular, the present invention utilizes ability of Vitamin E TPGS (Vitamin E ∂-alpha-Tocopheryl Polyethylene Glycol-1000 Succinate) to form liquid crystal structures with gastric fluid to enhance drug solubility, and further employs dosage forms with impermeable or semi-permeable barriers to control drug release over time, thereby providing a means for lowering dose within therapeutic window.

Vitamin E TPGS/drug compositions and methods were provided which obviate need for surfactants or non-evaporated co-solvents because active drug component was dissolved directly into Vitamin E TPGS to form a true molecular solution--not an emulsion or a micro-emulsion. The invention provides a slowly dissolving TPGS/Drug matrix that absorbs gastrointestinal fluid into matrix at the dosage form/fluid interface, where a gel-like liquid crystal is formed. This gel front forms a liquid crystal boundary where drug dissolution is highest. At this liquid crystal/GI fluid boundary, synchronization takes place in which rate of formation of liquid crystals equals dissolution rate of liquid crystals at the water interface, thereby giving controlled order release of drug into GI tract. The rate of dissolution is also controlled by geometry of the dosage form. The solid Vitamin E TPGS/drug matrices of invention can be solidified and compressed into tablets or filled into capsules, with other excipients, binders and/or fillers. The solid TPGS/drug solution of invention also can be made into an immediate release liquid formulation upon addition of water, or into a controlled release system solid tablet by use of impermeable or semi-permeable barriers or coatings surrounding portions of tablet.

8.5.1.3 Stability of Drug

Lyotropic liquid crystals have been used to make stable hydrocarbon foam. Hydrocarbon foams have been difficult to produce in the past because surface tension of hydrocarbon is low enough that adsorption to an oil-soluble surfactant would have no significant effect. Without adsorption, hydrocarbon simply behaves as liquid. When lyotropic liquid crystal molecules change from inverse micelles to lamellar sheets, they lower the surface tension enough for foam to form. The hydrocarbon and surfactant can dissolve in each other, and surfactant cannot dissolve in water, although water can dissolve in the surfactant and mix into the liquid crystal (61).

Example of Polymer Dispersed Liquid Crystals (PDLC)

Self-assembling block copolymers exhibiting wide range of morphologies and rheological behaviours have attracted significant attention because of their advantages, such as controlled drug release, bioadhesivity, and protection of sensitive drug molecules. Pluronic F127 exhibits thermoreversible gelation property in water, with high solubilising capacity. It can form liquid crystalline mesophases with high viscosity, which enhance protein stability and provide controlled drug release. Hence, this polymer is of special interest to the pharmaceutical industry (62, 63). The presence of drugs and excipients in Pluronic gel significantly affects properties of the system. Pisal et al have reported the effect of addition of vitamin B_{12} and other excipients on various pharmaceutical properties of a gel. Pandit et al have reported the effect of salt on Pluronic F127's gelation behaviour (64).The major limitation in Pluronic gels is high concentration of polymer required to obtain viscosity needed to achieve desired pharmaceutical performance. Therefore, attempts have been made to reduce polymer concentration by combining it with Carbopol, HPMC and CMC (65, 66). These polymers are added in concentration range of 1-5% w/w. Colloidal silicon dioxide is popular gelling agent that has been shown to gel in wide range of solvents (67). The addition of any other component to liquid crystalline system is expected to alter liquid crystalline phase and in turn the properties of system. Therefore, it is necessary to study effect of additive on liquid crystalline phase.

8.5.1.4 Effect of Base and Salts from Different Liquid Crystalline Structures on Drug Release Profile

This study was done for investigate the influence of two types of chlorhexidine species, chlorhexidine base and its salts, on physicochemical features of liquid crystalline systems and on drug transport through lipophilic membranes.For this non-ionic surfactant, Synperonic A7 (PEG7-C13-15) was selected for the liquid crystal formulation. Chlorhexidine species was modified liquid crystalline system. As a result of changes of liquid crystalline structures, drug release of various types of chlorhexidine could be also modified. The combination of base and salt forms of drug in one dosage form could eliminate drug release changes from liquid crystalline systems of dynamically changeable structures (68).

The acid, base or salt form of applied drug can interact differently with vehicle molecules, which might alter liquid crystalline structure and cause differences in release property (69). In case of drug components of similar chemical structure, liquid crystalline phase change and permeability coefficient were found to be solute and concentration dependent (70). Mueller-Goymann and Hamann studied drug release from reversed micellar and from lamellar liquid crystalline systems containing fenoprofen acid and fenoprofen sodium salt. Sustained release was achieved, when phase transformation occurred from micellar to lamellar phase (71).

8.5.1.5 Photopolymerization of Lyotropic Liquid Crystalline Systems: A New Route to Nanostructured Materials

A novel route to fabricating such materials is through the use of lyotropic liquid crystals (LLCs) that possess highly ordered nanostructures. However, LLC phases lack necessary physical robustness. So, templating LLC phase morphology onto other materials such as organic polymers would give a nanostructure retained as part of a robust polymeric matrix. This study focuses on photopolymerization behaviour and structure retention of hydroxyethyl acrylate (HEA)/dodecyltrimethyl ammonium bromide (DTAB)/water system in a select liquid crystal phase. The results suggest that polymerization behaviour is heavily dependent on type of LLC structure. Specifically, lamellar aggregates polymerize faster than either cubic or isotropic morphologies due to diffusional limitations on growing polymer chain. Monomer segregation also plays a role in determining the polymerization rates. Results also indicate that original LLC order of these systems is largely retained upon photopolymerization although some LLC phases do change upon cure. This order would be useful in applications such as ultrafiltration membranes, separation media, and drug delivery systems (Colleen D. Colson, Christopher L. Lester, and C. Allan Guymon).

8.5.1.6 Nanostructured Hexosome for Injectable Drug Delivery

A new class of amphiphiles with a glycerate headgroup, recently shown to form reverse hexagonal phase in excess water, have been dispersed to form Hexosome dispersions comprising sub-200 nm particles retaining the internal nanostructure of parent Hexosome phase. The application of these novel materials to development of a new injectable formulation of irinotecan was investigated. The formulation of irinotecan with a small percentage of oleic acid in oleyl glycerate permitted a clinically relevant dose of irinotecan to be dissolved in glycerate surfactant and dispersed in aqueous medium to form an injectable particle-based dose form of irinotecan. Importantly, incorporation of irinotecan into Hexosomes at neutral pH did not result in conversion from active lactone to inactive carboxylate form on storage, and is hence a promising alternative to current low pH formulation of irinotecan required to inhibit this conversion. Although release of irinotecan from Hexosomes was shown to be virtually instantaneous from Hexosomes on substantial dilution, the retention of the drug in lactone form at neutral pH demonstrates a potential application of these novel nanostructured particles in injectable drug delivery (18).

8.5.1.7 Temperature modulated drug permeation through liquid crystal embedded cellulose membranes

Stimuli-sensitive membranes may act as "on–off switches" or "permeability valves", producing patterns of pulsatile release, where the period and rate of mass transfer can be controlled by external or environmental triggers. Cellulose nitrate (CN) and cellulose acetate (CA) monolayer membranes containing thermotropic liquid crystals (LC) were developed as thermoresponsive barriers for drug permeation. A low molecular thermotropic liquid crystal, *n*-heptyl-cyanobiphenyl, with nematic to isotropic phase transition temperature of 41.5°C was chosen to modulate drug permeation. It was found that upon changing temperature of system around transition temperature, both cellulose membranes without liquid crystal showed no temperature sensitivity to drug permeation, whereas the results for liquid crystal entrapped membranes exhibited a distinct jump in permeability when temperature was raised to above transition temperature of liquid crystal for drug. On other hand, drug permeation through these liquid crystal embedded membranes can be thermally modulated. Thermoresponsive drug permeation through the membranes was reversible, reproducible and followed zero order kinetics. Liquid crystal embedded cellulose acetate membranes showed more temperature sensitivity than liquid crystal embedded cellulose nitrate membranes, apparently due to higher liquid crystal loading in their porous matrix compared to cellulose nitratemembranes (72). Another thermoresponsive membrane was developed by adsorbing the binary cholesteric liquid crystals, 36% cholesteryl oleyl carbonate and 64% cholesteryl nonanoate solved in an organic solvent into the cellulose nitrate membranes (15). Such thermoresponsive liquid crystal embedded membranes are much permeable to drug molecules at temperatures above 41.5 ◦C. This triggerable drug delivery system was released drug only above a certain temperature as a drug reservoir system. One of the rationales for such thermoresponsive system may be potential for their use in chemotherapeutics delivery under local hyperthermia.

8.5.1.8 Liquid Crystal for Treatment of Cancer and Other Disease

The most recent research involving LCPs has yielded new investigational anti-tumor drug called Tolecine, compound that also has antiviral and antibacterial applications. Created by Tsai, it has been shown to be even more effective than the current standard of care for herpes.

The patent application involves formulation that combines Tolecine and Apatone®, which attacks cancer cells via multiple pathways to offer improved efficacy. Apatone® has been successfully tested in more than 30 human tumor cell lines at Summa and in Phase I/IIa clinical trial, which demonstrated delaying effect in progression of end-stage cancer patients. In addition, FDA granted Apatone® orphan-drug status for treatment of metastatic, or locally advanced inoperable bladder cancer in August 2007 (73).

Unlike other chemotherapy drugs, Tolecine™ and Apatone® have low toxicity and do not target dividing cells. Instead, they are activated by inflammation that occurs in and around tumor cells, sparing healthy cells. We want to kill cancer cells specifically without killing surrounding tissues, says Jamison.

8.5.2 Non-Pharmaceutical Application

8.5.2.1 Liquid Crystal in Cosmetics

Liquid crystals are used today in cosmetics, both as eye shadow and lip gloss. The latter, marketed by Laura Mercier, is claimed to have "the brilliance of quartz & opal crystals by reflecting & refracting light giving lips fullness & depth". Lyotropic liquid crystals are used in cosmetic gels and emulsions to stabilize the structure and to retain moisture.

8.5.2.2 Liquid Crystal in Soap Industries

Lyotropic liquid crystalline behaviour is found in simple household soap. Soaps work better than pure water at removing dirt and grease because nonpolar insides of micelles are capable of dissolving nonpolar substances that will not dissolve in water. (This also works in reverse if the solvent is nonpolar and some of the substance to be removed is polar). Soaps also help water dissolve more because molecules tend to remain at surface, hydrocarbon tail away from water, thus lowering surface tension of water and allowing more material to enter it and be dissolved (74).

8.5.3.3 Liquid Crystals in Optics

Optical properties of liquid crystals make them suitable for many applications (75). Liquid crystal display technology employs electrically driven Liquid Crystal Spatial Light Modulators (LCSLMs). The commercial applications of this technology have increased research efforts and investment in new materials. Contrast ratio, viewing angle, polarization sensitivity, electronic consumption, temperature stability, operating speed and computer configuration are some of the characteristics of new devices. So, application of LCSLMs technology in photolithography, optical tweezers, optical processing, beam shaping, active and adaptive optics etc. (76-78).

8.6 Conclusion

The future of liquid crystals has just begun. There are still unsolved physical problems in this area. The need for liquid crystal applications in pharmaceutical field and other field are more important. Knowledge in the field of liquid crystals is crucial for an understanding of biological membranes, thus biologists, medical researchers and pharmacists are also interested in liquid crystal research. A strange form of matter which piqued the curiosity of scientists at the end of the last century has grown into an enormous industry with a great variety of applications and is still growing.

References

1. Sluckin TJ, Dunmur DA and Stegemeyer H. Crystals That Flow - classic papers from the history of liquid crystals, London: Taylor & Francis, (2004).
2. Vill V. Early History of Liquid Crystalline Compounds. Mol. Cryst. Liq. Cryst, 213: 67-71 (1992); Condens. Matter News, 1 (5): 25-28 (1992).
3. Fehr C, Goze-Bac C, Anglaret E, Benoit C, Hasmy A. Orientational order and dynamics of a nematic liquid crystal in porous media. Europhys. Lett. 73: 553-559 (2006).
4. Goodby JW, Curr. Opin. Solid State Mater. Sci.,4: 361 (1999)
5. Fairhurst CE, Fuller S, Gray J, Holmes MC, and Tiddy GJT. Lyotropic Surfactant Liquid Crystals in Handbook of Liquid Crystals, editors Demus D, Goodby JW, Gray GW, Spiess HW, and Vill V. Vol-3, High Molecular Mass Liquid Crystals, pp. 341 (1998).

6. Fazio D, Mongin C, Donnio B, Galerne Y, Guillon D, and Bruce DW. J. Mater. Chem, 11: 2852 (2001).

7. Leahuge-Ballesteros D, Abdul-Fattah A, Stevenson CL, Bennett DB. Properties and stability of liquid crystal form of cyclosporine-the first reported naturally occurring peptide that exists as thermotropic liquid crystal. J Pharm Sci., 9:1821-31(2003).

8. Tschierske C. J. Mater. Chem.,11: 2647 (2001).

9. Unger G. X-ray Studies of Nematic Systems in Physical Properties of Liquid Crystals: Nematics, editors Dunmur DA, Fukuda A, and Luckhurst GR. INSPEC, Institution of Electrical Engineers, London, pp. 177 (2001).

10. Shah MH, Paradkar A. Cubic liquid crystalline glyceryl monooleate matrices for oral delivery of enzyme. Int J Pharm., 294:161-171 (2005).

11. Sokol RJ. Improvement of Cyclosporin Absorption in Children after Liver Transplantation by Means of Water-soluble Vitamin E. The Lancet, pp. 212-215 (1991).

12. Argao EA. "d-.alpha.-Tocopheryl Polyethylene Glycol-1000 Succinate Enhances the Absorption of Vitamin D in Chronic Cholestatic Liver Disease of Infancy & Childhood". Pediatric Res., 31(2): 146-150 (1992).

13. Nylander T, Mattisson C, Razumas V, Miezis Y, Hakansson B. A study of entrapped enzyme stability and substrate diffusion in monoglyceride-based cubic liquid crystalline phase. Colloids Surf A: PhysicochemEng Aspects, 114: 311-320 (1996).

14. Watson SJ, Gleeson HF, D'Emanuele A. An examination of the drug transport properties of liquid crystal embedded membranes. Mol. Cryst. Liq. Cryst., 367: 3223–3231 (2001).

15. Lin SY, Lin HL, Li MJ. Adsorption of binary liquid crystals onto cellulose membrane for thermo-responsive drug delivery. Ads. J. Inter. Ads. Soc., 8: 197–202 (2002).

16. Osborne DW, Ward AJI. Lyotropic liquid crystals as topical drug delivery vehicles. Int. J. Pharm. Adv., 1: 38–45(1995).

17. Makai M, Csányi E, Németh Z, Pálinkás J, Er′os I. Structure and drug release of lamellar liquid crystals containing glycerol. International Journal of Pharmaceutics, 256: 95–107 (2003).

18. Boyd BJ, Whittaker DV, Khoob S, Davey G. Hexosomes formed from glycerate surfactants—Formulation as a colloidal carrier for irinotecan. International Journal of Pharmaceutics, 318: 154–162 (2006).

19. Vill V. Chemical Structures and Polymorphism in "Chirality in Liquid Crystals". Springer-Verlag, New York, (2001).

20. Meibner D, Grassert I, Oehme G, Holzhüter G, Vill V. Preparation and characterization of new micelle-forming cholesterol amphiphiles. Colloid Polym. Sci., 278, 364-368 (2000).

21. Chandrasekhar S. Liquid Crystals,Cambridge University Press, Cambridge, (1992).

22. De Gennes P, Prost J. The Physics of Liquid Crystals, Clarendon Press, Oxford, (1993).

23. Fukuda A, Takanishi Y, Isozaki T, Ishikawa K and Takezoe H. liquid crystal. J. Mater. Chem., 4: 997–1016 (1994).

24. Carlsson T, Stewart IW, 'Theoretical studies of smectic C liquid crystals confined in a wedge; stability considerations and Frederiks transitions.' Liq. Cryst., 9: 661-678, (1991).

25. Madsen LA, Dingemans TJ, Nakata M, and Samulski ET. Thermotropic Biaxial Nematic Liquid Crystals". Phys. Rev. Lett.,92: 2-13 (2004).

26. Joseph A. Liquid Gold: The Story of Liquid Crystal Displays and the Creation of an Industry,Castellano, (2005).

27. Kopp VI, Fan B, Vithana HK, Genack AZ. "Low threshold lasing at the edge of a photonic stop band in cholesteric liquid crystals". Opt. Lett.,23: 1707–1709 (1998).

28. Jakli A, Saupe A. One and Two Dimensional Fluids – Physical Properties of Smectic Lamellar and Columnar Liquid Crystals (Taylor and Francis), (2006).

29. Lydon J. Chromonic liquid crystal phases. Curr. Opin. Colloid Interface Sci., 3:458–466, (1998).

30. Lydon j. Chromonic mesophases. Curr. Opin. Colloid Interface Sci., 8: 480–490 (2004).

31. Kostko AF, Cipriano BH, Pinchuk OA, Ziserman L, Anisimov MA, Danino D, and Raghavan SR. Salt effects on the phase

behaviour, structure and rheology of chromonic liquid crystals, J. Phys. Chem. B, 109: 19126–19133 (2005).

32. Horowitz VR, Janowitz LA, Modic AL, Heiney PA, and Collings PA. Aggregation behaviour and chromonic liquid crystal properties of an anionic monoazo dye. Phys. Rev. E, 72, (2005).

33. Brown HG, Doan JW, Neff VD. A Review of Structure and Physical Properties of Liquid Crystals, CRC, Cleveland, OH, (1971).

34. Winsor PA. Binary and multicomponent solutions of amphiphilic compounds. Solubilization and the formation, structure and theoretical significance of liquid crystalline solutions, Chem. Rev.,68:1–40 (1968).

35. Blunk D, Praefcke K, Vill V. Amphotropic Liquid Crystals, Handbook of Liquid Crystals, pp. 305-340 (1998).

36. Rädler JO, Radiman S, de Vallera A, Toprakcioglu C. SANS studies of liquid crystalline microemulsion gels, Physica B: Physics of Condensed Matter, 156: 398-401 (1989).

37. Greenberg, Stephen M, Louis A. Ester based water-in-oil cosmetic microemulsions, US Patent 4940577 (1990).

38. Alexandridis P, Zhou D, Khan A. Lyotropic liquid crystallinity in amphiphilic block copolymers: temperature effect on phase behaviour and structure copolymers of different composition. Langmuir., 12: 2690-2700 (1996).

39. Peters D, Vill V. Prediction of phase transition temperatures of liquid crystals with help of similarity concept, Proc. SPIE-Int. Soc. Opt. Eng., 74-77 (1998).

40. Ibrahim HG, Sallam E, Takieddin M, Habboub M. Effects of solute characteristics and concentration on a lyotropic liquid crystal: solute induced phase change, pharmaceutical research, Vol.10, 5 (1993).

41. Nguyen HL, Nguyen HT, and Sigaud G. Synthesis and characterization of thermotropic amphiphilic liquid crystals: semiperfluoroalkyl-ßD-glucopyranosides, LiqCryst, 27(11): 1451-1456 (2000).

42. Nounesis G, Shi Y, Kumar S ,X-ray Diffraction and Calorimetric Studies of Nematic - Nematic and Smectic - Smectic Phase Transitions, ACS Meeting, Anaheim, CA, April, 2-7 (1995).

43. Acciacca A, Spong BR, Fleisher D, Hornedo NR. Mol. Pharmaceutics, 5: 956–967 (2008).
44. Drummond CJ, Fong C. Surfactant self-assembly objects as novel drug delivery vehicles. Curr. Opin. Colloid Interf. Sci., 4: 449–456 (1999).
45. Laughlin RG. The Aqueous Phase Behaviour of Surfactants. Academic Press Limited, London (1994).
46. Hyde ST, Andersson S, Larsson KZB, Landh TSL, Ninham BW. The language of shape. The role of curvature in condensed matter: physics, chemistry and biology (1997).
47. Briggs J, Caffrey M. The temperature-composition phase diagram and mesophase structure characterization of monopentadecenoin in water. Biophys. J., 67: 1594–1602 (1994).
48. Qui H, Caffrey M. Lyotropic and thermotropic phase behaviour of hydrated monoacylglycerols: structure characterisation of monovaccenin. J. Phys. Chem. B, 102: 4819–4829 (1998).
49. Engstrom S, Ljusberg-Wahren H, Gustafsson A. Bioadhesive properties of monoolein-water system. Pharm. Technol. Eur., 7:14-17 (1995).
50. Nielsen LS, Schubert L, Hansen J. Bioadhesive drug delivery system, I: characterization of mucoadhesive properties of systems based on glyceryl monooleate and glyceryl monolinoleate. Eur J Pharm Sci., 6: 231-239 (1998).
51. Wyatt DM, Dorschel D. A cubic phase delivery system composed of glyceryl monooleate and water for sustained release of water-soluble drugs. Pharm Technol., 16: 116 (1992).
52. Boyd BJ, Davey G, Drummond CJ, Hartley P, Fong C, Krodkiewska I, Murphy A, Tait R, Warr G, Wells D, Whittaker DV, Ye R. Surfactants and Lyotropic Phases Formed There from, International Patent WO 2004/022530, (2004).
53. Chang CM, Bodmeier R. Low viscosity monoglyceride-based drug delivery systems transforming into a highly viscous cubic phase. Int. J. Pharm., 173: 51–60 (1998).
54. Shah JC, Sadhale Y, Chilukuri DM. Cubic phase gels as drug delivery systems. Adv. Drug Del. Rev., 47: 229–250 (2001).
55. Larsson K. Cubic lipid-water phases—structures and biomembrane aspects. J. Phys. Chem., 93: 7304–7314 (1989).

56. Gustafsson J, Ljusberg-Wahren H, Almgren M, Larsson K. Submicron particles of reversed lipid phases in water stabilized by a non-ionic amphiphilic polymer. Langmuir, 13: 6964–6971 (1997).

57. Spicer PT, Hayden KL. Novel process for producing cubic crystalline nanoparticles. Langmuir, 17: 5748–5756 (2001).

58. Fong C, Krodkiewska I, Wells D, Boyd BJ, Booth J, Bhargava S, McDowall A, Hartley PG. Submicron dispersions of hexosomes based on novel glycerate surfactants. Aust. J. Chem., 58: 683–687 (2005).

59. Abraham T, Hato M, Hirai M. Glycolipid based cubic nanoparticles: preparation and structural aspects. Coll. Surf. B: Biointerf., 35: 107–117 (2004).

60. Fehe′r A, Urba′n E, Ero″s I, Szabo′-Re′ve′sz P, Csa′nyi E. Lyotropic liquid crystal preconcentrates for the treatment of periodontal disease, International Journal of Pharmaceutics, 358: 23–26 (2008).

61. Paavola A, Kilpelaine I, Yliruusi J, Rosenberg P. Controlled release injectable liposomal gel of ibuprofen for epidural analgesia. Int. J Pharm., 199: 85-93 (2000).

62. Veyries ML, Couarraze G, Geiger S, et al. Controlled release of vancomycin from poloxamer 407 gels. Int. J Pharm., 192: 183-193 (1999).

63. Ricci EJ, Lunardi LO, Nanclares DMA, Marchetti JM. Sustained release of lidocaine from Poloxamer 407 gels. Int. J Pharm., 288: 235-244 (2005).

64. Pandit N, Trygstad T, Croy S, Bohorquez M, Kock C. Effect of salts on the micellization, clouding and solubilization behavior of Pluronic F127 solutions. J Colloid Interface Sci., 222: 213-220 (2000).

65. El-Kamel AH. In vitro and in vivo evaluation of Pluronic F127-based ocular delivery system for timolol maleate. Int. J Pharm., 241: 47-55 (2002).

65. Lin H, Sung KC. Carbopol/pluronic phase change solution for ophthalmic drug delivery. J Control Release., 69: 379-388 (2000).

67. Raghavan SR, Walls HJ, Khan SA. Rheology of silica dispersions in organic liquids: new evidence for solvation forces dictated by hydrogen bonding. Langmuir., 16:7920-7930 (2000).

68. Farkas E, Kiss D, Zelk´o R. Study on the release of chlorhexidine base and salts from different liquid crystalline structures, International Journal of Pharmaceutics, 340: 71–75 (2007).

69. M¨uller-Goymann CC, Frank SG. Interaction of lidocaine and lidocaine-HCl with the liquid crystal structure of topical preparations. Int. J. Pharm., 29: 147–159 (1986).

70. Ibrahim HG. Release studies from lyotropic liquid crystal systems. J. Pharm. Sci., 78: 683–687 (1989).

71. M¨uller-Goymann CC, Hamann HJ. Sustained release from reverse micellar solutions by phase transformations into lamellar liquid crystals. J. Control. Rel., 23: 165–174 (1993).

72. Atyabi F, Khodaverdi E, Dinarvand R. Temperature modulated drug permeation through liquid crystal embedded cellulose membranes, International Journal of Pharmaceutics, 339: 213–221 (2007).

73. Tsai C, Jamison J, Miller T. New liquid crystal technology hope for treatment of cancer and other diseases, Kent State University, Kent state magazine, volume-7, (2007).

74. O'Lenick J A, Bilbo RE. Soap Cosmetics. Chem Specialities, 63(4): 52-65 (1987).

75. Khoo IC. Liquid Crystals: Physical properties and Nonlinear Optical Phenomena;Wiley-VCH: Hoboken, NJ, USA, (1994).

76. Hayashi T, Shibata T, Kawashima T, Makino E, Mineta T, Masuzawa T. Photolithography system with liquid crystal display as active gray-tone mask for 3D structuring of photoresist. Sens. Actuat. A: Phys., 144: 381-388 (2008).

77. Hands PJW, Tatarkova SA, Kirby AK, Love G. Model liquid crystal devices in optical tweezing: 3D control and oscillating potential wells. Opt. Expr., 14: 4525-4537 (2006).

78. Zmija J, Klosowicz SJ, Kedzierski J, Nowinowskt-Kruszelnicki E, Zielinski J, Raszewski Z, Walczak A, Parka J. Application of liquid crystals in optical processing of optical signals. Opto-Electron. Rev., 5: 93-106 (1997).

9
Chirality in Drug Development

Y. Madhusudan Rao and S. Jagan Mohan
University College of Pharmaceutical Sciences,
Kakatiya University, Warangal, India

9.1 History of Chirality

The discovery of isomers occurred about morethan 150 years ago when Louis Pasteur noted that some organic molecules existed as two non-superimposable mirror images. These two images were enantiomers and named (S) for left and (R) for right. Racemates are 1:1 mixture of two enantiomers. In most cases, one enantiomer provides the desired therapeutic benefits, while the other enantiomer may be inactive or may cause side effects. The growing drug industry demands for enantiomerically pure compounds and specialty chemicals are the driver for many companies to pursue biocatalytic technology (1).

In the August 25, 2003 issue of Chemical and Engineering News, the results of a survey of the journal's readers on ''the most beautiful experiments in the history of chemistry'' were reported. First place was given to Louis Pasteur's manual separation of tartrate enantiomers (1848). One prominent chemist (and historian of chemistry) characterized its scientific significance: Pasteur's separation of optical isomers opened up an area of chemical structure particularly important to organic chemistry and biochemistry.

In particular, Pasteur's achievement constituted the basis for unfolding the concept of chirality.

In this chapter, I will explore the background to Pasteur's first major experimental work, explicate what he did in studying tartrates and other organic crystalline compounds, and, finally, trace the development of the chemical concept of chirality from Pasteur's discoveries.

The word ''chirality'' did not exist in Pasteur's day. Pasteur used the word ''dissyme´ trie,'' or ''dissyme´ trie mole´ culaire.'' The words ''chiral'' and ''chirality'' were introduced by William Thompson, Lord Kelvin, in 1884: I call any geometrical figure, or group of points, chiral, and say that it has chirality, if its image in a plane mirror, ideally realized, cannot be brought to coincide with itself.

''Chiral'' and ''chirality'' have a classical derivation from the Greek word, ''cheir,'' which means ''hand.'' Pairs of human hands (and feet) are the most common and prominent examples of Kelvin's geometrical type; they are ''identical opposites,'' mirror-images of each other. The word ''enantiomorph,'' closely associated with chirality, captures in its Greek-derived meaning of ''opposite form'' (enantios morphe) just this essence.

This is what Pasteur meant by ''dissyme´trie mole´culaire.'' In a lecture that he delivered to the Socie´te´ chimique de Paris in 1860, titled, ''Recherches sur la dissyme´trie mole´culaire des produits organiques naturels,'' he speculated (almost uniquely for him) on the atomic arrangements in dextro- and levo-organic compounds that are optically active (2).

9.2 Nomenclature

(+) and (-): These symbols are used to specify the sign of rotation of plane-polarized light upon passage through a solution or solid; (+) denotes rotation to the right and (-) signifies rotation to the left.

d and l: These symbols are an alternative nomenclature of (+) and (-). d stands for dextrorotatory and l stands for levorotatory. The d, l nomenclature is not used in this handbook.

D and L: These symbols refer to the absolute configuration of a molecule. They are based on a convention which uses D-(+)-glyceraldehyde and L-(-)-glyceraldehyde as standards. The configuration notations D and L must not be confused with the rotational prefixes d and l to which they bear no immediate relationship. Today, the D,L system is used mainly for amino acids and carbohydrates but not for other compounds.

(R) and (S): The Cahn-Ingold-Perlog (R,S) system has replaced the D,L system for the designation of absolute configuration in stereochemistry.

The groups or atoms attached to the asymmetric atom are arranged according to a sequence rule in priority order of decreasing atomic number. If the atoms directly attached to the asymmetric center are the same, the priority order is determined by the nature of atoms further away. Multiple linkages are considered as multiple single bonds. The group of lowest priority is viewed through an imaginary triangle formed by joining the three other atoms or groups in the molecule. If the observed order of decreasing priority of groups forming the triangle is clockwise, then the configuration is called (R) which stands for *rectus* [e.g., (S)-(+)-amphetamine. If the order of decreasing priority of groups in the molecule is counterclockwise, then the configuration is called (S) which stands for *siniste*r [e.g., (S)-(+)-amphetamine. It is to be noted that the (R, S) designation bears no immediate relationship to the rotational prefixes (+) and (-).

***Erythro and Threo*:** These terms are used or refer to compounds having substituents on two adjacent carbon atoms.

Cis and Trans: *Cis* and *Trans* are used to describe the configuration of atoms in certain diastereomers. *Cis, Trans* isomers are superimposible on their mirror images and, therefore, they do not usually show optical activity. *Cis, Trans* isomerism is no restricted to molecules having carbon -carbon double bonds; If a molecule contains a chiral center in addition to a double bond or is equivalent (e.g., a cyclopropane ring), then it will exhibit both optical and geometrical isomerism (e.g., 2-phenylcyclopropylamines).

(*Z*) and (*E*): A problem arises with the *Cis, Trans* nomenclature when the four groups attached to the double bond of a molecule are all different. The (Z, E) nomenclature of the Cahn-Ingold-Perlog convention can describe such molecules accurately. According to this system, the two groups attached to each double-bonded atom are ranked using the sequence rules. That isomer which has the two highest ranking groups on the same side of the double bond is called (Z) (from the German word *Zusammen*, meaning together) and the other isomer is called (E) (from *entgagen*, meaning opposite).

α and β: This system is used for steroids and related compounds. An atom or group attached is termed α if it is below the plane of the ring and β if it lies above the plane [as in estradiol.

Chiral stationary phase: A stationary phase which incorporates a chiral selector. If not a constituent of the stationary phase as a whole, the chiral selector can be chemically bonded to (chiral bonded stationary phase) or immobilized onto the surface of a solid support or column wall (chiral coated stationary phase), or simply dissolved in the liquid stationary phase.

Chiral selector: The **chiral** component of the separation system capable of interacting enantioselectively with the enantiomers to be separated.

Chiral additive: The chiral selector which has been added as a component of a mobile phase or electrophoretic medium.

Chiral mobile phase: A mobile phase containing a chiral selector.

In addition to the terms mentioned above there are many other chiral technical terms which may be difficult to undestand for a biginner in chiral research area (Table 9.1)

Table 9.1 Glossary of selective terms in Stereochemistry

Chirality	It is a property of an object which is non-superimposable with its mirror image
Homochirality	It is the biological chirality in which all biologic compounds have the same chirality such as all amino acids are levorotary isomers.
Chiral switch	It is a procedure used to transform an old racemic drug into its single active enantiomer
Chiral molecules	Molecules whose mirror images are not superimposable upon each other
Achiral compounds	Molecules whose mirror images are superimposable on each other
Stereoisomers	Compounds that have the same atoms connected in same order but differ from each other in the way the atoms are oriented each other in the way are oriented in space.
Enantiomer	Stereoisomer whose mirror images cannot be superimposed
Enantiomers	Chiral molecules that are structurally different from each other only in the left and right-handedness of their orientations
Levo-	Isomer that rotates the plane of polarized light to the left

Contd...

Dextro-	Isomer that rotates the plane of polarized light to the right
Racemates	The two enantiomers that comprise a racemic mixture
Absolute configuration	Indicates the actual arrangements of the substituents in the chiral compound
Diastereomers	Stereoisomers with multiple chiral centers that are not Enantiomers
Meso compound	Diastereomer with two or more chiral centers where the four groups on each of the chiral carbon atoms contains a plane of symmetry with in the molecule
Chiral inversion	Conversion of one enantiomer into its mirror image
Distomer	Refers to the enantiomer with lower pharmacological affinity or activity
Eutomer	Refers to the enantiomer with higher pharmacological affinity or activity
Epimers	Diastereomers which have a different configuration at only one chiral center are called epimers.
Isomers	Compounds that have identical molecular formulae but differ in nature or sequence of bonding of their atoms or in the arrangement of their atoms in space
Constitutional isomers	Isomers that have the same number and kinds of atoms but differ in terms of the arrangement of atoms in the molecules.
Racemic mixture	A mixture of equimolar amounts of enantiomers
Racemization	Conversion of an enantiomer to its racemate
Conformation	This term is used to refer to any one of an infinite number of arrangements of atoms in space that result from the rotations about any of the bonds in the molecule.
Configuration	This term refers to the relative position or order of the arrangement of atoms in space that characterizes a certain stereoisomer.
Absolute configuration	This term refers to the actual order of the arrangement of atoms about a chiral center.

9.3 Introduction

The chirality that is inherent in the enzyme systems of living organisms results in an abundance of enantiopure organic molecules in the living world. In addition to the optical properties first noticed by Pasteur, stereospecific interactions at recognition sites result in differences in both biological and toxicological effects. This fact underlies the continuing growth in chiral chemistry, rooted as it is in fundamental biochemistry. The pharmaceutical industry has undergone a strategic shift and embraced the wide spectrum of asymmetrical synthetic methods now available. The use of these processes in developmental synthesis and large-scale manufacturing has provided new challenges in drug discovery, motivated by a desire to improve industrial efficacy and decrease the time from the conception of a new drug to the market. The economic impact of the industrial production of chiral drugs is now huge--more than 50% of the 500 top-selling drugs were single-enantiomers in 1997. Sales have continued to increase by more than 20% for the past 6 yr and worldwide annual sales of enantiomeric drugs exceeded US$100 billion for the first time in the year 2000, chiral drugs representing close to one-third of all sales worldwide. While some 'chiral switches' may be of less apparent benefit, or indeed detrimental in some cases, encouragement by the regulatory agencies and the ability to extend the life cycle of a drug coming off patent promotes the trend. However, it may turn out to be the ability to provide chiral templates, and thereby attack the key targets of selectivity and specificity, that will lead to the greatest benefits. Research into new chemical entities that can interact specifically with enzyme families may potentially lead to new therapies for complex disease processes (3).

Chirality is a fundamental characteristic of nature and pervades the living world. We have been under its constant influence throughout evolution as a result of the asymmet- rical nature of the environment and, while the origin of this phenomenon is a matter of speculation, the evidence surrounds us all. Most proteins, for example, are formed of L-amino acids while carbohydrates are composed of natural sugars, all D-isomers. Biological receptor systems comprise a complex structural organization of helices and sheets and display `handedness'. This results in a profound effect on drug-receptor interactions.

The subject has fascinated scientists since the middle of the 19th century, when Louis Pasteur65 demonstrated the stereoisomeric forms of tartaric acid. Following earlier work by de la Provostaye, he undertook crystallographic studies on tartaric acid and its salts, and demonstrated the presence of hemihedral facets. In some instances, these were orientated to the left and in others to the right. By handpicking the crystals, he divided them into two groups and found that the solutions rotated light in equal but opposite direction s. Pasteur recognized that the cause of this phenomenon lay in the molecular structure, and by extending these ideas he evolved the theory of the asymmetrical carbon atom.

Chirality is formally defined as the geometric property of a rigid object (like a molecule or drug) of not being superimposable with its mirror image. Molecules that can be superimposed on their mirror images are achiral (not chiral). Chirality is a property of matter found throughout biological systems, from the basic building blocks of life such as amino acids, carbohydrates, and lipids to the layout of the human body. Chirality is often illustrated with the idea of left- and right-handedness: a left hand and right hand are mirror images of each other but are not superimposable. The two mirror images of a chiral molecule are termed *enantiomers*. Like hands, enantiomers come in pairs. Both molecules of an enantiomer pair have the same chemical composition and can be drawn the same way in 2 dimensions (e.g., a drug structure on a package insert), but in chiral environments such as the receptors and enzymes in the body, they can behave differently. A *racemate* (often called a racemic mixture) is a mixture of equal amounts of both enantiomers of a chiral drug. Chirality in drugs most often arises from a carbon atom attached to four different groups, but there can be other sources of chirality as well. Single enantiomers are sometimes referred to as *single isomers* or *stereoisomers*. These terms can also apply to achiral drugs and molecules and do not indicate that a single enantiomer is present. For example, molecules that are isomers of each other share the same stoichiometric molecular formula but may have very different structures. However, many discussions of chiral drugs use the terms *enantiomer*, *single isomer*, and/or *single stereoisomer* interchangeably.

The two enantiomers of a chiral drug are best identified on the basis of their absolute configuration or their optical rotation. Other designations such as D and L (note the upper case) are used for sugars and amino acids but are specific to these molecules and are not generally applicable to other compounds. The terms *d*, or *dextro,* and *l*, or *levo*, are considered obsolete and should be avoided. Instead, the *R*/*S* system for absolute configuration and the +/− system for optical rotation should be used. The absolute configuration at a chiral center is designated as *R* or *S* to unambiguously describe the 3-dimensional structure of the molecule. *R* is from the Latin *rectus* and means to the right or clockwise, and *S* is from the Latin *sinister* for to the left or counterclockwise. There are precise rules based on atomic number and mass for determining whether a particular chiral center has an *R* or *S* configuration. A chiral drug may have more than one chiral center, and in such cases it is necessary to assign an absolute configuration to each chiral center. Optical rotation is often used because it is easier to determine experimentally than absolute configuration, but it does not provide information about the absolute configuration of an enantiomer. For a given enantiomer pair, one enantiomer can be designated (+) and the other as (−) on the basis of the direction they rotate polarized light. Optical rotations have also been described as *dextrorotatory* for (+) and *levorotatory* for (−). Racemates can be designated as (*R, S*) or (±).

FDA issued its guidelines governing the development of chiral drugs in 1992. In roughly the same period other regulatory agencies across the globe viz. European Union, Japan, Canada, Australia, etc. issued similar guidelines. Although the new regulations do not ban the introduction of new racemic therapeutics their overall effect is in fact the near-total disappearance of racemic substances as new drugs. Because of this majority of the new chiral drugs were marketed as single-enantiomer forms. The strong current trend towards enantiopure chiral drugs will also exert pressure on racemic therapeutics already on the market in the sense that a shift to a more potent or less toxic enantiomer could be well motivated from both medicinal and economical stand points. This is well reflected by the racemic switches on and near market.

Today, chiral drugs are coming under the increasing scrutiny among pharmaceutical manufacturers, pharmacologists, pharmacokinetists, medicinal chemists, separation scientists, agronomists, entomologists, regulatory agencies and others. This is because the left-and right-handed twins of mirror-image molecules often behave differently from each other in bio-environment. For the very same reason, while pursuing scientific investigations (pharmacokinetic, pharmcodynamic or toxicologic) on handed-drugs, it is mandatory that the aspect under study be viewed from both sides of the mirror. Clearly the ultimate goal of all these scientific endeavours is higher quality medicines to make drug therapy more effective, specific and safer for the benefit of mankind.

Importance of chirality: Although converting a molecule from one enantiomer to the other seems like only a small change in the structure, it can provide a significant impact on the way the molecule interacts with its surroundings, and especially other chiral compounds. Many of the molecules that are important in nature are chiral, these include proteins (and their constituent amino acids), which control most processes within biological systems, and the nucleic acids DNA and RNA which are responsible for holding the information necessary for proteins to be synthesized.

For this reason, if a chiral compound interacts with a protein to induce a specific response in a biological organism, it is likely that its enantiomer will either not interact or produce a completely different response. Some of these differences can be quite startling, for example limonene contains a chiral carbon atom. One enantiomer produces the smell of oranges whereas the other gives rise to the smell of lemons.

Understanding chirality is extremely important in the preparation of therapeutic drugs. For example, one enantiomer of penicillamine is a potent anti-arthritic agent whereas the other enantiomer is highly toxic. Perhaps the most startling example of the difference in activity between enantiomers is Thalidomide. This drug was seen as a panacea for the treatment of morning sickness in pregnant women, and indeed one enantiomer reliably has this effect. The other enantiomer, unfortunately, has been associated with the well-characterised birth defects that arose from use of Thalidomide.

One further difference between enantiomers is the way that they interact with light. Light is an electromagnetic radiation. This means that it consists of electronic and magnetic components. The electronic components of light interact with electrons, such as bonds, within a molecule. Changing the arrangement of the bonds changes the way that light interacts with the molecule. This difference in chiral molecules only becomes apparent when polarised light is shone through a solution of the molecule. In polarised light, all the electronic components are aligned. As the polarised light passes through the solution of a chiral compound the polarised light is twisted, with the plane of polarisation being rotated.

Two enantiomers rotate the plane of polarised light by equal amounts *but in opposite directions.* For this reason, stereoisomerism is also sometimes referred to as optical isomerism.

Examples of Chiral Drugs from Various Therapeutic Classes

Antiarrhythmics: Propafenone, tocainide

Antibiotics: Ofloxacin, moxalactam

Anticoagulants: Warfarin, acenocoumarol

Anaesthetics: Prilocaine, ketamine, pentobarbital

Antiemetics: Ondansetron

Antihistamines: Terfenadine, loratadine

Antihyperlipidemics: Atorvastatin

Antineoplastics: Cyclophosphamide, iphosphamide

Antimalarials: Chloroquine, halofantrine, mefl oquine

Muscle relaxants: Methocarbamol, baclofen

NSAIDS: Ibuprofen, ketorolac

β –blockers: Propranolol, metoprolol

β –adrenergics: Salbutamol, terbutaline

Calcium channel blockers: Verapamil, nimodipine

Opiate analgesics: Methadone, pentazocine

Proton pump inhibitors: Omeprazole, pantoprazole, lansoprazole

9.4 Pharmacological difference between enantiomers

Qualitatively and quantitatively enantiomers may have similar or different pharmacological effects. This may be related to stereoselective pharmacokinetics or pharmacodynamics. The terms "eutomer" for the more potent isomer and "distomer" for the less potent one have been suggested (4). It is important to consider which pharmacodynamic effects are being considered when using the terms eutomer or distomer as they are usually employed to describe only the single most obvious effects. For example, S (–) timolol is marketed as a stereochemically pure enantiomer for hypertension (5, 6). The S (–) enantiomer of timolol decreases intraocular pressure and is used in the treatment of glaucoma, although the R (+) enantiomer is also effective. The S (–) enantiomer applied to the eyes can cause β-adrenoreceptor bronchial blockade (7, 8). Hence, the terms "eutomer" and "distomer" apply to both R and S timolol depending on which pharmacological effect is being examined.

No generalizations can be made concerning enantiomers since they exhibit a wide variation in effects. Examples of these effects consist of the following and other are mentioned in **Table 9.2.**

1. Equipotent enantiomers (e.g., cyclophosphamide, flecainide).
2. One enantiomer with all or most of the activity (e.g., NSAIDs, β-blockers).

Table 9.2 Examples of few drugs showing different activities with different enantiomers.

Drug	Enantiomer	Activity
Thalidomide	S- Enantiomer R- Enantiomer	Teratogen Sedative
Ethambutol	(S,S)-Enantiomer (R,R)- Enantiomer	Tuberculostatic Blindness
Penicillamine	(S)- Enantiomer (R)- Enantiomer	Antiarthritic Mutagen
Asparagine	(S)- Enantiomer (R)- Enantiomer	Bitter Sweet
Carvone	(S)- Enantiomer (R)- Enantiomer	Caraway Spearmint

3. Both enantiomers active with similar therapeutic and toxic effects but differ in magnitude (e.g., warfarin).
4. Both enantiomers active but with quantitatively different therapeutic and toxic effects (eg, verapamil) (9-11).

Broccoli contains sulforaphane which increases the activity of certain enzymes capable of degrading toxic/carcinogenic compounds. But only (R)-stereoisomer is active and found in plants **(Fig. 9.1).**

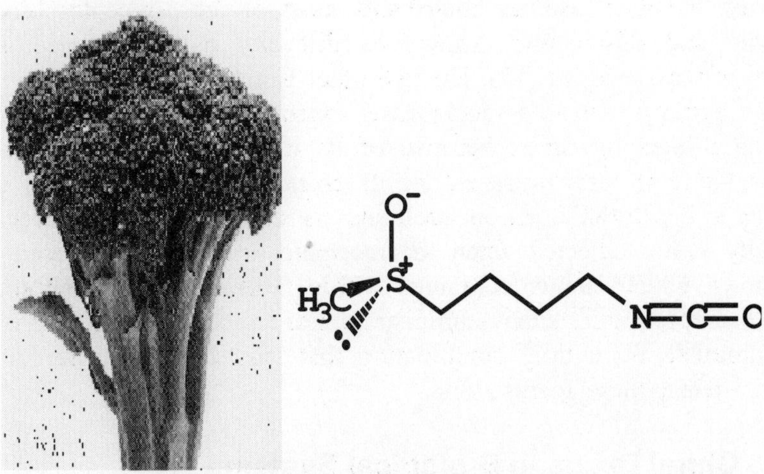

Figure 9.1 Broccoli plant.

Pharmacological Significance

Deoxyribonucleic acid (DNA) is composed of right-handed helices made of 2-deoxyribose monomer carbohydrates of the D- (or R-) confi guration.

Proteins and drug metabolizing enzymes, serum proteins, and transport proteins are composed of L- or usually S- (cysteine being the exception) amino acid monomers. The preferential interaction of one enantiomer of a racemate with chiral macromolecules within the body, leads to expressed pharmacokinetic and pharmacodynamic effects (12-14).

Enantiomers can be absorbed, distributed, metabolized, and excreted differently. Further, disease states, route of administration, genetic variability, and drug-interactions may be stereospecific. Enantiomers administered alone may have different actions within the body than a marketed racemate (15). It has been know for many years that opiate enantiomers can interact with different receptors, leading to different pharmacological effects (16).

The (–) enantiomers of opiates are potent narcotic analgesics, whereas their (+) enantiomers are useful anti-tussive agents (eg, propoxyphene, dextromethorphan). Labetalol has multiple isomers that differ in α- and β-receptor blocking activity.

The S, R isomer of labetalol is predominant for α-receptor blocking activity; the R, R isomer contributes most of the β-receptor blocking activity and this isomer known as dilevalol has been used as an antihypertensive agent (17). The two other isomers may contribute to the drug's activity, but to a much lesser extent. While labetalol is unique among β-blocking drugs because of its α-adrenoreceptor activities is know, there is less awareness of its chiral features, or that its (R,R) isomer's hepatotoxic action prohibits its clinical use. However, this toxicity is not detected when the racemàte drug is administered. It is unknown which isomer contained within labetalol is hepatoprotective. This example suggests that administration of a racemate, of a drug is akin to administering a drug combination that may be very different than administering these agents alone.

9.5 Chiral Drugs in Biological Systems

Enantiomers of a chiral drug have identical physical and chemical properties in an achiral environment. In a chiral environment, one enantiomer may display different chemical and pharmacologic behavior than the other enantiomer. Because living systems are themselves chiral, each of the enantiomers of a chiral drug can behave very differently *in vivo*. In other words, the *R*-enantiomer of a drug will not necessarily behave the same way as the *S*-enantiomer of the same drug when taken by a patient. For a given chiral drug, it is appropriate to consider the two enantiomers as two separate drugs with different properties unless proven otherwise.

The difference between two enantiomers of a drug is illustrated in **Figure 9.2** using a hypothetical interaction between a chiral drug and its chiral binding site. In this case, one enantiomer is biologically active while the other enantiomer is not. The portions of the drug labeled A, B, and C must interact with the corresponding regions of the binding site labeled a, b, and c for the drug to have its pharmacologic effect. The active enantiomer of the drug has a 3-dimensional structure that can be aligned with the binding site to allow A to interact with a, B to interact with b, and C to interact with c. In contrast, the inactive enantiomer cannot bind in the same way no matter how it is rotated in space. Although the inactive enantiomer possesses all of the same groups A, B, C, and D as the active enantiomer, they cannot all be simultaneously aligned with the corresponding regions of the binding site.

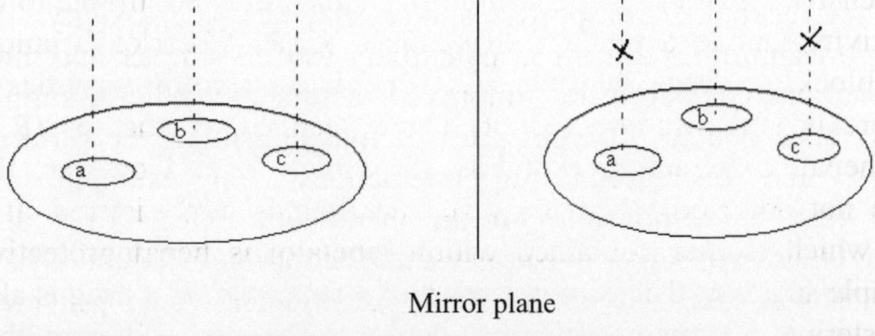

Figure 9.2 Hypothetical interaction between the two enantiomers of a chiral drug and its binding site.

This difference in 3-dimensional structure prevents the inactive enantiomer from having a biological effect at this binding site. In some cases, the portion of a molecule containing the chiral center(s) may be in a region that does not play a role in the molecule's ability to interact with its target. In these instances, the individual enantiomers may display very similar or even equivalent pharmacology at their target site. Even in these cases, the enantiomers may differ in their metabolic profiles as well as their affinities for other receptors, transporters, or enzymes.

9.6 Importance of Chirality in Drugs

Approximately 50% of marketed drugs are chiral, and of these approximately 50% are mixtures of enantiomers rather than single enantiomers. In this section, the potential advantages of using single enantiomers of chiral drugs are discussed and some specific examples of single-enantiomer drugs currently on the market are given. Single-enantiomer drugs will become increasingly more available to the practicing physician, and both the single-enantiomer form and the mixture of enantiomers of a given drug may be available at the same time. In these cases, it is critical to distinguish the single enantiomer from the racemic form because they may differ in their dosages, efficacies, side effect profiles, or even indicated use. It is also important to realize that the safety and efficacy data for a drug evaluated as a mixture of enantiomers are still valid. The introduction of a single-enantiomer preparation of a drug previously approved as a mixture of enantiomers does not necessitate that the single enantiomer should become the standard of care. The decision to use a single enantiomer versus a mixture of enantiomers of a particular drug should be made on the basis of the data from clinical trials and clinical experience (18, 19).

The two enantiomers of a chiral drug may differ significantly in their bioavailability, rate of metabolism, metabolites, excretion, potency and selectivity for receptors, transporters and/or enzymes, and toxicity. The use of single-enantiomer drugs can potentially lead to simpler and more selective pharmacologic profiles, improved therapeutic indices, simpler pharmacokinetics due to different rates of metabolism of the different enantiomers, and decreased drug interactions. For example, one enantiomer may be responsible for the therapeutic effects of a drug whereas the other enantiomer is inactive and/or contributes to undesirable effects. In such a case, use of the single enantiomer would provide a superior medication and may be preferred over the racemic form of the drug. Single-enantiomer formulations of (*S*)-albuterol, a β_2-adrenergic receptor agonist for treatment of asthma, and (*S*)-omeprazole, a proton pump inhibitor for treatment of gastroesophageal reflux, have been shown to be superior to their racemic formulations in clinical trials. In other cases, however, both enantiomers of a chiral drug may contribute to the therapeutic effects, and the use of a single enantiomer may be less effective or even less safe than the racemic form. For example, the (−)-enantiomer of sotalol has both β-blocker and antiarrhythmic activity, whereas the (+)-enantiomer has antiarrhythmic properties but lacks β-adrenergic antagonism (20, 21). In addition, the R-enantiomer of fluoxetine, at its highest administered dose, led to statistically significant prolongation of cardiac repolarization in phase II studies; the studies were subsequently stopped (22).

Sales of chiral drugs

Sales of enantiomeric intermediates and single-enantiomer drugs are increasing every year and a statistics mentioned in the **Tables 9.3(a) & 9.3(b)** will give us an idea about future market of chiral drugs.

Table 9.3 (a) Statistics indicating sale of chiral intermediates and drugs.

	Enantiomeric Intermediates			Bulk Enantiomeric Drugs		
$ MILLIONS	**1999**	**2000**	**2005**	**1999**	**2000**	**2005**
Anti-inflammatory/ analgesics	150	156	168	200	223	241
Antiviral	794	830	1,643	983	1,180	2,054
Cancer	892	1,073	1,297	1,783	2,146	2,593
Cardiovascular	1,133	2,281	3,269	1,889	3,802	5,449
Central nervous system	1,038	1,142	1,821	1,483	1,632	2,602
Dermatology	82	85	106	164	170	212
Gastrointestinal	251	331	649	413	567	1,082
Ophthalmic	238	284	401	340	405	573
Respiratory	576	656	914	1,151	1,511	2,287
Other	140	170	356	315	426	891
TOTAL	5,294	7,008	10,624	8,721	12,062	17,984
SOURCE: Technology Catalysts International						

Table 9.3 (b) Statistics indicating sale of chiral drugs.

	All Drugs		Single-Enantiomer Drugs		
$ MILLIONS	1999	2000	1999	2000	2005
Analgesics	21,500	23,000	1,173	1,291	1,395
Antibiotic/ antifungals	29,300	31,700	24,918	26,140	29,747
Antiviral	17,700	19,100	6,717	8,820	12,201
Cancer	13,700	15,600	8,891	10,690	13,605
Cardiovascular	42,700	46,600	24,895	27,650	34,627
Central nervous System	47,700	53,900	8,439	9,094	14,700
Dermatology	17,900	18,400	16,202	1,272	1,540
Gastrointestinal	43,900	47,200	1,970	4,033	6,590
Hematology	16,500	15,400	7,405	8,879	11,295
Hormone/ endocrinology	20,000	22,000	14,510	15,384	19,790
Ophthalmics	7,100	7,400	1,270	2,265	2,705
Respiratory	36,500	40,500	5,696	6,615	9,620
Vaccines	6,500	7,300	2,503	3,349	4,320
Other	39,000	41,900	6,248	7,032	9,730
TOTAL	360,000	390,000	117,763	132,514	171,865
SOURCE: Technology Catalysts International					

Chiral Drugs- Opportunities for Indian Pharmaceutical Industry

While discussing the Indian Pharmaceutical Industry and its compulsions to grow to global dimensions reaching out to global markets, it is often emphasized that from its present status of being a supplier of patented and off-patent bulk drugs, both for domestic and export markets, it has to move to become an R&D-based industrial segment. To-date, particularly since the implementation of the Indian Patents Act 1970, the Indian Pharma Industry has deployed its resources for developing process technology for bulk drugs and conventional formulations, made possible, by the absence of valid product patents in India.

With the advent of the post-2005 era, when India joins the major league of patent-strong countries with a harmonised patent system consistent with the provisions under TRIPS and WTO, the face of the

Indian industry needs to change, if it is to survive and grow. And that growth has to come not merely by being a supplier of generic (off-patent) products, but also from new patented products developed through Indian R&D efforts.

The majority of synthetic drugs developed in the past are not chiral; the ones developed from natural products are largely chiral. Although the different enantiomers (chirals) have the same chemical formula, they differ widely in their biological properties. This is primarily due to the fact that since chirality is related to three dimensional structures, one form may be more suitable for specific interaction with other biological molecules such as receptors, enzymes, etc.

Thus, there is obvious benefit in studying the properties of the two enantiomers of a molecule with respect to their therapeutic efficacy and safety. It is for this reason that the FDA (USA) and the European Committee For Proprietary Medicinal Products, since 1992, require that the properties of each enantiomer in a racemic molecule, should be studied separately before decisions are taken to market the drug as one of the enantiomers or as a mixture (racemate). In addition, there is an increasing awareness of the need for re-evaluating the properties of individual enantiomers of currently marketed racemic drug molecules.

The Regulatory Agencies, particularly, the US FDA, encourage such work and provide even incentives for developing chiral drugs of existing products in the market. Such requirements have led to a whole new field of new technologies to make chiral forms of existing drugs either from their racemic forms, or through direct synthesis and evaluating their biological properties. It has been predicted that 30% of the drugs under development will be marketed in single-enantiomeric forms and that in the coming years, three quarters of synthetic drugs will be sold as single enantiomers.

Chiral Technology Options

The development of a chiral molecule requires multiple skills and expertise, since the technology involved depends a great deal on the nature and properties of the molecule. The most favoured approaches which could be used to dovetail a technology for a specific product are based on two basic concepts of either introducing chirality in the synthetic route or by appropriate resolution of racemic mixtures. The best possible way of introducing chirality in a synthetic sequence would be to use a natural product with the desired chiral characteristics. This is particularly useful for complex natural products with built in chiral centres which then can be suitably derivatised. Taxol, Penicillins, Lovastatin, all of them subsequently used for conversion to more effective derivatives are all examples of this approach. In addition, small molecular weight naturally occurring amino acids can also be used as appropriate starting materials.

Resolution of Racemates using chiral reagents including acids and bases is a well-known and much used method. A number of commercially important products have been developed, such as S-Naproxen, S-Ibuprofen, S-Omeprazole, R-Fluoxetine, etc. Apart from classical crystallization including entrainment techniques, chromatographic separation using chiral stationary phase, as well as selective extraction systems, are useful techniques. Kinetic resolution which may involve selective derivatisation of one of the enantiomers in preference to the other, as well as processes known as dynamic resolution are also being increasingly used.

Asymmetric synthesis which involves the introduction of chirality through selective chemical transformation such as hydrogenation, oxidation, etc., have the advantage that the conversions can result in better yields with little loss of material, since the unwanted isomer is not involved. On the other hand, the synthesis of racemic molecules and the resolution to single enantiomers have the advantage of having an opportunity of evaluating both isomers. At the same time, the processes are often wasteful, particularly when kinetic resolution is not possible. However, the advent of combinatorial synthesis along with High Throughput Screening (HTS), would make the use of chiral building blocks as starting materials for synthesis attractive. Further down the line, as we understand the finer aspects of receptor bindings and three dimensional structural requirements through computer modeling, the required spatial requirements of the chiral isomer will be more apparent.

Major Players

Since the introduction of one of the first major chiral therapeutic agents, Alpha Methyl DOPA (Aldomet) by Merck in 1960, a number of synthetic drugs have been discovered and developed by leading pharmaceutical companies such as Abbott, Pfizer, Zeneca, etc. However, modern techniques for chiral synthesis and separation have been highly complex, sophisticated and specialised. The technology platforms developed for the purpose need to be adapted for each product. Due to these considerations, a number of small, high technology companies have come in the scene which primarily act as service companies or as contract manufacturing organisations, for large corporations to develop chiral drugs based on the latter"s products. Some of them are Sepracor, Syntheon, Celgene, Altus Biologicals, Chiral Technologies, Degussa, all in the U.S.A., Chiroscience in U.K., Sipsy in France, etc.

Both technology platforms used for chiral synthesis as well as separation technologies can be patented in most countries. The patents as well as the know-how are then licensed to large companies for their commercial exploitation on lumpsum licence fees and/or royalty payments on sales. For example, patents for a number of chiral molecules corresponding to marketed racemates are held by Chiral Companies, such as Sepracor. Many chiral reagents as well as the process technologies for their synthesis and applications have also been patented. For example, Sipsy has extensively used Sharpless" asymmetric epoxidation technology for production of glycidol, a key intermediate for chiral (-Blockers and Corey"s Oxaborolidine process for reduction of ketones. Patents in Sharpless Dihydroxylation and Jacobs epoxidation reaction have been acquired by Chirex for producing the single-enantiomer, Indene Oxide, an intermediate for Merck"s HIV inhibitor Indinavir (Crixivan). DSM Andeno, the Dutch major acquired Chemie Linz, the Austrian Chiral Technology Company, to produce chiral side-chains of amino acids. There are many more such examples of successful collaborations between academia, large corporations and relatively small chiral technology companies.

Market for Chiral Products

Even though the market value of global sales of chiral drugs is not available, the U.S. figures could be considered indicative of the world market. It has been estimated that the annual growth in sales of chiral drugs has been 20% p.a. In a recent report on U.S. Chiral Chemicals (for all applications), demand by 2005 is estimated at $ 15.1 billion growing annually at 9.4%. Of this, hormones and related products are $ 2.6 billion, anti-inflammatory $ 2.1 billion and Cardio-Vascular and Central Nervous System $ 1.2 and $ 1.36 billion respectively.

Pharmaceuticals which constitute the bulk are estimated to grow to $ 11.5 billion by 2005. Almost all new products in the pipeline for development having chiral centres will be marketed as single enantiomers, which will be a mandatory regulatory requirement. More than half of all new drugs entering the market will have optically pure active ingredients. This will be in addition to products developed from drug racemates currently in the market.

Indian Opportunities

It is difficult to predict whether a single enantiomer of a marketed drug will offer any advantages in terms of improvement in efficacy or reduction in side effects over the corresponding racemate. It is possible that once a pure chiral form is available, it may have a totally different pharmacological and therapeutic profile. Examples, such as that of the highly toxic, banned drug, Thalidomide, which led to several thousand mal-formed babies, resurging as an anti-cancer drug in its optically active form are not all that uncommon. Considering the potential for developing new and improved products from existing drugs at much lower costs, compared to discovery and development of new molecular entities, development of chiral molecules is an attractive proposition for Indian companies. The strategies that need to be evolved include identification of the right candidate drug for evaluation of its optically pure enantiomer, adequate technology strengths in synthesis, optical resolutions, separation techniques, biological evaluation methodologies, clinical trials and drug approvals by regulatory agencies, in a time frame much shorter and at much lower costs than what will be required for new drug discovery and development.

Whether, following the U.S. model, specialised companies to develop chiral products would be elevant to India is to be evaluated. Since different technology platforms are needed depending on the product, there is a case for specialised companies or laboratories providing contract research and services to large pharma companies. It is also important that India has in place appropriate policies on patenting of chiral enantiomers of off-patent drugs as well as clarity on regulatory requirements including provision for fast-track clearances.

The principal argument for Indian Companies to involve themselves in the development of new chiral products based on existing drugs, is that the developmental process would be shorter and much less expensive; factors which are very important considering the Indian Companies" constraints on resources. In leading Indian Companies which are in new drug research, the first option, if possible, should be to develop drugs with no chiral centers, and in case chirality is a necessity, to ensure that single enantiomers are developed right from the beginning, either through direct synthesis or resolution of the corresponding racemate. Overall, considering Indian strengths in chemical process technology, research on chiral drugs is an opportunity area for Indian Companies to make their presence in the global arena. (24)

9.7 Thalidomide a Typical Example The First Appearance of Thalidomide

Thalidomide first appeared in Germany on 1st October 1957 **(Figure 9.3)**. It was marketed as a sedative with apparently remarkably few side effects. The Drug Company who developed it believed it was so safe it was suitable for prescribing to pregnant women to help combat morning sickness (25).

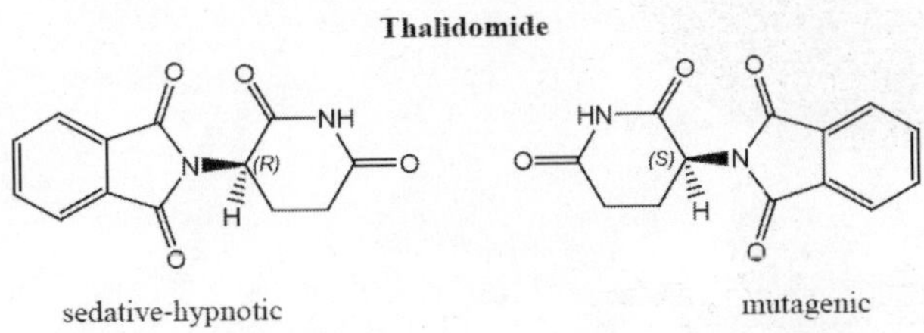

Figure 9.3 Thalidomide

Thalidomide is a sedative drug that was prescribed to pregnant women, from 1957 into the early 60's. It was present in at least 46 countries under different brand names. "When taken during the first trimester of pregnancy, Thalidomide prevented the proper growth of the foetus, resulting in horrific birth defects in thousands of children around the world". Why? The Thalidomide molecule is chiral. There are left and right-handed Thalidomides, just as there are left and right hands. The drug that was marketed was a 50/50 mixture. One of the molecules, say the left one, was a sedative, whereas the right one was found later to cause foetal abnormalities. The tragedy is claimed to have been entirely avoidable had the physiological properties of the individual thalidomide [molecules] been tested prior to commercialization (26).

It was quickly being prescribed to thousands of women and spread to most corners of the globe. Nobody had any idea of what was to follow. Drug testing procedures were far more relaxed at this time, and although tests had taken place on thalidomide, they didn't reveal any of its tetragenic (roughly meaning causing malformations) properties. In most countries, drug companies were not required to submit testing results to the appropriate government agencies. The tests on thalidomide were conducted on rodents which metabolise the drug in a different way to humans. Later tests on rabbits and monkeys produced the same horrific side effects as in humans.

Towards the end of the fifties, children began to be born with shocking disabilities. It was not immediately obvious what the cause of this was. Probably the most renowned is *Pharcomelia*, the name given to the flipper-like limbs which appeared on the children of women who took thalidomide. Babies effected by this tragedy were given the name 'Thalidomide Babies' **(Fig. 9.4(a) & 9.4(b)).**

Figure 9.4(a) Figures showing the victims of thalidomide tragedy.

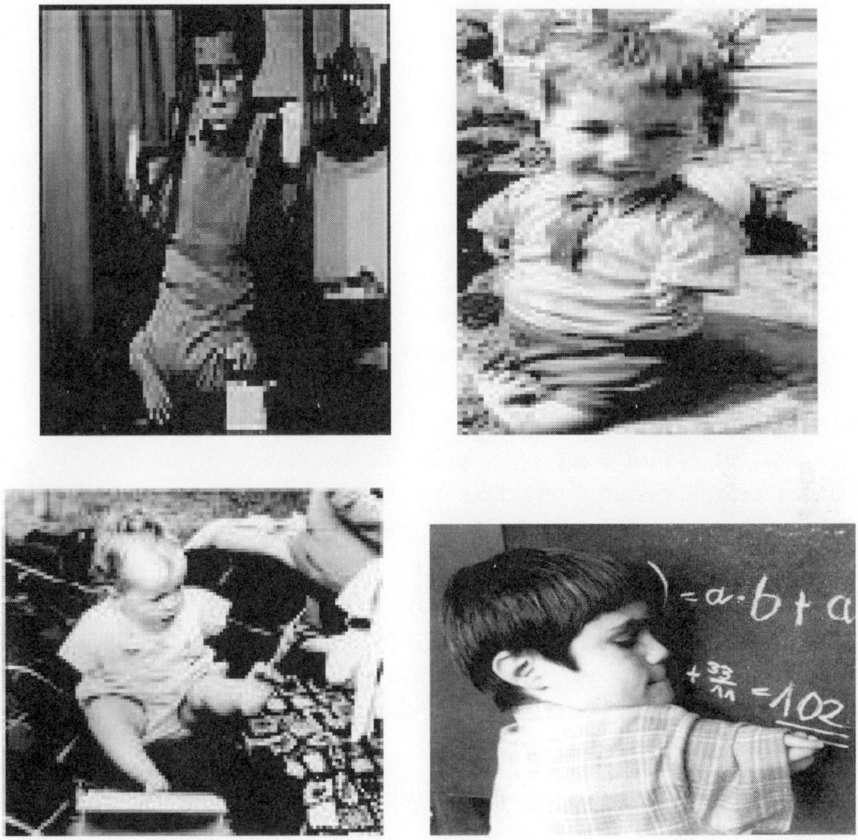

Figure 9.4(b) Figures showing the victims of thalidomide tragedy.

Some Effects of Thalidomide

Around 12,000 children were born with some kind of disability due to damage caused by thalidomide. Countless more miscarriages, which weren't recorded as caused by thalidomide, mean we can't assess correctly the true scale of the disaster. Estimates put the number of children affected at around 20,000. Even this doesn't take account of the families of the children affected (27, 28).

One effect of thalidomide is targeting blood vessels. It is able to block growth and so tends to target parts of the body undergoing growth. Pregnant women taking the drug could therefore feel the benefit of the drug in combating morning sickness, but damage was being done to the rapidly growing foetus. It is possible that so many of the thalidomide babies experienced phocomelia because morning sickness can appear around the time of foetal limb growth.

Many other forms of damage were caused to the children, including brain damage. Fortunately many children's disabilities were purely physical and they could learn to cope with disabilities and show normal intelligence.

Not all the effects of this drug were unpleasant, and a positive change has resulted from the disaster. As was mentioned before, at this time it was not usually necessary to submit research to government agencies before approval for sale was given. This was not the case in America. The case for Thalidomide was dealt with by reviewing medical officer Frances Kelsey, pictured left. This was her first case and was expected to be a simple approval, with the drug already widely available around the world. Kelsey was not convinced and withheld her approval while more research was conducted. After a while, reports of the drug's tetragenic effects began to emerge from Europe, and the drug was quickly withdrawn the world over. Thanks to her conviction, only 17 thalidomide babies were born in America, mostly due to thalidomide bought in other countries.

Now all countries require this information, and Drug tests have become infinitely more thorough. A lot of research has been put into optically active molecules. In most cases where only one enantiomer of a molecule is an effective drug, a stereo specific manufacturing method will be created.

The Re-Emergence of Thalidomide

After the 1960's tragedy and thalidomide being taken out of use, not all the drug was destroyed. A few years later, the story took a surprising turn by means of an Israeli doctor. In 1965 the doctor was treating leprosy patients. One in particular had developed a bacterial infection coupled with inflammation. After stumbling across some thalidomide, remembering it was a sedative, the doctor administered it in the hope it would help the patient sleep and ease his pain. Remarkably, in the morning the swelling had being brought under control. This extremely fortunate discovery lead to research on the effects of thalidomide on inflammatory diseases (29, 30).

Leprosy is a horrific desease which still causes suffering in parts of the world. Any new medicines which might help irradicate it are desirable. When our bodies encounter foreign matter cells they become injured and release a protein which increases blood flow. This can cause serious inflamation if the body overreacts. Thalidomide is able to block one of the proteins and stop the body over reacting, surpressing the defences.

This is not the only situation in which Thalidomide has proved effective. Since it reduces growth of blood vessels, it has been tested on cancer patients, as tumours require new blood vessels. It has been used to treat weight loss associated with AIDS/HIV and on Chron's disease patients. It has also been studied for its potential property of stopping HIV progression.

What Now For Thalidomide?

A drug with so many possible benefits would usually be heralded a wonder drug, but the dark side of this compound means some people will never be happy for it to be on the market. Former victims, although afraid to see it re-introduced, don't want to deprive those who could benefit from its positive side of a chance to do so (31).

It's a difficult situation to risk using this drug. No matter what precautions are taken there is always a high chance more mutated children will be born if it is being used. Some victims have said they would rather it was used properly and legally than banning it and forming a black market. At least then, every effort can be made to educate people and control the drug. They are determined however that the drug should always be called 'THALIDOMIDE' and not sold under a trade name.

The target for all should be to produce drugs with the positive effects of Thalidomide but without the side effects. Celgene Corporation has already produced new therapeutic products structurally related to Thalidomide. Some of these compounds show up to 4000 times the potency of thalidomide and many have lower side effects. Hopefully I'll be able to update this page soon with news of a replacement.

9.8 Chiral Chromatography

Importance of Chiral Separation

The separation of chiral compounds has been of great interest because the majority of bioorganic molecules are chiral. Living organisms, for example, are composed of chiral biomolecules such as amino acids, sugars, proteins and nucleic acids. In nature these biomolecules exist in only one of the two possible enantiomeric forms, e.g., amino acids in the L-form and sugars in the D-form. Because of chirality, living organisms show different biological responses to one of a pair of enantiomers in drugs, pesticides, or waste compounds, etc (32-34).

Chirality is a major concern in the modern pharmaceutical industry. This interest can be attributed largely to a heightened awareness that enantiomers of a racemic drug may have different pharmacological activities, as well as different pharmacokinetic and pharmacodynamic effects. The body being amazingly chiral selective, will interact with each racemic drug differently and metabolize each enantiomer by a separate pathway to produce different pharmacological activity. Thus, one isomer may produce the desired therapeutic activities, while the other may be inactive or, in worst cases, produce unwanted effects. Consider the tragic case of the racemic drug of n-phthalyl-glutamic acid imide that was marketed in the 1960's as the sedative Thalidomide. Its therapeutic acitivity resided exclusively in the R-(+)-enantiomer. It was discovered only after several hundred births of malformed infants that the S-(+)-enantiomer was teratogenic.

The U.S. Food and Drug Administration, in 1992, issued a guideline that for chiral drugs only its therapeutically active isomer be brought to market, and that each enantiomer of the drug should be studied separately for its pharmacological and metabolic pathways. In addition, a rigorous justification is required for market approval of a racemate of chiral drugs. Presently, a majority of commercially available drugs are both synthetic and chiral. However, a large number of chiral drugs are still marketed as racemic mixtures. Nevertheless, to avoid the possible undesirable effects of a chiral drug, it is imperative that only the pure, therapeutically active form be prepared and marketed (35-37). Hence there is a great need to develop the technology for analysis and separation of racemic drugs. Chiral compounds are also utilized for asymmetric synthesis, i.e., for the preparation of pure optically active compounds. They are also used in studies for determining reaction mechanisms, as well as reaction pathways. Chiral compounds are also important in the agrochemical industries.

Current methods of enantiomeric analysis include such non-chromatographic techniques as polarimetry, nuclear magnetic resonance, isotopic dilution, calorimetry, and enzyme techniques. The disadvantages of these techniques are the need for pure samples, and no separation of enantiomers is involved. Quantitation, which does not require pure samples and separation of enantiomers, can be done simultaneously by either gas chromatography (GC) or high performance liquid chromatography (HPLC). Chiral HPLC has proven to be one of the best methods for the direct separation and analysis of enatiomers. It is more versatile than chiral GC because it can separate a wide variety of nonvolatile compounds. It provides fast and accurate methods for chiral separation, and allows on-line detection and quantitation of both mass and optical rotation of enantiomers if appropriate detection devices are used. Current chiral HPLC methods are either direct, which utilizes chiral stationary phases (CSPs) and chiral additives in the mobile phase, or indirect, which involves derivatization of samples (39-40).

Direct chiral separations using CSPs are more widely used and are more predictable, in mechanistic terms, than those using chiral additives in the mobile phase.

To date nearly a hundred HPLC CSPs have been developed and are commercially available. However, there is no single CSP that can be considered universal, i.e., has the ability to separate all classes of racemic compounds. Choosing the right CSP for the enantioseparation of a chiral compound is difficult. Most chiral separations achieved on CSPs, however, were obtained based upon the accumulated trial-and-error knowledge of the analyst, intuition, and often simply by chance. An alternative way of choosing a CSP is by using predictive empirical rules that have been developed based on empirical structures. Neither scheme of choosing a right CSP offers a guarantee for a successful enantiomeric separation.

Although enantioseparation is hoped to be achieved by knowing the chemistry of the racemic analytes and the CSP sometimes, however, it does not work because the interactions of the mobile phase with both the racemic analyte and CSP have to be considered. All three components, analyte, CSP, and mobile phase, must be taken into consideration when developing a chiral separation method. The key, therefore, to a successful enantioseparation of a particular class of racemates on a given CSP is the understanding of the possible chiral recognition mechanisms.

Enantioselective HPLC analysis

There are basically two options for chiral HPLC analysis namely direct and indirect approach. In the indirect approach, drug enantiomers are derivatised with an enantiopure chiral reagent to form a pair of diastereomers, which may be then separated on a conventional chromatographic column, since diastereomers exhibit different physicochemical properties. In the direct method, transient rather than covalent diastereomeric complexes are formed between the drug enantiomers and a chiral selector present either added to the mobile phase (CMPA) or coated/bonded to the surface of a silica support (CSP). The technique relying on chiral stationary phases (CSPs) are preferred as they offer specific advantages over indirect methods. There is no need to chemically manipulate the analytes, interference with sample matrix, chiral purity of the chiral stationary phase (CSP) does not need be known, fast analysis, method can be readily scaled to commercial production, online coupling with MS or NMR permits structure identification (41).

Chiral Stationary Phases (CSPs)

The principle behind the functioning of a chiral stationary phase for HPLC is that an enantiopure chiral molecule (chiral derivatising agent, CDA) is coated or immobilised on the surface of a solid support (usually silica microparticles). This selectively retains the enantiomers of analytes by forming transient diastereomeric adsorbates of different stabilities with them during elution process (42).

Classification of CSPs

Wainer classified HPLC chiral stationary phases into five types. They are:

- Pirkle (Brush) CSPs
- Polysaccharide CSPs
- Cavity CSPs
- Ligand exchange CSPs
- Protein bound CSPs

Ligand exchange CSPs find limited application in pharmaceutical research and development environment.

Pirkle CSPs

π-Acidic and π-Basic Phases

The brush-type of CSP was introduced by Pirkle who was one of the pioneers of modern enantioselective liquid chromatography. The most frequently used π -acceptor phases are derived from the amino acids phenylglycine (DNBPG) or leucine (DNBLeu) covalently or ionically bonded to 3-aminopropyl silica gel (43,44). These CSPs are commercially available for analytical or preparative separation of enantiomers.

One of the most frequently used π-acceptor is a (R)- N- (3,5-Di nitro benzoyl) phenylglycine . Some drugs having π-donor groups can be directly resolved on π-acceptor phases. In most cases, however the introduction of π-donating groups by derivatization steps is necessary. Pirkle et al. have also developed a new generation of π-donor phases containing a napthyl moiety bound via a C_{11}- spacer to silica gel .These phases showed a remarkable enantioselectivity for the 3,5- dinitro benzoyl derivatives of amines , amino acids and di- and tripeptides (45,46).

Further CSPs based on amino acid or amine chiral selectors such as valine, phenylalanine, tyrosine (47) and 1,2-*trans*-diaminocyclohexane (DACH-DNB phase) (48) and 1,2-*trans*-diphenylethylene diamine (ULMO phase) were also developed. These CSPs have been applied for the preparative separation of the enantiomers of a few racemic compounds, but the number of reported preparative applications has remained very limited over the last 10 years.

These CSPs, developed and designed by Dr. W H Pirkle for HPLC separation, are based on ionic or covalent attachment of one enantiomer of an amino acid derivative (e.g., (R) -N-(3, 5-dinitrobenzoyl) phenylglycine) to aminopropyl silica. Chiral separation is based on preferential binding of one enantiomer to the CSP resulting in a diastereomeric complex. Transient diastereomeric complexes involve electron donor-acceptor (π- π) interactions, hydrogen bonding and dipole-dipole interactions.

Pirkle CSPs could be classified into three categories namely π-electron acceptor, π-electron acceptor phases have a dinitrobenzoyl derivatised phenylglycine or leucine group linked to a silica support. These will be separate compounds (e.g., aminoalochols, hydantoins, lactams), which possess an aromatic π-electron donating aromatic system. Conversely, π-electron donor phases incorporating, for example naphthylamine or naphthylurea phases, will separate π-electron acceptors such as suitably derivatised (e.g., with dinitrobenzoyl chloride or dinitrophenyl isocyanate) amines, alcohols and thiols. Derivatisation is generally required when separating analytes with strongly acidic or basic groups. Most revolutionary addition to Pirkle concept series is π-electron acceptor-donor phases. This concept is innovative incorporation of both π-electron acceptor-donor characteristics, resulting in a phase that can be used for the resolution of wide variety of compounds.

A number of Pirkle type CSPs are commercially available - Pirkle 1-J (π-electron acceptor phase), Naphthyl leucine (π-electron donor phase) and Whelk O-1 (π-electron acceptor-donor phases). They are used most often in the normal phase mode. The strengths of the Pirkle CSPS are:

- Robust and allows high sample load
- Better column durability due to covalent phase bonding
- Universal solvent. Compatibility
- Allows inversion of elution order
- Excellent chromatographic efficiency
- Separation of enantiomers of a wide variety of compounds and
- Ability to invert elution order

Polysaccharide CSPs

A wide range of powerful polysaccharide-based stationary phases have been developed during the last 20 years (49) and several of these CSPs have become available for preparative purposes.

These polymeric materials have been applied as pure polymers in a form adequate for chromatographic purposes or as a coating on an inert achiral support to confer mechanical stability.

For many years, the most widely used cellulose derivative for separations on a preparative scale was cellulose triacetate (CTA-I) introduced in its fully acetylated form by Hesse and Hagel in 1973 (50). The high versatility and the high loading capacity as well as the low preparation costs have certainly contributed to an extended use of this sorbent, even if there are some practical limitations. For CTA, it must be emphasized that the crystal structure of the polymeric material has a determining influence on the chromatographic properties and the chiral recognition ability (51). Indeed, cellulose triacetate exists in at least two different crystal polymorphic forms which can confidently be distinguished by X-ray diffraction (52). Only the so-called CTA I structure (cellulose triacetate, crystal form I) shows a large spectrum of applications. A broad variety of racemic structures have been resolved on CTA I on a preparative scale (53,54) but this phase suffers from the fact that it is readly soluble in most organic solvents which are usually applied for chromatographic separations.

However, Okamoto's group in Japan developed a technology consisting of coating macroporous silica gel with about 20 wt% of the polysaccharide derivatives, conferring a much higher mechanical stability to the chiral stationary phase (55-57). Although a wide range of polysaccharide-based CSPs have been described, only a few derived from cellulose and amylose have been commercialized. Among these phases, four have proved to be complementary and are capable together of covering about 90% of all analytical applications (58). As all analytical separations can potentially be scaled up, one might immediately realize the potential for preparative enantioselective applications. All analytical phases are also available in 20 µm particle size for preparative purposes.

Although the inert achiral support confers suitable mechanical stability, it considerably reduces the loading capacity owing to the presence of 75 to 80% of achiral material which does not contribute to chiral discrimination. The coated polysaccharide-based phases have mostly been used in normal phase conditions, but an increasing number of preparative applications have been reported in supercritical fluid chromatography (59) or reversed phase mode (60).

The broad applicability of the coated polysaccharide-based CSPs has made them very popular and they are now widely used for preparative separation of enantiomers and large-scale applications up to tonnes per year have been reported (61, 62). The success of these CSPs is documented in numerous papers and these CSPs are the most used phases for analytical and preparative applications. However, like all the non-immobilized phases, these CSPs have a non negligible drawback associated with the relatively good solubility of most polysaccharide derivatives in many common organic solvents like chlorinated alkanes, ethyl acetate, toluene, acetone, tetrahydrofuran, or dioxane.

Yoshio Okamoto invented amylose and cellulose CSPs but the technique was commercialised by Diacel Chemical Industries Ltd., Japan. To increase their chiral recognition capacity, these polysaccharides are converted into their triester and tricarbamate derivatives by modifying the 'R' group.

Few commercially available polysaccharide based on CSPs (coated on silica gel) are cellulose triacetate (chiralcel OA), cellulose tribenzoate (chiralcel OB) and cellulose triphenylcarbamate (chiralcel OC), amylose tris (3, 5-dimethylphenyl carbamate (chiralpak AD).

Polysaccharide derivatives coated on silica matrix have been extensively used as CSPs for their high selectivity and loading capacity in enantioseparation by HPLC. Immobilisation of the polymeric chiral selector on the support has been considered as direct approach to confer a universal solvent compatibility to this kind of CSP, thereby broadening the choice of solvents able to be used as mobile phases. In this context, Diacel Chemical Industries Ltd has recently developed a new generation of CSPs for HPLC using a novel immobilisation technology, Chiralpak IA, a 3,5-dimethylphenyl carbamate derivative of amylose, immobilised onto silica, is the first of this series of CSPs to become commercially available. The second member of this series is of the same nature as in chiralcel OD. Chiralpak IA has been employed for the HPLC separation of Hexobarbital, Bupivacaine, Naproxen, Suprofen, etc. and Chiralpak IB for the separation of Propranolol, Oxprenolol, Hydroxycine etc.

Cavity CSPs

Another strategy for chiral discrimination on a stationary phase is creation of chiral cavities, in which enantioselective guest-host interactions govern the resolution. The first important consideration for chiral distinction in this kind of CSPs is the proper fit of the analyte in the chiral cavity, structure of the analyte with reference to the stereogenic centre and the interaction between the analyte and chiral selector. This category of stationary phases includes cyclodextrins, macrocyclic antibiotics/glycopeptides and crown ethers.

Cyclodextrins: Commercially available cyclodextrins (CDs) are cyclic oligosaccharides of six, seven or eight glucose units. They are chiral due to inherent chirality of building glucose units. These cyclodextrins are chiral molecules that resemble truncated cone. The interior surface of the cone forms hydrophobic chiral cavity rimmed by the secondary 2- and 3-hydroxyl groups at the larger opening and by the primary 6-hydroxyl groups at the smaller orifice.

Cyclodextrins (CDs) are most frequently used chiral selectors in CE. The depth of the cavity and the solubility of native CDs can be increased by derivatization. The hydroxyl groups in positions 2, 3 and 6 are available for this process. Several neutral and charged CD derivatives have been synthesized. To date, various derivatives are commercially available and the enantioselectivity has been shown to vary drastically among them.

(a) ***Neutral CD derivatives:*** A great variety of neutral derivatives of CDs such as heptakis-O-methyl-CD (M-CD), heptakis (2,6-di-O-methyl) CD (DM-CD), heptakis (2,3,6-tri-O-methyl) CD (TM-CD), hydroxyethyl-CD (HE-CD) and hydroxyl propyl-CD (HP-CD) have been synthesized and applied to a great variety of compounds (63).

(b) ***Negatively charged CDs:*** Negatively charged CDs are suitable for the separation of basic and neutral drugs. The improved selectivity compared to neutral CDs mainly attributed to the counter – current mobility. Sulphated CDs, sulfobutyl and sulfoethyl ether-ß-CD are the most frequently used charged CDs (64, 65). Phosphated CDs are another group of negatively charged CDs successfully applied to the separation of some drug enantiomers.

(c) ***Positively charged CDs:*** Cationic CDs such as 6-[(3aminoethyl)amino]-6-deoxy -ß-CD, 6^A –methyl amino-ß-CD, 6^A,6^D – dimethylamino-ß-CD, were the first cationic CDs described that found application to the chiral separation of various and neutral compounds (66). CDs containing quaternary ammonium groups are strong bases and the electorphoretic mobility is pH dependent. Only few selector concentrations are required to resolve acidic concentrations because of the strong ionic interactions.

(d) ***Amphoteric CDs:*** Two new amphoteric CD derivatives, mono-(6-glutamylamino-6-deoxy)- β-CD (Glu- β-CD) and AM- β-CD of undefined structure have recently been introduced and shown to be applicable to neutral, acidic and basic compounds (67,68).

(e) ***CDs and non-chiral additives:*** The combination of CDs with chiral micelle forming surfactants such as sodium dodecyl sulphate (SDS) is utilized in the principle of CD-mediated micellar electrokinetic chromatography (69). While uncharged CDs migrate with same velocity as the EOF, the negatively charged micelles migrate in the direction opposite to the EOF. Partition of hydrophobic analytes between the bulk of solution, the CD and micelle phase takes place causing retention of analyte, which enables the separation of uncharged analytes with neutral CDs.

The strength of CD based CSPS include:

- Aqueous compatibility of CDs and its unique molecular structure make the CD based CSPs highly promising for use in chiral separation of drugs
- They are relatively cheaper than other CSPs
- Reversed phase conditions can be applied. The major weakness is that this kind of CSPs is limited to compounds that can enter into CD cavity

Cyclodextrin CSPs are marketed by Diacel Chemical Industries, Japan and Astec, USA.

Chiral Crown Ether Phases

Chiral crown ether phases on polystyrene or silica gel bases were first introduced by Cram and coworkers (70). Because of the chiral crown ether's ability to include guest molecules stereoselectivity, the authors called this principle "host-guest" chromatography. A crown ether phase is now commercially available from Daicel (CR (+)). This phase is suitable for the resolution of amino acids and primary amines. Norephedrine, amphetamine and baclofen and similar drugs have been successfully resolved on this new phase (71).

Protein bound CSPs

Proteins are complex and high molecular weight biopolymers. They are composed of L-amino acids and possess ordered three-dimensional structure. They are known to bind and interact stereoselectively with small molecules reversibly making them versatile CSPs for chiral separation of pharmaceuticals. Lots of CSPs have been developed by immobilising proteins or enzymes. Protein polymer remains in twisted form because of the different intramolecular bonding. These bonding are responsible for different chiral loops/grooves present in the protein molecule. Separation mechanism of proteins relies on unique combination of hydrophobic and polar interactions by which the analytes reoriented to the chiral surfaces. Hydrogen bonding and charge transfer may also contribute to enantioselectivity.

Albumin (Resolvosil), 1-Acid Glyco-Protein (enantiopac), Cellulose Bio-Hydrolase (Chiral CBH), Human Serum Albumin (Chiral HSA) and Vomucoid (Ultron Es -OVM) CSPs are commercially available. Protein based CSPs has been successfully employed for the chiral resolution of a number of pharmaceuticals viz. omeprazole, mephenytoin and atenolol.

Bovine Serum Albumin (BSA)

The ability of proteins to undergo enantioselective interactions with a great variety of drugs is utilized in the principle of chiral affinity chromatography. The primary interactions which are supposed to be responsible for the chiral recognition are hydrophobic and electrostatic interactions, BSA shows enantioselectivity especially for aromatic compounds. Direct analysis of drugs in serum samples is the combination of a Pinkerton internal surface reserved phase (ISRP) with a BSA column for the resolution of warfarin enantiomers (72).

Human Serum Albumin (HSA)

In addition to BSA a silica bonded HSA-phase has recently been developed by Domenici et al. (73). On this type of phase, drugs like benzodiazepines, warfarin and leucovorin can be resolved.

α_1-Glycoprotein (AGP)

Another type of protein phase, AGP, has been introduced by Hermansson (74,75). The first generation AGP column is based on AGP adsorbed on diethylaminoethyl silica, followed by crosslinking. The second generation AGP-column contains AGP immobilized by a crosslinking covalent attachment to the surface-modified silica gel. The latter phase shows a significantly higher stability and improved enantioselectivity.

Enzyme Phases

The use of enzymes as CSP's for chiral resolution has recently been examined by Wainer et al. (76, 77). The authors immobilized α-chymotrypsin (ACHT) on silica gel and resolved racemic O- and N, O-derivatized amino acids on this phase. Similar phases using trypsin and ACHT have been prepared by Kolbe et al. (78).

Major limitations of protein based CSPs are (i) protein phases are expensive (ii) low loading capacity (iii) extremely fragile and delicate to handle (iv) low efficiency and (v) cannot invert elution order.

Today, the significant difference in the pharmacokinetic, pharmacodynamic and toxicological profiles of the enantiomers of a racemic therapeutic is well appreciated. Further, regulatory agencies insist on enantiospecific data if the molecule under investigation is chiral. Hence there is a need for analytical tool to quantify enantiomers of chiral drug in formulations and biological fluids, study stereochemical stability during formulation and production, carry out enantiospecific bioavailability and bioequivalence assessment, control enantiomeric purity in chiral synthesis, and check for racemisation process and chiral impurity profiling. It is in this context chiral HPLC becomes a valuable tool in drug research analysis.

The market for HPLC chiral columns is about $30–50 million. The pharmaceutical industry is responsible for the majority of the market and is expected to drive future growth. Applications in agriculture and chemical markets also fuel demand.

Most commercially available chiral stationary phases and their manufacturers are given in **Table 9.4.**

Table 9.4 Most used commercially available preparative chiral stationary phases.

Packing name Chiral selector	Manufacturer
Cellulose and amylose derivatives	
Chiralcel OD. cellulose 3,5-dimethylphenylcarbamate	Daicel
Chiralcel OF. cellulose 4-chlorophenylcarbamate	Daicel
Chiralcel OG. cellulose 4-methylphenylcarbamate	Daicel
Chiralcel OJ. cellulose p-methylbenzoate	Daicel
Chiralcel OK. cellulose cinnamate	Daicel
Chiralpak IB. immobilized cellulose 3,5-dimethylphenylcarbamate	Daicel
Chiralpak AD. amylose 3,5-dimethylphenylcarbamate	Daicel
Chiralpak AS-V. amylose (S)-phenylethylcarbamate	Daicel
CTA-I cellulose triacetate	Daicel
Synthetic polymers	
ChiraSpher. poly[(S,-N-acryloylphenylalanine ethyl ester]	Merck
CHI-DMB. cross-linked O-3,5-dimethylbenzoyl tartramide	Eka Nobel
CHI-TTB. cross-linked O-4-tert-butylbenzoyl tartramide	Eka Nobel
Cyclodextrin-based CSPs	
ChiraDex. beta-cyclodextrin	Astec
Cyclobond. alpha-, beta-, gamma-cyclodextrin	Merck
Chiral Prep CD ST beta-cyclodextrin	YMC
Chiral Prep CD PM phenyl-beta-cyclodextrin	YMC
Chirobiotic	
Chirobiotic T. teicoplanin	Astec
Chirobiotic TAG. modified teicoplanin	Astec
Chirobiotic V. vancomycin	Astec
Chirobiotic R. ristocetin A	Astec
Brush-type CSPs	
DNBLeu 3,5-dinitrobenzoylleucine	Regis
DNBPG-co 3,5-dinitrobenzoylphenylglycine	Regis
DACH-DNB diaminocyclohexane 3,5-dinitrobenzamide	Regis
Whelk-01. 3,5-dinitrobenzoyl tetrahydrophenanthrene amine	Regis
ULMO. diphenylethylene diamine 3,5-dinitrobenzamide	Regis
Chiralpak QN-AX. quinine O-9-(tert-butylcarbamoyl)	Daicel
Chiralpak QD-AX. quinidine O-9-(tert-butylcarbamoyl)	Daicel

Application of chirality to pharmaceutical industries

Every living body contains amino acids, sugars, proteins and nucleic acids. An interesting feature of these chiral biomolecules is that in nature they usually exist in only one of the two possible enantiomeric forms. When a chemist synthesizes a chiral molecule in an achiral environment using achiral starting materials, an equal mixture of the two possible enantiomers (i.e. a racemic mixture) is produced. In order to make just one enantiomer, some enantioenriched starting material, reagent, catalyst, or template must be present in the reaction medium. Oftentimes, only a single enantiomer of a chiral molecule is desired, as is the case when the target molecule is a chiral drug that will be used in living systems. Drug molecules can be likened to tiny keys that fit into locks in the body and elicit a particular biological response. Since the 'locks' in living organisms are chiral, and exist in only one of the two possible enantiomeric forms, only one enantiomers of the 'key' molecule should be used (the mirror image of our car key will not start our car).

Future trends

With the move to green chemistry gaining more momentum every day, the "clean" technique of Chiral HPLC continues to grow in popularity. Once considered too costly to be practical in many laboratories, Chiral technologies are becoming cheaper and more effective than ever.

With the introduction of more and more sources of chiral phases, the costs are beginning to decline. Recently, there has been a trend in having immobilized chiral stationary phases (CSPs). Previously, most chiral stationary phases were coated which meant that there were limitations in the types of solvents that could be used. With the newer immobilized phases, widely varying solvent polarities can be used to develop and optimize chiral methods. There has also been a trend in the use of smaller particles analogous to other modes of HPLC. Although sub-two micron CSPs are not yet widely available, 3-μm columns are and they allow faster separations with good resolution. Another trend has been in the use of supercritical fluid chromatography for the separations of chiral compounds. This trend does put some stress on CSPs because supercritical fluids can play havoc on some of the coated phases. Another trend is the development of newer wide-range phases that allow most enantiomeric separations to occur on just a few CSPs. Only a few years ago, users had to purchase a dozen or so expensive chiral columns in order to find one that might do the job.

Future of Chiral HPLC

Although the market for biological drugs is growing rapidly, the traditional small-molecule drugs are entering clinical trials in much greater numbers than biologics. Clearly, use of chiral chromatography, especially for preparative separations, will continue to grow. Furthermore, chromatography, being the fastest route to market, will be viewed as part of drug production development in the future (79).

9.9 Chiral Drug Formulation

Chirality influences drug delivery because a single enantiomer or a non-racemic blend may have improved solubility, dissolution, and stability. In addition, many available pharmaceutical excipients (e.g., cellulose and its derivatives) either naturally occur as single enantiomers or are derivatives of the latter chiral molecules. These stereochemically pure molecules may interact with other chiral molecules (i.e., the active ingredient) and form stereoisomers. The latter will have physicochemical properties different from the original chiral molecule. For example, the presence of heptakis(2,6-di-O-ethy l)-beta-cyclodextrin results in stereoselective dissolution of tiaprofenic acid. While this stereoselective release did not result in stereoselective bioavailability, it highlights the potential implication of the effect of chirality on physicochemical properties of drugs. Similar to the solid dosage forms containing chiral excipients, biological membranes may provide chiral environments. Most drugs cross the gastrointestinal membrane through simple passive diffusion; thus, no stereoselectivity in the process is expected (80).

Chiral excipients: With the advent of stereospecific analytical methods in recent years, more attention has to be drawn to the influence of chiral excipients on the modification of *in vitro* release and *in vivo* disposition of chiral drugs. Chiral excipients have been widely used in pharmaceutical dosage forms.

Challenges and regulatory perspectives in the development of formulations containing chiral drugs

Chiral excipients have been widely used in pharmaceutical dosage forms. A selected list of commonly used excipients along with their major applications is shown in **Table 9.5.** The interactions of chiral excipients with drugs and their influence on the therapeutic outcomes have not been thoroughly investigated- This may be due to the misconception that excipients are "Inactive" materials and are not expected to affect the performance of dosage forms. Another reason may be a lack of proper understanding regarding the stereochemistry of these molecules and their implications on the performance of the dosage form, especially in the presence of chiral drugs (81).

Table 9.5 Chiral excipients and their major applications.

Excipients	Major Applications
A. Celluloses	
Carboxymethylcellulose calcium	Disintegrant, suspending agent, stabilizer
Carboxymethylcellulose sodium	Disintegrant, suspension stabilizer, coating agent
Elhylcellulose	Binder, coating material, viscosity builder
Hydroxyethylcellulose	Binder, film former, dissolution modifier, viscosity builder
Hydroxypropyl cellulose	Viscosity builder, binder, stabilizer
Hydroxypropyl methylcellulose	Dissolution modifier, suspension stabilizer coating agent, film former, viscosity builder
Hydroxypropyl methylcellulose phthalate	Enteric coating agent, taste masking
Microcrystalline cellulose	Binder, diluent, disintegrant, stabilizer
Cellulose acetate	Dissolution modifier, film coating agent
Cellulose acetate butyrate	Dissolution modifier, film former
Cellulose acetate phthalate	Dissolution modifier, film former
B. Starch, Sugars and Derivatives	
Alginic acid	Binder, disintegrant, viscosity builder
Carrageenan	Sustained release matrix, suspending agent
Dextrose	Diluent, sweetener
Fructose	Diluent, sweetener, dissolution enhancer
Guar gum or Galactomannan	Binder, suspending agent, viscosity builder, disintegrant
Lactose	Diluent, sweetener
Mannitol	Diluent, sweetener, bulking agent for lyophilized products
Maltose	Sweetener, diluent
Pectin	Dissolution modifier, dispersant
Scleroglucan	Dissolution modifier

Sorbitol	Diluent, humectant, sweetner
Starch	Binder, diluent
Sucrose	Sweetener
Xanthan_gum	Viscosity builder, .suspension stabilizer
C. Cyclodextrins	
Beta-cyclodextrin	Complexing agent, dissolution enhancer, Stabilizer.
Hydroxypropyl β-cyclodextrin	Complexing agent, dissolution enhancer, Stabilizer.
Heptakis-(2,6-di-O-ethyl) β-cyclodextrin	Complexing agent, dissolution enhancer, Stabilizer.
D. Acids	
Ascorbic acid	Antioxidant
Lactic acid	Acidifying agent, acidulant
Malic acid	Acidulant, antioxidant, flavor, buffering agent
Tartaric acid	Effervescent agent, diluent
E. Amino Acids, Peptides and Derivatives	
Arginine	Stabilizer
Aspartame	Sweetener
Bovine serum albumin	Solubility enhancer (for biomolecules)
Human scrum albumin	Solubility enhancer (for biomolecules)
Lysine	Stabilizer
Protamine	Stabilizer
F. Fats, Oils, Essentials Oils	
Medium chain triglycerides	Emulsifying agent, suspending agent in emulsion systems, absorption enhancer, dissolution modifier
Carvacrol	Flavor, Permeation enhancer
Carvone	Flavor, Permeation enhancer
1,8, cineole	Flavor, Permeation enhancer
± Linalool	Flavor, Permeation enhancer
D-limonene	Flavor, Permeation enhancer
Menthone	Flavor, Permeation enhancer
L-menthol	Flavor, Permeation enhancer
α-tocopherol	Antioxidant

Importance of Chirality in Pharmacokinetics

Enantiospecificity in pharmacokinetics arises because of enantioselectivity in one or more of the processes of drugs absorption, distribution, metabolism and excretion. In turn, enantioselectivity results from the interaction of chiral drugs with a chiral biological milieu.

Absorption

Stereoselectivity is not an issue with regard to the absorption of the vast majority of drugs, which diffuse passively through membrane matrices. This is because enantiomers do not differ in their lipid and aqueous solubilities. However, the aqueous solubilities and crystal forms of racemates can differ from those of the individual isomers, and this may give rise to corresponding differences in dissolution rates at sites of administration. Carrier transport systems involved in the absorption of natural amino acids, sugars and other endogenous chemicals are stereoselective. Consequently, drugs with analogous structures may also be absorbed stereoselectively by these mechanisms. Thus, L-Dopa passes across the intestinal wall more rapidly than the D-isomer, which is not carried by the amino acid pump (82). Similarly, some beta-lactam antibiotics are carried by the intestinal dipeptide transport mechanism. For example, in the rat the L-isomer of cephalexin was shown to have a higher affinity for the carrier site than the D-form. However, only the latter could be detected in plasma since the L-isomer is also more susceptible to hydrolysis by peptidases in the intestinal wall (83).

A greater oral availability of the more active (-)-isomer of terbutaline (0.15 vs 0.075 for the (+)-form) was accounted for partly by selectivity in first-pass metabolism but also by the suggestion that this enantiomer selectivity increases membrane permeability (84). An indirect cause of stereoselective absorption arises when enantiomers differ in their ability to constrict or dilate blood vessels at sites of injection. Thus, many local anaesthetics, such as mepivacaine and bupivacaine, are optically active; the isomers having different effects on local blood flow which, in turn, account for differences in rates of systemic absorption and hence in duration of anaesthesia (85,86).

Distribution

Plasma Binding

The extent of stereoselective plasma binding of drugs ranges up to a factor of about 1.5, reflecting a diastereoisomeric association with both albumin, the major binding protein for most acids, and with alpha$_1$-acid glycoprotein, which binds predominantly basic compounds. Propranolol illustrates a case where the stereoselectivity in binding occurs in opposite directions for the different proteins. Thus, whereas the R(+)-isomer binds less to human alpha$_1$-acid glycoprotein (free fraction, fu = 0.162) than the S(-)- form (fu = 0.127); its binding to human albumin is greater (fu = 0.607 vs 0.649). In whole plasma the binding to alpha$_1$-acid glycoprotein predominates such that the R (+)-isomer binds less than the S (-)-isomer (fu = 0.203 vs 0.176) (87).

Tissue Binding

A comparison of the volumes of distribution of propranolol isomers based on unbound drug in plasma is suggestive of stereoselective tissue binding of this beta-blocker. In dogs the values were greater for the S(-)-enantiomer whereas in man the opposite was indicated(88,89). Other data confirm that the more active S-enantiomers of propranolol and atenolol undergo selective storage and secretion by adrenergic nerve endings in cardiac and other tissue (90). An apparent stereoselectivity in the uptake of the active S-isomer of ibuprofen into human synovial fluid may be explained by the difference in plasma binding of the enantiomers as it affects the concentration gradient of diffusible free drug (91). It is now well-established that the R-forms of some NSAIDs are selectively taken up into fat and that a much greater uptake of both isomers is seen when the RS-and R-rather than the S-forms are administered (92,93). Technically, however, this is not really a distribution process since metabolism is also involved.

Metabolism

Enantioselectivity in metabolism may be manifest either as a difference in the bio-transformation of chiral drugs (Substrate Enantioselectivity) or in the production of chiral metabolites from pro-chiral drugs (Product Enantioselectivity). The latter phenomenon assumes greatest importance when the metabolic products are active. A special consideration is the metabolic inversion of some enantiomers.

Substrate Enantioselectivity

Most drugs are metabolised predominantly in the liver and their hepatic clearance (CL_H) is determined to different extents by hepatic blood flow, blood binding and intrinsic enzyme activity (expressed as CLu_{int}-intrinsic clearance based on unbound drug concentration). The first of these determinants will be of primary importance for high clearance drugs given parenterally; the latter two determinants will predominate for low clearance drugs given parenterally and for both high and low clearance drugs given orally (94). It is important, therefore, to appreciate that stereoselectivity in hepatic clearance may arise through one or more of these determinants and, consequently, that the extent of any difference may be route-dependent. For example, stereoselectivity in the systemic clearance of propranolol is the net result of stereoselectivity in the effects of the isomers on hepatic blood flow, differences in their plasma binding and in their enzymatic transformation. A lowering of hepatic blood flow by the active S(-)-isomer contributes to its lower systemic clearance (CL_{iv}) compared to that of the inactive R(+)-form (95,96). After oral administration clearance (CL_{po}) is largely independent of liver blood flow and the lower value for the S(-)-isomer now reflects the net balance of enantioselectivity in plasma binding and intrinsic clearance . As the systemic clearances of individual enantiomers both approach the limit of liver blood flow enantioselectivity after parenteral administration will become less apparent than after oral dosage when differences in first-pass metabolism magnify the influence of differences in binding and intrinsic clearance. The relative contributions of enantioselectivity in plasma binding and enzymatic transformation are evident when plasma clearance values are divided by free fraction to give clearances based upon unbound drug (CLu). These latter values are more relevant than clearances based on total drug in plasma when relating enantiospecific kinetics to pharmacological effect during continuous drug dosage.

Since drugs may be metabolised by more than one pathway stereoselectivity in intrinsic clearance will reflect the net balance of selectivity in different enzymes. For example, measurement of partial metabolic clearances *in vivo* shows that the higher oral clearance of R (+)–propranolol in man appears to be due largely to selective ring 4-hydroxylation. Although not statistically significant, difference between the isomers in side-chain oxidation tend to favour the R(+)–isomer also, whereas direct glucuronidation tends to favour the S(-)–isomer (97). A further possiblilty is that several isozymes may be involved in the formation of the same product and each may exhibit different steric preferences. Thus, studies with human liver microsomes indicate that the 4-hydroxylation of propranolol is mediated by two different enzymes, only one of which is stereoselective (98). Similarly metoprolol is O-demethylated by two isozymes which in this case show opposite stereoselectivity (99). In some cases differences in the metabolic clearance of enantiomers may be mediated by virtually specific preference for completely different pathways. For example, in man S(-)-warfarin undergoes mainly 7-hydroxylation whereas the R(+)-isomer is metabolised mainly by 6-hydroxylation and ketone reduction (100).

Product Enantioselectivity

Stereoselectivity in the formation of chiral products of drug metabolism may arise from reactions at pro-chiral centres or at positions remote from such centres. Examples of the first possibility include the 4-hydroxylation of debrisoquine, which is selective for the S-product, and the activation of polycyclic aromatic hydrocarbons to carciogenic dihydrodiol epoxides (101). In man, para-hydroxylation of one of the benzene rings of phenytoin creates a chiral centre indirectly, with preferential formation of the S-product (102). A further possibility is the formation of diastereoisomers from a chiral drug by the metabolic introduction of a second chiral centre. Examples include the ketone-reduction of R-warfarin to racemic diastereoisomeric alcohols, and the conjugation of chiral drugs with beta-D-glucuronic acid and L-glutathione (103).

Inversion

The phenomenon of essentially unidirectional inversion of the R-isomers of 2-aryl propionic acid NSAIDs, via coenzyme A thioester formation, to the therapeutically active S-antipodes has been reviewed extensively by (104, 105). Salient pharmacokinetic aspects can be summarised as follows:

1. It does not appear to be a general occurrence for all profens in man. Thus, although there is evidence for inversion of the R-forms of ibuprofen, benoxaprofen, cicliprofen, fenoprofen and thioxaprofene, those of indoprofen, tiaprofenic acid, flurbiprofen, ketoprofen and carprofen do not appear to invert significantly.
2. There are marked species differences in the extent of inversion.
3. Understanding of the phenomenon is complicated by and complicates the appreciation of stereoselectivtity in other routes of metabolism (oxidation, acyl glucuronidation and de-glucuronidation).
4. Dose- and absorption rate-dependent kinetics and kinetic interactions between isomers complicate assessment of the extent of inversion.
5. Renal impairment and the co-administration of drugs which block the active renal secretion of the acyl glucuronides of the profens both enhance the extent of chiral inversion. This arises as a result of a 'Futile cycle' whereby the glucuronides are hydrolysed back to the parent drug.

Excretion

Renal

Renal drug clearance is the net result of glomerular filtration, active secretion and passive and active reabsorption. Stereoselectivity in this process may be secondary to any difference in the plasma binding of the isomers as it affects glomerular filtration and passive reabsorption. The demonstration of an intrinsic stereoselectivity in active excretory processes requires the calculation of the renal clearance of unbound drug (Clu_R). Although it is likely that the differences arise in the process of active secretion, contributions from stereoselectivity in active reabsorption and renal drug metabolism cannot be ruled out.

Biliary

Little is known about the stereoselectivity of the active processes involved in the biliary secretion of drugs and their metabolites. Thus, although marked differences in biliary recovery of antipodes have been demonstrated, for example those of nicoumalone in the rat estimates of biliary clearance are necessary to eliminate contributions from other sources of stereoselectivity (106).

Gastrointestinal

In rabbits the oral administration of activated charcoal decreased the area under the plasma drug concentration-time curve after i.v. injection of R-disopyramide, but was without effect on that of the S-isomer (107). This does not reflect, however, any intrinsic difference in the ability of the charcoal to enhance the gastrointestinal clearance of the isomers. Rather it is a consequence of the difference in their hepatic extraction ratios and the fact that the gut and the liver are arranged in series.

Understanding Chiral Drug Interactions

Drug-Drug

The classic example of how an appreciaton of stereochemical aspects is essential to unravel the mechanism of an interaction relates to the warfarin-phenylbutazone combination. Coadministration of phenylbutazone enhances the anticoagulant effect of warfarin. Measurement of total (racemic) plasma warfarin concentrations indicates no change in kinetics in the presence of phenylbutazone, suggesting that the interaction is purely pharmacodynamic (108). However, if the enantiomers of warfarin are measured it is seen that phenylbutazone inhibits the clearance of the more active S-isomer, consistent with the enhanced pharmacological effect. The clearance of the R-isomer is apparently induced by phenylbutazone. The truth finally emerges when the plasma kinetics of the unbound isomers are examined. Inhibition of the (unbound) clearance of both enantiomers is now apparent, more so in the case of the S-form. Displacement of warfarin from plasma binding sites by phenylbutazone obscures the inhibitory effect on the metabolism of the R-isomer by increasing its elimination half-life. Clearly interactions involving those drugs, such as phenylbutazone, which selectively inhibit the metabolism of the more active S-isomer of warfarin are likely to be of greater clinical significance. Other drugs, such as cimetidine, selectively inhibit the metabolism of R-Warfarin; yet others, such as amiodarone, inhibit the clearance of both isomers equally. Enzyme-inducers also show differential effects on the clearance of warfarin isomers.

Enantiomer-Enantiomer

Since enantiomers should be considered as separate drugs, the possibility of enantiomer interactions must be anticipated. Thus, the kinetics of an isomer may differ when given as such and when administered as part of a racemic mixture. A differential effect of enantiomers on haemodynamics will be cancelled when they are administered together. Thus, the greater local vasoconstriction produced by the L-Forms of bupivacaine and mepivacaine would be expected to delay the absorption and prolong the anaesthetic action of their D-isomers (109). Similarly, the effect of the active S(-)isomer of propranolol in reduing liver blood flow will lower the hepatic clearance of its antipode (110). Competition for active tubular reabsorption was invoked to account for the observation that (+)-terbutaline seemed to enhance the renal clearance of (-)-terbutaline when given in a racemate (111).

The areas under the total plasma drug concentration-time curves of both isomers of ibuprofen were found to be less when administerted as the racemate than when given alone (112). This observation can be accounted by competition between the isomers for concentration-dependent binding to plasma protein (113). Competition for concentration-dependent plasma binding also appears to account for interactions between the isomers of disopyramide based upon measurement of their total (bound plus free) plasma concentrations. Total clearance and volume of distribution were higher for the R(-)-enantiomer when the S(+)-form was present, whereas the clearance and volume of distribution of the S(+)-isomer were higher when it was administered alone. The fu value of the R(-)-enantiomer is greater in the presence of the S(+)-form the fu vlaue of the latter is actually greater at a given total disopyramide concentration when present by itself (114,115) .

In *vitro* studies have demonstrated interactions between enantiomers at enzymatic binding sites. For example, the glucuronidation of R-propranolol by dog liver microsomes was inhibited by S-propranolol (116), whereas R-propranolol was found to be a non-competitive inhibitor of S-propranolol glucuronidation using an immobilised enzyme from rabbit (117). Corresponding examples of in *vivo* metabolic interactions between enantiomers are rare. In rats, Murphy et al. (118) found that the oral administration of (-)- propoxyphene significantly increased the plasma concentrations and analgesic effect of (+)-propoxyphene. There appear to be no examples of interaction between isomers for enzymatic transformation in man.

Discussion: Till now only one thing is well proved that "The stereospecific drug release and enantiospecific drug absorption is possible only when the mechanism of drug release or drug absorption is by diffusion process. But till now no body has proved molecular level driving force for this enantiomeric discrimination in the drug release as well as absorption. But the possible hypothetical assumed mechanism we want to propose is "There might be some steric or chiral interactions between some chiral enantiomers and chiral excipients." Further we want to assume that the chiral centre of a particular enantiomer might be susceptible for the chiral interaction by a particular chiral excipient. Hence that particular enantiomer as a result of chiral interaction is either driven from the dosage form fast or hindered or it may be absorbed fast or its absorption may be hindered.

Challenges: About 50% of the drugs what are present in the current market are racemic drugs with chirality and even today we do not have their complete pharmacokinetic and pharmacodynamic profiles as an individual enantiomer. In the currently existing racemic drugs, are both the enantiomers of all the racemic drugs are equally active? Or one enantiomer is highly active and other is less active? Or one of the two enantiomers is completely inactive? Or one is active and other is toxic? Or if both the enantiomers are active, are those enantiomers active in only racemic state or even when they are administered individually? If one enantiomer is active and other is inactive, is that active enantiomer active when it is formulated as single enantiomer? If it is active as single enantiomer is it stable or undergoes any invitro or invivo chiral inversion into its inactive form? Are all the chiral excipients compatible or not? Which chiral excipients produce which kind of stereospecific drug release with existing chiral drugs? Is this chiral discrimination possible with all the chiral excipients and chiral drugs? If only with few, what are those few chiral drugs and excipients? For howmany racemic drugs we have the solutions for the above list of questions? These are all the challenges which are present infront of all the chiral scientists all over the world expecting an immediate attention.

9.10 Conclusions

The study of chirality, chiral interactions and chiral inversions has a great significance. It is useful in the prevention of unwanted conversion of eutomer to distomer either *in vitro* or *in vivo*; however the scope for the conversion in *invitro* is less.

The study of chirality, chiral interactions and chiral inversions also gives us an opprtunity to select such chiral excipients which do not bring the chiral inversion from distomer to eutomer. Using suitable chiral excipients we can also enhance the eutomer/distomer ratio of enantiomer release versus time of a racemic drug.

In case of cardiovascular risks, the release of eutomer rather than the distomer is very essential to elicit required pharmacological effect. This is made possible with use of suitable chiral excipients. If the enantiomers are sufficiently different in pharmacological effects, it may be possible to get a patent on one or both. It is evident, however, that it is far better to use the specific active enantiomer in view of dosage and economic considerations to give a better pharmacological benefit to the patients.

As current regulatory guidelines also stress the importance of investigating Pharmacokinetic and Pharmacodynamic studies of drugs with a chiral centre because the enantiomers of drugs often display different pharmacological and toxicological profiles we feel that there is lot of scope need to extend research in this chiral area with an immediate attention.

References

1. Mccoy M. Biocatalysis: Making drugs with Little Bugs. Chemical & Engineering News-Northeast News Bureau, 37-43 (2001).
2. Busch KW, Busch MA. Chiral analysis. 3 (2006).
3. Burke D, Henderson DJ, Br J Anaesth. Chirality: a blueprint for the future. Br. J Anaesth, 88(4):563-76 (2002).
4. Lehmann FPA, Rodrigues de Miranda JF, Ariens EJ. Stereoselectivity and affinity in molecular pharmacology. Drug Res, 20:101–42 (1976).
5. Karhuvaara S, Kaila T, Huuponen R. β-adrenoceptor antagonist activities and binding affi nities of timolol enantiomers in rat atria. J Pharm Pharmacol, 41:649–50 (1989).
6. Richards R, Tattersfi eld AE. Undesired action of timolol racemates. Br J Clin Pharmacol, 20:459–62 (1985).
7. Richards R, Tattersfi led AE. Bronchial β-adrenoreceptor blockade following eye drops of timolol and it isomer L-714,465 in normal subjects. Br J Clin Pharmacol, 20:459–62 (1985).
8. Kente EU, Stove R. Effect of D-timolol on intraocular pressure in patients with ocular hypertension. Am J Opthamol, 98:73–8 (1984).
9. Hutt AJ, Tan SC. Drug Chirality and its Clinical Signifi cance. Drugs, 52(5):1–12 (1996).
10. Hutt AJ, Grady J. Drug chirality: a consideration of the significance of the stereochemistry of antimicrobial agents. J Antimicrob Chemother, 37(1):7–32 (1996).
11. Lien EJ. Chirality and drug targeting: pros and cons. J Drug Target, 2(6):527–32 (1995).
12. Caldwell J. The importance of stereochemistry in drug action and disposition. J Clin Pharmacol, 32:925–9 (1992).
13. Drayer DE. Pharmacodynamic and pharmacokinetic differences between drug enantiomers in humans: An overview. Clin Pharmacol Ther, 40(2):125–33 (1986).

14. Eichelbaum M. Pharmacokinetic and pharmacodynamic consequences of stereoselective drug metabolism in man. Biochem Pharmacol, 37:93–6 (1988).

15. Walle T, Walle UK. Pharmacokinetic parameters obtained with racemates. TIPS, 155(1986).

16. Beckett AH. Analgesics and their antagonists, some steric and chemical considerations. Int J Pharmacol, 8:848–59 (1956).

17. El-Ackad TM, Panidis IP, Michelson EL. The relative cardiac beta-blocking potency of oral labetalol and its isomer SCH 19927 in normotensive subjects. Clin Res, 32:330–5 (1984).

18. Suedee R, Brain KR, Heard CM. Differential permeation of propranolol enantiomers across human skin in vitro from formulations containing an enantioselective excipient. Chirality, 11(9): 680–683 (1999).

19. Dzgoev A, Haupt K. Enantioselective molecularly Imprinted polymer membranes. Chirality, 11: 465–469 (1999).

20. Kato R, Ikeda N, and Yabek S. Electrophysiologic effects of the levo- and dextrorotatory isomers of sotalol in isolated cardiac muscle and their *in vivo* pharmacokinetics. J Am Coll Cardiol, 7:116–125 (1986).

21. Advani SV, Singh BN. Pharmacodynamic, pharmacokinetic and antiarrhythmic properties of d-sotalol, the dextro-isomer of sotalol. Drugs, 49:664–679. (1995)

22. DeVane CL, Boulton DW. Great expectations in stereochemistry: focus on antidepressants. CNS Spectrums, 7:28–33 (2002).

23. Blaschke G, Kraft HP, Fickentscher K, Koehler F. Chromatographic racemic separation of thalidomide and teratogenic activity of its enantiomers. Arznei. Forschung, 29:1640–1642 (1979).

24. Eriksson,T., Bjorkman, S., Roth, B., Fyge, A., Hoglund, P. Stereospecific determination, chiral inversion in vitro and pharmacokinetics in humans of the enantiomers of thalidomide. Chirality, 7:44–52 (1995).

25. Fabro S, Smith RL, Williams RT. Toxicity and teratogenicity of optical isomers of thalidomide. Nature, 215: 269 (1967).

26. Eriksson T, Bjo rkman S, Roth B, Fyge A, Ho glund P. Stereospecific determination, chiral inversion in vitro and pharmacokinetics in humans of the enantiomers of thalidomide. Chirality, 7:44–52 (1995).

27. Eriksson T, Bjo rkman S, Roth B, Fyge A, Ho glund P. Enantiomers of thalidomide: blood distribution and the influence of serum albumin on chiral inversion and hydrolysis. Chirality, 10:223–228 (1998).

28. Reist M, Carrupt P, Francotte E, Testa B. Chiral inversion and hydrolysis of thalidomide: mechanisms and catalysis by bases and serum albumin, and chiral stability of teratogenic metabolites. Chem. Res. Toxicol, 11:1521–1528 (1998).

29. Gordon JN, Goggin PM. Thalidomide and its derivatives: emerging from the wilderness, Postgrad. Med. J, 79:127 -132 (2003).

30. Areins EJ, Wuis EW, Veringa EJ. Biochem pharmacol, 37:9 (1988).

31. Bojarski J, Aboul-Enein HY, Ghanem A. Curr Anal Chem, 1: 59 (2005).

32. Maier NM, Franco P, Lindner W. J Chromatogr A, 906: 3 (2001).

33. Nelson TD, Welch CJ, Rosen JD. Chirality, 16:609 (2004).

34. Shah RR, Midgley JM, Branch SK. Adv Drug React Toxical Rev, 17:145-190 (1998).

35. Caldwell J. J Chromatogr A, 39:694 (1995).

36. Ariens E. Trends Pharmacol Sci, 7: 200-205 (1986).

37. M. Kagawa, Y. Machida, H. Nishi, J Chromatogr A, 857:127 (1999).

38. Francotte ER, Ahuja S. Chiral Separations, Applications and Technology. American Chemical Society, 10:271 (1997).

39. Roussel C, Piras P, Heitmann I. Biome Chromatogr, 11:311 (1997).

40. Kaliszan R, Noctor TA, Wainer IW, Chromatographia 33:546 (1992).

41. Pirkle WH, Finn JM, Hamper BC, Schreiner J, Pribish JA. A useful and conveniently accessible chiral stationary phase for the liquid chromatographic separation of enantiomers, in Asymmetric Reactions and Processes in Chemistry. ACS, 185:245–260 (1982).

42. Pirkle WH, Pochapsky TC. Adv. Chromatogr. 27:73–127 (1987).

43. Pirkle WH, Pochapsky TC, Mahler GC, Corey DE, Remo R, Alessi DM. J OrgChem, 51:4991 (1985).

44. Pirkle WH, Alessi DM, Hyun MH, Pochapsky TC. J Chromatogr, 398:203 (1987).

45. Caude A. Tambute L. Siret J. Chromatogr, 550:357–382 (1991).

46. Gasparrini D, Misiti C, Villani. Chirality, 4:447–458 (1992).

47. Meurisse RL, De Ranter CJ. Chromatographia, 38:629 (1994).

48. Blackwell JA, Stringham RW, Xiang D, Waltermine RE. J. Chromatogr A, 852:383 (1999).

49. Francotte E. Chim. Nouv, 14:1541 (1996).

50. Pirkle WH, Welch CJ. J. Liq. Chromatogr, 15:1947 (1992).

51. Welch CJ, Szczerba T, Perrin SR. J. Chromatogr A, 758:93 (1997).

52. Murer P, Lewandowski K, Svec F, Frechet JM. Anal. Chem, 71:1278 (1999).

53. Welch CJ, Bhat G, Protopopova MN. J. Comb. Chem, 1:364 (1999).

54. Welch CJ, Protopopova MN, Bhat G. Enantiomer 3:471 (1998).

55. Wu Y, Wan A, Yang T. Anal. Chem, 7:1688 (1999).

56. Lewandowski K, Murer P, Svec F, Frechet JM. J. Comb. Chem, 1:105 (1999).

57. Francotte E. Chirality 10:492 (1998).

58. Stringham R, Blackwell JA. "Entropically Driven" Chiral Separations in Supercritical Fluid Chromatography. Confirmation of Isoelution Temperature and Reversal of Elution Order. Anal. Chem., 68:2179-2185 (1996).

59. Fornstedt T, Sajonz P, Guiochon G. Thermodynamic Study of an Unusual Chiral Separation. Propranolol Enantiomers on an Immobilized Cellulase. J. Am. Chem. Soc., 119:1254-1264 (1997).

60. Maftouh M, Granier-Loyaux C, Chavana E, Marini J, Pradines A, Vander Heyden Y, Picard C. Screening approach for chiral separation of pharmaceuticals. Supercritical fluid chromatography for analysis and purification in drug discovery. J. Chromatogr. A, 1088:67–81 (2005).

61. Koppenhoefer B, Zhu X, Jakob A, Wuerthner S. J Chromatogr A 875, 135-161 (2000).

62. Fanali S. J Chromatogr A, 792:227-267 (1997).

63. Gubitz. J Chromatogr A, 792:179-225 (1997).

64. Verlesyn K, Sandra P. Electrophoresis, 19:2798-2833 (1998).

65. Lelievre F, Guiet C, Gareil P, Bahaddi Y. Electrophoresis, 81:891-896 (1997).

66. Tanaka Y, Terebe S. J Chromatogr A, 781 (9):151-160 (1997).

67. Terabe S, Miyashita Y, Shibata O. J Chromatogr, 516:23-31(1990).

68. Sousa LB, Sogah Y, Hoffmann DH, Cram DJ. J Am Chem Soc, 100:4569(1978).

69. Application guide for chiral column selection, Daicel chem. Ind. (1989).

70. Wainer IW. J Pharm Biomed Anal, 7:1033 (1989).

71. Domenici E, Bertucci C, Salvadori P, Felix G, Cahagne I, Motellier G Wainer IW. Chromatographia, 29:170 (1990).

72. Hermansson J. J Chromatogr, 298:67 (1984).

73. Hermansson J, Eriksson M. J Liq Chromatogr, 9:621 (1986).

74. Wainer IW, Jadaud P, Schombaum G, Kadodkar GV, Henry MP. Chromatographia, 25: 903 (1988).

75. Thelohan S, Jadaud P, Wainer IW. Chromatographia, 28:551 (1989).

76. Kalbe J, Hocker H, Berndt H. Chromatographia, 28:193 (1989).

77. Saleem Shaikh MS, Muneera OA. Chiral chromatography and its application to the pharmaceutical industry. Pharmainfo.net, 7(1): (2009).

78. Eichelbaum M, and Gross AS. Stereochemical aspects of drug action and disposition. Advances in Drug Research, 1-64 (1996).

79. Indra K R, Reza M. Chirality in Drug Design and Development. Marcel Dekker, 10:50-51 (2004).

80. Wade D, Mearrick PT, Morris JL. Nature 242, 436-465 (1972).

81. Tamai I, Ling HY, Timbul SM, Nishikido J, Tsuji A. J Pharm Pharmac, 40:320-324 (1988).

82. Borgstrom L, Nyberg L, Jonsson S, Lindberg C, Paulson J. Br J Clin Pharmac, 27:49-56 (1987).

83. Akerman B, Persson H, Tegner C. Acta Pharamc Tox, 25:233-241 (1967).

84. Luduena FP. Ann Rev Pharmac, 9:503-520 (1969).

85. Walle T. Res Common. Chem Path Pharmac, 23:453-464 (1979).

86. Walle T, Webb J, Bgwell EE, Walle UK, Danill HB, Gaffney TE. Biochem. Pharmac, 37: 115-124 (1988).

87. Bai SA, Walle UK, Wilson MJ, Walle T. Drug Metab Disp, 11:394-395 (1983).

88. Webb JG, Street JA, Bagwell EE, Walle T, Gaffney TE. J Pharmac exp Ther, 247:958-964 (1988).

89. Day RO, Williams KM, Graham C, Lee EJ. Clin Pharmac Ther, 43:480-487 (1988).

90. Williams K, Day R, Knihinicki r, Duffield A. Biochem Pharmac, 35:3403- 3405 (1986).

91. Sallustio BC, Meffin P, Knights KM. Biochem Pharmac, 37:1919-1923 (1988).

92. Wilkinson GR, Shand DG. Clin Pharmac Ther, 18:377-390 (1975).

93. Branch RA, Nies AS, Shand DG. Drug Metab Disp, 1:687-690 (1973).

94. Nation RL, Evans AM, Shand DS. Amer Heart J, 85:97-102 (1973).

95. Ward SA, Walle T, Walle UK, Wilkinson GR, Branch RA. Clin Pharmac Ther, 45:72-79 (1983).
96. Otton SV, Lennard MS, Tucker GT, Woods HF. Eur J Clin Pharmac, 36 (1988).
97. Otton SV, Lennard MS, Tucker GT, Woods HF. J Pharmac exp Ther, 247:242-247 (1988).
98. Lewis RJ, Trager WF, Chan KK, Breckenridge AK, Orme M, Rowland M, Schary W.
99. J Clin Invest 53:1607-1617 (1974).
100. Yang SK. Biochem Pharmac, 37:61-70 (1988).
101. Butler TC, Dudley KH, Johnson D, Roberts SB. J Pharmac Exp Ther, 199:82-92 (1976).
102. Caldwell JC, Hutt AJ, Ournel-Gigleux SF. Biochem Pharmac, 37:105-114 (1988).
103. Caldwell JC, Winter SM, Hutt AJ. Xenobiotica, 18:59-70 (1988).
104. Jamali F. Eur J Drug Metab Pharmacokin, 13:1-9 (1988).
105. Thijssen W, Baars GM, Vervoort peters HM. Br J Pharmac, 95:675-682 (1988).
106. Huang JD. J Pharm Sci, 77: 959-962 (1988).
107. Reilly RA, Trager WF, Motley CH, Howald W. J Clin Invest, 65:746-753 (1980).
108. Aps C, Reynolds F. Br J Clin Pharmac, 6:63-68 (1978).
109. Branch RA, Nies AS, Shand DG. Drug Metab Disp I, 687-690 (1973).
110. Borgstrom L, Nyberg L, Jonsson S, Lindberg C, Paulson J, Br J Clin Pharmac, 27: 49-56 (1989).
111. Lee EJD, Williams K, Day R, Graham G, Champion D, Br J Clin Pharmac, 19:669-
112. 674 (1985).
113. Nation RL, Evans AM, Shand D. Amer Heart J, 85:97-102 (1973).
114. Giacomini M, Nelson WL, Pershe RA, Valdevieso L, Turneer-Tamayasu K,
115. Blaschke, TF. J Pharmacokin Biopharm, 14:335-356 (1986).

116. Valdivieso L, Giacomini KM, Nelson WL, Pershe R, Blaschke TF. Pharmaceut. Res, 5:316-318 (1988).

117. Wilson BK, Thompson JA. Drug Metab. Disp, 12:161-164 (1984).

118. Yost GS, Hohnson LP, Pallante S, Colvin M, Fenselua C. Fed Proc, 40:650 (1981).

119. Murphy PJ, Nickander RC, Bellamy GM, Kurtz WL. J Pharmac Exp Ther, 199:415-422 (1976).

Index

M

N

O

P

T

U

W

Z